Introduction to Optimizat[illegible] **The**[illegible] is
a comprehensiv[illegible]
all majo[illegible]
niqu[illegible]
[illegible]

- Uses examples requiring no special knowledge in the technical disciplines.
- Contains 114 illustrations and 70 numerical tables.

INTRODUCTION TO OPTIMIZATION THEORY

PRENTICE-HALL INTERNATIONAL SERIES
IN INDUSTRIAL AND SYSTEMS ENGINEERING

W.J. Fabrycky and J.H. Mize, Editors

Fabrycky, Ghare, and Torgersen *Industrial Operations Research*
Gottfried and Weisman *Introduction to Optimization Theory*
Mize, White, and Brooks *Operations Planning and Control*
Whitehouse *Systems Analysis and Design Using Network Techniques*

INTRODUCTION TO OPTIMIZATION THEORY

BYRON S. GOTTFRIED

Department of Industrial Engineering,
Systems Management Engineering and Operations Reseach
University of Pittsburgh

JOEL WEISMAN

Department of Chemical and Nuclear Engineering
University of Cincinnati

PRENTICE-HALL, INC., *Englewood Cliffs, New Jersey*

Library of Congress Cataloging in Publication Data

GOTTFRIED, BYRON S.
Introduction to optimization theory.

(Prentice-Hall international series in industrial and systems engineering)
Includes bibliographical references.
1. Mathematical optimization. 2. Programming (Mathematics) I. Wiesman, Joel, joint author. II. Title.
QA402.5.G68 519.7 72-8519
ISBN 0-13-491472-4

10 9 8 7 6 5 4 3 2 1

PRENTICE-HALL INTERNATIONAL, INC., *London*
PRENTICE-HALL OF AUSTRALIA, PTY. LTD., *Sydney*
PRENTICE-HALL OF CANADA, LTD., *Toronto*
PRENTICE-HALL OF INDIA PRIVATE LIMITED, *New Delhi*
PRENTICE-HALL OF JAPAN, INC., *Tokyo*

to our wives

Marcia and Bernice

CONTENTS

PREFACE

The recent proliferation of optimization techniques and the widespread availability of digital computers have encouraged most universities to offer a collection of courses in optimization theory, with applications to various disciplinary areas. These courses have been received with considerable enthusiasm by students, faculty, and, in many cases, by postgraduate professional people. The pioneer courses of this nature have been difficult to develop, however, owing to the lack of suitable text material which would bring the entire field into a proper perspective for introductory classroom instruction. We have attempted to write such a text, intending it for either the beginning graduate level or, with some modification, the advanced undergraduate level. Our intent is to present the basic ideas of each of the major classes of optimization methods, offering at the same time a basis for unification and some essence of comparison among them.

It was necessary for us to decide whether to orient the text toward certain disciplinary areas, or to write a more general textbook which a competent instructor could supplement with specialized comprehensive problems reflecting his particular area of interest. We have chosen the latter. In using early versions of this manuscript to teach senior- and graduate-level engineering optimization courses at Carnegie-Mellon University, the University of

Pittsburgh, and the University of Cincinnati, we found it very effective to supplement the text material with a few carefully chosen disciplinary problems requiring a computer solution. Thus our students received an exposure to some analytical model building in their own area of expertise. We strongly recommend that others consider this approach, particularly if a few general purpose computer codes (e.g., linear programming, hill-climbing, etc.) are available.

This text contains more than enough material for a two-semester (30-week) course at the introductory graduate level. The text has been organized so that linear programming is presented as an integrated part of optimization theory. Before linear programming is discussed (Chapter 4), optimization terminology (Chapter 1); the classical Calculus (Chapter 2); Kuhn-Tucker conditions (Chapter 2); and unconstrained optimization (Chapter 3) are presented. This sequence allows linear programming to be related to other optimization techniques as well as allowing duality theory to be derived on the basis of the Kuhn-Tucker conditions.

Chapters 1 through 4, plus Appendix B, can be used for the first semester course. The remaining chapters offer considerable latitude for the second semester. These chapters present nonlinear programming (Chapter 5); integer programming and the method of decomposition (Chapter 6); optimization of functionals (Chapter 7); dynamic programming and the discrete maximum principle (Chapter 8); and, finally, optimization under risk and uncertainty (Chapter 9). If the curriculum calls for a first semester course more heavily oriented towards linear programming, this can be provided by Chapters 1, 2, 4, 6, and Appendix B. A one-semester course in nonlinear programming can be structured from Chapters 3, 5, 6, and a part of Chapter 9. Also, Chapters 7 and 8, and a portion of Chapter 9, can be used for a one-semester course in dynamic systems optimization.

Each chapter is supplemented with an extensive set of problems for student solution. The problems are designed to illustrate and, in some cases, extend the text material. In a number of problems, some seemingly quite simple, the student will find it difficult to arrive at a numerical solution by hand. These problems are meant to illustrate the real complexity of most optimization tasks. The student should be encouraged to set up such problems for solution but to carry the numerical solution procedure only as far as seems reasonable. When the class composition is such that development of special problems in a given discipline is not feasible, it is suggested that a few complex problems be selected from those provided by the text and assigned for solution on a digital computer.

The required mathematical background for this text does not extend beyond elementary calculus and differential equations, though an introductory exposure to linear algebra is desirable. Appendix B contains the basic concepts of linear algebra for those who lack background in this important area.

Finally, we wish to thank several people for their assistance and encouragement during the course of this work. To R. H. Curran, who proofread much of the manuscript; D. L. Keefer, who proofread portions of the manuscript and developed some of the problems; and to M. J. Lempel, who made a number of useful suggestions, we extend our sincere appreciation. Our special thanks to Dr. A. G. Holzman who, as a teacher and a colleague, contributed to this project in numerous ways. Last, but not least, we wish to acknowledge the patience and understanding shown by our wives and children for the many hours of preoccupation, which we hope will serve a worthwhile purpose.

BYRON S. GOTTFRIED
Pittsburgh, Pennsylvania

JOEL WEISMAN
Cincinnati, Ohio

NOMENCLATURE

Vector and Scalar Notation

E, e	Symbols in standard type refer to scalar quantities
$\mathbf{E}$	Symbols in bold face type indicate vectors or matrices
$\mathbf{E}'$	Primed, bold face symbol indicates transpose of matrix
$\mathbf{E}^{-1}$	Bold face symbol to power of (-1) indicates inverse of matrix
$\lvert\mathbf{E}\rvert$	Bold face symbol in $\lvert\ \ \rvert$ indicates determinant of matrix within $\lvert\ \ \rvert$

Symbol	Definition
a	constant, lower bound on x
a_{ij}	coefficients of decision variables in constraint relationships
A	constant
$\mathbf{A}$	a matrix, matrix where components are a_{ij}, vector where components represent lower bounds of $\mathbf{X}$

b	constant, upper bound of x
b_i	constraint requirement
B	constant
$\mathbf{B}$	a column vector, vector where components are b_i, vector where components represent upper bound of $\mathbf{X}$, matrix where columns are the basis vectors $\mathbf{P}_1 \ldots \mathbf{P}_m$
c_j	coefficient of x_j in objective function
$\mathbf{C}$	vector whose components are c_j — the objective function coefficients
d_j	coefficient of x_j in Gomory constraint or cut, or in any equation defining a hyperplane
D_k	demand during period k
$\mathbf{D}$	vector whose components are d_j
$\lvert\mathbf{D}_i\rvert$	an $i \times i$ determinant whose elements are second partials of y with respect to the x_j
$E(\ )$	expected value of quantity within brackets
$f(x)$	frequency function of x
F_n	n'th member of Fibonacci sequence
$F(x)$	distribution function of x; $F(x) = \int_{-\infty}^{x} f(s)\, ds$, where s is a dummy variable
$g_i(\mathbf{X})$	constraint relationship
G	constraint equation
$G_i(\mathbf{X}, \mathbf{X}^\circ)$	hyperplane approximating $g_i(\mathbf{X})$ at $\mathbf{X}^\circ$
$\mathbf{G}$	constraint set, $[\mathbf{G}(\mathbf{u}, \mathbf{x}, \mathbf{u}', \mathbf{x}', t) = \mathbf{0}]$
H	Hamiltonian function $H(\mathbf{u}, \mathbf{x}, \boldsymbol{\lambda}) = \sum_{j=0}^{n} \lambda_j(t)\, g_j(\mathbf{u}, \mathbf{x})$
$H^{(\alpha)}$	stagewise Hamiltonian function $H^{(\alpha)} = \sum_{j=0}^{n} \lambda_j^{(\alpha)}\, T_j^{(\alpha)}$
$\mathbf{H}_i$	matrix used to determine search direction in Fletcher-Powell algorithm
$\mathbf{H}$	Hessian matrix
I	integral of function with respect to t
I(0)	$\int_{t_0}^{t_f} \phi(u, u', t)\, dt$
I(ϵ)	$\int_{t_0}^{t_f} \phi(u + \epsilon\eta, u' + \epsilon\eta', t)\, dt$
$\mathbf{I}$	identity matrix
k	constant

K	positive constant, penalty factor
l	constant,
	augmented integral $= \int_{t_0}^{t_f} (\phi + \lambda G)\, dt$
L_n	interval of uncertainty after n search points
L	Lagrangian function
m	a constant,
	number of constraint equations,
	number of search points
$\mathfrak{M}(\mathbf{u}, \lambda)$	function defined by $\mathfrak{M}(\mathbf{u}, \boldsymbol{\lambda}) = \operatorname*{Max}_{\mathbf{x}} H(\mathbf{u}, \mathbf{x}, \boldsymbol{\lambda})$
n	number of decision variables,
	number of terms in series
N	total number of subregions
p	fraction of points in given region
P	probability
$\mathbf{P}$	projection of vector on hyperplane
$\mathbf{P}_j$	search vector,
	vector consisting of jth column of constraint matrix coefficients
r	net worth
r_k	return function for stage k
r_{ij}	fractional part of number
R	set of points
R_j	remainder term
s	distance
$\mathbf{S}$	feasible direction vector
t	time or any quantity for which a functional relationship, $u(t)$, exists,
	standard normal variate $= (x - \mu)/\sigma$
t_i	a polynomial expression with only positive coefficients
$T_j^{(\alpha)}$	transformation function—function which transforms state variable $u_j^{(\alpha-1)}$ into state variable $u_j^{(\alpha)}$
$\mathbf{T}$	matrix whose columns consist of search vectors P_i
u	state variable
u'	first derivative of $u(t)$
u_i	simplex multiplier for row i
u_j	technological coefficient, state variable
u_{ij}	integer part of number
$u_j^{(\alpha)}$	value of j'th state variable at stage α
$u_0^{(\alpha)}$	contributions to objective function from stages 1 through α
U	utility, value
v	value of game
v_j	simplex multiplier for column j,
	fractional advance in stack prices for year j

$v(\boldsymbol{\delta})$	dual function in geometric programming
$\mathbf{V}$	projection matrix
w	dual variable, weighting factor
$\mathbf{W}$	vector of dual variables
x_j	decision variable
$\mathbf{X}$	vector of decision variables
x_{ij}	weight to be given to basis vector $\mathbf{P}_i$ in expression determining $\mathbf{P}_j$
$\mathbf{X}_o$	vector whose components are values of the decision variables at extremum
$\mathbf{X}_0$	vector whose components are values of decision variables at some arbitrarily specified starting point
$\mathbf{X}^*$	minimum feasible solution
y, $y(\mathbf{X})$	objective function
y_j	sum of the products of $c_i\, x_{ij}$ for all the x_{ij} in the jth column of the Simplex tableau
$y_k(u_k)$	optimum value of objective function for k stages $= \underset{x_k, x_{k-1} \cdots x_1}{\max(\min)} \{r_k(u_k, x_h) + r_{k-1}(u_{k-1}, x_{k-1}) + \cdots + r_1(u_1 x_1)\}$
$y'(x)$	first derivative of $y(x)$ with respect to x
$y''(x)$	second derivative of $y(x)$ with respect to x
z	Lagrangian function
α	constant, index indicating stage or subregion number
α_{ij}	weights which replace variables and/or nonlinear functions in separable programming procedure
β	scalar constant between 0 and 1
β_i	scalars determining direction of search vector $\mathbf{P}_i$
δ, δ_{ij}	quantity whose value is 0 or 1
δ_i	change in cost coefficient c_i
δ_i, δ'_{il}	dual variables of geometric programming problem
$\delta \mathrm{I}$	first variation of integral $\mathrm{I}(\epsilon)$ $= \lim_{\epsilon \to 0} \left(\frac{\mathrm{I}(\epsilon) - \mathrm{I}(0)}{\epsilon}\right)$
$\boldsymbol{\Delta}_i$	initial pattern search vector
$\Delta^+ x_j$	magnitude of positive change in x_j
$\Delta^- x_j$	magnitude of negative change in x_j
ϵ	small increment
$\eta(t)$	arbitrary function with continuous second derivatives and vanishing boundary conditions
η'	first derivative of $\eta(t)$

θ	scalar constant; in some cases a scalar restricted to be between 0 and 1
λ	Lagrange multiplier
$\lambda_j^{(\alpha)}$	adjoint variable for state variable u_j at stage α
$\boldsymbol{\lambda}$	vector of Lagrange multipliers
$\boldsymbol{\lambda}_o$	vector whose coordinates are values of Lagrange multipliers at extremum
μ	scalar weighting factor, mean value
μ_1	weight, in geometric programming problem for constraint 1
μ_{ij}	weights applied to extremal point x_{ij}
ξ_j	arbitrary point
π	geometrical constant, failure cost sum
π_k, $\hat{\pi}_k$	dual variables in decomposition algorithm
π_i	simplex multipliers or shadow prices
ρ_j	change in j'th constraint requirement
σ	standard deviation (σ^2 = variance)
τ	time
ϕ	function of u, u' and t
$\phi(\mathbf{X})$	function of $\mathbf{X}$, augmented objective function
ψ	differential or integral constraint function
∇y	gradient of y [vector whose components are $\partial y/\partial x_j$]

INTRODUCTION TO OPTIMIZATION THEORY

INTRODUCTION

All of us must make many decisions in the course of our day-to-day events in order to accomplish certain tasks. Usually there are several, perhaps many, possible ways to accomplish these tasks, although some choices will generally be better than others. Consciously or unconsciously, we must therefore decide upon the *best*—or *optimal*—way to realize our objectives.

For example, all of us at one time or another find it necessary to drive through city traffic. We could attempt to find the shortest possible route from point A to point B without concern for the time required to traverse this route. Alternatively, we could seek out the quickest, though not necessarily the shortest, route between A and B. As a compromise, we might attempt to find the shortest path from A to B subject to the *auxiliary condition* that the transit time not exceed some prescribed value. Here we have examples of three similar, but different, optimization problems.

The stock market is another example of a fascinating, if not always successful, endeavor to form an optimal strategy. There are several objectives from which to choose when "playing the market," such as maximum rate of growth of capital, maximum rate of return from a fixed amount of capital, minimum chance of loss of capital, and so on. Thus one must first formulate carefully an appropriate objective, and then develop some strategy which

1

will allow that objective to be satisfied in the best possible manner. Unfortunately, this problem is extremely difficult to analyze mathematically, owing to the significant uncertainties in the behavior of individual stock prices.

For most of us, personal decisions are based upon a mixture of logic, common sense, intuition, and emotion. The proportion of each ingredient varies from individual to individual. Sometimes we must make decisions spontaneously, which excludes the possibility of formulating rational decisions based upon logic. However, the results of a single poor decision are rarely disastrous, so that we can usually tolerate a lack of precise strategy in our everyday lives.

On the other hand, in our professional capacities we must sometimes plan strategies and make decisions that may have serious and far-reaching consequences. The transcontinental truck driver must preplan his travel route carefully; the investment analyst must advise his clients how and where to invest large sums of money to best meet certain objectives; the plant engineer must determine the optimum operating parameters (and there are many) for a large oil refinery; the military commander must decide how to distribute his retaliatory forces so as to minimize the damage resulting from a surprise enemy attack.

Decisions of this nature obviously require extensive study, and specific answers can sometimes be obtained only through the use of sophisticated mathematical methods and the use of large, high-speed computers. In attempting to optimize problems of this magnitude, a variety of specialized mathematical techniques have been developed. These techniques—their logical foundations, their characteristics, and their methods of application—comprise the subject of optimization theory. It is the exposition of these ideas to which this book is addressed.

1.1 Optimization—Present and Past

In a classical sense, optimization can be defined as the art of obtaining best policies to satisfy certain objectives, at the same time satisfying fixed requirements. It would be presumptuous at this time to suggest that optimization has attained the status of a science rather than an art; however, recent advances in applied mathematics, operations research, and digital-computer technology enable many complex industrial problems in engineering and economics to be optimized successfully by the application of logical and systematic techniques. At present, not all problems can be solved by existing methodology. However, the development of new and increasingly powerful optimization techniques is proliferating rapidly. It can be said with reasonable assurance that the field of optimization is rapidly moving in the direction of becoming an established science.

The idea of selecting a best way to accomplish a task that involves fixed requirements is as old as civilization. For instance, the Egyptians, according to the Greek historian Herodotus, were concerned with finding the best way to impose taxes on farmlands, taking into account any change in the value of each plot of land resulting from the annual flooding of the Nile River. This problem led to the creation of certain elementary concepts of plane geometry, which were required as measurement and decision-making tools.

The creation of calculus by Newton and Leibnitz in the seventeenth century was a milestone in the development of optimization theory. Calculus allows one to obtain not only an optimal condition in terms of the maximum or minimum of a mathematical function, but also the value of the independent variables that cause the function to be maximized or minimized. The use of calculus is restricted to certain well-behaved functions, however, and even then the algebraic problems that arise from the application of calculus frequently are extremely formidable. Consequently, the calculus is not a sufficiently powerful tool to handle realistic optimization problems of practical interest.

After World War II a new class of optimization techniques became available. These could be applied successfully to the problems of a sophisticated and complex society. Two factors made this possible—the development of high-speed electronic digital computers, and the application of mathematical analysis to the development of numerical techniques for obtaining maxima or minima. These numerical procedures (which are largely the basic tools in the field of *operations research*) bypass many of the difficulties associated with the calculus. Moreover, the modern digital computer, with its large memory and its extremely rapid calculational ability, enables the practical utilization of these numerical techniques in a reasonable amount of time and at a tolerable expense.

1.2 Systems and Mathematical Models

Reference is frequently made to the terms *systems* and *systems engineering*. Yet these terms are rarely defined in an explicit manner. There is a partial explanation for the vagueness of these terms, since the techniques of systems engineering (of which optimization techniques constitute an important subclass) are applicable to a very wide variety of physical problems. Let us think of a system in the thermodynamic sense as some portion of the universe, which, for purposes of study, is isolated from the remainder of the universe. Hence we may choose as a system a fixed mass, a collection of fixed masses (e.g., a machine having many moving parts), a process, or a collection of people, machines, and processes engaged in one or more common activities (e.g., a corporation, a national economy, a war involving several nations). Thus the size and scope of our system may vary over several orders of magnitude.

To study a system quantitatively it is necessary to establish mathematical cause-and-effect relationships—commonly known as a *mathematical model.* Such a model must accurately describe the laws of economics and human behavior, as well as those of chemistry, physics, and biology, as they pertain to a given system. It is important to realize that the information which results from the solution of a mathematical model, no matter how sophisticated, can be no better than the mathematical model which describes the system. Furthermore, it is necessary to model an *entire system* in the broadest possible sense, since the optimal conditions for several subsystems usually do not result in the optimal conditions for the overall system.

The great utility of the techniques of systems engineering lies in the universal application of these techniques to a wide variety of mathematical models. Such models may describe complex physical, chemical, biological, and economic systems in considerable detail. Of course, such models are not easily solved, even with high-speed computers. In our subsequent discussions we shall, however, assume that the system of interest is reasonably well understood and that an appropriate mathematical model is available.

1.3 Variables and Objective Functions

In order to consider attaining certain goals in an optimal manner, we must first define our objective quantitatively. We recognize that the desired objective must depend upon the proper choice of certain *variables* $x_1, x_2, \ldots, x_n$, which are sometimes referred to as *policy variables* or *decision variables.* In an ordinary optimization problem, we define an *objective function* (sometimes called a *profit function, performance index,* or *return function*) $y(x_1, x_2, \ldots, x_n)$ which represents some quantity, such as profit or cost, that we wish to optimize. More specifically, we seek a particular point $(x_{10}, x_{20}, \ldots, x_{n0})$ in a closed region (i.e., a domain plus its boundaries) that causes $y(x_1, x_2, \ldots, x_n)$ to be either a maximum or a minimum.

For example, y may represent the cost of producing an automobile in a proposed assembly plant. This may depend upon the fixed cost of the proposed plant, the amortization rate of the plant, the number of automobiles to be manufactured per month, the rate of production, the degree of quality control to be imposed, the costs of labor, materials, utilities, and transportation, etc. The objective function, in this case, expresses the automobile production cost as a function of the above variables. Our goal is to define the objective function quantitatively and realistically, and then to select those values of the variables which minimize the production cost.

The function $y(x_1, x_2, \ldots, x_n)$ is of a known form in an ordinary optimization problem. However, there are many problems in which this formulation is not valid. For example, consider the problem of finding the path that will provide the shortest distance between two points on the surface of some object. For simplicity, assume that we are considering a plane. To solve this problem, a function must be found that will minimize an integral, I, of the form

$$I = \int_{x_1}^{x_2} ds, \tag{1.3.1}$$

where s represents distance and

$$ds = \sqrt{(dx)^2 + (dy)^2}. \tag{1.3.2}$$

Hence

$$I = \int_{x_1}^{x_2} \sqrt{1 + \left(\frac{dy}{dx}\right)^2}\, dx. \tag{1.3.3}$$

Such an integral is called a *functional*. Geometrically, a problem of this type may be interpreted as finding a curve in space, specified by parametric equations or functions, that maximizes or minimizes a functional of the functions. This is in contrast with the previous description of an ordinary optimization problem as one in which we seek a point in n-dimensional space that maximizes or minimizes the value of our known objective function.

In a mathematical sense the objective function is not restricted to any particular kind of expression. That is, the objective function may be an algebraic function of the independent variables, an integral involving the independent variables, or a function requiring the solution of a system of ordinary or partial differential equations for specific evaluation. The objective function may require that probabilities be included in its formulation. We shall simply treat the objective function as a functional relationship involving the independent variables, without regard for its specific form.

1.4 Multifactor Objectives

In many instances the objective of an ordinary optimization problem cannot be stated easily, since it involves the minimization or maximization of some function of dissimilar entities. Suppose, for example, that we wish to purchase a used automobile. If we are only concerned with obtaining the unit with the lowest initial cost, our objective is clear. Usually this is not the case, however, as we shall probably be concerned about other features such as styling and safety. It is not immediately obvious how we can express our feelings about style and safety on the same scale as our concern about cost.

To handle such multifactor situations, we can utilize an approach known as *value theory*. For each factor (style, safety, and cost) we set up a scale going from 0 to 1. We then examine each automobile model being considered for purchase and assign values, between 0 and 1, to our evaluation of safety, cost, and style. An automobile that was perfectly safe would have a safety value of 1, and a totally unsafe automobile would be assigned a value of 0. An automobile that cost nothing would have a cost value of 1, while the most expensive car being considered would be assigned a cost value of 0. (Note that an automobile having a cost one half that of the most expensive unit need not have a cost value of 0.5. That is, our value for money is not necessarily linear with money.) Once having done this we may write an equation for the overall value, U, of the automobile as

$$U = w_1 U_1 + w_2 U_2 + w_3 U_3, \tag{1.4.1}$$

where w_i is a weighting factor expressing the relative importance of an item i, $0 < w_i \leq 1$; U_1 is the cost value; U_2 is the safety value; and U_3 is the style value. When the weighting factors, w_i, are not known a priori, iterative methods have been proposed for their evaluation (cf. Guilford [1954]).

In most multifactor optimization problems, the factors entering into the objective function will be functions of our decision variables. Consider the optimization of an aircraft component. Generally, we would like to keep both component cost and weight low. Our objective function could take the form of

$$y = U = w_1 U_1(M) + w_2 U_2(C), \tag{1.4.2}$$

where $U_1(M)$ is the value assigned to the component weight, M; $U_2(C)$ is the value assigned to the component cost, C; and w_1 and w_2 are the respective weighting factors.

Since the component weight and cost are functions of the decision variables, x_i, we have

$$U_1(M) = f_1(x_1, x_2, \ldots, x_n) \tag{1.4.3}$$

and

$$U_2(C) = f_2(x_1, x_2, \ldots, x_n). \tag{1.4.4}$$

Our objective function can then be rewritten as

$$y = w_1 f_1(x_1, x_2, \ldots, x_n) + w_2 f_2(x_1, x_2, \ldots, x_n). \tag{1.4.5}$$

We would seek that set of decision variables which would maximize $y(x_1, x_2, \ldots, x_n)$. Again we should observe that neither U_1 nor U_2 need be linear functions.

A particularly important two-factor objective function arises in the assessment of "risky" ventures. Often, the more risk that is associated with a venture, the greater will be the expected return. A prudent investor will not only consider expected return but will choose an investment that does not expose him to an unacceptably high risk. The methods available for balancing expected return and risk will be discussed in Chapter 9.

Another type of problem which may arise is that of determining the coefficients of a given function which, in some sense, provide a "best" fit to a set of observations. This is an optimization problem, since we seek to minimize a measure of the difference between the predicted values, ϕ_i, and the observed values, ϕ_i'. The procedure which is used frequently is to minimize the weighted sum of the differences squared. Our objective function then becomes

$$y = \sum_{i=1}^{n} w_i(\phi_i - \phi_i')^2, \tag{1.4.6}$$

where the w_i are weighting factors, $0 < w_i \leq 1$, as before. When each of the weighting factors is set equal to 1, this reduces to the classical "least-squares" problem. Often the various ϕ_i represent different physical entities, however, so that the corresponding w_i must take on appropriate fractional values.

1.5 Stationary Values and Extrema

In Chapter 2 we shall discuss necessary and sufficient conditions for the existence of a maximum or a minimum in the neighborhood of a given point in an open region R. The point at which a function satisfies the necessary conditions for a maximum or minimum is called a *stationary point*. The value of the function at that point is called a *stationary value*. Not all stationary values represent optimal conditions. As we shall see, maxima, minima, inflection points, and saddle points are all examples of stationary points.

The value of a function that satisfies both the necessary and sufficient conditions at a given point is called an *extremum*. The sufficiency conditions rule out the possibility that an extremum be a stationary point which is not a maximum or a minimum. In later chapters we shall develop certain optimization techniques that enable us to find stationary values, but do not necessarily require that the stationary values be extrema. In such cases we must be careful not to assume that we have found a maximum or a minimum when in reality we may have an inflection point or a saddle point.

1.6 Relative and Absolute Extrema

Many functions possess more than one extremum of a given kind (e.g., more than one maximum). Hence we must recognize that the maximum or minimum which we may have found at some particular point is not necessarily the largest maximum or the smallest minimum in the entire region of interest. Contrast, for example, the problem of climbing Pike's Peak with that of finding the highest mountain in Colorado (which happens to be Mt. Elbert). An extremum that exists at a point within some subregion R' of the region R is called a *relative* or *local extremum;* the greatest maximum (or the smallest minimum) in R is called an *absolute* or *global maximum* (or an *absolute* or *global minimum*.)

Notice that an absolute extremum may exist either at an interior point of R or on the region boundaries. (In fact, in Chapter 4 we shall discuss a class of optimization problems whose extrema *always* occur on the boundaries of the region.) Figure 1.1 shows a continuous, one-dimensional function $y(x)$, $a \leq x \leq b$, which has relative maxima at a, x_2, and b, and relative minima at x_1 and x_3. The absolute maximum occurs on the boundary at $x = b$, whereas the absolute minimum occurs at the interior point x_3.

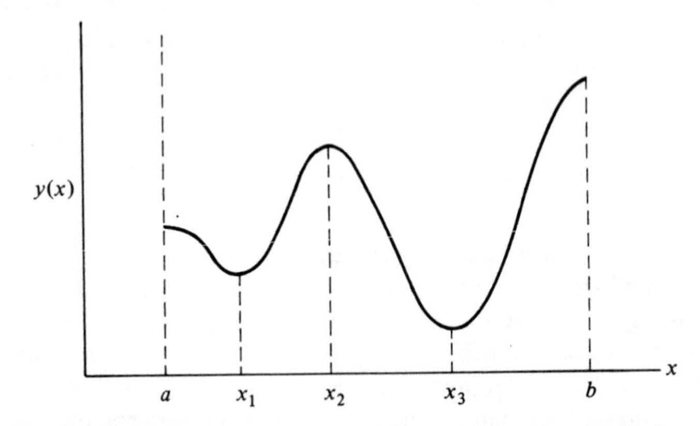

Figure 1.1. A One-Dimensional Function with Several Relative Maxima and Minima

The principal objective of optimization theory is the development of systematic and efficient techniques for determining the locations and values of all the relative extrema, with the subsequent determination of the absolute extrema. Unfortunately, this goal has not been attained for all types of optimization problems. In many cases an absolute extremum can be found only after extensive and relatively inefficient searching procedures. Even then, there may be the possibility that one of the relative extrema, which may, in fact, be the desired absolute extremum, has been overlooked.

1.7 Equivalence of Minimum and Maximum

A minimization problem can be converted into an equivalent maximization problem, and vice versa, as can be seen from Figures 1.2a and 1.2b. Clearly, the minimum of $y(x)$ and the maximum of $-y(x)$ both occur at x_0. Moreover, we can see that

$$\min\{y(x)\} = -\max\{-y(x)\}. \tag{1.7.1}$$

Thus we can restrict our attention to the characteristics of relative maxima if we wish, knowing that we can always establish an analogous discussion for the characteristics of relative minima through Equation (1.7.1).

1.8 Convex, Concave, and Unimodal Functions

Some functions have the particularly well-behaved characteristic that

$$y\{ax_1 + (1 - a)x_2\} \leq ay(x_1) + (1 - a)y(x_2), \qquad 0 < a < 1. \tag{1.8.1}$$

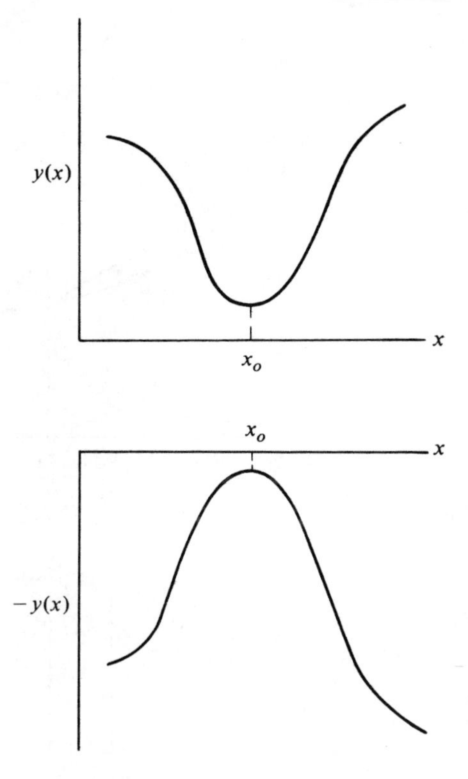

Figure 1.2. Conversion of Minimization Problem to Maximization Problem

In other words, given two points x_1 and x_2, the value of the function at some intermediary point $[ax_1 + (1 - a)x_2]$ is less than or equal to the weighted arithmetic average of $y(x_1)$ and $y(x_2)$. If Equation (1.8.1) holds for all points x_1 and x_2, then the function $y(x)$ is said to be a *convex function*. Furthermore, $y(x)$ is said to be *strictly convex* if Equation (1.8.1) is a strict inequality. An example of a one-dimensional convex function is shown in Figure 1.3. It is intuitively obvious (and can indeed be proved rigorously—cf. Hadley [1964]) that a convex function has one and only one relative minimum. Moreover, a strictly convex function takes on its absolute minimum at one and only one point.

The above discussion also applies to multidimensional functions $y(x_1, x_2, \ldots, x_n)$, providing Equation (1.8.1) is written as*

$$y\{a\mathbf{X}_1 + (1 - a)\mathbf{X}_2\} \leq ay(\mathbf{X}_1) + (1 - a)y(\mathbf{X}_2), \qquad 0 < a < 1, \qquad (1.8.2)$$

where $\mathbf{X}_1$ and $\mathbf{X}_2$ are points in an n-dimensional space, represented by the vectors

---

* The basic mathematical concepts of n-dimensional Euclidean space are presented in Appendix B.

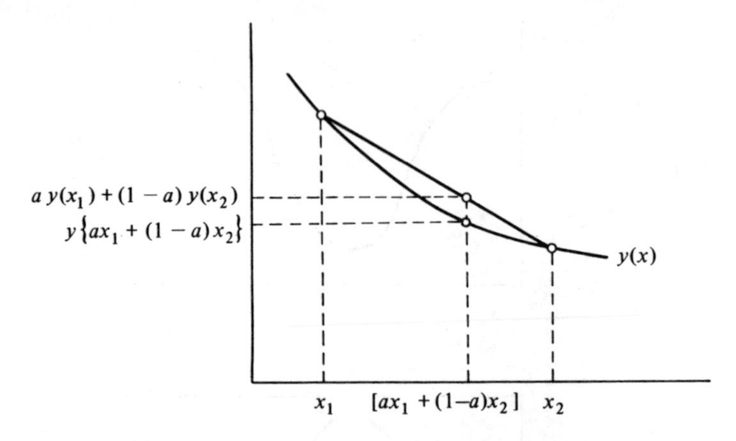

Figure 1.3. One-Dimensional Convex Function

$$\mathbf{X}_1 = \begin{bmatrix} x_{11} \\ x_{21} \\ \cdot \\ \cdot \\ \cdot \\ x_{n1} \end{bmatrix} \quad \text{and} \quad \mathbf{X}_2 = \begin{bmatrix} x_{12} \\ x_{22} \\ \cdot \\ \cdot \\ \cdot \\ x_{n2} \end{bmatrix}. \tag{1.8.3}$$

Similarly, a function $y(x_1, x_2, \ldots, x_n)$ is said to be a *concave function* if

$$y\{a\mathbf{X}_1 + (1 - a)\mathbf{X}_2\} \geq ay(\mathbf{X}_1) + (1 - a)y(\mathbf{X}_2), \qquad 0 < a < 1, \tag{1.8.4}$$

where $\mathbf{X}_1$ and $\mathbf{X}_2$ are any two points in the n-dimensional space as defined by Equation (1.8.3). The function $y(\mathbf{X})$ is said to be *strictly concave* if Equation (1.8.4) is a strict inequality. Figure 1.4 shows a one-dimensional concave function. It can be seen that a concave function has one and only one relative maximum (cf. Hadley [1964]), and that a strictly concave function takes on its absolute maximum at a unique point. In Chapter 2 we shall develop a criterion, in terms of second partial derivatives, that will assist us in recognizing concavity or convexity for any given function $y(\mathbf{X})$.

A function that has only one relative maximum is said to be *unimodal.* Stated differently, consider the function $y(x_1, x_2, \ldots, x_n)$ which takes on a maximum at $(x_{1o}, x_{2o}, \ldots, x_{no})$ or, using vector notation, at $\mathbf{X}_o$. If $\mathbf{X}_1$ and $\mathbf{X}_2$ are any two points in a neighborhood of $\mathbf{X}_o$ with $\|\mathbf{X}_1 - \mathbf{X}_o\| < \|\mathbf{X}_2 - \mathbf{X}_o\|$, then a path passing successively through $\mathbf{X}_o$, $\mathbf{X}_1$, and $\mathbf{X}_2$ is unimodal if $y(\mathbf{X}_2) < y(\mathbf{X}_1) < y(\mathbf{X}_o)$. If any point $\mathbf{X}$ can be connected to $\mathbf{X}_o$ by a unimodal path defined by $y(\mathbf{X})$ for all points $\mathbf{X} \neq \mathbf{X}_o$, then the function $y(\mathbf{X})$ is

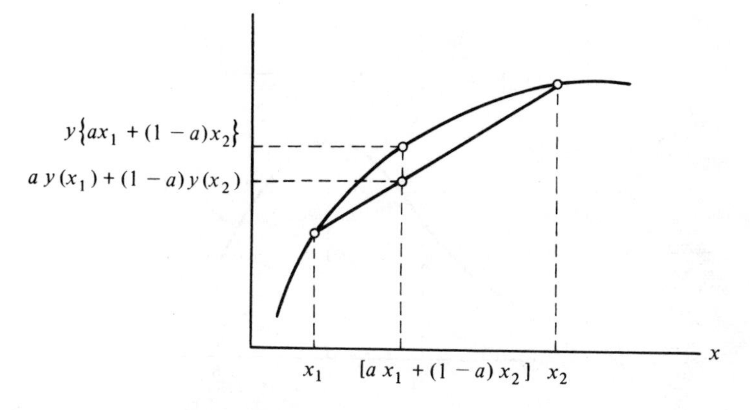

Figure 1.4. One-Dimensional Concave Function

unimodal. If **X** and $\mathbf{X}_o$ are connected by unimodal *straight-line* paths for all **X**, then the function $y(\mathbf{X})$ is said to be *strongly unimodal.* An example of a one-dimensional unimodal function is shown in Figure 1.5.

Notice that all concave functions are unimodal, although not all unimodal functions are concave. Hence concavity is a stronger condition than unimodality. Figure 1.6 shows a one-dimensional unimodal function that is convex for $a \leq x \leq x_1$ and concave for $x_1 \leq x \leq b$.

Finally, in our definitions of concave, convex, and unimodal functions we have said nothing about continuity or differentiability. A convex or a concave function need not be differentiable, although such functions must be continuous. A unimodal function need not even be continuous. In later chapters we shall relate the kinds of functions defined above to specific optimization techniques.

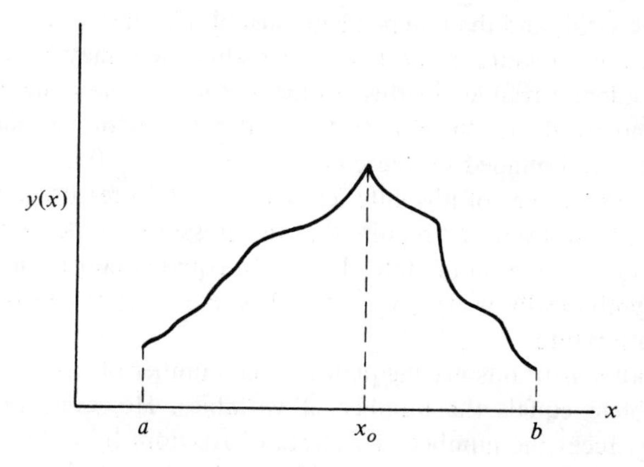

Figure 1.5. One-Dimensional Unimodal Function

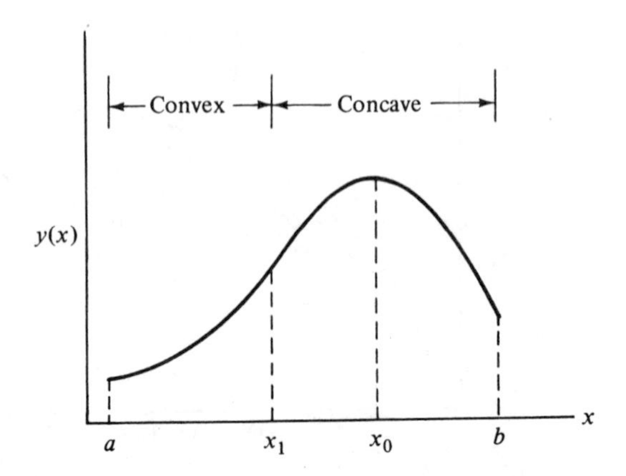

Figure 1.6. Unimodal Function with Both Convex and Concave Regions

1.9 Constraints

Sometimes the problem variables within a region are not entirely free, but must satisfy certain bounds or functional relationships. These auxiliary conditions are known as *constraints.* Usually the constraints represent either system limitations or the physical and economic laws that the variables must satisfy.

For example, the optimal design of a chemical reactor might involve the determination of some absolute temperature and absolute pressure that would maximize the yield of product. The objective function $y(x_1, x_2)$ would represent the yield, and the independent variables x_1 and x_2 would represent temperature and pressure, respectively (in reality there may be other significant independent variables in this problem, such as the concentrations of reactants and products, the flow rates, and the reactor volume). Clearly, the variables are bounded by the conditions that $x_1 > 0$ and $x_2 > 0$, since negative or zero values of absolute temperature and pressure are physically meaningless. In addition, we require that the pressure, temperature, and density satisfy some equation of state. This is an equality constraint, which can be written symbolically as $g(x_1, x_2, x_3) = 0$, where x_3 represents the density of the reaction mixture.

When all constraints are inequalities, the number of degrees of freedom of the problem equals the number of variables. However, each equality constraint reduces the number of degrees of freedom by 1. Each such constraint means that one of the variables can be expressed in terms of the others. Hence the number of *independent* variables equals the total number of variables less the number of equality constraints. In some numerical techniques it is desirable to deal only with the independent variables in determination of the optimum.

The presence of constraints in an optimization problem sometimes introduces considerable restrictions on the value of the solution. For example, Figure 1.7 shows the level curves of a unimodal, two-dimensional function

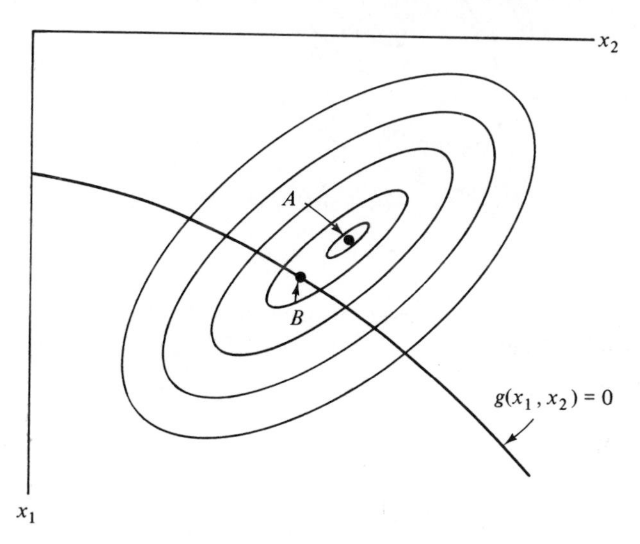

Figure 1.7. Contour Curves of Unimodal Function Subject to a Single Equality Constraint

subject to a single equality constraint $g(x_1, x_2) = 0$. Notice that the locus of the extremum lies along the curve defined by the constraint rather than in the entire x_1x_2 quadrant; hence the introduction of the equality constraint has greatly decreased the allowable region in which we seek an optimum. The *constrained optimum* for this problem is, of course, at point B rather than at point A. Notice that point B has a less favorable value than point A, but point B does not violate the constraint, whereas point A does.

Figure 1.8 shows the same unimodal function as Figure 1.7. Now, however, the constraint is expressed as an inequality $g(x_1, x_2) \leq 0$ rather than as an equality. The locus of the extremum remains a reasonably large, two-dimensional region, although it is a smaller region than in the case of the unconstrained problem. The extremum in this case is equal to the unconstrained extremum, which is found at point A. Notice that the inequality constraint has the effect of forming a part of the region boundary.

In Figure 1.9 we see the same problem as before except that the inequality is reversed; i.e., $g(x_1, x_2) \geq 0$. Again the inequality constraint forms a part

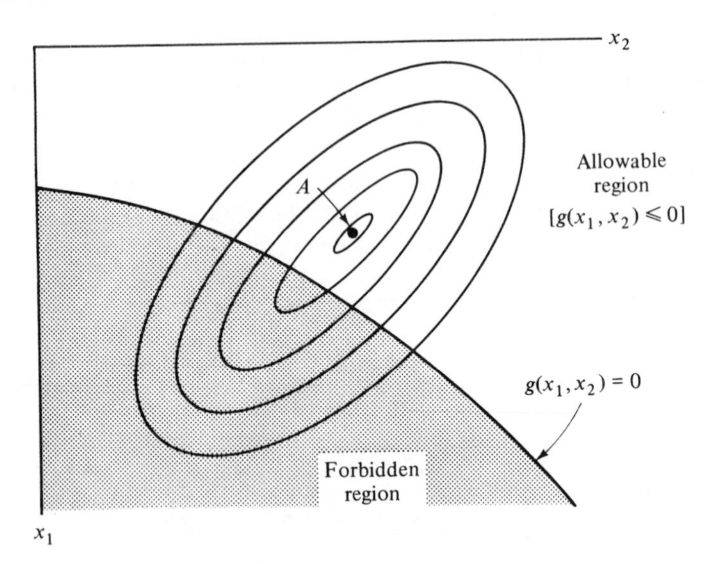

Figure 1.8. Contour Curves of Unimodal Function Subject to Inequality Constraint

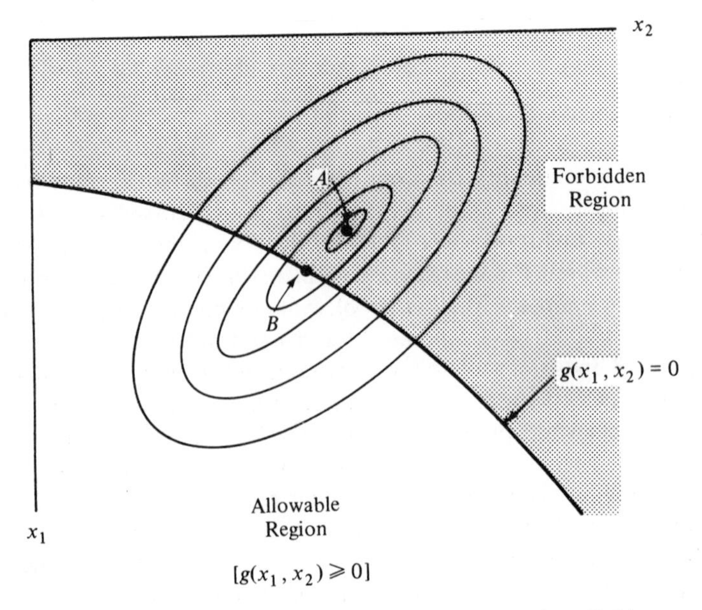

Figure 1.9. Contour Curves of Unimodal Function Subject to Reversed Inequality Constraint

of the region boundary. Now, however, the unconstrained extremum lies in the forbidden region, so that the constrained optimum again lies at point *B*.

Figure 1.10 shows the level curves of the same unimodal function that we have seen before. Now, however, we have imposed a different equality constraint. The shape of this particular constraint introduces *two* constrained optima, at points *B* and *C*. This further complicates the determination of an absolute optimum, since both points must be recognized and compared.

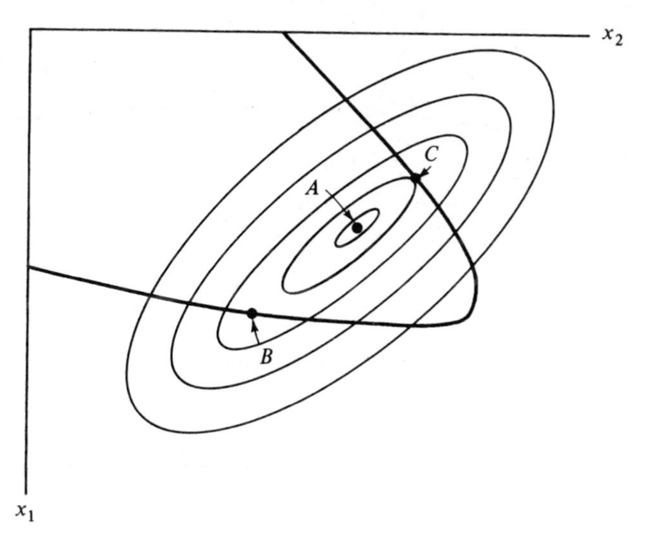

Figure 1.10. Contour Curves of Unimodal Function Showing Two Constrained Optima

A given optimization technique would, in general, find only one of the two local optima. Which point would be selected is dependent upon both the technique being used and the choice of the initial point. It is sometimes necessary to solve a problem of this nature many successive times in order to have some reasonable assurance that all the local optima have been found.

It is, of course, most significant that the presence of this constraint negates our original advantage in working with a unimodal objective function.

Figure 1.11 shows a unimodal function subject to two equality constraints, $g_1(x_1, x_2) = 0$ and $g_2(x_1, x_2) = 0$. Since the objective function is two dimensional and the constraints are represented by two equations in two unknowns, point *B* is the only point that can satisfy the constraints, and, hence, is the optimal point. Notice that in this problem we have no freedom in the value we may assign to the objective function; the single value that the objective function can take on is determined uniquely by the constraints.

Finally, in Figure 1.12 we see the level curves of a unimodal objective

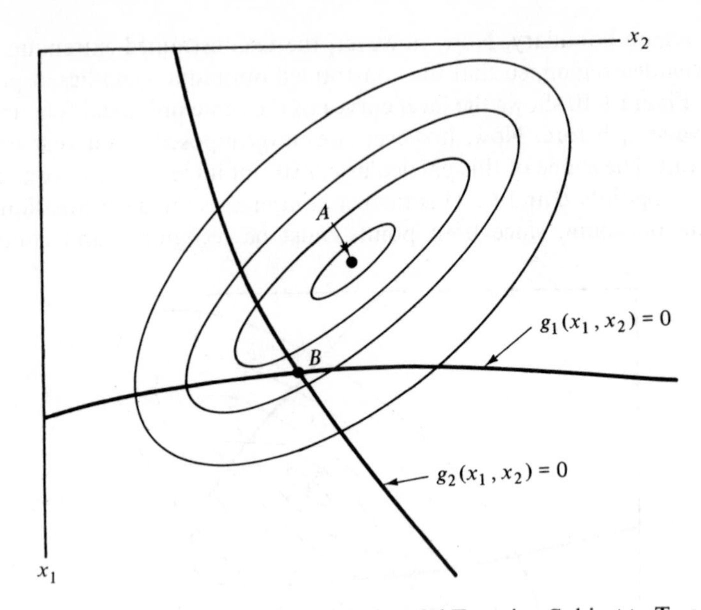

Figure 1.11. Contour Curves of Unimodal Function Subject to Two Constraints

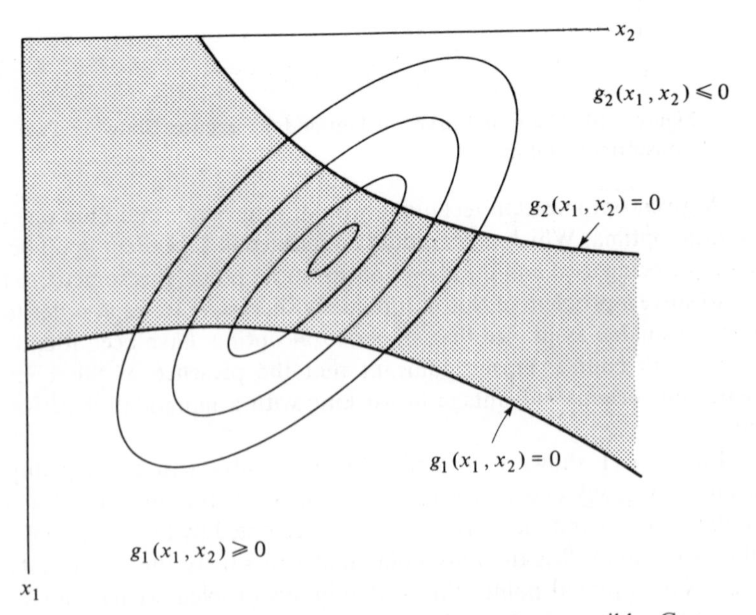

Figure 1.12. Unimodal Function Subject to Incompatible Constraints

function that is subject to two indequality constraints, $g_1(x_1, x_2) \leq 0$ and $g_2(x_1, x_2) \geq 0$. Although this situation is in general allowable, the particular constraints shown here are such that one constraint cannot be satisfied without violating the other. Hence the constraints can never be satisfied simultaneously, and the problem cannot be solved.

The difficulty with problems of this type lies in their formulation, as the imposed constraints are really asking that an impossible situation be satisfied. For example, it is obviously senseless to ask that a foundry produce no more than 1000 automobile engine castings a day when the assembly plants supplied solely by that foundry require a minimum of 1200 castings simply to maintain an 8-hour shift. Thus the analyst must exercise considerable care in order to formulate an optimization problem that is both meaningful and solvable.

In succeeding chapters we shall discuss further the characteristics of constrained optimization problems, and a variety of methods will be presented for solving such problems. For now, let us simply take note of the difficulties introduced into an optimization problem by the presence of constraints.

1.10 Mathematical Programming Problems

To summarize much of our earlier discussions, most optimization problems require the maximization or minimization of a given objective function subject to a set of constraints and a set of bounds on the policy variables. Mathematically, we wish to:

$$\text{Extremize } y(\mathbf{X}) \text{ with respect to } \mathbf{X} \tag{1.10.1}$$

subject to

$$\begin{aligned} g_i(\mathbf{X}) &\leq 0, \quad i = 1, 2, \ldots, k, \\ g_i(\mathbf{X}) &\geq 0, \quad i = k+1, k+2, \ldots, l, \\ g_i(\mathbf{X}) &= 0, \quad i = l+1, l+2, \ldots, m, \end{aligned} \tag{1.10.2}$$

and

$$\mathbf{A} \leq \mathbf{X} \leq \mathbf{B}^*, \tag{1.10.3}$$

where **A** and **B** are vectors whose components represent the lower and upper bounds, respectively, of the components of **X**.

* This condition is more frequently written in the less general form $\mathbf{X} \geq \mathbf{0}$.

Notice that we need not distinguish between the two types of inequality constraints, since a ($\geq$) constraint can always be converted to a ($\leq$) constraint by simply multiplying both sides of the expression by a minus sign. Observe also that the number of equality constraints must be less than the number of variables; otherwise the variables will be either uniquely determined by the constraints (if $m - l = n$) or overspecified ($m - l > n$). The number of inequality constraints is unrestricted.

Problems of this type are known as *mathematical programming problems.** Any vector **X** that satisfies all the constraints and bounds is called a *feasible solution* to the programming problem. There will in general be an infinite number of feasible solutions to a properly posed mathematical programming problem. The *particular* feasible solution that *extremizes* the objective function is called the *optimal feasible solution* (often referred to specifically as the *maximum feasible solution* or the *minimum feasible solution*). The optimal feasible solution vector $\mathbf{X}_o$ is frequently referred to as the *solution vector* or the *policy vector*.

1.11 Dimensionality

In principle it is just as easy to discuss the optimization of an n-dimensional function (where n is a large number) as a two-dimensional function. However, as is the case with solving simultaneous linear equations, the actual work involved in obtaining solutions increases enormously with a modest increase in the number of independent variables (i.e., with increasing *dimensionality*). Therefore, we must recognize high dimensionality as a practical difficulty, and we should realize that the utility of a given optimization technique may depend strongly on the dimensionality of the problem. Certain powerful optimization techniques, such as dynamic programming, become practically useless when the dimensionality of the problem exceeds four or five independent variables—even with the largest available electronic computers!

1.12 Simultaneous and Sequential Optimization

Causal reference is frequently made to *simultaneous* versus *sequential* methods of optimization. This distinction actually has two meanings. We shall consider each meaning separately.

---

* The term "programming" as used above referred originally to the scheduling of events or activities. There is no immediate connection with computer programming, although most mathematical programming problems are solved on a digital computer.

Most optimization techniques are designed such that the objective function is successively improved as we move from some point $\mathbf{X}_i$ to the next (i.e., some *better*) point $\mathbf{X}_{i+1}$. In choosing the point $\mathbf{X}_{i+1}$ we make use of past information obtained at points $\mathbf{X}_i, \mathbf{X}_{i-1}, \ldots$, etc. This is known as a *sequential* technique. In contrast, some methods (in particular, *random* methods) choose the current search point $\mathbf{X}_{i+1}$ independently of the history of previous search points. Such methods are called *nonsequential* or *simultaneous.* Although nonsequential techniques may appear to be wasteful, since they ignore previously determined information, we shall see that they can be very useful when sequential methods fail.

In a different sense, simultaneous optimization techniques solve a problem by determining the optimal value of all variables at the same time. Some of these techniques, such as gradient methods and linear programming, employ iterative algorithms*, which gradually converge to the desired optimum conditions. All the variables are evaluated during each iteration, even though none of these variables may attain its optimal value. Systematic procedures are then applied to successive iterations to move closer to the desired optimum. Other simultaneous techniques, such as classical calculus, are noniterative, so that they yield an optimum after a fixed procedure of predetermined length.

Sequential techniques obtain the desired optimum by means of a step-by-step procedure (sometimes called a *stagewise* procedure). The stages are traversed systematically in such a manner that information obtained at a given stage is used to determine further information at the next stage, and so on. The optimal values of certain of the independent variables are determined at each stage—in contrast to simultaneous tecniques, where all the optimal independent variables are obtained at once. Dynamic programming is a powerful, well-known optimization technique of this type.

REFERENCES

Guilford, J. P., *Psychometric Methods*, 2nd ed., McGraw-Hill, New York, 1954.

Hadley, G., *Nonlinear and Dynamic Programming*, Addison-Wesley, Reading, Mass., 1964.

Wilde, D. J., and C. S. Beightler, *Foundations of Optimization*, Prentice-Hall, Englewood Cliffs, N.J., 1967.

---

* An *algorithm* is a mathematical strategy or computational scheme.

PROBLEMS

1.1. Determine whether or not the following functions are convex or concave:

(a) $y = (x_1 - 2)^2 + 3(x_2 + 1)^2,\ x_1, x_2 \geq 0$
(b) $y = 4x_1 - 3x_2 + 6x_3,\ x_1, x_2, x_3 \geq 0$
(c) $y = 2x_1^2 + x_2^2 + 3x_3^2,\ x_1, x_2, x_3 \geq 0$
(d) $y = 2x_1^2 + x_2^2 - 3x_3^2,\ x_1, x_2, x_3 \geq 0$
(e) $y = x_1 - x_2^4 - 2x_3 - x_4^2,\ x_1, x_2, x_3, x_4 \geq 0$

1.2. If $f_1(\mathbf{X})$ and $f_2(\mathbf{X})$ are convex functions within the region $\mathbf{a} \leq \mathbf{X} \leq \mathbf{b}$, show that

$$y = f_1(\mathbf{X}) + f_2(\mathbf{X})$$

is also convex within the same region. Is this also true of the product

$$z = f_1(\mathbf{X}) \cdot f_2(\mathbf{X})?$$

Can the results be generalized to

$$y = \sum_{i=1}^{n} f_i(\mathbf{X}), \qquad \mathbf{a} \leq \mathbf{X} \leq \mathbf{b},$$

and

$$z = \prod_{i=1}^{n} f_i(\mathbf{X}), \qquad \mathbf{a} \leq \mathbf{X} \leq \mathbf{b}?$$

1.3. Given the functions

$$f_1(x_1, x_2) = x_1 x_2,$$
$$f_2(x_1, x_2) = x_1^2 + 3x_2^2,$$

solve the following constrained optimization problems graphically:

(a) Maximize $f_1(x_1, x_2)$, subject to

$$f_2(x_1, x_2) \leq 9,$$
$$x_1, x_2 \geq 0$$

(b) Maximize $f_2(x_1, x_2)$ subject to

$$f_1(x_1, x_2) \leq 1,$$
$$x_1, x_2 \geq 0$$

(c) Maximize $f_1(x_1, x_2)$, subject to

$$f_2(x_1, x_2) \geq 9,$$
$$x_1, x_2 \geq 0$$

(d) Minimize $f_2(x_1, x_2)$, subject to

$$f_1(x_1, x_2) \geq 1,$$
$$x_1, x_2 \geq 0$$

(e) Minimize $f_2(x_1, x_2)$, subject to

$$f_1(x_1, x_2) \leq 1,$$
$$x_1, x_2 \geq 0$$

(f) Maximize $f_1(x_1, x_2)$, subject to

$$f_2(x_1, x_2) \leq 3,$$
$$x_1 - x_2 \geq 5,$$
$$x_1, x_2 \geq 0$$

1.4. Solve problems 1.3(a) and (b) as equivalent minimization problems. In each case write out the problem statement and solve graphically.

1.5. A common problem in numerical analysis is the determination of the real roots of two simultaneous, nonlinear algebraic equations, $f_1(x_1, x_2) = 0$ and $f_2(x_1, x_2) = 0$. Suggest how this problem might be solved by transforming it into an equivalent optimization problem.

1.6. An *equation of state* is an algebraic equation that relates the pressure, specific volume (i.e., volume per unit mass or volume per unit mole), and absolute temperature of a gaseous substance. The pressure, volume, and temperature of hydrocarbon mixtures can often be accurately related by means of the expression

$$P = \frac{10.73a(T + 460)}{(V - b)^m} - \frac{c}{V^n},$$

where P represents pressure in pounds per square inch (psi), V represents specific volume in cubic feet per pound mole, and T represents temperature in degrees Fahrenheit. The constants $a, b, c, m, n \geq 0$ must be determined for each particular substance.

Suppose that a set of constants is to be determined which will "fit" the above equation of state to the following experimental data:

P (psi)	V (ft³/lb mole) 100°F	160°F	220°F	280°F
1000		5.40	6.35	7.21
1500	2.44	3.28	4.02	4.66
2000	1.77	2.36	2.92	3.43
2500	1.49	1.91	2.33	2.74
3000	1.36	1.65	1.98	2.31
3500	1.27	1.49	1.76	2.03
4000	1.21	1.39	1.60	1.83
4500	1.16	1.32	1.49	1.69
5000	1.12	1.26	1.41	1.58
6000	1.07	1.18	1.30	1.43
7000	1.03	1.11	1.21	1.32
8000	0.995	1.07	1.16	1.25

Develop a mathematical model of an optimization problem whose solution will yield the desired curve fit.

1.7. Restructure the mathematical model of problem 1.6 so that different weights can be assigned arbitrarily to each of the two terms on the right-hand side of the equation of state.

1.8. A farmer wishes to determine the optimum winter diet for his cattle. For good health his cattle should each receive at least 3 lb of protein, 10 lb of carbohydrates, and 30 lb of roughage per day. The total consumption of food must be less than 65 lb/day. He has available for feed oats, corn, and hay. The prices and composition of these feeds are

Feed	*Composition (%)* *Protein*	*Carbohydrate*	*Roughage*	*Water*	*Cost/lb*
Oats	8	20	40	32	0.09
Corn	4	20	45	31	0.07
Hay	1	5	65	29	0.015

Derive a mathematical model that, when solved, will yield the minimum-cost diet subject to the required constraints.

1.9. A college admissions officer believes that high school grades, intelligence (based on IQ test), and leadership ability (based on an interview) are the three characteristics which should be considered for college admission. He feels that grades are as important as intelligence, but that leadership ability is twice as important as grades. How would you establish a function that would allow each candidate to be rated with a single numerical score?

1.10. A company manufactures three different kinds of products, utilizing two scarce raw materials. The profit derived from the scale of each product is proportional to the number of units of that product manufactured in a given period of time. Since the raw materials are scarce, no more than some specified amount of each raw material can be utilized during a given time period. Derive a mathematical model whose solution will yield an optimal distribution of products. Will this model yield two or more local optima? Explain.

1.11. Suppose that the profit per unit for each kind of product in problem 1.10 is dependent upon the number of units manufactured in a given period of time. How will this alter the mathematical model? Will this affect the situation with regard to local optima?

1.12. A rocket is to be fired vertically from a flat surface. If atmospheric resistance is negligible, the acceleration of gravity is constant, the thrust is vertical, and the exit velocity of the propellant gases is constant, derive a model that will allow determination of the propellant mass flow rate as a function of time such that the altitude attained by the rocket is maximized. Assume that the propellant mass flow rate is bounded by $0 \leq \dot{m} \leq \dot{m}_0$, where $\dot{m}$ represents the mass flow rate and $\dot{m}_0$ is a constant.

The movement of the rocket is governed by Newton's laws of motion:

$$v = -\frac{\alpha \dot{m}}{m} - g,$$

where v represents the rocket velocity, given by

$$v = \dot{x},$$

and x represents the height of the rocket above the firing surface, α is the exit gas velocity, m is the mass of the rocket casing and propellant (which decreases with time), and g is the gravitational acceleration. Assume that the initial mass (including propellant) of the rocket is known, the initial velocity is zero, and the final mass and velocity are specified, nonzero quantities. (Cf. G. Leitmann, *Optimization Techiques with Applications to Aerospace Systems*, Academic Press, New York, 1962, pp. 127–143.)

Note that the term $\dot{m}$ and $\dot{x}$ used in this problem are shorthand notations for dm/dt and dx/dt, respectively, where t represents time.

OPTIMIZATION FUNDAMENTALS

To understand why there is a need for specialized optimization techniques, let us first review the classical aspects of optimization theory—the application of calculus to problems involving maxima and minima. We shall see that the use of calculus is restricted to the optimization of functions which are piecewise continuous and differentiable. Even then, the application of calculus to multidimensional problems frequently involves much tedious and burdensome mathematical manipulation. Under these conditions the use of the classical calculus becomes impractical even if possible.

We first ask whether or not a continuous and differentiable function $y(x_1, x_2, \ldots, x_n)$ has a maximum or a minimum (extremum) at some point $(x_{1o}, x_{2o}, \ldots, x_{no})$ in a closed region R. The answer to this question is provided in the affirmative by a well-known theorem of Weierstrass, which we state without proof (cf. Courant, Vol. I [1936]): *Every function which is continuous in a closed region R of the variables $(x_1, x_2, \ldots, x_n)$ possesses a largest and a smallest value within the interior or on the boundary of that region.* Notice that the theorem assures us of the *existence of extrema* within a region or on its boundaries. A function that is piecewise continuous within a region R can be broken down into subregions R' in which the function is continuous

2

and in which the Weierstrass theorem applies. The Weierstrass theorem can even be applied to functions that are constant within a given region; this, of course, is a trivial case for which every point in R and on its boundaries is both a maximum and a minimum.

Given the existence of extrema for a continuous function in R or on its boundaries, we then ask how many extrema exist, where they exist, and whether a given extremum represents a relative or absolute maximum or minimum. Unfortunately, these questions cannot be easily answered. Moreover, the existence of inflection points and saddle points makes it necessary to distinguish between stationary values and extrema—at the expense of additional mathematical complexity.

To answer at least some of the above questions, let us proceed to examine the behavior of first unconstrained and then constrained functions that are continuous and differentiable in a closed region R. We shall also assume that the constraints, when present, are continuous and differentiable. We shall consider both the classical treatment of equality constraints and the more recent extension to inequality constraints.

2.1 One-dimensional Functions

Let us consider the function $y(x)$ to be continuous and differentiable within an open interval $a < x < b$. (This interval may actually be a subinterval within a larger interval $a_o \leq a < x < b < b_o$ or $a_o < a < x < b \leq b_o$, in which the function is piecewise continuous.) Suppose that at some interior point x_o, $a < x_o < b$, y takes on a greater value than at neighboring points; i.e., $y(x_o) > y(x)$. If $x < x_o$, then $y(x)$ must decrease as x decreases, so that $y'(x_o) \geq 0$. Similarly, if $x > x_o$, $y(x)$ must decrease as x increases, so that $y'(x_o) \leq 0$. Since $y(x)$ is differentiable, $y'(x_o)$ must be single valued, and hence it must vanish at x_o; i.e., $y'(x_o) = 0$. This situation is shown schematically in Figure 2.1.

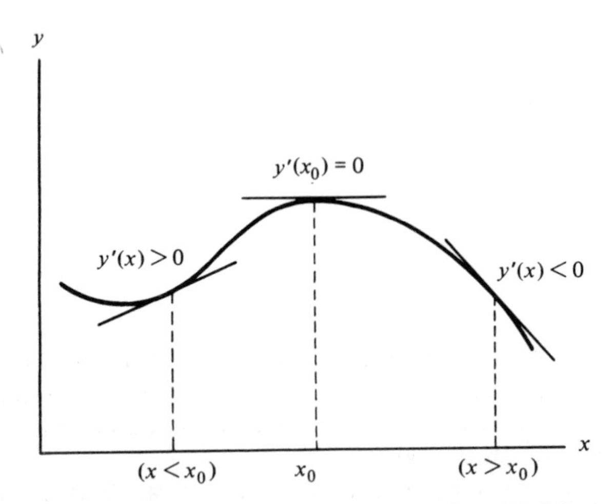

Figure 2.1. Derivatives of Continuous Function with Maximum

We can apply similar reasoning to some other point $x_1 \neq x_o$, $a < x_1 < b$, at which y takes on a lesser value than at neighboring points; i.e., $y(x_1) < y(x)$. If $x > x_1$, then $y(x)$ must increase as x decreases, so that $y'(x_1) \leq 0$. If $x > x_1$, then $y(x)$ must increase as x increases, so that $y'(x_1) \geq 0$. Again, $y'(x_1)$ must be single valued, so that $y'(x_1) = 0$. This case is illustrated in Figure 2.2.

So far we have established that $y'(x_o) = 0$ is a necessary condition for the existence of an extremum at x_o. The vanishing first derivative is not a sufficient condition, however, as can be seen in Figure 2.3. Clearly, $y(x_o)$ is neither a relative maximum nor a relative minimum; yet the curve is horizontal at x_o, and $y'(x_o) = 0$. This is called a *horizontal inflection point*. Notice that if $x < x_o$, then $y(x) > y(x_o)$ for x sufficiently close to x_o. Hence y in-

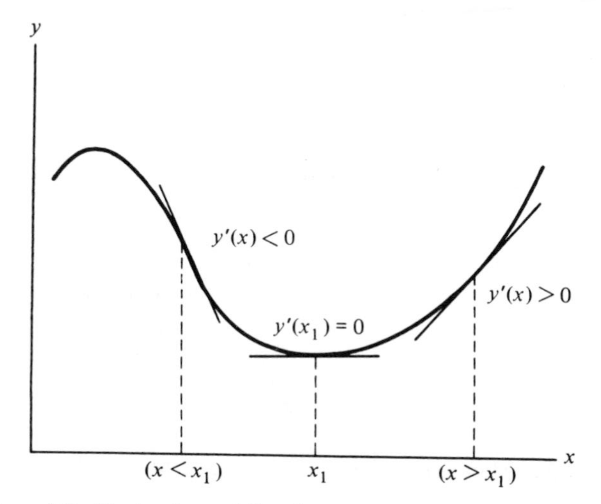

Figure 2.2. Derivatives of Continuous Function with Minimum

creases as x decreases, so that $y'(\xi) < 0$, $x < \xi < x_o$. Similarly, if $x_o > x$, then $y(x) < y(x_o)$, so that y decreases as x increases. Thus $y'(\xi) < 0$, $x_o < \xi < x$. In contrast to the relative maximum or minimum, $y'(x)$ is of the same sign on either side of a horizontal inflection point, which means that $y'(x_o)$ is either a relative maximum or a relative minimum. This is illustrated in Figure 2.4, where the derivative of the function shown in Figure 2.3 is plotted against x.

We can obtain further insight into the differences between relative maxima, minima, and horizontal inflection points by considering the higher derivatives. In the vicinity of the maximum shown in Figure 2.1, for example, the slope decreases monotonically, which implies that $y''(x_o) \leq 0$. For the minimum shown in Figure 2.2, the slope increases monotonically in the vicinity of x_1; hence $y''(x_1) \geq 0$. On the other hand, Figure 2.4 shows us that $y'(x)$ passes through an extremum at a horizontal inflection point of y, which requires that $y''(x_o) = 0$. Thus, *for a function which varies continuously over an open interval $a < x < b$, if $y'(x_o) = 0$ and $y''(x_o) < 0$, $a < x_o < b$, y passes through a relative maximum at x_o; if $y'(x_o) = 0$ and $y''(x_o) > 0$, $y(x_o)$ is a relative minimum; if, however, both $y'(x_o)$ and $y''(x_o)$ vanish, the behavior of y at x_o is inconclusive.*

If both $y'(x)$ and $y''(x)$ (and perhaps some successive higher derivatives) vanish at an interior point x_o, we can determine the behavior of $y(x_o)$ by application of Taylor's theorem about the point x_o:

$$y(x) = y(x_o) + y'(x)_o[x - x_o] + \frac{1}{2!}y''(x_o)[x - x_o]^2 + \cdots + \frac{1}{n!}y^{(n)}(x_o)[x - x_o]^n + R_n, \tag{2.1.1}$$

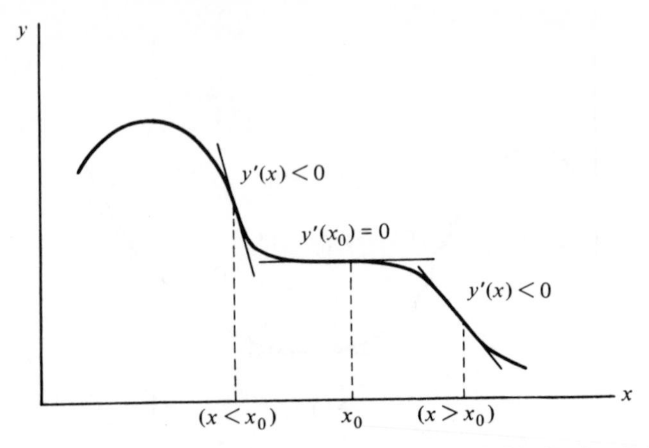

Figure 2.3. Continuous Function with Horizontal Inflection Point

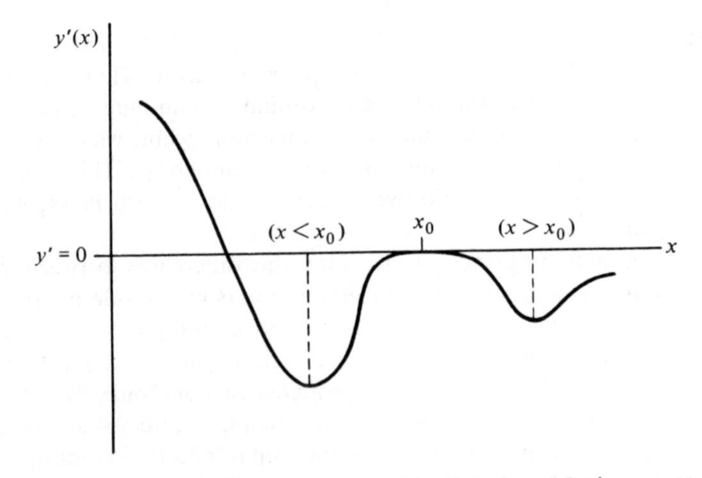

Figure 2.4. Continuous Function with Relative Maximum at Horizontal Inflection Point

where R_n, the remainder term, is given by

$$R_n = \frac{1}{(n+1)!} y^{(n+1)}(\xi)[x - x_o]^{(n+1)} \tag{2.1.2}$$

and $x < \xi < x_o$ or $x_0 < \xi < x$. Let $y'(x_o) = y''(x_o) = 0$ and let $R_n = R_2$, where

$$R_2 = \frac{1}{3!} y'''(\xi)[x - x_o]^3, \qquad |y'''(\xi)| > 0. \tag{2.1.3}$$

Now R_2 will be of different sign for $x < x_o$ and $x > x_o$, providing the sign of $y'''(\xi)$ remains fixed. In particular, as ξ approaches x_o, $y(x)$ will be *less* than $y(x_o)$ as x_o is approached from one side, and *greater* than $y(x_o)$ as x_o is approached from the other side. We therefore must have a horizontal inflection point at x_o for the particular case $y'(x_o) = y''(x_o) = 0$, $|y'''(x_o)| > 0$. Furthermore, the same argument applies whenever the remainder term involves an odd power of $[x - x_o]$, i.e., whenever the highest vanishing derivative at x_o is even. If, however, the highest vanishing derivative at x_o is odd, the power of $[x - x_o]$ in the remainder term will be even and the term $[x - x_o]^{(n+1)}$ will always be positive. Hence $y(x)$ will vary in the same direction on either side of x_o, and $y(x_o)$ will necessarily be either a relative maximum or a relative minimum.

We can summarize our discussion of necessary and sufficient conditions for relative extrema and inflection points of one-dimensional functions as follows:

Let $y(x)$ be defined over the open interval $a < x < b$. If $y'(x_o) = y''(x_o) = \ldots = y^{(n)}(x_o) = 0$ but $y^{(n+1)}(x_o) > 0$ at some interior point x_o, then $y(x)$ has a relative maximum at x_o if n is odd and $y^{(n+1)}(x_o) < 0$, and $y(x)$ has a relative minimum at x_o if n is odd but $y^{(n+1)}(x_o) > 0$. If n is even, then $y(x)$ has a horizontal inflection point at x_o.

So far our discussion has been concerned only with relative maxima and minima at interior points. We can evaluate these extrema by solving for those values of x which yield a vanishing first derivative, and then determining the corresponding values of y. We have no way of knowing *a priori* how many extrema exist in a given closed interval. To determine the *absolute* maximum or minimum in the closed interval, it is necessary to evaluate *all* the relative extrema at the interior points and compare them with the values that the function takes on at the boundaries.

EXAMPLE 2.1.1

Determine the location of the stationary points for the function

$$y = x \cos x$$

within the interval $-\pi \leq x \leq \pi$. Identify the stationary points and locate the absolute maximum and the absolute minimum.

The first and second derivatives can easily be written as

$$\frac{dy}{dx} = \cos x - x \sin x,$$

$$\frac{d^2y}{dx^2} = -2 \sin x - x \cos x.$$

Equating the first derivative to zero and solving numerically, we see that the first derivative vanishes at $x = \pm 0.860$. Hence we have stationary values at these points.

Evaluating the second derivatives at the stationary points, we obtain $y'' = -2.077$ at $x = 0.860$ and $y'' = 2.077$ at $x = -0.860$; thus we have a relative maximum at $x = 0.860$ and a relative minimum at $x = -0.860$. The values of these relative extrema are $y = 0.561$ and $y = -0.561$, respectively. At the bounds $x = \pm\pi$, the function takes on the values $y = \mp\pi$. Therefore, the function has its absolute maximum at the boundary $x = -\pi$ and its absolute minimum is at the other boundary, $x = \pi$.

The characteristics of the function $y = x \cos x$ can be seen quite clearly from the graph of the function, which is shown in Figure 2.5.

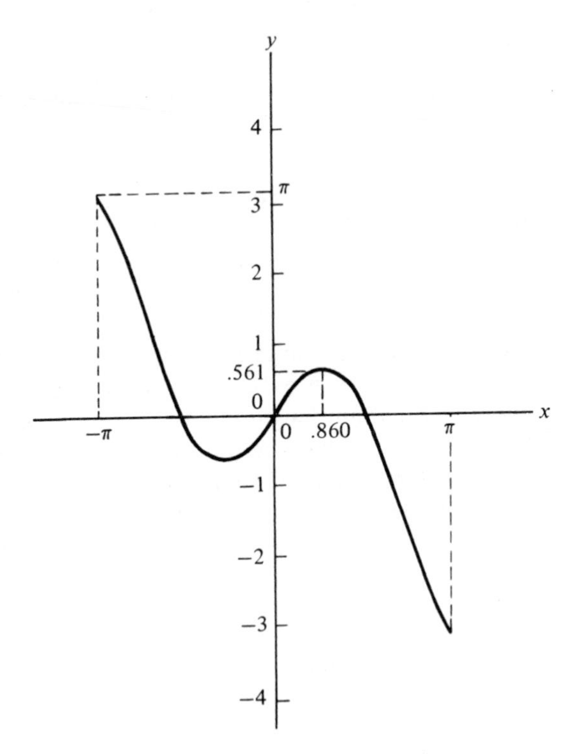

Figure 2.5. Characteristics of the Function $y = x \cos x$

2.2 Multidimensional Functions

Before attempting to develop necessary and sufficient conditions for the existence of multidimensional extrema, let us first inquire as to the kind of stationary values we might expect to find. For simplicity in visualization, we shall consider the surface generated by a function of two independent variables; i.e., $y = y(x_1, x_2)$.

Figures 2.6 and 2.8 show a relative maximum and a relative minimum for a continuous function $y(x_1, x_2)$. The corresponding level curves (i.e., curves of constant y) are shown in Figures 2.7 and 2.9. If we pass planes through the extremum points that are parallel to the x_1—y and x_2—y planes, their intersections with the surface $y(x_1, x_2)$ produce curves having straight-line projections aa' and bb' in the x_1x_2 plane. These curves and their projections are shown in the figures. Since the curves themselves pass through a maximum or a minimum at the point (x_{1o}, x_{2o}), where the surface $y(x_1, x_2)$ is maximized or minimized, we see that $\partial y(x_1, x_{2o})/\partial x_1 = 0$ at $x_1 = x_{1o}$, and

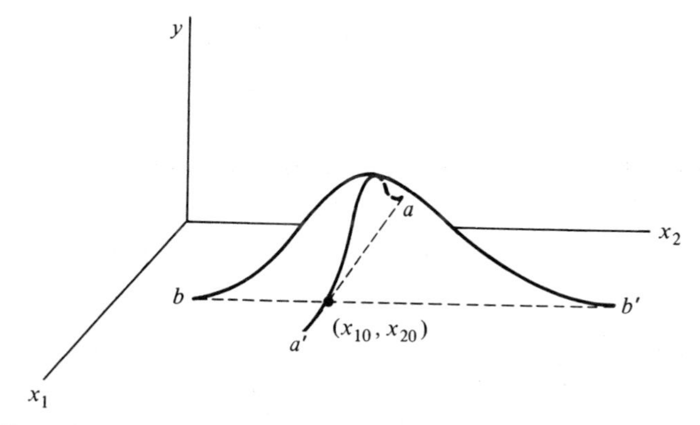

Figure 2.6. Surface with Relative Maximum Generated by Function of Two Variables

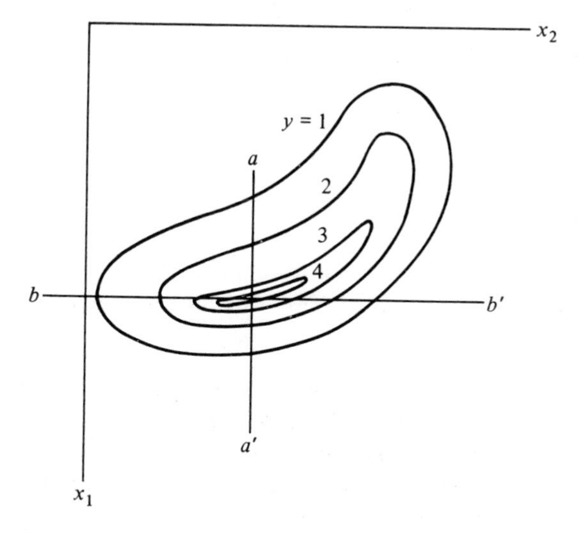

Figure 2.7. Contour Lines for Two-Variable Function Shown in Previous Figure

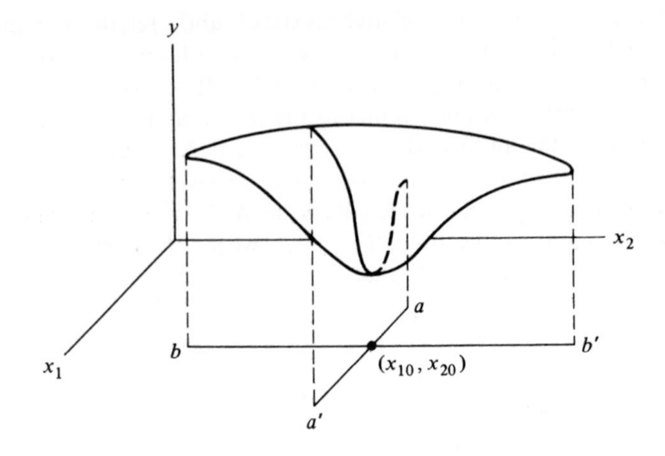

Figure 2.8. Surface with Relative Minimum Generated by Function of Two Variables

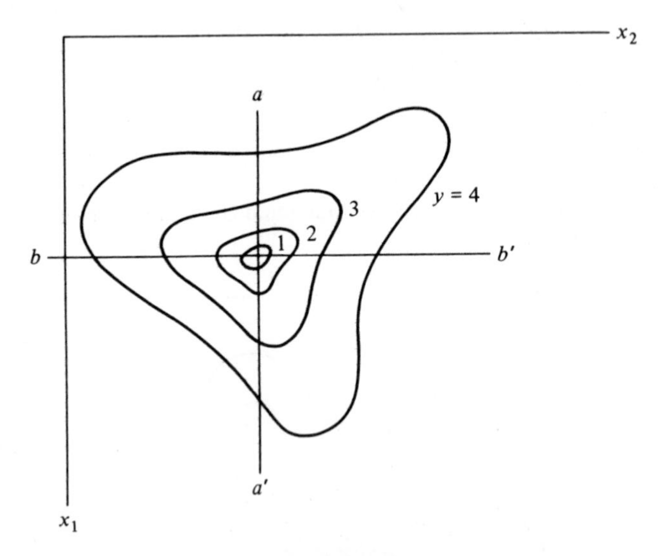

Figure 2.9. Contour Lines for Two-Variable Function of Previous Figure

$\partial y(x_{1o}, x_2)/\partial x_2 = 0$ at $x_2 = x_{2o}$. Hence, *for a function $y(x_1, x_2)$ which varies continuously within an open region R, it is necessary that $\partial y/\partial x_1 = \partial y/\partial x_2 = 0$ at (x_{1o}, x_{2o}) in order that the function pass through an extremum at (x_{1o}, x_{2o}).*

As in the case of one-dimensional functions, the necessary vanishing of the first partial derivatives is not a sufficient condition to guarantee the existence of a relative maximum or minimum. This can be seen from the sur-

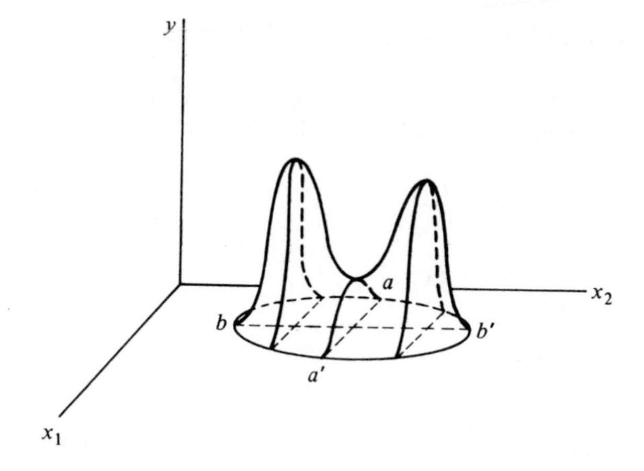

Figure 2.10. Two-Dimensional Function Showing Saddle Point

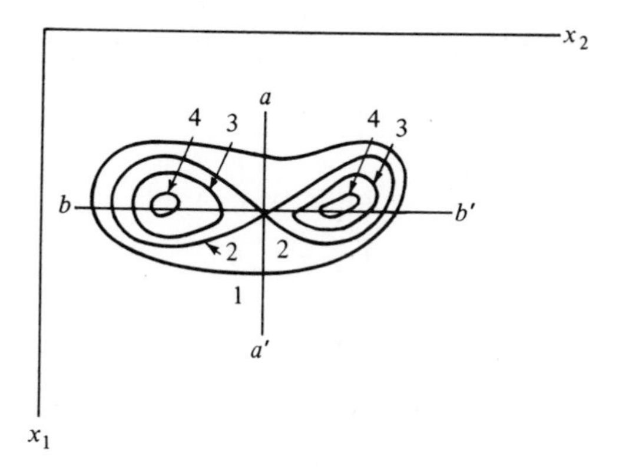

Figure 2.11. Contour Lines for Function Showing Saddle Point

face shown in Figure 2.10 and the corresponding level curves in Figure 2.11. Again we pass through the surface planes that are parallel to the x_1–y and the x_2–y planes. The planes intersect the surface to form curves that exhibit extrema at (x_{1o}, x_{2o}) and whose projections in the x_1–x_2 plane are aa' and bb'. Thus we see that both $\partial y(x_1, x_{2o})/\partial x_1$ and $\partial y(x_{1o}, x_2)/\partial x_2$ vanish at (x_{1o}, x_{2o}). However, the curve $y(x_{1o}, x_2)$ passes through a *minimum* at (x_{1o}, x_{2o}), whereas $y(x_1, x_{2o})$ exhibits a *maximum* at the same point. A surface exhibiting this characteristic does *not* itself pass through an extremum at (x_{1o}, x_{2o}). However, a different type of stationary point exists at (x_{1o}, x_{2o}); this is called a *saddle point*.

Sufficient conditions for two-dimensional functions

We can learn more about the behavior of $y(x_1, x_2)$ at a stationary point (x_{1o}, x_{2o}) by expanding $y(x_1, x_2)$ in a Taylor series about (x_{1o}, x_{2o}):

$$
\begin{aligned}
y(x_1, x_2) = y(x_{1o}, x_{2o}) &+ \left.\frac{\partial y}{\partial x_1}\right|_{(x_{1o},\, x_{2o})} (x_1 - x_{1o}) + \left.\frac{\partial y}{\partial x_2}\right|_{(x_{1o},\, x_{2o})} (x_2 - x_{2o}) \\
&+ \frac{1}{2!}\left[\left.\frac{\partial^2 y}{\partial x_1^2}\right|_{(\xi_1, \xi_2)} (x_1 - x_{1o})^2\right. \\
&+ 2\left.\frac{\partial^2 y}{\partial x_1\, \partial x_2}\right|_{(\xi_1, \xi_2)} (x_1 - x_{1o})(x_2 - x_{2o}) \\
&+ \left.\left.\frac{\partial^2 y}{\partial x_2^2}\right|_{(\xi_1\, \xi_2)} (x_2 - x_{2o})^2\right] + \cdots,
\end{aligned} \tag{2.2.1}
$$

where ξ_1 and ξ_2 lie on the line connecting (x_1, x_2) and (x_{1o}, x_{2o}). Let us rewrite this equation as

$$
y(x_1, x_2) = y(x_{1o}, x_{2o}) + \left.\frac{\partial y}{\partial x_1}\right|_{(x_{1o},\, x_{2o})} \Delta x_1 + \left.\frac{\partial y}{\partial x_2}\right|_{(x_{1o},\, x_{2o})} \Delta x_2 + \frac{(\Delta x_1)^2}{2!}\phi(u), \tag{2.2.2}
$$

where

$$\Delta x_1 = (x_1 - x_{1o}), \tag{2.2.3}$$

$$\Delta x_2 = (x_2 - x_{2o}), \tag{2.2.4}$$

$$u = \frac{\Delta x_2}{\Delta x_1}, \tag{2.2.5}$$

$$\phi(u) = A + 2Bu + Cu^2 \tag{2.2.6}$$

$$A = \left.\frac{\partial^2 y}{\partial x_1^2}\right|_{(\xi_1,\, \xi_2)}, \tag{2.2.7}$$

$$B = \left.\frac{\partial^2 y}{\partial x_1\, \partial x_2}\right|_{(\xi_1,\, \xi_2)}, \tag{2.2.8}$$

$$C = \left.\frac{\partial^2 y}{\partial x_2^2}\right|_{(\xi_1,\, \xi_2)}. \tag{2.2.9}$$

Since (x_{1o}, x_{2o}) is a stationary point, the first derivatives must vanish, leaving

$$y(x_1, x_2) = y(x_{1o}, x_{2o}) + \frac{(\Delta x_1)^2}{2!}\phi(u). \tag{2.2.10}$$

In order that $y(x_{1o}, x_{2o})$ be a relative maximum, $\phi(u) < 0$ for all u; if $y(x_{1o}, x_{2o})$ is a relative minimum, then $\phi(u) > 0$ for all u.

If the function $\phi(u)$ has real roots [i.e., real values of u for which $\phi(u) = 0$], then ϕ can be either positive, negative, or zero, depending upon the particular value of u chosen. If $\phi(u)$ has no real roots, then ϕ must always be either positive or negative for all values of u.

Let us now investigate the conditions under which $\phi(u)$ has no real roots.

Setting $\phi(u) = 0$, we have

$$\phi(u) = A + 2Bu + Cu^2 = 0 \tag{2.2.11}$$

whose roots are

$$u = \frac{-B \pm \sqrt{B^2 - AC}}{C} \tag{2.2.12}$$

The function will be complex conjugate for $B^2 - AC < 0$, so that $\phi(u)$ will be either always positive or always negative for all u. This corresponds to an extremum. If $B^2 - AC > 0$, then $\phi(u)$ can be either positive or negative, depending on the particular value of u. This is the case of a saddle point. If $B^2 - AC = 0$, however, there is one repeated real root and the analysis is inconclusive.

Carrying the analysis one step farther, let $B^2 - AC < 0$. Therefore, both A and C must have the same sign (if not, AC would be negative and $B^2 - AC$ would be positive, violating our original assertion). If A is negative and $u = 0$, then $\phi(0) = A < 0$ and, therefore, $\phi(u) < 0$ for all u. Moreover, if $A < 0$, then $A + C < 0$. On the other hand, if $A > 0$, then $A + C > 0$ and $\phi(u) > 0$ for all u. These two cases correspond to a relative maximum and a relative minimum, respectively.

Recall that A, B, and C are evaluated at (ξ_1, ξ_2), a point arbitrarily close to (x_{1o}, x_{2o}). In the limits as (ξ_1, ξ_2) approaches (x_{1o}, x_{2o}), the above conditions apply at (x_{1o}, x_{2o}). Hence we can summarize the results of our analysis as follows:

Let $y(x_1, x_2)$ very continuously in an open region R. Let

$$A_o = \left.\frac{\partial^2 y}{\partial x_1^2}\right|_{(x_{1o}\ x_{2o})}, \tag{2.2.13}$$

$$B_o = \left.\frac{\partial^2 y}{\partial x_1\, \partial x_2}\right|_{(x_{1o},\ x_{2o})}, \tag{2.2.14}$$

$$C_o = \left.\frac{\partial^2 y}{\partial x_2^2}\right|_{(x_{1o},\ x_{2o})}. \tag{2.2.15}$$

If $\partial y/\partial x_1$ and $\partial y/\partial x_2$ vanish at (x_{1o}, x_{2o}), then

1. $B_o^2 - A_o C_o < 0$ *and* $A_o + C_o < 0$ *indicate the presence of a relative maximum at* (x_{1o}, x_{2o}).
2. $B_o^2 - A_o C_o < 0$ *and* $A_o + C_o > 0$ *indicate the presence of a relative minimum at* (x_{1o}, x_{2o}).
3. $B_o^2 - A_o C_o > 0$ *indicates the presence of a saddle point at* (x_{1o}, x_{2o}).
4. $B_o^2 - A_o C_o = 0$ *is a special case for which the analysis is inconclusive.*

Sufficient conditions for multidimensional functions

From the foregoing discussion it is readily apparent that the two-dimensional case is considerably more complex than the one-dimensional case. The complexity of the analysis increases with increasing dimensionality. We shall simply state without proof the following sufficient conditions for an extremum of an n-dimensional function (cf. Kiepert [1910]):

Let $y(x_1, x_2, \ldots, x_n)$ *vary continuously in an open region* R. *Consider the set of determinants* $|\mathbf{D}_i|$, $i = 1, 2, \ldots, n$, *where*

$$(2.2.16) \qquad |\mathbf{D}_i| = \left| \begin{matrix} \dfrac{\partial^2 y}{\partial x_1^2} & \dfrac{\partial^2 y}{\partial x_1\, \partial x_2} & \cdots & \dfrac{\partial^2 y}{\partial x_1\, \partial x_i} \\ \dfrac{\partial^2 y}{\partial x_2\, \partial x_1} & \dfrac{\partial^2 y}{\partial x_2^2} & \cdots & \dfrac{\partial^2 y}{\partial x_2\, \partial x_i} \\ \vdots & \vdots & & \vdots \\ \dfrac{\partial^2 y}{\partial x_i\, \partial x_1} & \dfrac{\partial^2 y}{\partial x_i\, \partial x_2} & \cdots & \dfrac{\partial^2 y}{\partial x_i^2} \end{matrix} \right|_{(x_{1o}, x_{2o}, \ldots, x_{no})}.$$

If $\partial y/\partial x_1 = \partial y/\partial x_2 = \cdots = \partial y/\partial x_n = 0$ *at* $(x_{1o}, x_{2o}, \ldots, x_{no})$, *then*

1. $|\mathbf{D}_i| < 0$ *for* $i = 1, 3, 5, \ldots$ *and* $|\mathbf{D}_i| > 0$ *for* $i = 2, 4, 6, \ldots$ *indicate the presence of a relative maximum at* $(x_{1o}, x_{2o}, \ldots, x_{no})$.
2. $|\mathbf{D}_i| > 0$ *for* $i = 1, 2, \ldots, n$ *indicates the presence of a relative minimum at* $(x_{1o}, x_{2o}, \ldots, x_{no})$.

Failure to satisfy either of these two conditions indicates the presence of a saddle point.

Vector–matrix formulation

It is often convenient to write Equation (2.2.1) in vector form. Thus

$$y(\mathbf{X}) = y(\mathbf{X}_o) + \nabla y|_{(\mathbf{X}_o)}(\mathbf{X} - \mathbf{X}_o) + \tfrac{1}{2}(\mathbf{X} - \mathbf{X}_o)'\mathbf{H}|_{(\mathbf{X}_o)}(\mathbf{X} - \mathbf{X}_o) + \cdots, \tag{2.2.17}$$

where $\mathbf{H}$ is the *Hessian* matrix (see Appendix B) given by

$$\mathbf{H} = \begin{bmatrix} \frac{\partial^2 y}{\partial x_1^2} & \frac{\partial^2 y}{\partial x_1\,\partial x_2} & \cdots & \frac{\partial^2 y}{\partial x_1\,\partial x_n} \\ \frac{\partial^2 y}{\partial x_2\,\partial x_1} & \frac{\partial^2 y}{\partial x_2^2} & \cdots & \frac{\partial^2 y}{\partial x_2\,\partial x_n} \\ \vdots & & & \\ \frac{\partial^2 y}{\partial x_n\,\partial x_1} & \frac{\partial^2 y}{\partial x_n\,\partial x_2} & \cdots & \frac{\partial^2 y}{\partial x_n^2} \end{bmatrix}. \tag{2.2.18}$$

Since $\nabla y|_{(\mathbf{X}_o)} = \mathbf{0}$ at a stationary point, we have, neglecting the higher-order terms, a vectorial expression analogous to Equation (2.2.10); i.e.,

$$y(\mathbf{X}) = y(\mathbf{X}_o) + \tfrac{1}{2}(\mathbf{X} - \mathbf{X}_o)'\mathbf{H}|_{(\mathbf{X}_o)}(\mathbf{X} - \mathbf{X}_o). \tag{2.2.19}$$

From this expression we see that $y(\mathbf{X}_o)$ will be a maximum if the quadratic form $(\mathbf{X} - \mathbf{X}_o)'\mathbf{H}|_{(\mathbf{X}_o)}(\mathbf{X} - \mathbf{X}_o) \leq 0$ for all $\mathbf{X} \neq \mathbf{X}_o$, and $y(\mathbf{X}_o)$ will be a minimum for $(\mathbf{X} - \mathbf{X}_o)'\mathbf{H}|_{(\mathbf{X}_o)}(\mathbf{X} - \mathbf{X}_o) \geq 0$ for all $\mathbf{X} \neq \mathbf{X}_o$. (The reader is referred to Appendix B for a discussion of quadratic forms and their properties.)

The positive (negative) definiteness of the Hessian matrix can establish whether or not a stationary point is an extremum, and is also related to the convexity (concavity) of the given function. Unless the given function is quadratic, however, the positive (negative) definiteness of its Hessian matrix may be difficult to establish for all $\mathbf{X}$.

The determination of an *absolute* extremum of a continuous, multidimensional function $y(x_1, x_2, \ldots, x_n)$ frequently involves a great deal of computation. First we must determine the location of the interior stationary points. When we set the first partial derivatives equal to zero, however, we sometimes obtain a system of n simultaneous, nonlinear algebraic equations that we must solve to determine the stationary points $(x_{1o}, x_{2o}, \ldots, x_{no})_j$, $j \geq 1$. This in itself may be a most arduous task, even with the aid of a digital computer. Having determined the stationary points, it is then necessary to evaluate the corresponding stationary values and select from these the greatest maximum or the least minimum.

Region boundaries

We then proceed to evaluate the function on the boundaries of the region. To illustrate the procedure, let us assume that the independent variables are restricted to non-negative values; i.e., $x_i \geq 0$, $i = 1, 2, \ldots, n$, so that the set of points defined by one or more $x_i = 0$ will constitute the region boundaries. We begin by letting one of the independent variables equal zero, and then extremize the function with respect to the remaining $n - 1$ independent variables. (Again, we may be confronted with solving a complicated system of nonlinear algebraic equations.) The procedure is repeated n times, each time letting a different x_i equal zero. Next we let two of the variables equal zero, and extremize the function with respect to the remaining $n - 2$ independent variables. This procedure is continued $n!/(2)!\,(n-2)!$ times, using different combinations of the two vanishing variables each time [the quantity $n!/(2)!\,(n-2)!$ is the number of different ways n variables can be chosen two at a time]. We then proceed to set any three variables equal to zero, giving us $n!/(3)!\,(n-3)!$ evaluations for stationary points, and so on. Finally we reach a point where we simply evaluate our objective function with all the independent variables equal to zero. At the conclusion of this procedure all the extrema, including those obtained at interior points, are compared and an absolute extremum is selected.

In practice the procedure is easily carried out if the partial derivatives yield algebraic equations that are easily solved and if the dimensionality is low. However, some reflection on the method shows that it becomes impractical, or even totally useless, when the dimensionality of the problem becomes modestly high.

EXAMPLE 2.2.1

Analyze the stationary points of the quadratic form

$$y = \mathbf{X}'\mathbf{A}\mathbf{X} + \mathbf{B}'\mathbf{X},$$

where

$$\mathbf{A} = \begin{bmatrix} 1 & 3 \\ -1 & 2 \end{bmatrix}, \qquad \mathbf{B} = \begin{bmatrix} 2 \\ 1 \end{bmatrix}.$$

Carrying out the matrix multiplication, we have

$$\mathbf{A}\mathbf{X} = \begin{bmatrix} (\ \ x_1 + 3x_2) \\ (-x_1 + 2x_2) \end{bmatrix}, \qquad \mathbf{B}'\mathbf{X} = 2x_1 + x_2,$$

$$\begin{aligned} \mathbf{X}'\mathbf{A}\mathbf{X} &= x_1^2 + 3x_1x_2 - x_1x_2 + 2x_2^2 \\ &= x_1^2 + 2x_1x_2 + 2x_2^2, \\ y &= x_1^2 + 2x_1x_2 + 2x_2^2 + 2x_1 + x_2. \end{aligned}$$

The first partial derivatives are

$$\frac{\partial y}{\partial x_1} = 2x_1 + 2x_2 + 2 = 0,$$

$$\frac{\partial y}{\partial x_2} = 2x_1 + 4x_2 + 1 = 0,$$

which yield a stationary point at $x_1 = -\frac{3}{2}$, $x_2 = \frac{1}{2}$.

The second partial derivatives are

$$\frac{\partial^2 y}{\partial x_1^2} = 2 = A_o,$$

$$\frac{\partial^2 y}{\partial x_2^2} = 4 = C_o,$$

$$\frac{\partial^2 y}{\partial x_1\,\partial x_2} = 2 = B_o.$$

Hence we have $B_o^2 - A_oC_o = -4$ and $A_o + C_o = 6$. Therefore, we have a relative minimum at $x_1 = -\frac{3}{2}$, $x_2 = \frac{1}{2}$, whose value is $y = -\frac{5}{4}$.

EXAMPLE 2.2.2

Determine the extrema of the function

$$y = x_1^2 - x_2^2 + x_3^2 - 2x_1x_3 - x_2x_3 + 4x_1 + 12.$$

Also locate the absolute maximum in the region defined by $x_1, x_2, x_3 \geq 0$.

Taking the first derivatives, we have

$$\begin{aligned}
\frac{\partial y}{\partial x_1} &= 2x_1 - 2x_3 + 4 = 0,\\
\frac{\partial y}{\partial x_2} &= -2x_2 - x_3 = 0,\\
\frac{\partial y}{\partial x_3} &= -2x_1 - x_2 + 2x_3 = 0.
\end{aligned}$$

Solving the system of three linear equations in three unknowns, we obtain a stationary point at $x_1 = -10$, $x_2 = 4$, and $x_3 = -8$.

The second derivatives are

$$\begin{aligned}
\frac{\partial^2 y}{\partial x_1^2} &= 2 & \frac{\partial^2 y}{\partial x_1\,\partial x_2} &= 0,\\
\frac{\partial^2 y}{\partial x_2^2} &= -2 & \frac{\partial^2 y}{\partial x_1\,\partial x_3} &= -2,\\
\frac{\partial^2 y}{\partial x_3^2} &= 2 & \frac{\partial^2 y}{\partial x_2\,\partial x_3} &= -1.
\end{aligned}$$

Forming the determinant given by Equation (2.2.16), we have

$$|\mathbf{D}_1| = \left|\frac{\partial^2 y}{\partial x_1^2}\right| = 2,$$

$$|\mathbf{D}_2| = \begin{vmatrix} \dfrac{\partial^2 y}{\partial x_1^2} & \dfrac{\partial^2 y}{\partial x_1\,\partial x_2} \\[2ex] \dfrac{\partial^2 y}{\partial x_2\,\partial x_1} & \dfrac{\partial^2 y}{\partial x_2^2} \end{vmatrix} = \begin{vmatrix} 2 & 0 \\ 0 & -2 \end{vmatrix} = -4,$$

$$|\mathbf{D}_3| = \begin{vmatrix} 2 & 0 & -2 \\ 0 & -2 & -1 \\ -2 & -1 & 2 \end{vmatrix} = -2.$$

Since we do not satisfy the necessary extremum conditions associated with Equation (2.2.16), we conclude that the stationary point at $(-10, 4, -8)$ is a saddle point.

To solve the second part of the problem, we first let $x_1 = 0$ and examine the function

$$y = -x_2^2 + x_3^2 - x_2 x_3 + 12.$$

We see that this function possesses a saddle point at $(0, 0, 0)$.

Now let $x_1 \neq 0$ and $x_2 = 0$, which gives us

$$y = x_1^2 + x_3^2 - 2x_1 x_3 + 4x_1 + 12.$$

This function does not have a stationary point.

We now let x_1 and $x_2 \neq 0$ but let $x_3 = 0$, which gives us

$$y = x_1^2 - x_2^2 + 4x_1 + 12.$$

Again we obtain a saddle point, at $(-2, 0, 0)$.

If we let x_1 and $x_2 = 0$, we have

$$y = x_3^2 + 12,$$

which possesses a relative minimum with respect to x_3 at $(0, 0, 0)$, whose value is $y = 12$.

We now let x_1 and $x_3 = 0$ with $x_2 \neq 0$. This gives us

$$y = -x_2^2 + 12,$$

which has a relative maximum with respect to x_2 at $(0, 0, 0)$, whose value is $y = 12$.

If we now let $x_1 \neq 0$ but let x_2 and $x_3 = 0$, we obtain

$$y = x_1^2 + 4x_1 + 12,$$

which possesses a relative minimum with respect to x_1 at $(-2, 0, 0)$.

From this analysis we conclude that the function does not have an absolute maximum within the region $x_1, x_2, x_3 \geq 0$. The value of y can be made to increase or decrease without bounds for the proper choice of x_1, x_2, and x_3.

2.3 Optimization with Constraints—Lagrange Multipliers

Let us now turn our attention to the optimization of a function whose independent variables must satisfy one or more equality constraints. For example, suppose that $y(x_1, x_2)$ is a continuous function which we wish to optimize in an open region R, subject to the auxiliary condition $g(x_1, x_2) = 0$. The function $g(x_1, x_2)$ prevents x_1 and x_2 from varying independently; thus we seek the value and location of a *constrained optimum*, which may differ from an unconstrained optimum.

Solution by direct substitution

One way to sove a constrained optimization problem is by direct substitution. Suppose, for example, that we can solve $g(x_1, x_2)$ for one variable in terms of the remaining variable; i.e., $x_2 = h(x_1)$. We can then rewrite $y\ (x_1, x_2)$ as $y(x_1, h)$. Since a necessary condition for the existence of an extremum is $dy = 0$, we can write

$$dy = \frac{\partial y}{\partial x_1} dx_1 + \frac{\partial y}{\partial h} dh = 0, \tag{2.3.1}$$

where

$$dh = \frac{dh}{dx_1} dx_1. \tag{2.3.2}$$

Combining equations, we obtain

$$dy = \frac{\partial y}{\partial x_1} dx_1 + \frac{\partial y}{\partial h}\frac{dh}{dx_1} dx_1 = 0 \tag{2.3.3}$$

or

$$\frac{\partial y}{\partial x_1} + \frac{\partial y}{\partial h}\frac{dh}{dx_1} = 0. \tag{2.3.4}$$

The values of x_1 and x_2 satisfying the above equations give the location of the constrained stationary point. This procedure is very straightforward, and is well suited to simple problems. However, we cannot always solve the auxiliary equation for one variable in terms of the others. Even if we can, the procedure often becomes cumbersome and tedious for problems of higher dimensionality.

Preferred solution—method of Lagrange

We can proceed somewhat differently, however, by writing

(2.3.5) $$g(x_1, x_2) = 0,$$

(2.3.6) $$dg = \frac{\partial g}{\partial x_1} dx_1 + \frac{\partial g}{\partial x_2} dx_2 = 0$$

along the constraint path (since g remains constant), so that

(2.3.7) $$\frac{dx_2}{dx_1} = -\frac{\partial g/\partial x_1}{\partial g/\partial x_2}.$$

Combining this result with Equation (2.3.4) gives

(2.3.8) $$\frac{\partial y}{\partial x_1}\frac{\partial g}{\partial x_2} - \frac{\partial y}{\partial x_2}\frac{\partial g}{\partial x_1} = 0.$$

Equations (2.3.5) and (2.3.8) represent the necessary conditions for $y(x_1, x_2)$ to have a stationary point subject to the auxiliary condition $g(x_1, x_2) = 0$.

Now let us consider the *unconstrained* function

(2.3.9) $$z(x_1, x_2, \lambda) = y(x_1, x_2) + \lambda g(x_1, x_2),$$

where λ is some undetermined constant. The necessary conditions for $z(x_1, x_2, \lambda)$ to have a stationary value are

(2.3.10) $$\frac{\partial z}{\partial x_1} = \frac{\partial y}{\partial x_1} + \lambda\frac{\partial g}{\partial x_1} = 0,$$

(2.3.11) $$\frac{\partial z}{\partial x_2} = \frac{\partial y}{\partial x_2} + \lambda\frac{\partial g}{\partial x_2} = 0,$$

(2.3.12) $$\frac{\partial z}{\partial \lambda} = g = 0.$$

If we eliminate λ from Equations (2.3.10) and (2.3.11), we obtain

(2.3.13) $$\frac{\partial y}{\partial x_1}\frac{\partial g}{\partial x_2} - \frac{\partial y}{\partial x_2}\frac{\partial g}{\partial x_1} = 0,$$

which is identical to Equation (2.3.8). Thus we see that the constrained optimization problem is equivalent to obtaining the stationary values of the unconstrained function $z(x_1, x_2, \lambda) = y(x_1, x_2) + \lambda g(x_1, x_2)$. The function $z(x_1, x_2, \lambda)$ is called the *Lagrangian function*, and the constant λ, which is determined in the course of the solution, is known as a *Lagrange multiplier*.

Multivariable, multiconstraint problems

The Lagrange-multiplier concept also applies to the optimization of a multivariable function $y(x_1, x_2, \ldots, x_n) = 0$, subject to the constraints $g_i(x_1, x_2, \ldots, x_n) = 0$ for $i = 1, 2, \ldots, m$, with $m < n$. (Notice that m, the number of constraint equations, must be less than n; otherwise we would have a system of n equations in n unknowns, which would result in a unique solution for $x_1, x_2, \ldots, x_n$ when $m = n$, or in an overspecified system of equations when $m > n$.) The procedure is to construct the unconstrained Lagrangian function

$$z(x_1, x_2, \ldots, x_n, \lambda_1, \lambda_2, \ldots, \lambda_m) = y(x_1, x_2, \ldots, x_n) + \sum_{i=1}^{m} \lambda_i g_i(x_1, x_2, \ldots, x_n), \qquad (2.3.14)$$

where the unspecified constants λ_i are the *Lagrange multipliers*, which are determined in the course of the extremization.

The necessary conditions for z to possess an extremum are

$$\frac{\partial z}{\partial x_k} = \frac{\partial y}{\partial x_k} + \sum_{i=1}^{m} \lambda_i \frac{\partial g_i}{\partial x_k} = 0, \qquad k = 1, 2, \ldots, n \qquad (2.3.15)$$

and

$$\frac{\partial z}{\partial \lambda_i} = g_i = 0, \qquad i = 1, 2, \ldots, m. \qquad (2.3.16)$$

Notice that Equation (2.3.16) simply restates the original constraints acting on $y(x_1, x_2 \ldots, x_n)$.

Equations (2.3.15) and (2.3.16) are a system of $n + m$ equations in $n + m$ unknowns. Hence their solution will yield stationary values for x_1, $x_2, \ldots, x_n$ and $\lambda_1, \lambda_2, \ldots, \lambda_m$. The values $x_1, x_2, \ldots, x_n$ that satisfy the necessary conditions for $z(x_1, x_2, \ldots, x_n, \lambda_1, \lambda_2, \ldots, \lambda_m)$ also satisfy the necessary constrained extremum conditions for $y(x_1, x_2, \ldots, x_n)$. We have no assurance, however, that the values of y obtained at this point is an extremum rather than a saddle point. Moreover, in the next section we shall see that the Lagrangian function exhibits a saddle point when the original function is extremized subject to the given constraints. The increased dimensionality, which is characteristic of the Lagrange-multiplier method, is generally more than compensated for by the relative simplicity of the computation.

It should be pointed out that a derivation of the above conditions involves division by an $m \times m$ determinant whose elements are first partial derivatives of the constraining equations. The derivation requires that this determinant be nonzero. It is shown in more advanced texts (cf. Hadley [1964]) that the Lagrange-multiplier concept is valid under certain conditions when this determinant does vanish; however, the Lagrange multipliers cannot be uniquely determined.

Region boundaries

To obtain an *absolute* extremum, we are again confronted with the task of extremizing the given objective function subject to the given constraints, both within the region of interest and on its boundaries. To do this we follow the same procedure as in Section 2.2. The Lagrange-multiplier technique can be used to extremize the given objective function on a region boundary as well as within the interior of the region. When exactly $n - m$ of the variables take on fixed values, however, the constraints will uniquely determine the values of the objective function and the remaining m variables. Consequently, we cannot simultaneously fix the values of more than $n - m$ independent variables.

EXAMPLE 2.3.1

Determine the extrema of the function

$$y = x_1^2 + x_2^2 - 3x_1x_2$$

on the circle $x_1^2 + x_2^2 = 4$.

We form the Lagrangian

$$z(x_1, x_2, \lambda) = x_1^2 + x_2^2 - 3x_1x_2 + \lambda(x_1^2 + x_2^2 - 4)$$

and obtain the first partial derivatives

$$\frac{\partial z}{\partial x_1} = 2x_1 - 3x_2 + 2\lambda x_1 = 0,$$

$$\frac{\partial z}{\partial x_2} = -3x_1 + 2x_2 + 2\lambda x_2 = 0,$$

$$\frac{\partial z}{\partial \lambda} = x_1^2 + x_2^2 - 4 = g(x_1, x_2) = 0.$$

Eliminating λ from the first two partial derivatives results in

$$x_1^2 - x_2^2 = 0.$$

Combining this result with the third partial derivative (i.e., the original constraint) gives

$$2x_1^2 = 4$$

or

$$x_1 = \pm\sqrt{2}.$$

We also see that $x_2 = \pm\sqrt{2}$. Hence we have stationary points at $(\sqrt{2}, \sqrt{2})$, $(\sqrt{2}, -\sqrt{2})$, $(-\sqrt{2}, -\sqrt{2})$, and $(-\sqrt{2}, \sqrt{2})$. From the original form of the constrained objective function, it is clear that the constrained maxima correspond to the points where x_1 and x_2 differ in sign, whereas the constrained minima occur where x_1 and x_2 are of the same sign. Therefore, we have relative maxima of $y = 10$ at $(\sqrt{2}, -\sqrt{2})$ and $(-\sqrt{2}, \sqrt{2})$, and relative minima of $y = -2$ at $(\sqrt{2}, \sqrt{2})$ and $(-\sqrt{2}, -\sqrt{2})$.

It is intersting to observe that the *unconstrained* function $y = x_1^2 + x_2^2 - 3x_1x_2$ does not have any relative extrema, but it exhibits a saddle point at the origin.

2.4 Behavior of the Lagrangian Function

Suppose that a constrained function $y(\mathbf{X})$ takes on a relative maximum at the point $\mathbf{X}_o$, subject to the constraints $g_i(\mathbf{X}_o) = 0$, $i = 1, 2, \ldots, m$. We shall require that $\mathbf{X}_o$ be an interior point within the given region. Let the corresponding set of Lagrange multipliers be represented by the point $\boldsymbol{\lambda}_o$. [Notice that $\mathbf{X}_o$ and $\boldsymbol{\lambda}_o$ are n-dimensional and m-dimensional vectors, respectively, in an $(m + n)$-dimensional space.] Suppose also that the Lagrangian function $z(\mathbf{X}, \boldsymbol{\lambda})$ takes on a relative maximum with respect to $\mathbf{X}$ at $\mathbf{X}_o$; this condition will be assumed to hold for *any* set of constants $\boldsymbol{\lambda}$ that satisfies the expression

$$\frac{\partial z}{\partial x_j} = \frac{\partial y}{\partial x_j} + \sum_{i=1}^{m} \lambda_i \frac{\partial g_i}{\partial x_j} = 0, \qquad j = 1, 2, \ldots, n. \tag{2.4.1}$$

In other words, we can say that for any $\boldsymbol{\lambda}$,

$$\max_{\mathbf{X}} z(\mathbf{X}, \boldsymbol{\lambda}) = f(\boldsymbol{\lambda}). \tag{2.4.2}$$

Observe that Equation (2.4.1) does not represent all of the necessary conditions for $y(\mathbf{X})$ to have a constrained extremum, since the constraints are not included [i.e., we have not specified that $\partial z/\partial \lambda_i = g_i(\mathbf{X}) = 0$, $\mathbf{i} = 1, 2, \ldots, m$]. Hence any point $\boldsymbol{\lambda}$ that satisfies (2.4.1) is not unique and is therefore not necessarily equal to $\boldsymbol{\lambda}_0$.

Let us examine the behavior of $f(\boldsymbol{\lambda})$ as $\boldsymbol{\lambda}$ approaches $\boldsymbol{\lambda}_o$. If we consider $\mathbf{X}$ to take on only those values which satisfy the constraints, i.e., $g_i(\mathbf{X}) = 0$, $i = 1, 2, \ldots, m$, then

$$\max_{\mathbf{X}} z(\mathbf{X}, \boldsymbol{\lambda}) = \max_{\mathbf{X}} y(\mathbf{X}) = y(\mathbf{X}_o) \tag{2.4.3}$$

From equation (2.4.2), however, we can write

$$y(\mathbf{X}_o) = z(\mathbf{X}_o, \boldsymbol{\lambda}) = f(\boldsymbol{\lambda}_o) \tag{2.4.4}$$

since $(\mathbf{X}_o, \boldsymbol{\lambda}_o)$ is that point in the $(m + n)$-dimensional space at which z is maximized with respect to $\mathbf{X}$ *and* the given constraints are satisfied. Now satisfaction of the constraints imposes restrictions on the allowable values of $\mathbf{X}$. Hence we can write

$$f(\boldsymbol{\lambda}) \geq f(\boldsymbol{\lambda}_o) \tag{2.4.5}$$

since the expression for $f(\boldsymbol{\lambda})$ is not restricted to those values of $\mathbf{X}$ which satisfy the constraint equations. Thus we see that

$$z(\mathbf{X}, \boldsymbol{\lambda}_o) \leq z(\mathbf{X}_o, \boldsymbol{\lambda}_o) \leq z(\mathbf{X}_o, \boldsymbol{\lambda}), \tag{2.4.6}$$

or

$$\max_{\mathbf{X}} z(\mathbf{X}, \boldsymbol{\lambda}_o) = \min_{\boldsymbol{\lambda}} z(\mathbf{X}_o, \boldsymbol{\lambda}) = f(\boldsymbol{\lambda}_o). \tag{2.4.7}$$

We therefore conclude that *the Lagrangian function $z(\mathbf{X}, \boldsymbol{\lambda})$ exhibits a saddle point at $(\mathbf{X}_o, \boldsymbol{\lambda}_o)$ where the objective function $y(\mathbf{X})$ is maximized subject to the given constraints.*

Note that Equation (2.4.7) is a general expression of a saddle-point condition. In our particular problem the second inequality in Equation (2.4.7) should be written as a strict equality since $\mathbf{X}_o$ satisfies the given constraints and therefore $z(\mathbf{X}_o, \boldsymbol{\lambda}) = y(\mathbf{X}_o)$ for any vector $\boldsymbol{\lambda}$.

By similar arguments it can be shown that the Lagrangian function exhibits a saddle point at $(\mathbf{X}_1, \boldsymbol{\lambda}_1)$ where the constrained function $y(\mathbf{X})$ takes on a minimum subject to the given constraints. Under these conditions, however, the above conditions are reversed; i.e.,

$$\min_{\mathbf{X}} z(\mathbf{X}, \boldsymbol{\lambda}_1) = \max_{\boldsymbol{\lambda}} z(\mathbf{X}_1, \boldsymbol{\lambda}) = f(\boldsymbol{\lambda}_1) \tag{2.4.8}$$

and

$$z(\mathbf{X}, \boldsymbol{\lambda}_1) \geq z(\mathbf{X}_1, \boldsymbol{\lambda}_1) \geq z(\mathbf{X}_1, \boldsymbol{\lambda}). \tag{2.4.9}$$

We see, then, that *the conditions for extremizing an objective function subject to a set of constraints are equivalent to the conditions for a saddle point of the corresponding Lagrangian function.* We shall shortly find it convenient to consider the maximization or minimization of a given function with constraints in terms of the equivalent saddle-point problem.

The problem of maximizing (minimizing) $z(\mathbf{X}, \boldsymbol{\lambda})$ with respect to $\boldsymbol{\lambda}$, subject to the constraints

$$\frac{\partial y}{\partial x_j} + \sum_{i=1}^{m} \lambda_i \frac{\partial g_i}{\partial x_j} = 0, \qquad j = 1, 2, \ldots, n \tag{2.4.10}$$

is called the *dual* of the problem that minimizes (maximizes) $z(\mathbf{X}, \boldsymbol{\lambda})$ with respect to **X**, subject to the constraints

$$g_i(x_1, x_2, \ldots, x_n) = 0, \qquad i = 1, 2, \ldots, m. \tag{2.4.11}$$

The λ_i are the *dual variables.* We shall see that the concept of duality has important application in both linear and geometric programming.

2.5 Inequality Constraints and Lagrange Multipliers

The use of Lagrange multipliers for the optimization of a constrained continuous function can be extended to the case in which some or all of the constraints are inequalities. Consider extremizing the function $y(x_1, x_2, \ldots, x_n)$, subject to the constraints

$$\begin{aligned} g_i(x_1, x_2, \ldots, x_n) &\leq 0, \qquad i = 1, 2, \ldots, k, \\ g_i(x_1, x_2, \ldots, x_n) &\geq 0, \qquad i = k+1, k+2, \ldots, l, \\ g_i(x_1, x_2, \ldots, x_n) &= 0, \qquad i = l+1, l+2, \ldots, m, \end{aligned} \tag{2.5.1}$$

where $(m - l) < n$. We can express the inequality constraints as equalities if we introduce l fictitious nonnegative variables x_{si} as follows:

$$\begin{aligned} g_i(x_1, x_2, \ldots, x_n) + x_{si} &= 0, \qquad i = 1, 2, \ldots, k, \\ g_i(x_1, x_2, \ldots, x_n) - x_{si} &= 0, \qquad i = k+1, k+2, \ldots, l. \end{aligned} \tag{2.5.2}$$

The variables x_{si} are called *slack* variables for $i = 1, 2, \ldots, k$, and *surplus* variables for $i = k + 1, k + 2, \ldots, l$. We then form the Lagrangian function

$$\begin{aligned} &z(x_1, x_2, \ldots, x_n, x_{s1}, x_{s2}, \ldots, x_{sl}, \lambda_1, \lambda_2, \ldots, \lambda_m) \\ &= y(x_1, x_2, \ldots, x_n) + \sum_{i=1}^{k} \lambda_i[g_i(x_1, x_2, \ldots, x_n) + x_{si}] \\ &\quad + \sum_{i=k+1}^{l} \lambda_i[g_i(x_1, x_2, \ldots, x_n) - x_{si}] + \sum_{i=l+1}^{m} \lambda_i g_i(x_1, x_2, \ldots, x_n). \end{aligned} \tag{2.5.3}$$

We know that a necessary condition for the existence of an extremum of z at an interior point is the vanishing of all the first partial derivatives of z with respect to the nonzero variables. Taking derivatives with respect to x_{si},

$$\begin{aligned} \frac{\partial z}{\partial x_{si}} &= \lambda_i = 0, \qquad i = 1, 2, \ldots, k, \\ \frac{\partial z}{\partial x_{si}} &= -\lambda_i = 0, \qquad i = k+1, k+2, \ldots, l. \end{aligned} \tag{2.5.4}$$

Now if $x_{si} > 0$, then the vanishing of λ_i implies that the ith constraint does not apply at an interior point and therefore applies only at the region boundaries. Stated differently, *the inequality constraints constitute a part of the boundary of the region in which we wish to optimize* $y(x_1, x_2, \ldots, x_n)$. If, on the other hand, $x_{si} = 0$, then Equations (2.5.4) do not apply. Hence we conclude that

$$\lambda_i x_{si} = 0, \qquad i = 1, 2, \ldots, l. \tag{2.5.5}$$

We can make use of the these results to construct a computational scheme as follows: First ignore the inequality constraints and proceed to extremize the objective function, subject to the equality constraints, using the Lagrange-multiplier technique. If the solution so obtained also satisfies the inequality constraints, then the extremum lies within the boundaries of the constrained region and the problem is solved. If one or more inequality constraints are violated, however, then the solution we seek will fall on a region boundary. We then allow one of the inequality constraints to become an equality and repeat the procedure.

This is continued successively, treating each different inequality constraint as an equality, until an extremum is obtained that satisfies all the remaining inequality constraints. If such a solution cannot be found with one active inequality constraint, then two of the inequality constraints are treated as equalities and the above procedure is repeated. We then allow three inequalities to become active, if necessary, solving all combinations of this problem, and so on. The problem is solved when some unique combination of inequalities is treated as a set of equalities, and the resulting absolute extremum satisfies the remaining inequalities. Note that the *number* of inequalities treated as equalities, together with the number of original equality constraints, cannot exceed n, the number of independent variables.

Again notice that the amount of computation required to obtain such a solution may be prohibitive if the dimensionality is modestly high.

EXAMPLE 2.5.1

Minimize the function

$$y = 2x_1^2 + 2x_1x_2 + x_2^2 - 20x_1 - 14x_2,$$

subject to the constraints

$$x_1 + 3x_2 \leq 5,$$
$$2x_1 - x_2 \leq 4,$$
$$x_1, x_2 \geq 0.$$

The problem is shown graphically in Figure 2.12. The allowable region is the shaded area formed by the x_1 and x_2 axes and the lines $x_1 + 3x_2 = 5$ and $2x_1 - x_2 = 4$.

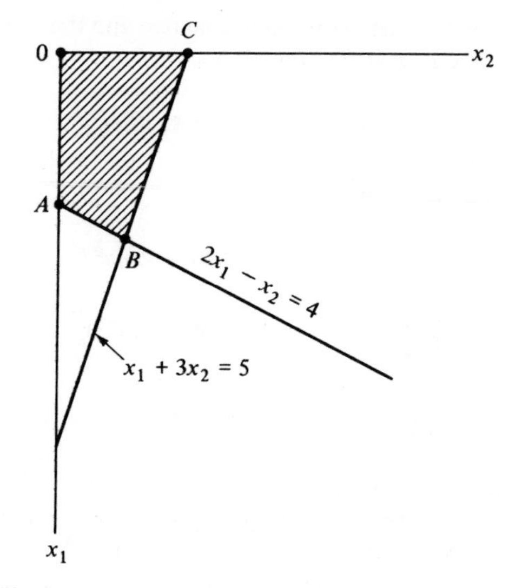

Figure 2.12. Feasible Region for Optimization of Example 2.5.1

Extremizing the unconstrained objective function, we have

$$\frac{\partial y}{\partial x_1} = 4x_1 + 2x_2 - 20 = 0,$$

$$\frac{\partial y}{\partial x_2} = 2x_1 + 2x_2 - 14 = 0,$$

which yield a minimum of $y = -58$ at $x_1 = 3$, $x_2 = 4$. However, $x_1 + 3x_2 = 15$, so that the first inequality constraint is violated.

Allowing the first inequality to become active, we form the Lagrangian function

$$z = 2x_1^2 + 2x_1x_2 + x_2^2 - 20x_1 - 14x_2 + \lambda(x_1 + 3x_2 - 5).$$

The necessary conditions for a stationary point are

$$\frac{\partial z}{\partial x_1} = 4x_1 + 2x_2 - 20 + \lambda = 0,$$

$$\frac{\partial z}{\partial x_2} = 2x_1 + 2x_2 - 14 + 3\lambda = 0,$$

$$\frac{\partial z}{\partial \lambda} = x_1 + 3x_2 - 5 = 0,$$

which result in a constrained minimum at $x_1 = \frac{59}{13}$, $x_2 = \frac{2}{13}$. This point violates the second inequality constraint, since $2x_1 - x_2 = 8\frac{12}{13}$.

We now allow the first inequality to be inactive and the second equality to be active. We thus have the Lagrangian function

$$z = 2x_1^2 + 2x_1x_2 + x_2^2 - 20x_1 - 14x_2 + \lambda(2x_1 - x_2 - 4).$$

The necessary conditions now are

$$\frac{\partial z}{\partial x_1} = 4x_1 + 2x_2 - 20 + 2\lambda = 0,$$

$$\frac{\partial z}{\partial x_2} = 2x_1 + 2x_2 - 14 - \lambda = 0,$$

$$\frac{\partial z}{\partial \lambda} = 2x_1 - x_2 - 4 = 0.$$

These expressions result in a relative minimum at $x_1 = \frac{18}{5}$, $x_2 = \frac{16}{5}$. However, since $x_1 + 3x_2 = 13\frac{1}{5}$, we see that the first inequality constraint is now violated.

Let us now allow both constraints to become active. This is shown as point *B* in Figure 2.12. The Lagrangian function is

$$z = 2x_1^2 + 2x_1x_2 + x_2^2 - 20x_1 - 14x_2 + \lambda_1(x_1 + 3x_2 - 5) + \lambda_2(2x_1 - x_2 - 4),$$

and the necessary conditions for an extremum are

$$\frac{\partial z}{\partial x_1} = 4x_1 + 2x_2 - 20 + \lambda_1 + 2\lambda_2 = 0,$$

$$\frac{\partial z}{\partial x_2} = 2x_1 + 2x_2 - 14 + 3\lambda_1 - \lambda_2 = 0,$$

$$\frac{\partial z}{\partial \lambda_1} = x_1 + 3x_2 - 5 = 0,$$

$$\frac{\partial z}{\partial \lambda_2} = 2x_1 - x_2 - 4 = 0.$$

Solving, we obtain $x_1 = \frac{17}{7}$, $x_2 = \frac{6}{7}$, which are the coordinates of point *B* in Figure 2.12. (Notice that we could have obtained these values simply from the two equality constraints, without recourse to the Lagrangian function.) The value of y at this point is $y = -43.8$.

We must now determine the values of y at the other three corners of the region (i.e., at the origin and at points *A* and *C* in Figure 2.12). At the origin, we have the trivial result that $y = 0$. When $x_1 = 0$, $x_2 = \frac{5}{3}$ and $y = -20\frac{5}{9}$ (point *C*). At the remaining corner (point *A*), we have $x_2 = 0$, $x_1 = 2$, and $y = -32$.

Hence we conclude that the given function exhibits a constrained absolute minimum of $y = -43.8$ within the allowable region at $x_1 = \frac{17}{7}$, $x_2 = \frac{6}{7}$. This is represented by point *B* in Figure 2.12.

Note that as an alternative to the foregoing procedure we could have accepted the increased dimensionality brought to the problem by the introduction of two slack variables to convert our inequalities to equalities. To avoid non-negativity restrictions on the original and slack variables, we can use a technique originally suggested by Klein [1955]. We may replace each x_i by $(y_i)^2$ and each x_{si} by $(y_{si})^2$. Thus we would write the constraint equations as

$$(y_1)^2 + 3(y_2)^2 + (y_{s1})^2 - 5 = 0,$$
$$2(y_1)^2 - (y_2)^2 + (y_{s2})^2 - 4 = 0.$$

To obtain a solution we would have to solve simultaneously these two equations and the four equations for the partial derivatives with respect to each of the variables. The original procedure would appear to be somewhat simpler in this instance.

EXAMPLE 2.5.2

Find an equation for the largest circle whose center is at some undetermined point (x_{1o}, x_{2o}); $x_{1o}, x_{2o} > 0$; such that $x_1 + x_2 \leq 5$ and $x_1, x_2 \geq 0$. In other words, we wish to maximize the function

$$y = (x_1 - x_{1o})^2 + (x_2 - x_{2o})^2, \qquad x_1, x_2 \geq 0, \qquad x_{1o}, x_{2o} > 0,$$

subject to

$$x_1 + x_2 \leq 5.$$

A graphical interpretation of this problem is shown in Figure 2.13. Notice that the

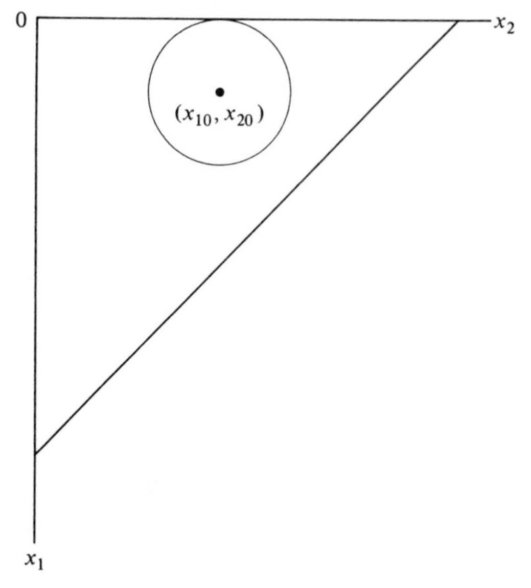

Figure 2.13. Graphical Representation of Example 2.5.2

inequality constraint is satisfied for all values of x_1 and x_2 which are within the triangle formed by the lines $x_1 = 0$, $x_2 = 0$, and $x_1 + x_2 = 5$.

By inspection we see that the unconstrained function has a minimum at $x_1 = x_{1o}$ and $x_2 = x_{2o}$, but that the function does not have a bounded maximum. We therefore allow the inequality constraint $x_1 + x_2 \leq 5$ to become active, and we form the Lagrangian function

$$z(x_1, x_2, \lambda) = (x_1 - x_{1o})^2 + (x_2 - x_{2o})^2 + \lambda(x_1 + x_2 - 5).$$

Taking the first partial derivatives, we obtain

$$\frac{\partial z}{\partial x_1} = 2(x_1 - x_{1o}) + \lambda = 0,$$

$$\frac{\partial z}{\partial x_2} = 2(x_2 - x_{2o}) + \lambda = 0.$$

Eliminating λ from the above two equations results in

$$(x_1 - x_{1o}) = (x_2 - x_{2o}). \qquad \text{(a)}$$

Combining the above expression with the constraint (which is equivalent to $\partial z/\partial \lambda$) yields

$$x_{1o} + 2x_2 - x_{2o} = 5$$

or

$$x_2 - x_{2o} = \tfrac{1}{2}(5 - x_{1o} - x_{2o}). \qquad \text{(b)}$$

On the boundary $x_1 = 0$, the unconstrained objective function becomes

$$y = x_{1o}^2 + (x_2 - x_{2o})^2.$$

If we extremize this function along the boundary $x_1 = 0$, we obtain

$$\frac{\partial y}{\partial x_2} = 2(x_2 - x_{2o}) = 0$$

or

$$x_2 = x_{2o}. \qquad \text{(c)}$$

Similarly, if we extremize the unconstrained function

$$y = (x_1 - x_{1o})^2 + x_{2o}^2$$

along the boundary $x_2 = 0$, we obtain

$$x_1 = x_{1o}. \qquad \text{(d)}$$

From Figure 2.13 it is obvious that the largest circle that will satisfy the given conditions is one which is tangent to all three boundary lines; i.e., the constraints and non-negativity conditions become equalities. This requires that y take on the same value at the three points which correspond to Equations (a) and (b), (c), and (d). Equations (a), (c), and (d) give us

$$2(x_2 - x_{2o})^2 = x_{1o}^2 = x_{2o}^2,$$

from which we conclude that $x_{1o} = x_{2o}$. Substituting Equation (b) into the above expression, with $x_{1o} = x_{2o}$, gives us

$$x_{1o}^2 - 10x_{1o} + 25/2 = 0.$$

Solving for x_{1o}, we obtain

$$\begin{aligned} x_{1o} = x_{2o} &= 5\left(1 \pm \frac{\sqrt{2}}{2}\right) \\ &= 5(1 \pm 0.707). \end{aligned}$$

Since x_{1o} must be positive and less than 5, we conclude that

$$x_{1o} = x_{2o} = 5(0.293) = 1.46.$$

Hence the equation of the desired circle is

$$(x_1 - 1.46)^2 + (x_2 - 1.46)^2 = 2.13.$$

The solution is shown graphically in Figure 2.14.

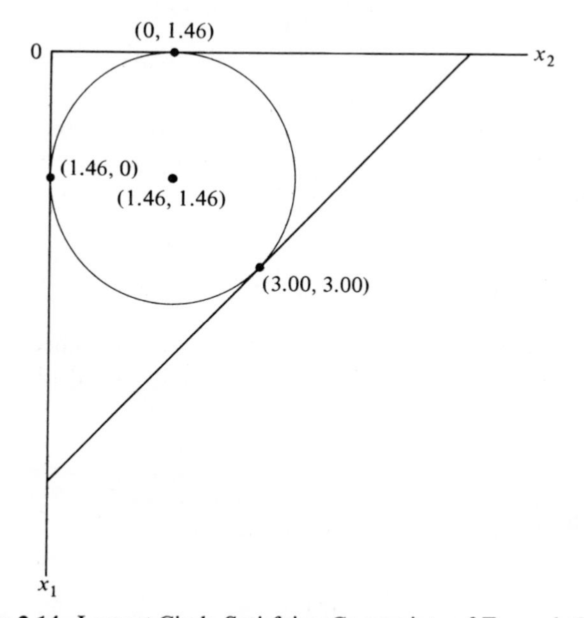

Figure 2.14. Largest Circle Satisfying Constraints of Example 2.5.2

2.6 Necessary and Sufficient Conditions for an Equality Constrained Optimum with Bounded Independent Variables

We have previously seen that the maximum of a given objective function subject to one or more equality constraints corresponds to a saddle point of the Lagrangian function, providing the maximum occurs at an interior point of the given region. Let us now consider the behavior of the Lagrangian function $z(\mathbf{X}, \boldsymbol{\lambda})$ near some point $(\mathbf{X}_o, \boldsymbol{\lambda}_o)$ where $y(\mathbf{X})$ is maximized, and $z(\mathbf{X}, \boldsymbol{\lambda})$ exhibits a saddle point; i.e.,

$$z(\mathbf{X}, \boldsymbol{\lambda}_o) \leq z(\mathbf{X}_o, \boldsymbol{\lambda}_o) \leq z(\mathbf{X}_o, \boldsymbol{\lambda}), \tag{2.6.1}$$

and $\mathbf{X}$ and $\boldsymbol{\lambda}$ are restricted in sign. Notice from Equation (2.6.1) that $z(\mathbf{X}, \boldsymbol{\lambda})$ is maximized with respect to $\mathbf{X}$ and minimized with respect to $\boldsymbol{\lambda}$ at $(\mathbf{X}_o, \boldsymbol{\lambda}_o)$.

Necessary conditions

Suppose that we restrict $\mathbf{X}$ to be nonnegative; i.e., $\mathbf{X} \geq \mathbf{0}$. If $\mathbf{X}_o$ is an interior point, then $x_j > 0$ for all j and

$$\left.\frac{\partial z}{\partial x_j}\right|_{(\mathbf{X}_o, \boldsymbol{\lambda}_o)} = \left.\frac{\partial y}{\partial x_j}\right|_{(\mathbf{X}_o, \boldsymbol{\lambda}_o)} + \sum_{i=1}^{m} \lambda_i \left.\frac{\partial g_i}{\partial x_j}\right|_{(\mathbf{X}_o, \boldsymbol{\lambda}_o)} = 0, \qquad j = 1, 2, \ldots, n, \tag{2.6.2}$$

as before. If, however, one of the x_{jo}, say x_{ko}, equals zero, then Equation (2.6.2) does not necessarily hold. The Lagrangian function cannot increase as x_k increases, though, so that we can write

$$\left.\frac{\partial z}{\partial x_k}\right|_{(\mathbf{X}_o, \boldsymbol{\lambda}_o)} = \left.\frac{\partial y}{\partial x_k}\right|_{(\mathbf{X}_o, \boldsymbol{\lambda}_o)} + \sum_{i=1}^{m} \lambda_i \left.\frac{\partial g_i}{\partial x_k}\right|_{(\mathbf{X}_o, \boldsymbol{\lambda}_o)} \leq 0. \tag{2.6.3}$$

Equation (2.6.3) is valid for all k for which $x_{ko} = 0$. Since either $x_j = 0$ or $\partial z/\partial x_j = 0$ at $(\mathbf{X}_o, \boldsymbol{\lambda}_o)$, we can combine the above results by writing

$$x_{jo} \left.\frac{\partial z}{\partial x_j}\right|_{(\mathbf{X}_o, \boldsymbol{\lambda}_o)} = 0, \qquad j = 1, 2, \ldots, n. \tag{2.6.4}$$

By similar reasoning, we can conclude that

$$\left.\frac{\partial z}{\partial \lambda_i}\right|_{(\mathbf{X}_o, \boldsymbol{\lambda}_o)} \geq 0 \tag{2.6.5}$$

and

$$\lambda_{io}\frac{\partial z}{\partial \lambda_i}\bigg|_{(\mathbf{X}_o, \boldsymbol{\lambda}_o)} = 0 \tag{2.6.6}$$

for each $\lambda_{io} \geq 0$; moreover,

$$\frac{\partial z}{\partial \lambda_i}\bigg|_{(\mathbf{X}_o, \boldsymbol{\lambda}_o)} \leq 0 \tag{2.6.7}$$

and

$$\lambda_{io}\frac{\partial z}{\partial \lambda_i}\bigg|_{(\mathbf{X}_o, \boldsymbol{\lambda}_o)} = 0 \tag{2.6.8}$$

for each $\lambda_{io} \leq 0$.

Sufficient conditions

Equations (2.6.3)–(2.6.8) are *necessary* conditions for the Lagrangian function to possess a saddle point at $(\mathbf{X}_o, \boldsymbol{\lambda}_o)$, where the point lies either within the region or on a boundary formed by one or more $x_j = 0$. If, in addition to Equations (2.6.3)–(2.6.8), the following two expressions

$$z(\mathbf{X}, \boldsymbol{\lambda}_o) \leq z(\mathbf{X}_o, \boldsymbol{\lambda}_o) + \sum_{j=1}^{n} \frac{\partial z}{\partial x_j}\bigg|_{(\mathbf{X}_o, \boldsymbol{\lambda}_o)} (x_j - x_{jo}) \tag{2.6.9}$$

$$z(\mathbf{X}_o, \boldsymbol{\lambda}) \geq z(\mathbf{X}_o, \boldsymbol{\lambda}_o) + \sum_{i=1}^{m} \frac{\partial z}{\partial \lambda_i}\bigg|_{(\mathbf{X}_o, \boldsymbol{\lambda}_o)} (\lambda_i - \lambda_{io}) \tag{2.6.10}$$

are satisfied at some point $(\mathbf{X}_o, \boldsymbol{\lambda}_o)$, then we are assured that $(\mathbf{X}_o, \boldsymbol{\lambda}_o)$ will be a saddle point. The above expressions are valid for all $\mathbf{X}$ and $\boldsymbol{\lambda}$, providing the Lagrangian function possesses only one saddle point; otherwise, these conditions hold only in the neighborhood of each local saddle point.

From the properties of concave and convex functions, one can easily show that Equation (2.6.9) is satisfied for any function which is concave with respect to $\mathbf{X}$, and that Equation (2.6.10) holds for any function which is convex with respect to $\boldsymbol{\lambda}$. Hence any Lagrangian function that satisfies Equations (2.6.3)–(2.6.8) and possesses the necessary concavity and convexity conditions in the vicinity of point $(\mathbf{X}_o, \boldsymbol{\lambda}_o)$ will exhibit a saddle point at $(\mathbf{X}_o, \boldsymbol{\lambda}_o)$.

2.7 Necessary and Sufficient Conditions for an Inequality Constrained Optimum

Now let us again turn our attention to the problem of extremizing an inequality constrained function. Assume that we are to maximize $y(x_1, x_2, \ldots, x_n)$, subject to the inequality constraints

$$g_i(x_1, x_2, \ldots, x_n) \leq b_i \qquad i = 1, 2, \ldots, \alpha,$$
$$(2.7.1) \qquad g_i(x_1, x_2, \ldots, x_n) \geq b_i \qquad i = \alpha + 1, \alpha + 2, \ldots, \beta,$$
$$g_i(x_1, x_2, \ldots, x_n) = b_i \qquad i = \beta + 1, \beta + 2, \ldots, m,$$

where $(m - \beta) < n$. We shall restrict $\mathbf{X}$ to be non-negative and assume that $z(\mathbf{X}, \boldsymbol{\lambda})$ takes on an absolute or relative maximum with respect to $\mathbf{X}$ at $(\mathbf{X}_o, \boldsymbol{\lambda}_o)$. We shall further require that the Lagrange multipliers exist and are unique for each binding constraint.

After addition of slack and surplus variables, we obtain the equivalent problem of maximizing $y(x_1, x_2, \ldots, x_n)$, subject to the m equality constraints

$$g_i(x_1, x_2, \ldots, x_n) + x_{si} = b_i, \qquad i = 1, 2, \ldots, \alpha,$$
$$(2.7.2) \qquad g_i(x_1, x_2, \ldots, x_n) - x_{si} = b_i, \qquad i = \alpha + 1, \alpha + 2, \ldots, \beta,$$
$$g_i(x_1, x_2, \ldots, x_n) = b_i, \qquad i = \beta + 1, \beta + 2, \ldots, m.$$

If we transpose the b_i to the left side of Equation (2.7.2), our problem is identical to that considered in Section 2.6 and the results obtained therein are valid. We thus may immediately apply Equations (2.6.3)–(2.6.8).

Sign of the Lagrange multiplier

We can now relate the sign of λ_{io} to the nature of the original constraints, i.e., whether $g_i(\mathbf{X}) \leq b_i$ or $g_i(\mathbf{X}) \geq b_i$. To do so, we shall first establish an additional interpretation for λ_{io}. We shall do so by computing the derivative $y(\mathbf{X}_o)$ with respect to b_i. By the chain rule,

$$(2.7.3) \qquad \left.\frac{\partial y}{\partial b_i}\right|_{(\mathbf{X}_o, \boldsymbol{\lambda}_o)} = \sum_{j=1}^{N} \left.\frac{\partial y}{\partial x_{jo}}\right|_{(\mathbf{X}_o, \boldsymbol{\lambda}_o)} \left.\frac{\partial x_{jo}}{\partial b_i}\right|_{(\mathbf{X}_o, \boldsymbol{\lambda}_o)},$$

where $N = n + \beta$ (the total number of independent variables, including slack and surplus variables).

To evaluate the terms on the right-hand side of Equation (2.7.3), we observe that at the optimum of our *equivalent* problem (all equality constraints) we can write

$$(2.7.4) \qquad g_i(\mathbf{X}_o) = b_i, \qquad i = 1, 2, \ldots, m,$$

and therefore we may write

$$(2.7.5) \qquad \sum_{j=1}^{N} \left.\frac{\partial g_k}{\partial x_{jo}}\right|_{(\mathbf{X}_o, \boldsymbol{\lambda}_o)} \left.\frac{\partial x_{jo}}{\partial b_i}\right|_{(\mathbf{X}_o, \boldsymbol{\lambda}_o)} = 1 \qquad \text{if } i = k,$$

and

$$(2.7.6) \qquad \sum_{j=1}^{N} \left.\frac{\partial g_k}{\partial x_{jo}}\right|_{(\mathbf{X}_o, \boldsymbol{\lambda}_o)} \left.\frac{\partial x_{jo}}{\partial b_i}\right|_{(\mathbf{X}_o, \boldsymbol{\lambda}_o)} = 0 \qquad \text{if } i \neq k.$$

These equations may be rewritten as

$$\delta_{ik} - \sum_{j=1}^{N} \frac{\partial g_k}{\partial x_{jo}}\bigg|_{(\mathbf{X}_o, \boldsymbol{\lambda}_o)} \frac{\partial x_{jo}}{\partial b_i}\bigg|_{(\mathbf{X}_o, \boldsymbol{\lambda}_o)} = 0, \qquad i = 1, \ldots, m, \tag{2.7.7}$$

where δ_{ik} is the *Kronecker delta.** If we multiply (2.7.7) by λ_{ko}, sum over all k's, and add the sum to (2.7.3), we obtain

$$\frac{\partial y}{\partial b_i}\bigg|_{(\mathbf{X}_o, \boldsymbol{\lambda}_o)} = -\sum_{k=1}^{m} \lambda_{ko}\delta_{ik} + \sum_{j=1}^{N} \left[\frac{\partial y}{\partial x_{jo}}\bigg|_{(\mathbf{X}_o, \boldsymbol{\lambda}_o)} + \sum_{k=1}^{m} \lambda_{ko} \frac{\partial g_k}{\partial x_{jo}}\bigg|_{(\mathbf{X}_o, \boldsymbol{\lambda}_o)}\right] \frac{\partial x_{jo}}{\partial b_i}. \tag{2.7.8}$$

We have previously shown the terms in the brackets to be zero. Hence†

$$\frac{\partial y}{\partial b_i}\bigg|_{(\mathbf{X}_o, \boldsymbol{\lambda}_o)} = -\sum_{k=1}^{m} \lambda_{ko}\delta_{ik} = -\lambda_{io}. \tag{2.7.9}$$

We are now in a position to consider for a constraint $g_i(\mathbf{X}_o, \boldsymbol{\lambda}_o) \leq b_i$, $b_i > 0$, what will happen if we increase b_i by some small amount. This has the effect of increasing the size of the feasible region. Hence all solutions that were previously feasible remain feasible and $y(\mathbf{X}_o)$ can only increase or remain unchanged. This means that

$$\frac{\partial y}{\partial b_i}\bigg|_{(\mathbf{X}_o, \boldsymbol{\lambda}_o)} \geq 0 \quad \text{and} \quad \lambda_{io} \leq 0 \quad \text{for} \quad i = 1, 2, \ldots, \alpha. \tag{2.7.10}$$

Now consider a constraint of the form $g_i(\mathbf{X}_o, \boldsymbol{\lambda}_o) \geq b_i$, $b_i > 0$. If we decrease b_i slightly, we again increase the size of the feasible region so that $y(\mathbf{X}_o)$ can only increase or remain unchanged. Therefore,

$$\frac{\partial y}{\partial b_i}\bigg|_{(\mathbf{X}_o, \boldsymbol{\lambda}_o)} \leq 0 \quad \text{and} \quad \lambda_{io} \geq 0 \quad \text{for} \quad i = \alpha + 1, \alpha + 2, \ldots, \beta. \tag{2.7.11}$$

For an equality constraint, the λ_{io} may have either sign.

Necessary conditions

Let us now summarize the results of the last two sections. Under the stated conditions the following expressions must be satisfied by the Lagran-

---

* $\delta_{ik} = 1, i = k; \quad \delta_{ik} = 0, i \neq k.$

† Suppose that b_i represents the capacity of some resource and y represents a profit (which we wish to maximize). Under these circumstances λ_i has the dimensions of dollars per unit of capacity of resource i and can therefore be interpreted as a value per unit resource. It is usual to define $\pi_i = -\lambda_i$. These π_i are referred to as *shadow prices* since they tell us how much our objective function is changed if b_i is increased by one unit without violating any of the other constraints. Thus we know the maximum price we should be willing to pay for an additional unit of b_i.

gian at a saddle point:

$$\left.\frac{\partial z}{\partial x_j}\right|_{(\mathbf{X}_o, \boldsymbol{\lambda}_o)} \leq 0, \qquad j = 1, 2, \ldots, N, \tag{2.7.12}$$

with strict equality holding if $x_j > 0$, and

$$x_{jo}\left.\frac{\partial z}{\partial x_j}\right|_{(\mathbf{X}_o, \boldsymbol{\lambda}_o)} = 0, \qquad j = 1, 2, \ldots, N. \tag{2.7.13}$$

Furthermore, it has been shown that

$$\lambda_{io} \leq 0 \tag{2.7.14}$$

and

$$\left.\frac{\partial z}{\partial \lambda_i}\right|_{(\mathbf{X}_o, \boldsymbol{\lambda}_o)} \leq 0 \tag{2.7.15}$$

for those $g_i(\mathbf{X}) \leq b_i$, and that

$$\lambda_{io} \geq 0 \tag{2.7.16}$$

and

$$\left.\frac{\partial z}{\partial \lambda_i}\right|_{(\mathbf{X}_o, \boldsymbol{\lambda}_o)} \geq 0 \tag{2.7.17}$$

for those $g_i(\mathbf{X}) \geq b_i$. Finally, we can write

$$\lambda_{io}\left.\frac{\partial z}{\partial \lambda_i}\right|_{(\mathbf{X}_o, \boldsymbol{\lambda}_o)} = 0, \qquad i = 1, 2, \ldots, m. \tag{2.7.18}$$

Notice that Equation (2.7.18) is not dependent upon the sign of λ_{io}.

Equations (2.7.12)–(2.7.18) are *necessary* conditions for the Lagrangian function to possess a saddle point at $(\mathbf{X}_o, \boldsymbol{\lambda}_o)$ when $y(\mathbf{X})$ is maximized at $\mathbf{X}_o$. These are the well-known *Kuhn–Tucker* conditions (cf. Kuhn and Tucker [1951]).

Kuhn and Tucker originally derived these necessary conditions under somewhat more general conditions than used above. They introduced the notion of a *constraint qualification.* We shall designate the gradients of the constraints as $\nabla g_i(\mathbf{X})$. With this notation we can state the constraint qualification as the requirement that the $\nabla g_i(\mathbf{X})$ for all active constraints (slack or surplus variables zero) be positively linearly independent.*

* A set of vectors are positively linearly independent if $\alpha_1 \mathbf{X}_1 + \alpha_2 \mathbf{X}_2 + \cdots + \alpha_k \mathbf{X}_k = \mathbf{0}$ implies that $\alpha_1 = \alpha_2 = \cdots = \alpha_k = 0$ (see Appendix B).

The constraint qualification is designed to rule out unusual behavior or singularities on the boundaries of the set of feasible solutions. The Kuhn–Tucker conditions can be shown to hold providing only that the constraint qualification is met.

Sufficient conditions

It is easy to shown that a *sufficient* set of conditions for a saddle point at $(\mathbf{X}_o, \boldsymbol{\lambda}_o)$ is that the expressions

$$z(\mathbf{X}, \boldsymbol{\lambda}_0) \leq z(\mathbf{X}_o, \boldsymbol{\lambda}_o) + \sum_{j=1}^{n} \left.\frac{\partial z}{\partial x_j}\right|_{(\mathbf{X}_o, \boldsymbol{\lambda}_o)} (x_j - x_{jo}) \tag{2.7.19}$$

$$z(\mathbf{X}_o, \boldsymbol{\lambda}) \geq z(\mathbf{X}_o, \boldsymbol{\lambda}_o) + \sum_{i=1}^{m} \left.\frac{\partial z}{\partial \lambda_i}\right|_{(\mathbf{X}_o, \boldsymbol{\lambda}_o)} (\lambda_i - \lambda_{io}) \tag{2.7.20}$$

apply, in addition to Equations (2.7.12)–(2.7.18). Thus we are assured that any point $(\mathbf{X}_o, \boldsymbol{\lambda}_o)$ which satisfies Equations (2.7.12)–(2.7.20) will cause the Lagrangian function to take on a maximum with respect to $\mathbf{X}$ and a minimum with respect to $\boldsymbol{\lambda}$ in the neighborhood of $(\mathbf{X}_o, \boldsymbol{\lambda}_o)$. We shall refer to this set of conditions as the Kuhn–Tucker sufficiency conditions.

If in the vicinity of $(\mathbf{X}_o, \boldsymbol{\lambda}_o)$ the Lagrangian function is concave with respect to $\mathbf{X}$, convex with respect to those λ_i for which $g_i(\mathbf{X}) \geq 0$, and concave for those λ_i for which $g_i(\mathbf{X}) \leq 0$, then the Kuhn–Tucker sufficiency conditions will apply, guaranteeing a saddle point at $(\mathbf{X}_o, \boldsymbol{\lambda}_o)$. Moreover, the saddle point will be a *global* saddle point if $z(\mathbf{X}, \boldsymbol{\lambda})$ possesses the above concavity and convexity conditions for all $\mathbf{X} \geq 0$ and all $\boldsymbol{\lambda}$.

An objective function $y(\mathbf{X})$ that is minimized at $\mathbf{X}_1$ can also be analyzed in terms of the equivalent Lagrangian saddle-point problem

$$z(\mathbf{X}, \boldsymbol{\lambda}_1) \geq z(\mathbf{X}_1, \boldsymbol{\lambda}_1) \geq z(\mathbf{X}_1, \boldsymbol{\lambda}). \tag{2.7.21}$$

Notice that the Lagrangian function is now minimized with respect to $\mathbf{X}$ and maximized with respect to $\boldsymbol{\lambda}$ at $(\mathbf{X}_1, \boldsymbol{\lambda}_1)$. Necessary and sufficient conditions can be developed for this problem which are parallel to those for the maximization of $y(\mathbf{X})$, the differences being a reversal in the inequalities and the concavity–convexity requirements.

Although the Kuhn–Tucker conditions themselves do not lead directly to computational techniques, they are significant theoretically and they form the basis for computational techniques that we shall consider in Chapter 5.

The Kuhn–Tucker conditions are treated in detail by Hadley [1964]. Additional necessary and sufficient conditions for extremizing a nonlinear objective function subject to linear constraints are presented by Dorn [1961].

2.8 Applicability of Classical Methods to Practical Optimization Problems

As we have seen in the earlier sections of this chapter, classical methods can, in principle, be applied to the optimization of continuous and differentiable functions within a given closed region. The methods can be applied to either unconstrained or constrained functions as long as the number of equality constraints is less than the number of independent variables. The theory is not as complete as we would like, however, as we have no way of knowing how may extrema exist in a given region without solving for the location of the stationary points.

Severe computational difficulties may be encountered in solving the algebraic equations that result from equating to zero the first derivatives of the objective function. This limits the practical utility of the classical approach. Futhermore, having solved these equations, we must be careful to distinguish between extrema and saddle points. Also, we must always evaluate the function on the region boundaries in order to obtain an absolute extremum; this may involve very tedious computation if the dimensionality of the function becomes high. Thus we see that, except for particularly simple and well-behaved functions, the classical theory does not provide us with efficient techniques for the optimization of practical problems. Nevertheless, classical optimization theory has its value in that it offers us considerable insight into the characteristics and problems associated with extremizing continuous functions. Moreover, the theory to some extent serves as a basis for the development of certain computational algorithms which we shall encounter in the next several chapters.

REFERENCES

Buck, R. C., *Advanced Calculus*, McGraw-Hill, New York, 1956.

Courant, R., *Differential and Integral Calculus*, Vols. I and II, Wiley-Interscience, New York, 1936. (Translation by E. J. McShane.)

Dorn, W. S., "On Lagrange Multipliers and Inequalities," *Operations Res.*, *9* (1961), 95.

Hadley, G., *Nonlinear and Dynamic Programming*, Addison-Wesley, Reading, Mass., 1964.

Kaplan, W., *Advanced Calculus*, Addison-Wesley, Reading, Mass., 1952.

Kiepert, L., *Grundiss der Differential- und Integral-Rechnung*, Hannover, 1910.

Klein, B., "Direct Use of Extremal Principles in Solving Certain Optimizing Problems Involving Inequalities," *Operations Res.*, *3* (1955), 168.

Kuhn, H. W., and A. W. Tucker, "Nonlinear Programming," *Proceedings of Second*

Berkeley Symposium on Mathematical Statistics and Probability, University of California Press, Berkeley, 1951, pp. 481–92.

PROBLEMS

2.1. Determine the nature and location of the stationary points for each of the following functions. Determine the absolute maximum and the absolute minimum.

(a) $y = 2x(1 + x^2)^{-1}, x \geq 0$
(b) $y = x^2 \sin x, -\pi \leq x \leq \pi$
(c) $y = x^{-1} \sin x, 0 \leq x \leq 2\pi$
(d) $y = x(x + \cos x), -\pi \leq x \leq \pi$
(e) $y = \cos 2x + 2 \cos x, 0 \leq x \leq \pi/2$
(f) $y = x_1^2 + x_2^2 - x_1x_2$
(g) $y = x_1^2 + x_2^2 - 3x_1x_2$
(h) $y = x_1^3 - 3x_1x_2 + 3x_2^2$
(i) $y = 2x_1 - x_2 + 3x_3$
(j) $y = 2(x_1 - 3)^2 - (x_2 + 1)^2 + 3(x_3 - 2)^2, x_1, x_2, x_3 \geq 0$
(k) $y = x_1^2 + 2x_2^2 + 2x_3^2 + x_4^2 - x_1x_2 + x_1x_3 - 2x_2x_4 + x_1x_4$
(l) $y = x_1^2 - 2x_2^2 - 2x_3^2 + x_4^2 - x_1x_2 + x_1x_3 - 2x_2x_4 + x_1x_4$
(m) $y = -x_1^2 - 0.5x_2^2 + x_1x_2 + x_1$
(n) $y = -x_1^2 - 0.5x_2^2 + x_1x_2 + x_1, x_i \geq 0, i = 1, 2$

2.2. What is the interpretation of the condition where $y(x_1, x_2)$ has a stationary point at (x_{1o}, x_{2o}), and B^2 in Equation (2.2.12) is equal to the product AC? How can the nature of the stationary point be identified?

2.3. Resolve the exercises in problem 1.1, making use of the material presented in Section 2.2.

2.4. Solve, using Lagrange multipliers:

(a) Maximize $y = x_1x_2$,
subject to $x_1 + 2x_2 = 4$.
(b) Minimize $y = (x_1 - 2)^2 + 2(x_2 - 1)^2 + (x_3 - 3)^2$,
subject to $2x_1 + x_2 + 2x_3 \geq 4$,
$x_1^2 + 2x_2^2 + 3x_3^2 \geq 48$.
(c) Minimize $y = (x_1 - 2)^2 + 2(x_2 - 1)^2 + (x_3 - 3)^2$,
subject to $2x_1 + x_2 + 2x_3 \leq 4$,
$x_1^2 + 2x_2^2 + 3x_3^2 \leq 48$.
(d) Maximize $y = (x_1 - 1)^2 + (x_2 - 3)^2 + (x_3 + 1)^2$,
subject to $x_1^2 + 4x_3^2 = 36$,
$x_1, x_2, x_3 \geq 0$.
(e) Minimize $y = x_1^2 + x_2^2$,
subject to $(x_1 - 1)^3 - x_2^2 = 0$.
(Cf. Courant, Vol. II [1936].)

Verify the solutions graphically wherever possible.

2.5. By use of the Lagrange multiplier method, find the extremum of $y = \mathbf{X}'\mathbf{A}\mathbf{X} - \mathbf{B}'\mathbf{X}$, subject to $\mathbf{C}'\mathbf{X} + 5 = 0$,

where

$$\mathbf{A} = \begin{bmatrix} 2 & 0 \\ -1 & 1 \end{bmatrix}, \quad \mathbf{B} = \begin{bmatrix} 3 \\ 1 \end{bmatrix}, \quad \mathbf{C} = \begin{bmatrix} 2 \\ -1 \end{bmatrix}, \quad \mathbf{X} = \begin{bmatrix} x_1 \\ x_2 \end{bmatrix}.$$

Identify the nature of the extremum.

2.6. Determine the minimum distance from the origin to the plane $x_1 + 3x_2 + 2x_3 = 12$.

2.7. Derive an equation for the largest sphere that can be enclosed by the ellipsoid $x_1^2 + 4x_2^2 + 3x_3^2 = 12$.

2.8. A rectangular parallelepiped has a diagonal of length d. Determine the dimensions of this parallelepiped such that its volume is maximized.

2.9. If x_1 and x_2 are related by $2x_1 + x_2 = 1$, find those points on the ellipsoid $2x_1^2 + x_2^2 + x_3^2 = 1$ which are, respectively, the closest and farthest from the origin.

2.10. A company manufactures three different kinds of products, utilizing two scarce raw materials. The net profit derived from the sale of the product line is expressed as

$$y = (40 + 6x_1)x_1 + (60 - 4x_2)x_2 + (50 + 2x_3)x_3,$$

where x_i is the number of units of the ith product manufactured per week. (Note that the profitability of products 1 and 3 increases as the production rate increases, whereas the profitability of product 2 decreases.) The availability of each of the scarce resources can be expressed as $12x_1 + x_2 + 8x_3 \leq 48$ and $9x_1 + 12x_2 + x_3 \leq 36$. Determine a production policy that will maximize net profit.

2.11. Show that the solution of the following problem is a saddle point in $(x_1, x_2, x_3, \lambda_1, \lambda_2)$ space: Maximize $y = x_1x_2x_3$, subject to

$$\begin{aligned} x_1^2 + x_2^2 + x_3^2 &= 9, \\ x_1 + 2(x_2^2 + x_3^2) &= 4 \\ x_1, x_2, x_3 &\geq 0. \end{aligned}$$

2.12. In problem 2.10 *estimate* the maximum profit if

(a) The availability of the first scarce resource is increased by two units/week, while the availability of the second resource is held constant.

(b) The availability of the second scarce resource is decreased by one unit/week, with the availability of the first resource kept at its original value.

Do *not* resolve the problem.

2.13. Solve the following problems and determine whether or not the Kuhn–Tucker conditions apply. Explain any discrepancies that may be encountered.

(a) Maximize $y = x_1^2 + x_2^2$, subject to

$$3x_1^2 + x_2^2 + 2x_1x_2 \leq 6,$$
$$x_1, x_2 \geq 0.$$

(b) Minimize $y = (x_1 - 2)^2 + 2(x_2 - 1)^2 + (x_3 - 3)^2$, subject to

$$2x_1 + x_2 + 2x_3 \geq 4,$$
$$x_1^2 + 2x_2^2 + 3x_3^2 \geq 48$$
$$x_1, x_2, x_3 \geq 0.$$

(c) Minimize $y = x_1^2 + x_2^2$, subject to

$$x_1^2 - (x_2 - 1)^3 \leq 0,$$
$$x_1, x_2 \geq 0.$$

2.14. Derive a set of sufficient conditions analogous to Equations (2.7.12)–(2.7.20) for the Lagrangian function $z(\mathbf{X}, \boldsymbol{\lambda})$ to have a saddle point at $(\mathbf{X}_o, \boldsymbol{\lambda}_o)$ if $y(\mathbf{X})$ is to be minimized, subject to the constraint set (2.7.1).

2.15. Show that Equations (2.7.19) and (2.7.20) are valid for a function $z(\mathbf{X}, \boldsymbol{\lambda})$ that is concave with respect to $\mathbf{X}$, concave with respect to those $\lambda_i \geq 0$, and convex with respect to those $\lambda_i \leq 0$.

UNCONSTRAINED OPTIMIZATION

In the last chapter we saw that the use of calculus in determining extrema is restricted to objective functions which are continuous and differentiable. Even then, the amount of computation required to locate an extremum is frequently excessive, making the use of calculus impractical. Consequently, we shall turn our attention to the development of numerical searching techniques that will allow us to extremize a given objective function relatively quickly and easily.

A variety of numerical techniques for extremizing both one-dimensional and multidimensional objective functions is presented in this chapter. Certain of these methods will borrow from classical theory to the extent that they apply to functions which are continuous and differentiable. These methods, known as *gradient techniques*, require the computation of certain partial derivatives (or approximations thereof). However, these methods yield an extremum much more readily than the classical techniques of Chapter 2. Other methods that we shall encounter do not make use of partial derivatives and therefore do not require that the objective function be differentiable. Some of these latter techniques can be applied to functions that are not even continuous.

3

Most of the techniques presented in this chapter are sequential in the sense that past information is used to generate future search points. The extrema which they yield are dependent upon the choice of the initial search point. Unless the objective function is unimodal, the sequential methods will yield an extremum that is not necessarily the absolute extremum in the entire region of interest. Many problems (in fact, most practical engineering problems) give rise to objective functions that are multimodal; however, we can in principle always find some subregion within the original region in which the objective function is unimodal and in which sequential searching methods are useful. At the end of the chapter we shall consider some random-search techniques that are not sequential and are less restrictive in their use.

For now we shall be concerned only with extremizing objective functions whose variables are not constrained. In later chapters we shall see that the addition of constraints adds considerable complexity to the task of numerical optimization. At the expense of simplicity, however, many of the techniques that apply to unconstrained functions serve as a foundation for the more general programming problem which we shall encounter later.

3.1 One-dimensional Optimization by Search Methods

Conceptually, direct-search techniques are perhaps the simplest of the optimization techniques. However, they are among the most powerful tools available. In a search procedure we evaluate the function to be extremized at a number of points and use these data to obtain an estimate of the location of the extremum.

Interval of uncertainty

Let us restrict our attention to finding the maximum of a unimodal, one-dimensional function $y(x)$ within the closed interval $a \leq x \leq b$. We shall adopt the sequential approach described above, of evaluating the function at several predetermined search points and then using the resulting information to establish a new set of search points. Unless we happen to be particularly lucky in the placement of one of the search points, we will not locate the position of the true maximum (in fact, we would lack sufficient information to recognize the true maximum even if we had determined its location). Rather, we can establish that the true maximum lies within the interval either to the left or the right of the calculated maximum, as shown in Figure 3.1. We cannot determine from the search points to which side of x_i the true maximum lies. Since the function is unimodal, however, and since $y(x_i) > y(x_j)$, $j \neq i$, we can conclude with certainty that the true maximum lies within the interval $(x_{i+1} - x_{i-1})$.

We shall designate the *largest* interval $(x_{j+1} - x_{j-1})$, not necessarily

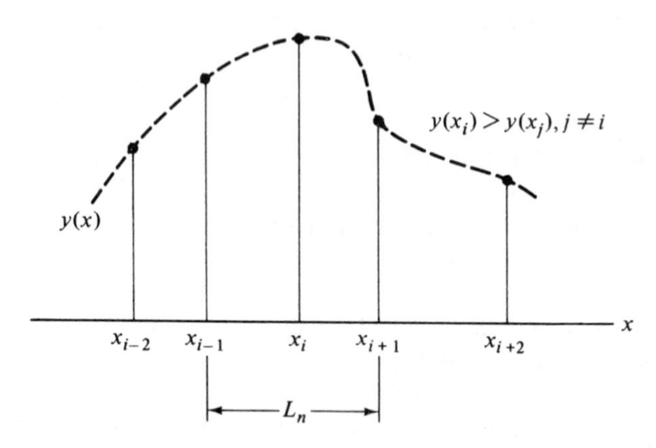

Figure 3.1. Interval of Uncertainty Obtained by Evaluation of Function at Predetermined Points

containing the maximum, as L_n, the *interval of uncertainty*. The subscript n will refer to the number of search points required to obtain L_n. Our objective in the next few sections will be the development of search schemes that will minimize L_n, and, consequently $(x_{j+1} - x_{j-1})$, with a given number of search points. Stated differently, we may wish to reduce L_n to some small predetermined value with as few search points as possible.

Notice that our strategy will be to minimize the worst of several possible occurrences. This requires that we cause the *maximum* possible value of L_n to be as small as possible. One way to accomplish this is to play a game of "equal odds," and force all possible L_n to be of the same size. A strategy of this kind, where one minimizes the maximum attainable condition, is known as a *minimax* strategy.

Symmetrical two-point search

To carry out our search procedure, we must locate at least two search points within a given search interval, evaluate the function at these search points, find the maximum value, and, using this information, define a new subinterval and repeat the procedure. Suppose that we wish to locate two new search points x_1 and x_2 within a given interval $a \leq x \leq b$, allowing for the fact that the search points must have at least some minimal separation.

Let ϵ be the minimum allowable separation between the search points such that the two values of y remain distinguishable. Note that there is always some such $\epsilon > 0$. The selection of an appropriate value for ϵ depends upon the sensitivity of the objective function to changes in x (i.e., dy/dx) over the entire search interval, however, so that ϵ can usually only be estimated in advance.

We can see that, by locating x_1 and x_2 symmetrically with respect to the center of the interval at a distance ϵ apart, our interval of uncertainty L_2 will equal $(b - x_1) = (x_2 - a) = (b - a + \epsilon)/2$. This situation is shown in Figure 3.2. Notice that L_2 must increase if x_1 and x_2 remain symmetrical

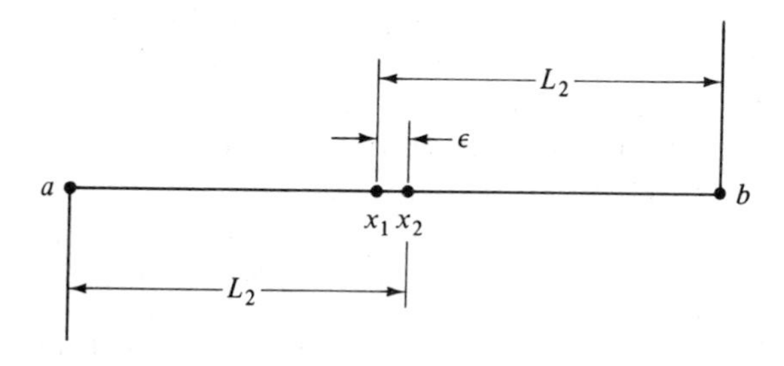

Figure 3.2. Interval of Uncertainty Obtained by Symmetric Two-Point Search

but move in opposite directions from the center of the interval. Moreover, if x_1 remains fixed but x_2 moves to the right, then $(b - x_1)$ remains equal to $(b - a + \epsilon)/2$, but $(x_2 - a)$, and hence L_2, must increase. The interval of uncertainty L_2 also increases if x_2 remains fixed but x_1 moves to the left. Hence we conclude that, *in placing two new search points within an interval, the interval of uncertainty is minimized by placing the search points symmetrically about the center of the given interval and as close together as possible.*

If, in this example, $y(x_2) > y(x_1)$, then the interval $x_1 \leq x \leq b$ could be explored by the symmetrical placement of two new search points within that interval, and so on. Such a scheme, which approximately cuts the search interval in half after the placement of every two new search points, is called a *dichotomous search.* We can easily see that, after the placement of $2k$ search points, the remaining interval of uncertainty for a dichotomous search is

$$L_{2k} = \frac{b-a}{2^k} + \left(1 - \frac{1}{2^k}\right)\epsilon. \tag{3.1.1}$$

If we let $k = 7$ and neglect ϵ, then Equation (3.1.1) tells us that the interval of uncertainty is reduced to less than 1 per cent of $(b - a)$.

The following question now arises: Given some $\epsilon > 0$, how large can we allow k to become before the numerical values of $y(x)$ become indistinguishable? Notice from Equation (3.1.1) that L_{2k} approaches ϵ asymptotically as k becomes large. Once L_{2k} becomes smaller than 2ϵ, however, the values of y evaluated within L_{2k} and on its boundaries become indistinguishable. Suppose, then, that we carry out a dichotomous search using $2k$ search points, such that

$$L_{2k} \geq 2\epsilon, \tag{3.1.2}$$

coming as close as possible to satisfying the equality. We shall then place a $(k + 1)$st search point at the center of L_{2k}, so that the single interior point x_{k+1} will be separated from each end point by a distance equal to or slightly greater than ϵ. The best value of $y(x)$ can then be chosen as closely as possible.

The procedure is illustrated in Figure 3.3 for the case in which (3.1.2) is satisfied as a strict equality. In this case the final search point x_{k+1} coincides with one of the previously placed search points.

The question remains as to the correct numerical value for k. If we combine Equations (3.1.2) and (3.1.1), we obtain

$$(2^k + 1)\epsilon \leq (b - a), \tag{3.1.3}$$

so that k, the total number of search pairs, becomes the largest positive integer that satisfies Equation (3.1.3).

Although the dichotomous-search technique is highly efficient from the

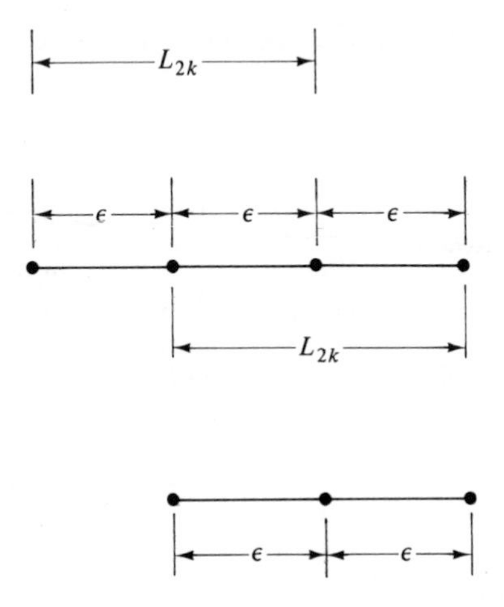

Figure 3.3. Uncertainty Interval at Conclusion of Dichotomous Search

standpoint of reducing the interval of uncertainty with a small number of search points, we shall develop still more efficient techniques in the next few sections.

EXAMPLE 3.1.1

Maximize the function $y = x \cos x$ within the interval $0 \leq x \leq \pi$ using a 14-point dichotomous search ($k = 7$) and $\epsilon = 0.0001$ radian. (Recall that this problem was solved in Example 2.1.1 using classical methods. The function is shown graphically in Figure 2.5.)

The first two search points are calculated as follows:

$$x_{s1} = \tfrac{1}{2}(\pi - \epsilon) = \tfrac{1}{2}(3.14149) = 1.570745,$$
$$x_{s2} = \tfrac{1}{2}(\pi + \epsilon) = \tfrac{1}{2}(3.14169) = 1.570845.$$

Since $y(x_{s1}) = 0.000085$ and $y(x_{s2}) = -0.000075$, the new search interval becomes $0 \leq x \leq x_{s2}$, or $0 \leq x \leq 1.570845$, and

$$x_{s1} = \tfrac{1}{2}(1.570745) = 0.785373,$$
$$x_{s2} = \tfrac{1}{2}(1.570945) = 0.785473,$$

and so on.

The interval bounds (called x_l and x_r), the interior search points (called x_{s1} and x_{s2}), and the values of y at the search points (called y_1 and y_2) are shown in Table 3.1.

The final solution is $y_{\max} = 0.561096$, which occurs at $x = 0.859001$. Notice that the final interval of uncertainty is 0.024643, which is 0.8 per cent of the original interval.

Table 3.1. MAXIMIZATION OF ($y = x \cos x$) BY DICHOTOMOUS SEARCH

x_l	x_r	x_{s1}	x_{s2}	y_1	y_2
0	3.141593	1.570745	1.570845	0.000085	−0.000075
0	1.570845	0.785373	0.785473	0.555356	0.555373
0.785373	1.570845	1.178059	1.178159	0.450863	0.450797
0.785373	1.178159	0.981716	0.981816	0.545436	0.545414
0.785373	0.981816	0.883544	0.883644	0.560535	0.560529
0.785373	0.883644	0.834458	0.834558	0.560406	0.560411
0.834458	0.883644	0.859001	0.859101	0.561096	0.561095

Three-point search

Now let us suppose that we wish to place *three* new search points, x_1, x_2, and x_3, in the interval $a \leq x \leq b$. Let x_2 be located at the center of the interval, and let x_1 and x_3 be separated by a distance of $(b - a)/2$, as shown in Figure 3.4. Notice that x_1 and x_3 are not placed symmetrically. Under these

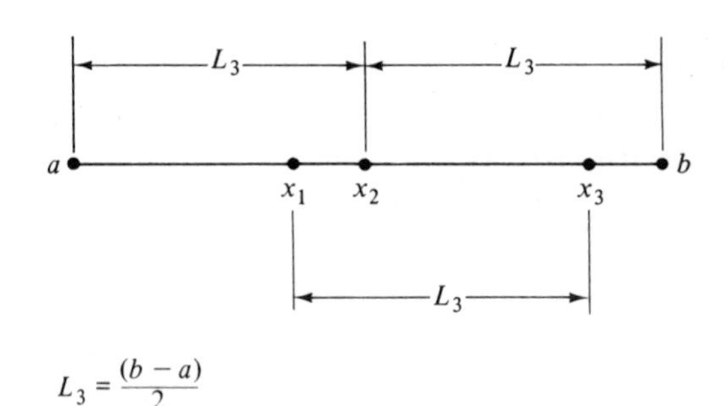

Figure 3.4. Uncertainty Interval in a Three Point Search

conditions $(b - x_2) = (x_3 - x_1) = (x_2 - a) = L_3 = (b - a)/2$. The interval $(x_3 - x_1)$ can be reduced without affecting L_3, as shown in Figure 3.5; of course, $(x_3 - x_1)$ must remain greater than ϵ. If, however, $(x_3 - x_1)$ exceeds $(b - a)/2$, or if x_2 moves in either direction, then the interval of uncertainty L_3 must increase, Hence, *in placing three new search points within an interval, the inteval of uncertainty is minimized by placing x_2 at the center of the interval and by preventing $(x_3 - x_1)$ from exceeding half the interval length.*

Notice that the minimum interval of uncertainty in a two-point search is $(b - a + \epsilon)/2$, whereas the minimum interval of uncertainty for a three-point search is $(b - a)/2$. Thus *the advantage of a three-point search over a two-point search is very slight.* In almost all cases the additional computation

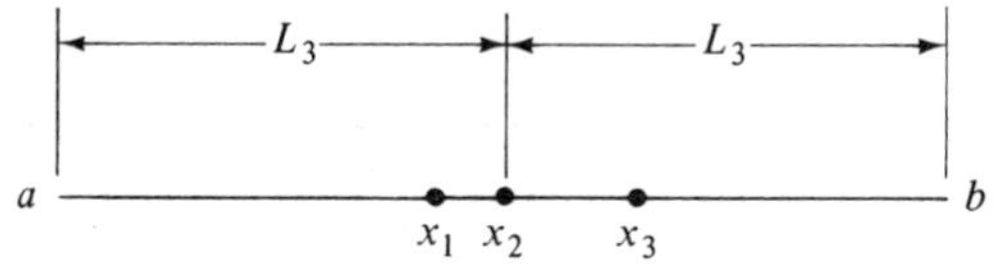

Figure 3.5. Minimum Uncertainty Interval in a Three Point Search

required for the third point makes it inadvisable to use a three-point rather than a two-point search.

We could investigate the possibility of placing pairs of search points symmetrically within a given interval. However, this begins to defeat our purpose of locating a few search points within an interval sequentially and then using the information obtained to define a new and smaller search interval. The placement of pairs of search points within an interval is discussed by Wilde [1964] and is shown to be inefficient.

Fibonacci search

Suppose that we place two points x_1 and x_2 symmetrically within the interval $a \leq x \leq b$, as shown in line (I) of Figure 3.6. If $y(x_2) > y(x_1)$, then we form a new search interval $x_1 \leq x \leq b$, as in line (II) of Figure 3.6. We could, if we wish, place two new search points within the new interval and repeat the above search procedure. Instead, however, let us take advantage of the fact that x_2 has already been placed in our new search interval from the previous search interval, and that $y(x_2)$ has already been evaluated. Therefore, let us place only one new search point x_3 within our new interval, locating it symmetrically with respect to x_2. The arrangement of the search points within the new search interval is shown in line (III) of Figure 3.6. Notice that by locating the search points symmetrically within any interval we shall always have subintervals of the same length, regardless of which search point yields the maximum; i.e., $(x_2 - x_1) = (b - x_3) = L_3$. Observe also that the search points rapidly move toward each other in successive search intervals.

Let us now require that L_n, the interval of uncertainty which results from the sequential placement of n search points, be minimized. We have already seen that the last two search points x_{n-1} and x_n must be symmetrically spaced and a distance ϵ apart within the last search interval, as in line (III) of Figure 3.7. If in the previous search interval $y(x_{n-1}) > y(x_{n-2})$, however, then x_{n-2} forms the new boundary of the last search interval, and x_{n-1} appears within the last search interval. This is showh in lines (II) and (III) in Figure 3.7.

Notice that the interval of uncertainty L_{n-1} becomes equal to $(b - x_{n-2})$,

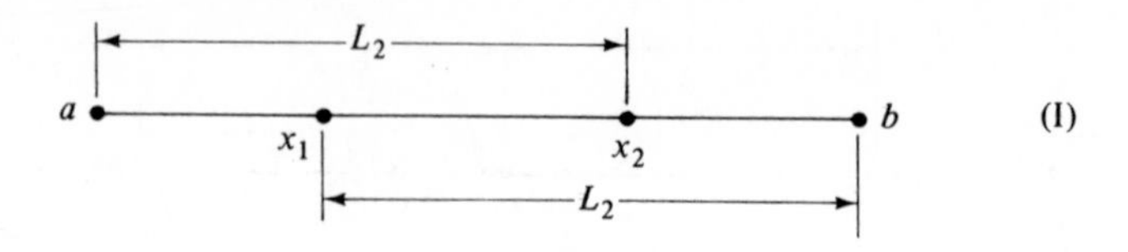

$y(x_2) > y(x_1)$:

x_1 x_2 b (II)

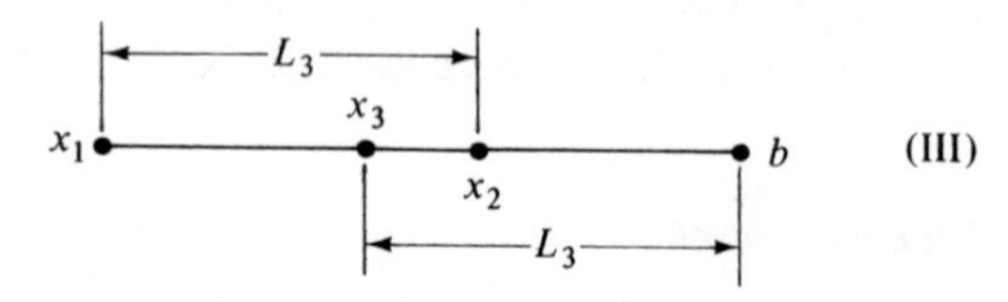

Figure 3.6. Search Point Arrangement in a Fibonacci Search

the new search interval. Hence we can write

$$L_n = \frac{(b - x_{n-2} + \epsilon)}{2} \tag{3.1.4}$$

or

$$L_n = \frac{(L_{n-1} + \epsilon)}{2}. \tag{3.1.5}$$

Notice also that the second-last search interval [i.e., line (II), Figure 3.7] can be written

$$(b - x_{n-3}) = (b - x_{n-1}) + (x_{n-1} - x_{n-3}), \tag{3.1.6}$$

which, from Figure 3.7. can be expressed as

$$(b - x_{n-3}) = L_n + L_{n-1}. \tag{3.1.7}$$

However, from line (I) of Figure 3.7 we can see that $(b - x_{n-3})$, the second-last search interval, is simply equal to L_{n-2} providing $y(x_{n-2}) > y(x_{n-3})$.

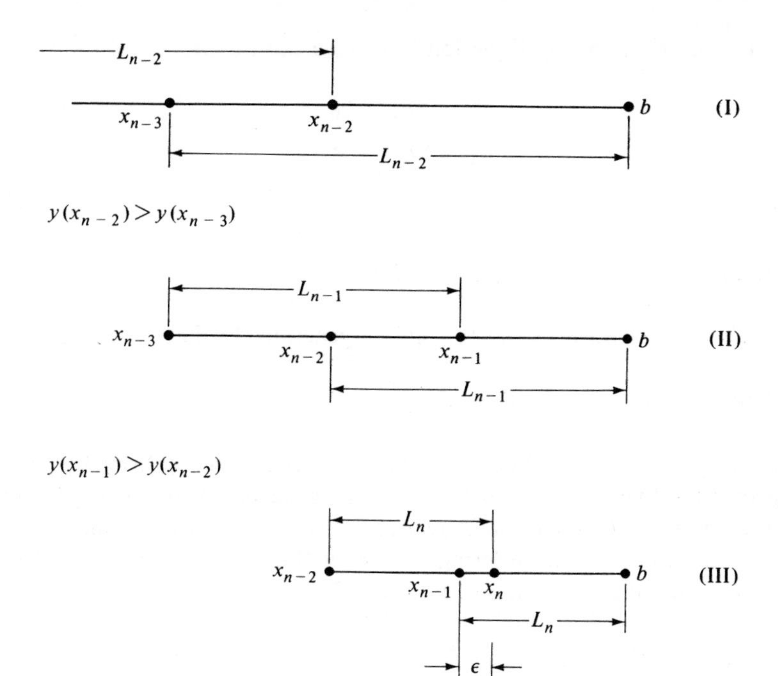

Figure 3.7. Sequence of Uncertainty Intervals in a Fibonacci Search

Thus we have shown that

$$L_{n-2} = L_{n-1} + L_n, \tag{3.1.8}$$

which can be generalized to read

$$L_i = L_{i+1} + L_{i+2}, \qquad i = 1, 2, \ldots, n-2. \tag{3.1.9}$$

The arguments that led to the above equations are not dependent upon which of the two search points in a given search interval yields the greater value for y. In Figure 3.7, for example, we could have interchanged the positions of x_{n-2} and x_{n-3} in line (I), and, again working our way back from the final interval, obtained Equations (3.1.5) and (3.1.9). All that we require for Equations (3.1.5) and (3.1.9) to be valid are the symmetrical placement of the search points within each search interval and the minimum separation of the search points within the last search interval.

Combining Equations (3.1.5) and (3.1.8) to eliminate L_{n-1} gives

$$L_{n-2} = 3L_n - \epsilon. \tag{3.1.10}$$

By successively applying Equation (3.1.9) we can obtain

(3.1.11) $$L_{n-3} = 5L_n - 2\epsilon,$$

(3.1.12) $$L_{n-4} = 8L_n - 3\epsilon,$$

$$\vdots$$

(3.1.13) $$L_{n-k} = F_{k+1}L_n - F_{k-1}\epsilon.$$

The coefficients F_{k+1} and F_{k-1} can be generated by the recurrence formula

(3.1.14) $$F_{k+1} = F_k + F_{k-1}, \qquad k = 1, 2, 3, \ldots,$$

providing $F_0 = F_1 = 1$. In other words, each coefficient is formed by adding together the two previous coefficients. The sequence of numbers so formed is the interesting and curious *Fibonacci number sequence*, which appears numerous times in the ordering of certain biological forms. The first 40 Fibonacci numbers are shown in Table 3.2.

Table 3.2. COMPILATION OF FIBONACCI NUMBERS

n	F_n	n	F_n
0	1	21	17,711
1	1	22	28,657
2	2	23	46,368
3	3	24	75,025
4	5	25	121,393
5	8	26	196,418
6	13	27	317,811
7	21	28	514,229
8	34	29	832,040
9	55	30	1,346,269
10	89	31	2,178,309
11	144	32	3,524,578
12	233	33	5,702,887
13	377	34	9,227,465
14	610	35	14,930,352
15	987	36	24,157,817
16	1,597	37	39,088,169
17	2,584	38	63,245,986
18	4,181	39	102,334,155
19	6,765	40	165,580,141
20	10,946		

Consider the special case of Equation (3.1.13) when $k = n - 1$. We have

$$L_1 = F_n L_n - F_{n-2}\epsilon \tag{3.1.15}$$

or

$$L_n = \frac{1}{F_n} L_1 + \frac{F_{n-2}}{F_n}\epsilon. \tag{3.1.16}$$

Letting L_1 represent the length of the original search interval, Equation (3.1.16) allows us to determine the interval of uncertainty remaining after the placement of n Fibonacci search points. The interval of uncertainty is reduced to less than 1 per cent of the original search interval for $n = 11$, as compared to $n = 14$ for a dichotomous search.

Once a Fibonacci search has been started, it is a simple matter to continue the placement of successive search points because of the symmetry that we require within each interval. The question remains as to the placement of the first two search points. We can answer this question by considering the relationship between L_1 and L_2.

If we evaluate Equation (3.1.13) for the case of $k = n - 2$, we obtain

$$L_2 = F_{n-1} L_n - F_{n-3}\epsilon. \tag{3.1.17}$$

Combining this result with Equation (3.1.15) to eliminate L_n gives

$$L_2 = \frac{F_{n-1}}{F_n} L_1 + \frac{F_{n-1}F_{n-2} - F_n F_{n-3}}{F_n}\epsilon, \tag{3.1.18}$$

which can be written as

$$L_2 = \frac{F_{n-1}}{F_n} L_1 + \frac{(-1)^n}{F_n}\epsilon. \tag{3.1.19}$$

The equivalence between Equations (3.1.18) and (3.1.19) is shown by Wilde [1964].

Thus from Equation (3.1.19) we can compute L_2 once we specify the total number of search points n and a value for ϵ. To begin a Fibonacci search within an interval $L_1 = (b - a)$, we simply locate x_1 a distance of L_2 units from b, and x_2 is placed L_2 units from a.

It should be emphasized that the placement of the first two search points is based upon the condition that L_n, the interval of uncertainty remaining after n search points have been explored, is to be minimized. If $n = 2$, the Fibonacci search simply reduces to a symmetrical two-point search, which minimizes L_2. When $n = 3$, as shown in Figure 3.5. we ask if the search interval $L_2 = (b - x_1)$ can be reduced, with x_2 and x_3 still symmetric in L_2 and separated by the minimum distance ϵ. For this to be so, x_1, and consequently x_2, must be located closer to the center of $L_1 = (b - a)$. Thus x_2 would appear in a different location in $L_2 = (b - x_1)$. Locating x_3 symmetrically in L_2, we would not have x_2 and x_3 separated by ϵ, which violates our original premise and either makes $y(x_2)$ and $y(x_3)$ indistinguishable or else increases L_3. Hence for $n = 3$ we see that the Fibonacci search does indeed minimize the remaining interval of uncertainty. We can continue this type of logic for higher n if we wish, and thus we can show by induction that the Fibonacci search technique minimizes L_n. A rigorous proof is presented by Kiefer [1953], who first devised the Fibonacci search technique.

Let us now consider what happens when ϵ is allowed to vanish. With this condition the last two search points will coincide, and L_n will equal $L_{n-1}/2$. From Equation (3.1.16), we have, for $\epsilon = 0$,

$$L_n = \frac{1}{F_n} L_1. \tag{3.1.20}$$

Since $x_n = x_{n-1}$, however, we obtain L_n with only $(n - 1)$, rather than n, search points.

We now consider the possibility that x does not vary continuously in $L_1 = (b - a)$, but rather takes on discrete values. We can still conduct a Fibonacci search to locate the maximum value of $y(x)$, providing the number of discrete points is one less than a Fibonacci number. If this is not the case, we can choose some n^* discrete x's and we can add a sufficient number of dummy points $x_1^*, x_2^*, \ldots,$ at one end, with $y(x_1^*) = y(x_2^*) = \ldots = u^*$, until we have one less discrete point than a Fibonacci number (e.g., we could have 7, 12, 20, 33, . . . , etc., discrete points). We then order the points. This is shown in line (I) of Figure 3.8 for the case of five discrete values of x, two dummy points, and $u^* = 0$. Notice that our search interval L_1 now contains F_n intervals (where $F_n = F_5 = 8$ for this example), with the spacing between adjacent search points corresponding to one interval. From Equation (3.1.19) we have

$$L_2 = \frac{F_{n-1}}{F_n} L_1 = \frac{F_{n-1}}{F_n} F_n = F_{n-1}. \tag{3.1.21}$$

Hence we place our search points F_{n-1} units from each end of L_1, which corresponds to the points x_3 and x_5 in Figure 3.8.

The Fibonacci search is continued in lines (II), (III), and (IV) of Figure 3.8 for the case in which $y(x_3) > y(x_5)$ and $y(x_2) < y(x_3) < y(x_4)$. Notice that L_5 is obtained after evaluating y at only four search points (x_3, x_5, x_2, and x_4) because $\epsilon = 0$.

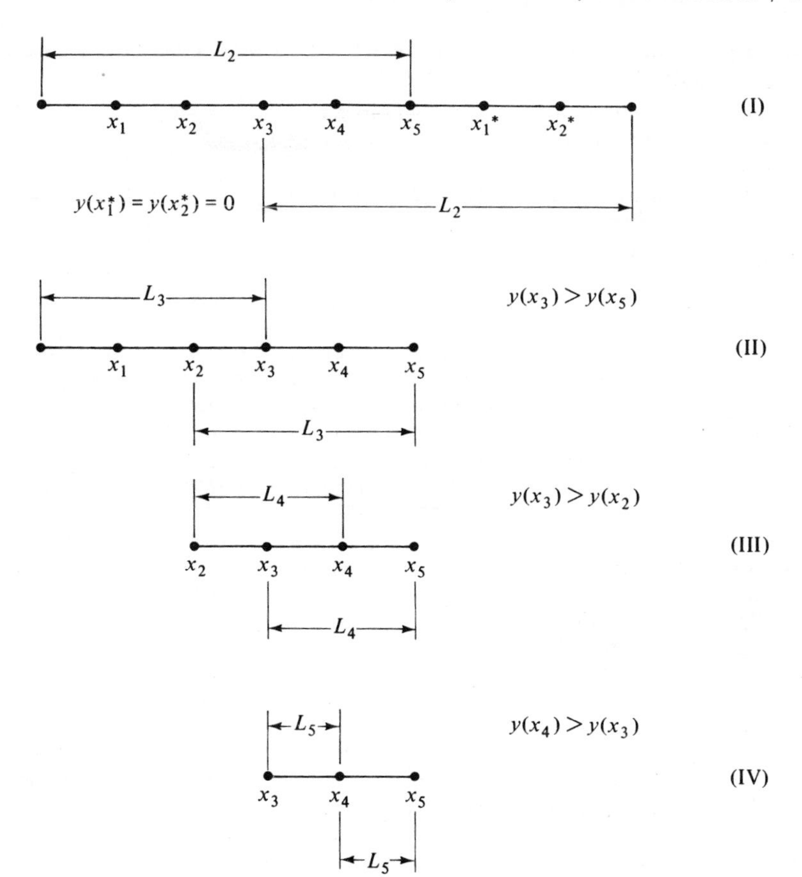

Figure 3.8. Fibonacci Search Over Set of Discrete Points

EXAMPLE 3.1.2

Maximize the function $y = x \cos x$ within the interval $0 \leq x \leq \pi$ using a 14-point Fibonacci search with $\epsilon = 0.0001$ radian (cf. Figure 2.5).

From Table 3.1 we see that $F_{13} = 377$ and $F_{14} = 610$. Substituting these results into Equation (3.1.19), we have

$$L_1 = \frac{377}{610}\pi + \frac{10^{-4}}{610} = 1.941606.$$

Hence $x_{s2} = 1.941606$ and $x_{s1} = (\pi - 1.941606) = 1.199984$. Evaluating y at these points, we see that $y(x_{s1}) = 0.434839$ and $y(x_{s2}) = -0.703580$. Therefore, x_{s2} becomes the upper bound of the new search interval and we locate the next search point within the new interval, symmetrically with respect to x_{s1}; i.e., $x_{s3} = 0 + (1.941606 - 1.199984) = 0.741622$.

The interval bounds, the interior search points, and the values of y at the search points are shown in Table 3.3. The notation is defined in Example 3.1.1.

Table 3.3. MAXIMIZATION OF ($y = x \cos x$) BY FIBONACCI SEARCH

x_l	x_r	x_{s1}	x_{s2}	y_1	y_2
0	3.141593	1.199984	1.941606	0.434839	−0.703580
0	1.941606	0.741622	1.199984	0.546853	0.434839
0	1.199984	0.458363	0.741622	0.411050	0.546853
0.458363	1.199984	0.741622	0.916725	0.546853	0.557755
0.741622	1.199984	0.916725	1.024881	0.557755	0.532120
0.741622	1.024881	0.849777	0.916725	0.560980	0.557755
0.741622	0.916725	0.808570	0.849777	0.558345	0.560980
0.808570	0.916725	0.849777	0.875519	0.560980	0.560855
0.808570	0.875519	0.834312	0.849777	0.560397	0.560980
0.834312	0.875519	0.849777	0.860053	0.560980	0.561096
0.849777	0.875519	0.860053	0.865242	0.561096	0.561070
0.849777	0.865242	0.854965	0.860053	0.561066	0.561096
0.854965	0.865242	0.860053	0.860153	0.561096	0.561095

The final solution is $y_{\max} = 0.561096$, which occurs at $x = 0.860053$. We see that the final interval of uncertainty has been reduced to 0.005188, and that the final search points are a distance $0.0001(\epsilon)$ units apart.

Golden-ratio search

Consider the rectangle shown in Figure 3.9. Let us choose the dimensions a and b in such a manner that if we remove a square of area b^2 from the

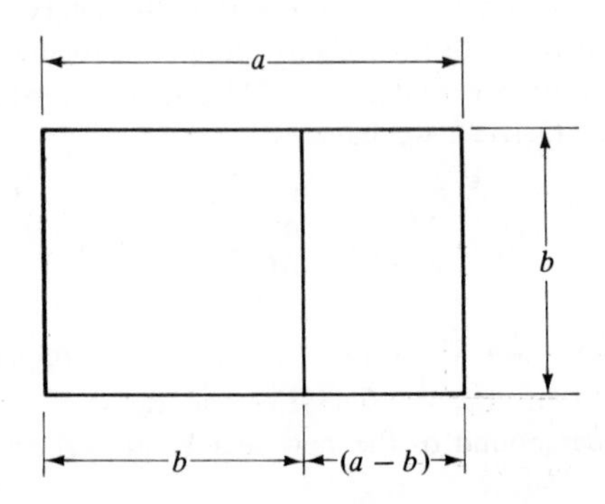

Figure 3.9. Rectangles Conforming to Golden Ratio

original rectangle, the remaining rectangle will have the same length-to-width ratio as the original rectangle. In other words,

$$\frac{a}{b} = \frac{b}{a - b} \tag{3.1.22}$$

or

$$a^2 - ab - b^2 = 0. \tag{3.1.23}$$

Solving for a in terms of b, we obtain

$$a = \frac{b}{2}(1 \pm \sqrt{5}). \tag{3.1.24}$$

Selecting the positive root and forming the ratio a/b,

$$\frac{a}{b} = \frac{1}{2}(1 + \sqrt{5}) = 1.618034.\ldots \tag{3.1.25}$$

[Notice that $a/b < 0$ if we had selected the negative root of Equation (3.1.24); in the geometric sense discussed above, this is meaningless.]

The ratio of a/b given by Equation (3.1.25) is known as the *golden ratio*, which we shall designate by the symbol ρ. The golden ratio has interesting aesthetic characteristics, which are frequently exploited in architectural design and in the fine arts.

If we form the ratio of Fibonacci numbers F_n/F_{n-1} for various $n \geq 1$, we obtain the values shown in Table 3.4. As n becomes large, we see that the

Table 3.4. RATIO OF FIBONACCI NUMBERS

n	F_n/F_{n-1}	n	F_n/F_{n-1}
1	1.0000	9	1.617647
2	2.0000	10	1.618181
3	1.5000	11	1.617978
4	1.6667	12	1.618056
5	1.6000	13	1.618026
6	1.6250	14	1.618037
7	1.615385	15	1.618033
8	1.619048	16	1.618034

ratio F_n/F_{n-1} approaches the golden ratio ρ. We can make use of this relationship between the Fibonacci numbers and the golden ratio as follows.

Let us evaluate Equation (3.1.13) for values of k equal to $n - i - 1$ and $n - i$ and, setting $\epsilon = 0$, let us eliminate L_n from the resulting two expressions. This gives

$$L_{i+1} = \frac{F_{n-i}}{F_{n-i+1}} L_i. \tag{3.1.26}$$

For $n - i$ reasonably large, the ratio of Fibonacci numbers in Equation (3.1.26) approaches $1/\rho$, and we can write

$$L_{i+1} \sim \frac{1}{\rho} L_i. \tag{3.1.27}$$

When $i = 1$, Equation (3.1.27) can be written

$$L_2 = \frac{1}{\rho} L_1. \tag{3.1.28}$$

Thus we could, if we wished, perform a Fibonacci-type search with the ratio of successive search intervals equal to $1/\rho$. In particular, we could begin the search by locating the first two search points a distance L_1/ρ units from each end of the original search interval. The search would then continue by simply locating successive search points symmetrically within each search interval, as described in the preceding section. This procedure is called a *golden-ratio search.*

Since we have essentially substituted the constant $1/\rho$ for the quantity F_{n-1}/F_n in Equation (3.1.19), we see that *in a golden-ratio search the location of the first two search points is independent of the total number of search points employed.* Therefore, the golden-ratio search technique does not minimize the interval of uncertainty remaining after a prespecified number of search points, so that the method is slightly less efficient than a Fibonacci search. Quite often, however, the difference in successive values of y is a more convenient measure of search effectiveness than L_n, the interval of uncertainty. With this type of search criterion, the number of search points required to obtain a satisfactory answer is not known in advance; rather, the search is continued until the computed values of y suggest that the procedure be halted. Under such conditions a golden-ratio search offers decided computational advantages over a Fibonacci search.

From Table 3.2 one can easily establish that F_{n-2}/F_n approaches a constant value of 0.382 as n increases. Thus as n becomes large, Equation (3.1.16) can be written

$$L_n \sim 0.382\epsilon, \tag{3.1.29}$$

since L_1/F_n approaches zero. Recall, however, that nothing is gained by allowing L_n to become smaller than 2ϵ, since the search points then become indistinguishable. Hence we can establish some finite bound on n, the number of golden-ratio search points.

If we rearrange Equation (3.1.16), we obtain

$$F_n L_n - F_{n-2}\epsilon = L_1. \tag{3.1.30}$$

Since we require that $2\epsilon \leq L_n$, this expression becomes

$$(2F_n - F_{n-2})\epsilon \leq L_1. \tag{3.1.31}$$

Thus for fixed values of L_1 and ϵ, the total number of golden-ratio search points must satisfy Equation (3.1.31).

We can make use of this result by proceeding as we did with the dichotomous search; i.e., we perform a golden-ratio search to obtain an interval of uncertainty L_n such that

$$L_n \geq 2\epsilon, \tag{3.1.32}$$

with n chosen so as to come as close as possible to satisfying the strict equality. We can then place an $(n + 1)$st search point at the center of L_n and select, as closely as possible within the ϵ restriction, the optimal value of y.

The Fibonacci and golden-ratio search techniques are among the best available one-dimensional algorithms. Both are easily programmed for a digital computer. We shall not discuss single-dimensional search any further, except to note that certain multidimensional gradient methods reduce essentially to a sequence of one-dimensional searches. We shall discuss this concept at length in the next section. The reader interested in additional single-dimensional search techniques is referred to Wilde and Beightler [1967].

EXAMPLE 3.1.3

Maximize the function $y = x \cos x$, $0 \leq x \leq \pi$, using a 14-point golden-ratio search with $\epsilon = 0.0001$ radian (cf. Figure 2.5).

The first search point is given by Equation (3.1.28) as

$$x_{s2} = \frac{\pi}{1.6180} = 1.941650,$$

and the second search point is

$$x_{s1} = (\pi - x_{s2}) = 1.199940.$$

Hence we have a symmetrical placement of the first two search points within the given interval $0 \leq x \leq \pi$. Notice the similarity in the values of x_{s1} and x_{s2} in this example and Example (3.1.2.). Having located the first two search points, the remaining procedure is identical to a Fibonacci search.

The tabular data shown in Table 3.5 correspond to the information given in Examples 3.1.1 and 3.1.2.

Table 3.5. MAXIMIZATION OF ($y = x \cos x$) BY GOLDEN-RATIO SEARCH

x_l	x_r	x_{s1}	x_{s2}	y_1	y_2
0	3.141593	1.199940	1.941650	0.434876	−0.703672
0	1.941650	0.741710	1.199940	0.546874	0.434876
0	1.199940	0.458229	0.741710	0.410957	0.546874
0.458229	1.199940	0.741710	0.916459	0.546874	0.557789
0.741710	1.199940	0.916459	1.025191	0.557789	0.532009
0.741710	1.025191	0.850443	0.916459	0.560995	0.557789
0.741710	0.916459	0.807726	0.850443	0.558256	0.560995
0.807726	0.916459	0.850443	0.873743	0.560995	0.560911
0.807726	0.873743	0.831026	0.850443	0.560209	0.560995
0.831026	0.873743	0.850443	0.854326	0.560995	0.561059
0.850443	0.873743	0.854326	0.869859	0.561059	0.561003
0.850443	0.869859	0.854326	0.865976	0.561059	0.561064
0.854326	0.869859	0.858209	0.865976	0.561092	0.561064

We see that $y_{max} = 0.561092$ at $x = 0.858209$. The final interval of uncertainty is 0.011650, which is smaller than that obtained with the dichotomous search but larger than that resulting from the Fibonacci search (cf. Examples 3.1.1 and 3.1.2). Notice also that the last two search points are separated by a distance greater than ϵ, which is in contrast with the Fibonacci search.

3.2 Multi-dimensional Optimization by Gradient Methods

Let us now consider a multidimensional function $y(x_1, x_2, \ldots, x_n)$ that is continuous and differentiable. In Chapter 2 we cited two difficulties associated with the use of classical methods in extremizing such problems: the solution of the algebraic equations resulting from setting all the first partial derivatives to zero, and the evaluation of the function on the boundaries. Often, however, we wish to extremize a unimodal, multidimensional function that is continuous and differentiable and whose maximum lies at an interior point of an n-dimensional region. Here the only deterrent to using classical calculus is the problem of solving a system of complicated algebraic equations. Under these circumstances we can use one of several very powerful *hillclimbing* techniques, known as *gradient* methods. These techniques are based upon classical ideas but they circumvent the algebra problem.

Hillclimbing along a gradient direction: steepest ascent

We wish to follow some path s that will lead us from an initial, non-stationary point to the maximum. For a two-dimensional function, the path we seek is shown in Figure 3.10, and its projection in the x_1x_2 plane appears in Figure 3.11.

Notice that $y(x_1, x_2, \ldots, x_n)$ will increase along s, providing $dy/ds > 0$. Hence let us choose that *particular* path which maximizes dy/ds at any given point within the region.

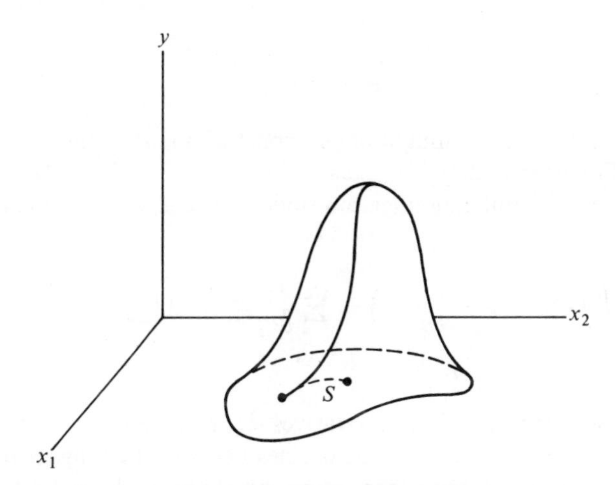

Figure 3.10. Shortest Path to Maximum

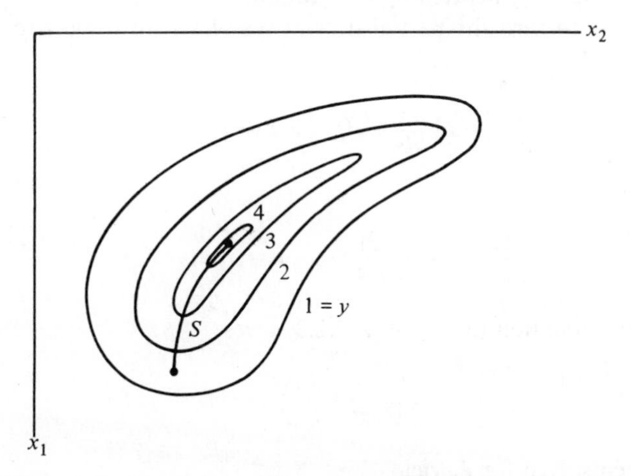

Figure 3.11. Projection of Shortest Path in the $x_1 - x_2$ Plane

Employing the chain rule for partial differentiation, we can write

$$\frac{dy}{ds} = \sum_{i=1}^{n} \frac{\partial y}{\partial x_i} \frac{dx_i}{ds}, \tag{3.2.1}$$

where s can be expressed in terms of the independent variables by the generalized Pythagorean theorem,

$$(ds)^2 = \sum_{i=1}^{n} (dx_i)^2, \tag{3.2.2}$$

or

$$\sum_{i=1}^{n} \left(\frac{dx_i}{ds}\right)^2 = 1. \tag{3.2.3}$$

Equation (3.2.3) can be thought of as a constraint that applies to the maximization of Equation (3.2.1).

Making use of the Lagrange-multiplier concept, we form the Lagrangian function

$$z\left(\frac{dx_1}{ds}, \frac{dx_2}{ds}, \ldots, \frac{dx_n}{ds}, \lambda\right) = \sum_{i=1}^{n} \frac{\partial y}{\partial x_i} \frac{dx_i}{ds} + \lambda\left\{\sum_{i=1}^{n} \left(\frac{dx_i}{ds}\right)^2 - 1\right\}, \tag{3.2.4}$$

where we have considered the derivatives dx_i/ds to be independent variables. These derivatives are the direction cosines between the tangent to s and the various x_i axes. Once these derivatives are known, the desired path $s(x_1, x_2, \ldots, x_n)$ can be generated by integration.

Equating to zero the partial derivatives of z with respect to dx_i/ds, we obtain

$$\frac{\partial y}{\partial x_i} + 2\lambda \frac{dx_i}{ds} = 0, \qquad i = 1, 2, \ldots, n, \tag{3.2.5}$$

which gives

$$\frac{dx_i}{ds} = -\frac{1}{2\lambda} \frac{\partial y}{\partial x_i}, \qquad i = 1, 2, \ldots, n. \tag{3.2.6}$$

Substituting Equation (3.2.6) into (3.2.3) gives

$$\frac{1}{4\lambda^2} \sum_{i=1}^{n} \left(\frac{\partial y}{\partial x_i}\right)^2 = 1, \tag{3.2.7}$$

which, when solved for λ, yields

$$\lambda = \pm\frac{1}{2}\left\{\sum_{i=1}^{n} \left(\frac{\partial y}{\partial x_i}\right)^2\right\}^{1/2}. \tag{3.2.8}$$

Combining this result with Equation (3.2.6) gives

$$\frac{dx_i}{ds} = \pm\left\{\sum_{i=1}^{n}\left(\frac{\partial y}{\partial x_i}\right)^2\right\}^{-1/2}\frac{\partial y}{\partial x_i}, \qquad i = 1, 2, \ldots, n, \tag{3.2.9}$$

which is our desired result.

When the positive value of Equation (3.2.9) is selected and substituted into Equation (3.2.1), we determine the *rate of steepest ascent* to be

$$\frac{dy}{ds} = \left\{\sum_{i=1}^{n}\left(\frac{\partial y}{\partial x_i}\right)^2\right\}^{1/2}. \tag{3.2.10}$$

Had we chosen the negative value of Equation (3.2.9), we could determine the *rate of steepest descent*, from which we could obtain the path that would lead us to a *minimum* of $y(x_1, x_2, \ldots, x_n)$.

We can construct a numerical hillclimbing procedure for maximizing $y(\mathbf{X})$, based upon the foregoing analysis, as follows: By rewriting Equation (2.3.9) in finite-difference form and using vector notation, we have

$$\mathbf{X}_{k+1}^{*} = \mathbf{X}_k + \{(\nabla y)'(\nabla y)\}^{-1/2}\,\nabla y \Delta s, \tag{3.2.11}$$

where

$$\mathbf{X} = \begin{bmatrix} x_1 \\ x_2 \\ \cdot \\ \cdot \\ \cdot \\ x_n \end{bmatrix}, \qquad \nabla y = \begin{bmatrix} \frac{\partial y}{\partial x_1} \\ \frac{\partial y}{\partial x_2} \\ \cdot \\ \cdot \\ \cdot \\ \frac{\partial y}{\partial x_n} \end{bmatrix}, \tag{3.2.12}$$

and $(\nabla y)'$ is the transpose of ∇y. Equation (3.2.11) can be rewritten as simply

$$\mathbf{X}_{k+1}^{*} = \mathbf{X}_k + \nabla y \Delta\tau, \tag{3.2.13}$$

where we have incorporated the $(\nabla y)'(\nabla y)$ term into $\Delta\tau$. This is permissible since Δs, and hence $\Delta\tau$, are arbitrarily large.

The algorithm begins by first choosing some point $\mathbf{X}_0$ within the region and then evaluating ∇y at point $\mathbf{X}_0$. The components of ∇y can be evaluated analytically if $y(\mathbf{X})$ is relatively uncomplicated. In many cases of practical interest, however, it will be easier to approximate the partial derivatives numerically; i.e.,

$$\frac{\partial y}{\partial x_i} \sim \frac{y[x_1, x_2, \ldots, x_i + (\Delta x_i/2), \ldots, x_n] - y[x_1, \ldots, x_i - (\Delta x_i/2), \ldots, x_n]}{\Delta x_i}, \qquad i = 1, 2, \ldots, n, \tag{3.2.14}$$

where Δx_i is an arbitrarily chosen small number.

A new point $\mathbf{X}_1$ is then obtained from Equation (3.2.13), using some small, arbitrarily specified value for $\Delta\tau$. The function ∇y is then reevaluated at $\mathbf{X}_1$, and another point $\mathbf{X}_2$ is determined from Equation (3.2.13), and so on. The procedure is continued until some point is found where ∇y either vanishes (which is the condition for a stationary point) or becomes sufficiently small. The stationary value of y is then determined at this point. This technique is called the *method of steepest ascent*. The method is shown schematically in Figure 3.12.

An analogous numerical technique, known as the *method of steepest descent*, can be developed if one chooses the negative root of Equation (3.2.9).

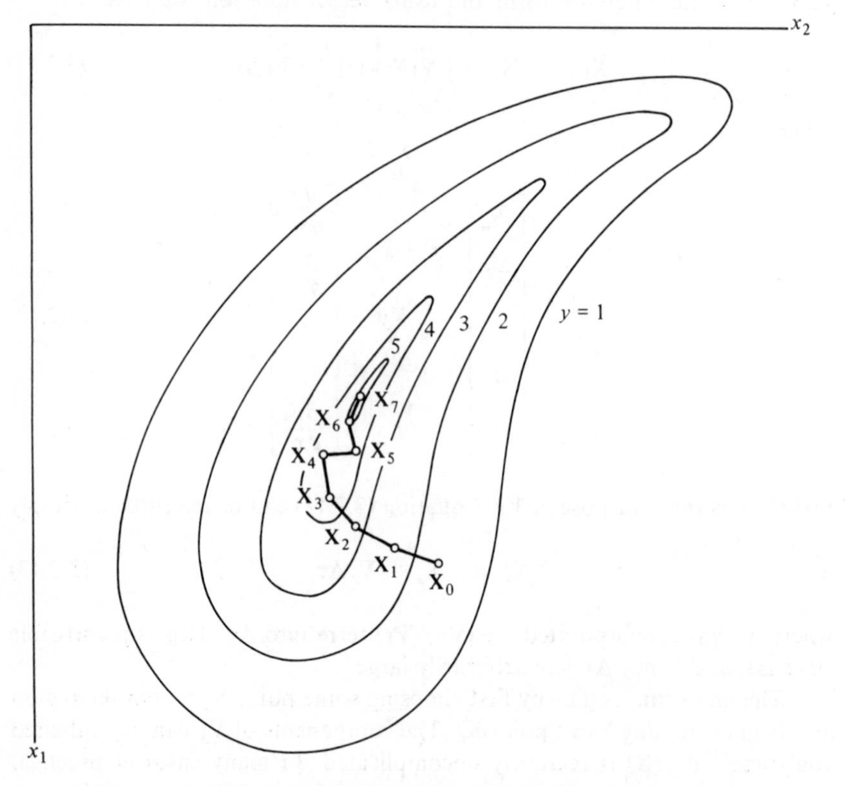

Figure 3.12. Schematic Representation of Steepest Ascent Procedure

The resulting expression, which parallels Equation (3.2.13), is

$$\mathbf{X}_{k+1}^* = \mathbf{X}_k - \nabla y \Delta\tau. \tag{3.2.15}$$

As its name implies, the method of steepest descent is used for minimizing $y(\mathbf{X})$ by numerically descending into a "valley" or "trough."

Two different procedures are available for selecting $\Delta\tau$. In the first, $\Delta\tau$ is simply held constant for all steps. The alternative to this is to select a new value for $\Delta\tau$ that is proportional to the magnitude of the gradient vector at each step. This latter method tends to take a big step on steep surfaces and a small step on relatively flat surfaces. It is, however, possible to "overshoot" the extremum in such a manner that the computation will continue to oscillate about the extremum point. The constant-step method may therefore be the safer of the two methods, although it may require more function evaluations than the variable-step method.

Although it is possible that the steepest ascent–descent algorithm will lead to a saddle point rather than an extremum, this is highly unlikely (cf. Wilde [1964]). In any event, the nature of the stationary point can and should be determined by numerical exploration. (Some random exploratory techniques will be discussed in a later section of this chapter.) If by chance the stationary point does turn out to be a saddle point, some nearby point in the direction of increasing y can be arbitrarily chosen and the steepest ascent–descent method again initiated.

The method of steepest ascent can often be improved upon if $\Delta\tau$ can be chosen such that $y(\mathbf{X})$ has a relative maximum along the line connecting $\mathbf{X}_k$ and $\mathbf{X}_{k+1}^*$ (experience with the range of independent variables frequently enables us to do this). The procedure is to pick some point $\mathbf{X}_0$, evaluate ∇y at this point, and then determine a point $\mathbf{X}_1^*$ using Equation (3.2.13) and a sufficiently large value of $\Delta\tau$. We then perform a one-dimensional search (e.g., a Fibonacci or golden-ratio search) along the line connecting $\mathbf{X}_0$ and $\mathbf{X}_1^*$ until we locate some point $\mathbf{X}_1$ where $y(\mathbf{X})$ is maximized. For a proper choice of $\Delta\tau$, $\mathbf{X}_1$ can be written

$$\mathbf{X}_1 = \theta_1\mathbf{X}_0 + (1 - \theta_1)\mathbf{X}_1^*, \qquad 0 < \theta_1 < 1, \tag{3.2.16}$$

so that the search proceeds with respect to the one independent variable θ_1. Having located $\mathbf{X}_1$ in this manner, we then evaluate ∇y at $\mathbf{X}_1$, choose a new value of $\Delta\tau$ (which may or may not be different from the old value), and establish the point $\mathbf{X}_2^*$ with Equation (3.2.13). Again a one-dimensional search is performed to locate the point $\mathbf{X}_2$, where

$$\mathbf{X}_2 = \theta_2\mathbf{X}_1 + (1 - \theta_2)\mathbf{X}_2^*, \qquad 0 < \theta_2 < 1, \tag{3.2.17}$$

and so on, until ∇y again vanishes or becomes sufficiently small. This procedure, known as *optimal steepest ascent*, is shown schematically in Figure 3.13. Notice that the algorithm reduces a multidimensional search to a sequence of one-dimensional searches.

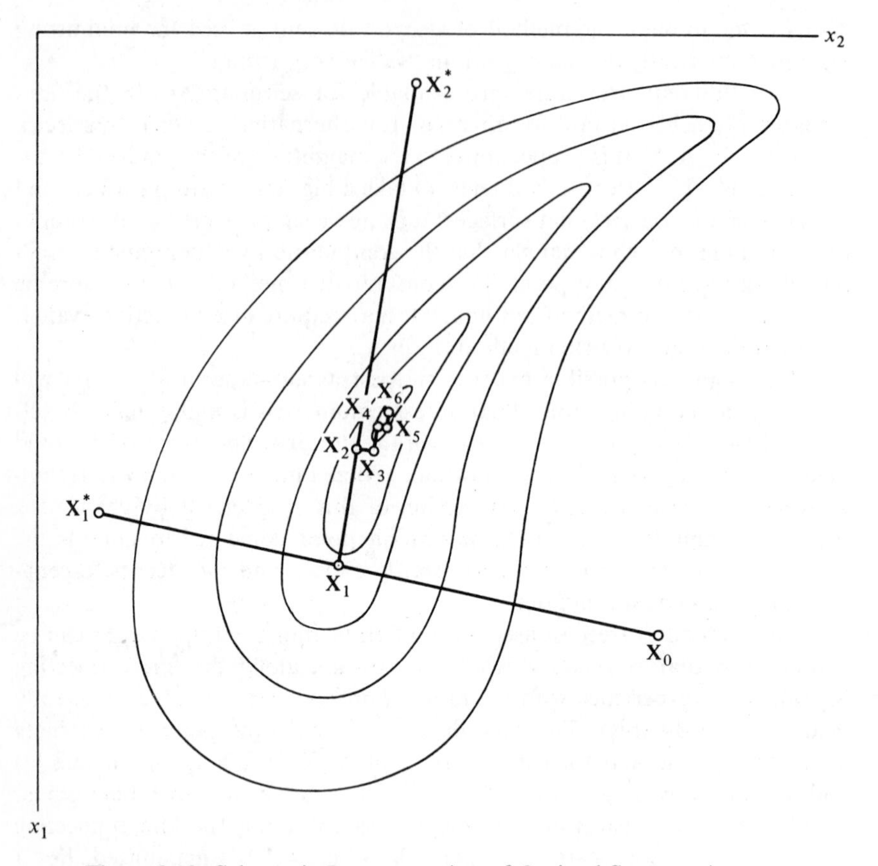

Figure 3.13. Schematic Representation of Optimal Steepest Ascent Procedure

The method of steepest ascent is the simplest and most straightforward of a rather large collection of gradient techniques. All these techniques are related intha t the ascent proceeds along a path which is at least in part defined by the partial derivatives of the objective function (hence the name "gradient" techniques). In subsequent sections of this chapter we shall examine some of the more sophisticated gradient methods. For now, however, let us recognize that there are certain numerical difficulties associated with the use of the steepest-ascent method. These difficulties limit its value as a practical optimization tool.

One difficulty associated with the use of most gradient techniques is that the amount of computation required to extremize a function depends on the sensitivity of the function to changes in each of the independent variables. By scaling the independent variables it is often possible to change the value of certain of the partial derivatives, thus increasing the computational efficiency of a particular method. Although there is no universal technique for doing this, the general procedure is to reduce the geometric eccentricity of the objective function at the optimum as much as possible. We might therefore attempt to introduce new independent variables such that a one-unit change in any one of the new independent variables will, at the optimum, affect the objective function to the same extent as a one-unit change in any other of the new independent variables.

For many problems of practical interest we cannot, unfortunately, find a new set of variables that will significantly reduce the computational effort. Quite often the method of steepest ascent–descent will "zigzag" toward an optimum, requiring more and more steps of a smaller and smaller size as the optimum is approached. The number of iterations required to locate an extremum accurately may thus be very large. This is a significant disadvantage to the use of steepest-ascent and steepest-descent methods.

Another difficulty, which is common to all gradient methods, occurs when the search region is bounded; i.e., we impose the condition that

$$\mathbf{a} \leq \mathbf{X} \leq \mathbf{b}. \tag{3.2.18}$$

As long as the computation is restricted to the interior of the region, there is no difficulty. If the gradient vector is directed out of the region, however, and the one-dimensional search proceeds to move out of the bounded region, then that particular move should terminate at the region boundary. This much is clear. The difficulty arises when a new gradient vector is calculated at the boundary and one or more components are still directed out of the bounded region. When this occurs a generally satisfactory procedure is to set equal to zero those components which point out of the bounded region, and then search along the modified search vector. The entire procedure terminates at an optimum when either all the components of the gradient vector point out of the bounded region or those components which do not point out of the bounded region are sufficiently close to zero.

Another method for handling a bounded region is to erect a "barrier" at the region boundary, so that the gradient vector never points out of the boundary. For example, if we wanted to maximize $y(\mathbf{X})$, subject to the bounds given by (3.2.18), then we could construct the auxiliary function

$$\phi(\mathbf{X}) = y(\mathbf{X}) - K \sum_{i=1}^{n} \left(\frac{1}{x_i - a_i} + \frac{1}{b_i - x_i} \right) \tag{3.2.19}$$

and then proceed to maximize the z function. In this expression K is a positive constant that determines how much weight to attach to the barrier term. The barrier term is chosen so that the desired maximum will always be found interior to the boundary (though perhaps very close). We only require that the computation be initiated at an interior point.

Barrier methods are susceptible to numerical difficulties, and the choice of a value for K is sometimes critical. We shall say more about barrier methods in Chapter 5, where we discuss nonlinear optimization in the presence of constraints.

EXAMPLE 3.2.1

Minimize the function

$$y = (x_1 - 3)^2 + 9(x_2 - 5)^2$$

using the method of steepest descent, with $\Delta\tau = 0.10$. Choose as an initial starting point $x_{10} = 1$, $x_{20} = 1$. Hence

$$\mathbf{X}_0 = \begin{bmatrix} 1 \\ 1 \end{bmatrix}.$$

Evaluating the derivatives analytically,

$$\left.\frac{\partial y}{\partial x_1}\right|_{(1,1)} = 2(x_1 - 3)\Big|_{(1,1)} = -4,$$

$$\left.\frac{\partial y}{\partial x_2}\right|_{(1,1)} = 18(x_2 - 5)\Big|_{(1,1)} = -72.$$

From Equation (3.2.15),

$$\mathbf{X}_1 = \mathbf{X}_1^* = \begin{bmatrix} 1 \\ 1 \end{bmatrix} - 0.10\begin{bmatrix} -4 \\ -72 \end{bmatrix} = \begin{bmatrix} 1.4 \\ 8.2 \end{bmatrix}.$$

Again evaluating the derivatives,

$$\left.\frac{\partial y}{\partial x_1}\right|_{(1.4,\,8.2)} = 2(x_1 - 3)\Big|_{(1.4,\,8.2)} = -3.2,$$

$$\left.\frac{\partial y}{\partial x_2}\right|_{(1.4,\,8.2)} = 18(x_2 - 5)\Big|_{(1.4,\,8.2)} = 57.6.$$

Again utilizing Equation (3.2.15),

$$\mathbf{X}_2 = \mathbf{X}_2^* = \begin{bmatrix} 1.4 \\ 8.2 \end{bmatrix} - 0.10\begin{bmatrix} -3.2 \\ 57.6 \end{bmatrix} = \begin{bmatrix} 1.72 \\ 2.44 \end{bmatrix}.$$

Continuing the procedure, we finally arrive at the point

$$\mathbf{X}_n = \begin{bmatrix} 3 \\ 5 \end{bmatrix},$$

where

$$\left.\frac{\partial y}{\partial x_1}\right|_{(3,5)} = \left.\frac{\partial y}{\partial x_2}\right|_{(3,5)} = 0$$

and $y = 0$. This is, of course, the desired minimum.

EXAMPLE 3.2.2

Resolve the previous example using scaling techniques and the method of optimal steepest descent. Choose the starting point $x_{10} = 1$, $x_{20} = 1$, and let $\Delta\tau = 1.0$.

Let us define a new variable

$$u = 3(x_2 - 5).$$

Hence the objective function becomes

$$y = (x_1 - 3)^2 + u^2,$$

and $x_{10} = 1$, $u_0 = 3(1 - 5) = -12$. Hence

$$\mathbf{X}_0 = \begin{bmatrix} 1 \\ -12 \end{bmatrix}.$$

Evaluating the derivatives,

$$\left.\frac{\partial y}{\partial x_1}\right|_{(1,-12)} = 2(x_1 - 3)\Big|_{(1,-12)} = -4,$$

$$\left.\frac{\partial y}{\partial u}\right|_{(1,-12)} = 2u\Big|_{(1,-12)} = -24.$$

From Equation (3.2.15),

$$\mathbf{X}_1^* = \begin{bmatrix} 1 \\ -12 \end{bmatrix} - 1.0\begin{bmatrix} -4 \\ -24 \end{bmatrix} = \begin{bmatrix} 5 \\ 12 \end{bmatrix}.$$

From Equation (3.2.16), we have

$$\mathbf{X}_1 = \theta_1\begin{bmatrix} 1 \\ -12 \end{bmatrix} + (1 - \theta_1)\begin{bmatrix} 5 \\ 12 \end{bmatrix}, \qquad 0 < \theta_1 < 1.$$

The objective function $y(\mathbf{X}_1)$ can be written

$$y = (5 - 4\theta_1 - 3)^2 + (12 - 24\theta_1)^2.$$

From a one-dimensional search with respect to θ_1, we see that $y(\mathbf{X}_1)$ takes on a minimum value of zero at $\theta_1 = \frac{1}{2}$. Hence the problem is solved with a single one-dimensional search.

Notice that, in scaling this problem, we have transformed a family of ellipses to a family of circles. In general, an objective function will not represent a family of circles (or hyperspheroids in n-dimensional space), even after scaling. When the objective function does not represent a family of circles (or hyperspheroids), the function cannot be extremized with a single one-dimensional search. In fact, an infinite number of iterations may be required to extremize the function exactly. We shall encounter such a situation in the next example.

EXAMPLE 3.2.3

Solve the following system of equations by minimizing the sum of the errors squared:

$$x_1 = 0.1136(x_1 + 3x_2)(1 - x_1), \qquad 0 \leq x_1 \leq 1,$$
$$x_2 = -7.50(2x_1 - x_2)(1 - x_2), \qquad 0 \leq x_2 \leq 1.$$

The corresponding objective function can be written

$$y = [x_1 - 0.1136(x_1 + 3x_2)(1 - x_1)]^2 + [x_2 + 7.50(2x_1 - x_2)(1 - x_2)]^2.$$

Notice that y is always nonnegative because of the manner in which it is defined.

Using the method of steepest descent, the partial derivatives can be evaluated numerically in accordance with Equation (3.2.14) as

$$\frac{\partial y}{\partial x_1} \sim \frac{y(x_1 + \epsilon/2, x_2) - y(x_1 - \epsilon/2, x_2)}{\epsilon},$$

$$\frac{\partial y}{\partial x_2} \sim \frac{y(x_1, x_2 + \epsilon/2) - y(x_1, x_2 - \epsilon/2)}{\epsilon},$$

where ϵ is some prespecified small number.

This problem has been programmed for a digital computer, using both the method of steepest descent and optimal steepest descent. Both problems began at $x_{10} = x_{20} = 0.5000$, and both problems set $\epsilon = 10^{-4}$. A value of $\Delta\tau = 0.001$ was used in the steepest-descent computation, and $\Delta\tau = 0.10$ was chosen for the optimal steepest-descent calculation. Both problems were to be terminated when the magnitude of each partial derivative became less than 10^{-5}. Some representative numerical values are shown in Tables 3.6 and 3.7.

Both of these problems required in excess of 500 iterations to attain the desired degree of accuracy. However, the method of optimal steepest descent converged toward the correct answer of $x_1 = \frac{1}{5}$, $x_2 = \frac{2}{3}$ much more rapidly than the ordinary method of steepest descent. In both cases the convergence slowed down consider-

Table 3.6. STEEPEST DESCENT RESULTS

n	x_{1n}	x_{2n}	$\partial y/\partial x_1$	$\partial y/\partial x_2$	$y_{(n+1)}$
0	0.5000	0.5000	36.53	−31.01	5.790
1	0.4635	0.5310	27.86	−21.23	3.815
2	0.4356	0.5522	22.51	−15.54	2.726
3	0.4131	0.5678	18.86	−11.86	2.049
5	0.3780	0.5889	14.15	−7.464	1.2706
10	0.3209	0.6152	8.350	−2.959	0.5027
50	0.2098	0.6368	5.393×10^{-1}	-1.918×10^{-2}	0.2440×10^{-2}
100	0.2015	0.6372	2.200×10^{-2}	-5.348×10^{-3}	0.958×10^{-4}
500	0.2011	0.6393	1.682×10^{-5}	-5.088×10^{-3}	0.822×10^{-4}

Table 3.7. OPTIMAL STEEPEST DESCENT WITH GOLDEN-RATIO SEARCH

n	x_{1n}	x_{2n}	$x^*_{1(n+1)}$	$x^*_{2(n+1)}$	$x_{1(n+1)}$	$x_{2(n+1)}$	$y_{(n+1)}$
0	0.5000	0.5000	−3.1529	3.6007	0.1704	0.7798	7.367×10^{-3}
1	0.1704	0.7798	0.1500	0.7472	0.1640	0.7696	5.059×10^{-3}
2	0.1640	0.7696	0.1762	0.7621	0.1762	0.7621	3.916×10^{-3}
3	0.1762	0.7621	0.1668	0.7448	0.1728	0.7559	3.217×10^{-3}
5	0.1803	0.7518	0.1736	0.7394	0.1778	0.7471	2.339×10^{-3}
10	0.1837	0.7345	0.1898	0.7311	0.1872	0.7326	1.290×10^{-3}
50	0.1964	0.6922	0.1980	0.6914	0.1969	0.6919	1.188×10^{-4}
100	0.1987	0.6784	0.1992	0.6780	0.1989	0.6782	2.119×10^{-5}
500	0.2000	0.6669	0.2000	0.6669	0.2000	0.6669	1.641×10^{-9}

ably as the minimum solution was approached; this is a common (and unfortunate) characteristic of most gradient methods.

The method of optimal steepest descent is more complicated logically than ordinary steepest descent, and the computer time required for a given iteration is therefore greater for optimal steepest descent than ordinary steepest descent. Consequently, the computer time required to solve these two problems was about the same for each problem, although the number of iterations performed in the ordinary steepest-descent version was 50 per cent greater than optimal steepest descent.

Conjugate gradients

Although the method of steepest ascent is a well-known procedure based upon a very plausible idea, we have seen that the method is computationally not very effective. We therefore turn our attention to some more advanced methods in which the path taken is related to, though not identical with, the gradient direction.

The first of these is known as the method of *conjugate gradients*. This method was originally devised by Hestenes and Steifel [1952] for solving the system of linear algebraic equations

$$\mathbf{AX} = \mathbf{B}, \tag{3.2.20}$$

where $\mathbf{A}$ is a real, symmetric, positive-definite matrix, by minimizing the corresponding quadratic form

$$y = \tfrac{1}{2}\mathbf{X}'\mathbf{AX} - \mathbf{B}'\mathbf{X}. \tag{3.2.21}$$

The equivalence of these two problems is established by the relationship

$$\nabla y = \mathbf{0} = \mathbf{AX} - \mathbf{B}. \tag{3.2.22}$$

Hence the vector $\mathbf{X}_o$ that minimizes Equation (3.2.21) will also be the solution vector of equation (3.2.20). We shall proceed to develop the method based upon the minimization of the quadratic form given by Equation (3.2.21). The extension of the algorithm to the optimization of a more general objective function will then be discussed.

The idea behind the conjugate-gradient procedure is similar to that of steepest descent in that a sequence of one-dimensional searches is carried out in directions which are determined by the partial derivatives of the objective function. Unlike the method of steepest descent, however, the search vectors are not equal to the negative gradient vectors; rather, a sequence of search vectors is determined in such a manner that each search vector is a function of both the current gradient vector and the previous search vector. The algorithm is guaranteed to minimize a quadratic function of n independent variables with no more than n iterations—a condition known as *quadratic convergence*.

Proceeding from an arbitrary initial search point $\mathbf{X}_0$, we locate a sequence of points that are successively closer to the minimum as follows:

$$\mathbf{X}_{i+1} = \mathbf{X}_i + \alpha_i \mathbf{P}_i, \tag{3.2.23}$$

where α_i is a positive scalar that defines the distance between $\mathbf{X}_i$ and $\mathbf{X}_{i+1}$ along the search vector $\mathbf{P}_i$. Notice that the minimum along $\mathbf{P}_i$ will occur where $\mathbf{P}_i$ is tangent to the family of contours given by Equation (3.2.21). Stated differently, the gradient of y at $\mathbf{X}_{i+1}$ will be normal to $\mathbf{P}_i$; i.e.,

$$\nabla y'_{i+1}\mathbf{P}_i = \mathbf{P}'_i \nabla y_{i+1} = 0. \tag{3.2.24}$$

If we apply Equation (3.2.23) recursively, we obtain

$$\mathbf{X}_k = \mathbf{X}_i + \sum_{j=i}^{k-1} \alpha_j \mathbf{P}_j, \qquad k = 1, 2, \ldots, n. \tag{3.2.25}$$

In particular,

$$\mathbf{X}_n = \mathbf{X}_i + \sum_{j=i}^{n-1} \alpha_j \mathbf{P}_j. \tag{3.2.26}$$

Subtracting $\mathbf{X}_o$ from each side of Equation (3.2.26) and premultiplying by **A**, we obtain

$$\mathbf{A}(\mathbf{X}_n - \mathbf{X}_o) = \mathbf{A}(\mathbf{X}_i - \mathbf{X}_o) + \sum_{j=i}^{n-1} \alpha_j \mathbf{A}\mathbf{P}_j. \tag{3.2.27}$$

Forming the gradient of Equation (3.2.21), however, we see that

$$\nabla y = \mathbf{A}(\mathbf{X} - \mathbf{X}_o). \tag{3.2.28}$$

Hence Equation (3.2.27) becomes

$$\nabla y_n = \nabla y_i + \sum_{j=i}^{n-1} \alpha_j \mathbf{A}\mathbf{P}_j. \tag{3.2.29}$$

As a special case of the above expression we have

$$\nabla y_{i+1} = \nabla y_i + \alpha_i \mathbf{A}\mathbf{P}_i. \tag{3.2.30}$$

Let us now develop a criterion for defining the search vector $\mathbf{P}_j$. Premultiplying Equation (3.2.29) by $\mathbf{P}_{i-1}$ gives

$$\mathbf{P}'_{i-1}\nabla y_n = \mathbf{P}'_{i-1}\nabla y_i + \sum_{j=i}^{n-1} \alpha_j \mathbf{P}'_{i-1}\mathbf{A}\mathbf{P}_j. \tag{3.2.31}$$

The first term on the right-hand side vanishes because of Equation (3.2.24). If we now choose the $\mathbf{P}_j$ such that

$$\mathbf{P}'_i\mathbf{A}\mathbf{P}_j = 0 \tag{3.2.32}$$

for $i \neq j$, then the summation term in Equation (3.2.31) will also vanish, so that

$$\mathbf{P}'_{i-1}\nabla y_n = 0. \tag{3.2.33}$$

The condition expressed by Equation (3.2.32) is known as **A**-*conjugacy*, and the set of vectors $\mathbf{P}_i$ is said to be **A**-*conjugate*.

It can easily be shown that the **A**-conjugacy condition is sufficient to ensure that the vectors $\mathbf{P}_0, \mathbf{P}_1, \ldots, \mathbf{P}_{n-1}$ are linearly independent. Therefore, the vector set $\mathbf{P}_i$ is a basis in the n-dimensional vector space, so that ∇y_n can be expressed in terms of one or more of the $\mathbf{P}_i$. Thus Equation (3.2.33) will be nonzero for at least one i unless $\nabla y_n = 0$. We see, then, that the condition that the vectors $\mathbf{P}_i$ be **A**-conjugate causes ∇y_n to vanish identically. Since this condition exists only where the quadratic form is minimized, we conclude that *the quadratic form is thus minimized after no more than* n *one-dimensional searches in the directions* $\mathbf{P}_0, \mathbf{P}_1, \ldots, \mathbf{P}_{n-1}$.

We still have not specified exactly how the $\mathbf{P}_i$ are chosen. Let us arbitrarily let $\mathbf{P}_0 = -\nabla y_0$, and then let

$$\mathbf{P}_{i+1} = -\nabla y_{i+1} + \beta_i \mathbf{P}_i, \qquad i = 0, 1, \ldots, n-1, \tag{3.2.34}$$

where the β_i are positive scalars that must be determined. This choice of $\mathbf{P}_i$ can be shown to satisfy the **A**-conjugacy condition expressed by Equation (3.2.32). For further development of this point the reader is referred to Beckman [1960].

From Equation (3.2.32), we can write

$$\mathbf{P}_i' \mathbf{A} \mathbf{P}_{i+1} = 0. \tag{3.2.35}$$

Conbining this result with Equation (3.2.34) gives

$$-\mathbf{P}_i' \mathbf{A} \nabla y_{i+1} + \beta_i \mathbf{P}_i' \mathbf{A} \mathbf{P}_i = 0, \tag{3.2.36}$$

or

$$\beta_i = \frac{\mathbf{P}_i' \mathbf{A} \nabla y_{i+1}}{\mathbf{P}_i' \mathbf{A} \mathbf{P}_i}. \tag{3.2.37}$$

Equation (3.2.37) can be used to compute the β_i if we so desire. In a practical sense, however, notice that Equation (3.2.37) requires an explicit knowledge of the matrix **A**. If a large problem (i.e., a problem of high dimensionality) is being solved on a digital computer, then the $n \times n$ matrix **A** must be stored in the computer. This can occupy a significant portion of core storage and hence limit the size of a problem that can be solved by the conjugate-gradient method. For this reason we will not use Equation (3.2.37) to determine β_i; rather, let us obtain an expression for β_i that does not contain the matrix **A**.

From Equation (3.2.30), we can write

$$\mathbf{P}_i' \nabla y_{i+1} = \mathbf{P}_i' \nabla y_i + \alpha_i \mathbf{P}_i' \mathbf{A} \mathbf{P}_i. \tag{3.2.38}$$

From Equation (3.2.24), however, we see that the left-hand side of Equation (3.2.38) vanishes, so that

$$\mathbf{P}_i' \mathbf{A} \mathbf{P}_i = -\frac{1}{\alpha_i} \mathbf{P}_i' \nabla y_i. \tag{3.2.39}$$

Now let us premultiply Equation (3.2.34) by ∇y_{i+1}. This gives

$$\nabla y'_{i+1}\mathbf{P}_{i+1} = -\nabla y'_{i+1}\nabla y_{i+1} + \beta_i \nabla y'_{i+1}\mathbf{P}_i. \tag{3.2.40}$$

Again referring to Equation (3.2.24), we see that the last term in Equation (3.2.40) vanishes. Hence,

$$\nabla y'_i\mathbf{P}_i = \mathbf{P}'_i\nabla y_i = -\nabla y'_i\nabla y_i. \tag{3.2.41}$$

Substituting this result into Equation (3.2.39) gives

$$\mathbf{P}'_i\mathbf{A}\mathbf{P}_i = \frac{1}{\alpha_i}\nabla y'_i\nabla y_i. \tag{3.2.42}$$

Let us again make use of Equation (3.2.30) to write

$$\nabla y'_{i+1}\nabla y_{i+1} = \nabla y'_{i+1}\nabla y_i + \alpha_i\nabla y'_{i+1}\mathbf{A}\mathbf{P}_i, \tag{3.2.43}$$

which can be rearranged to give

$$\nabla y'_{i+1}\mathbf{A}\mathbf{P}_i = \frac{1}{\alpha_i}(\nabla'_{i+1}\nabla y_{i+1} - \nabla y'_{i+1}\nabla y_i). \tag{3.2.44}$$

Recall that **A** is assumed to be symmetric; hence

$$\nabla y'_{i+1}\mathbf{A}\mathbf{P}_i = \mathbf{P}'_i\mathbf{A}\nabla y_{i+1}, \tag{3.2.45}$$

so that

$$\mathbf{P}'_i\mathbf{A}\nabla y_{i+1} = \frac{1}{\alpha_i}(\nabla y'_{i+1}\nabla y_{i+1} - \nabla y'_{i+1}\nabla y_i). \tag{3.2.46}$$

Utilizing Equation (3.2.30) one more time we can write

$$\mathbf{P}'_{i-1}\nabla y_{i+1} = \mathbf{P}'_{i-1}\nabla y_i + \alpha_i\mathbf{P}'_{i-1}\mathbf{A}\mathbf{P}_i = 0 \tag{3.2.47}$$

because of Equations (3.2.24) and (3.2.32). From Equation (3.2.34), however, we have

$$\mathbf{P}_{i-1} = \frac{1}{\beta_{i-1}}(\mathbf{P}_i + \nabla y_i). \tag{3.2.48}$$

Combining this result with Equation (3.2.47) gives

$$\nabla y'_{i+1}\mathbf{P}_{i-1} = \frac{1}{\beta_{i-1}}(\nabla'_{i+1}\mathbf{P}_i + \nabla y'_{i+1}\nabla y_i) = 0. \tag{3.2.49}$$

Since $\nabla y'_{i+1}\mathbf{P}_i$ vanishes because of Equation (3.2.24), we conclude that

$$\nabla y'_{i+1}\nabla y_i = 0. \tag{3.2.50}$$

Hence Equation (3.2.46) becomes

$$\mathbf{P}'_i\mathbf{A}\nabla y_{i+1} = \frac{1}{\alpha_i}\nabla y'_{i+1}\nabla y_{i+1}. \tag{3.2.51}$$

A simplified expression for β_i can now be obtained by substituting Equations (3.2.42) and (3.2.51) into Equation (3.2.37). This yields

$$\beta_i = \frac{\nabla y'_{i+1}\nabla y_{i+1}}{\nabla y'_i\nabla y_i}, \tag{3.2.52}$$

which does not require explicit knowledge of the matrix **A**.

Several other mathematical relationships can be shown to exist among the various vectors that appear in the algorithm. For a detailed account of these relationships, as well as a more rigorous exposition of the conjugate-gradient method, the reader is referred to a paper by Beckman [1960].

Let us now summarize the algorithm, including a generalization to the maximization problem. Choose an arbitrary point $\mathbf{X}_0$ and evaluate the gradient vector ∇y_0. Let $\mathbf{P}_0 = \pm\nabla y_0$ (the plus sign corresponding to a maximization problem, the minus to a minimization). Obtain

$$\mathbf{X}_{i+1} = \mathbf{X}_i + \alpha_i\mathbf{P}_i \tag{3.2.23}$$

as the point on the $\mathbf{P}_i$ vector where the objective function is extremized. This point is located by conducting a one-dimensional search along $\mathbf{P}_i$. Determine ∇y_{i+1}, the gradient vector, at $\mathbf{X}_{i+1}$. Compute β_i in accordance with Equation (3.2.52), and determine a new search vector

$$\mathbf{P}_{i+1} = \pm\nabla y_{i+1} + \beta_i\mathbf{P}_i. \tag{3.2.53}$$

Again the plus sign is chosen for a maximization problem and the minus sign corresponds to a minimization.

Although the method is supposed to find the optimum of a quadratic form with no more than n iterations, it is in fact rather sensitive to roundoff error. Fletcher and Reeves [1964] suggest that the computation be restarted with $\mathbf{P}_i = \pm\nabla y_i$ after every $n + 1$ iterations as an effective way to minimize the problem of cumulative roundoff.

Our discussion thus far has been concerned exclusively with the optimization of quadratic forms. We can see that the method is applicable to a broader class of functions, however, by expanding the objective function in a

Taylor series:

$$y(\mathbf{X}) = y(\mathbf{X}_o) + \nabla y|_{X_o}(\mathbf{X} - \mathbf{X}_o) + \tfrac{1}{2}(\mathbf{X} - \mathbf{X}_o)'\mathbf{H}(\mathbf{X} - \mathbf{X}_o) + \ldots, \tag{3.2.54}$$

where **H** is the Hessian matrix consisting of the second partial derivatives of y evaluated at $\mathbf{X}_o$; i.e.,

$$\mathbf{H} = \begin{bmatrix} \frac{\partial^2 z}{\partial x_1^2} & \frac{\partial^2 z}{\partial x_1 \partial x_2} & \cdots & \frac{\partial^2 z}{\partial x_1 \partial x_n} \\ \frac{\partial^2 z}{\partial x_2 \partial x_1} & \frac{\partial^2 z}{\partial x_2^2} & \cdots & \frac{\partial^2 z}{\partial x_2 \partial x_n} \\ \vdots & & & \\ \frac{\partial^2 z}{\partial x_n \partial x_1} & \frac{\partial^2 z}{\partial x_n \partial x_2} & \cdots & \frac{\partial^2 z}{\partial x_n^2} \end{bmatrix}_{\mathbf{X}_o}. \tag{3.2.55}$$

Notice that **H** is a real, symmetric matrix, providing the objective function is not linear.

If $\mathbf{X}_o$ represents an extremum of $y(\mathbf{X})$, then

$$\nabla y|_{X_o} = \mathbf{0}, \tag{3.2.56}$$

and Equation (3.2.54) becomes

$$y(\mathbf{X}) - y(\mathbf{X}_o) = \tfrac{1}{2}(\mathbf{X} - \mathbf{X}_o)'\mathbf{H}(\mathbf{X} - \mathbf{X}_o) + \ldots. \tag{3.2.57}$$

Hence the quantity $y(\mathbf{X}) - y(\mathbf{X}_o)$ can be expressed by Equation (3.2.57), providing **X** is sufficiently close to $\mathbf{X}_o$ so that the higher-order terms in the Taylor-series expansion vanish. In the neighborhood of an optimum, then, we would expect the conjugate-gradient method to perform quite efficiently when applied to any continuous and differentiable nonlinear function.

Implementation of the method does not require explicit knowledge of the matrix **H**. The first partial derivatives of the objective function must be known, however, in order to determine the gradient vector ∇y. These partial derivatives can be determined by finite differences. Unfortunately, the method is susceptible to the accumulation of roundoff errors—a shortcoming that becomes particularly troublesome when the partial derivatives are approximated numerically. Therefore, the use of a restart after every $n + 1$ iterations is strongly advised.

An appreciation for quadratically convergent methods can be obtained from Figure 3.14 where a quadratic function is minimized by the method of steepest descent and by a conjugate-gradient search. The simple gradient (steepest-descent) procedure, which is shown by the solid lines, tends to zigzag back and forth near the optimum, requiring many search iterations. This

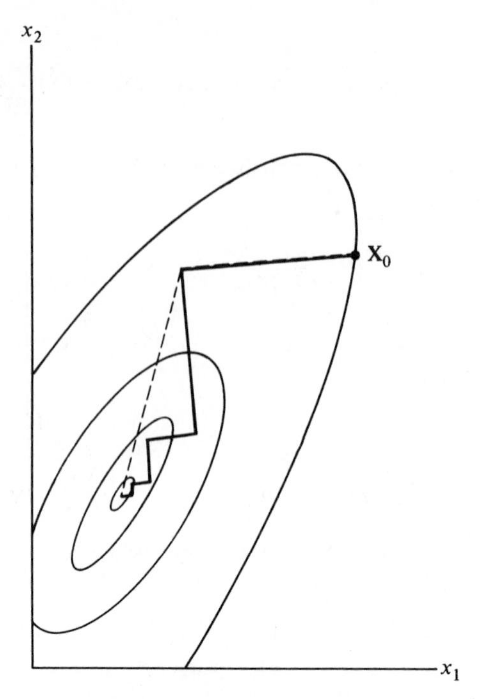

Figure 3.14. Comparison of Behavior of Steepest Descent and Conjugate Gradient Search in Minimization of Quadratic Function

type of oscillation near an optimum is very characteristic of steepest-descent techniques. On the other hand, the conjugate-gradient procedure, shown by dashed lines, finds the minimum of the function in only two iterations, thus avoiding the problem of oscillations near an optimum. Quadratically convergent gradient methods are sometimes referred to as *second-order methods* (cf. Crockett and Chernoff [1955]).

EXAMPLE 3.2.4

Reslove Example 3.2.1 using the method of conjugate gradients, with $x_{10} = 1$ and $x_{20} = 1$ as an initial point.

In vector notation,

$$\mathbf{X}_0 = \begin{bmatrix} 1 \\ 1 \end{bmatrix}$$

and, from Example 3.2.1,

$$\nabla y|_{\mathbf{x}_0} = -\begin{bmatrix} 4 \\ 72 \end{bmatrix},$$

so that

$$\mathbf{P}_0 = -\nabla y|_{\mathbf{x}_0} = \begin{bmatrix} 4 \\ 72 \end{bmatrix}.$$

From Equation (3.2.23), we have

$$\mathbf{X}_1 = \begin{bmatrix} 1 \\ 1 \end{bmatrix} + \alpha_0 \begin{bmatrix} 4 \\ 72 \end{bmatrix}, \qquad \alpha_0 > 0.$$

The objective function can be expressed as a function of α_0 as follows:

$$y(\alpha_0) = (4\alpha_0 - 2)^2 + 9(72\alpha_0 - 4)^2.$$

Minimizing $y(\alpha_0)$, we obtain $y = 3.1594$ at $\alpha_0 = 0.0555$. Hence

$$\mathbf{X}_1 = \begin{bmatrix} 1.223 \\ 5.011 \end{bmatrix}.$$

The gradient can now be determined as

$$\nabla y|_{\mathbf{x}_1} = \begin{bmatrix} -3.554 \\ 0.197 \end{bmatrix}$$

and β_0 can be computed as

$$\beta_0 = \frac{(3.554)^2 + (0.197)^2}{(4)^2 + (72)^2} = 0.00244.$$

Making use of Equations (3.2.34) and (3.2.23), we obtain

$$\mathbf{P}_1 = \begin{bmatrix} 3.554 \\ -0.197 \end{bmatrix} + 0.00244 \begin{bmatrix} 4 \\ 72 \end{bmatrix} = \begin{bmatrix} 3.564 \\ -0.022 \end{bmatrix}$$

and

$$\mathbf{X}_2 = \begin{bmatrix} 1.223 \\ 5.011 \end{bmatrix} + \alpha_1 \begin{bmatrix} 3.564 \\ -0.022 \end{bmatrix}.$$

Solving for α_1 as before [i.e., expressing $y(\mathbf{X}_2)$ as a function of α_1 and minimizing with respect to α_1] yields $y = 5.91 \times 10^{-10}$ at $\alpha_1 = 0.4986$. Hence

$$\mathbf{X}_2 = \begin{bmatrix} 3.0000 \\ 5.0000 \end{bmatrix},$$

which is, for all practical purposes, the desired result.

Notice that this two-dimensional function has been minimized with the determination of only two points. This is, of course, to be expected, since the objective function is a quadratic and the conjugate-gradient algorithm is quadratically convergent.

EXAMPLE 3.2.5

Resolve Example 3.2.3 using the method of conjugate gradients and a golden-ratio search. Let $\epsilon = 10^{-5}$ be the largest permissible value of each of the partial derivatives for which the problem will be considered to have converged to an optimum. The partial derivatives are to be evaluated numerically.

Some select results of this problem, obtained with a digital computer, are given in Table 3.8.

Table 3.8. RESULTS OF CONJUGATE GRADIENT OPTIMIZATION

n	x_{1n}	x_{2n}	y_n
0	0.5000	0.5000	5.7899
1	0.1800	0.7716	7.1772×10^{-3}
2	0.1703	0.7618	3.8202×10^{-3}
3	0.1922	0.7264	1.2420×10^{-3}
5	0.1977	0.6984	2.9656×10^{-4}
10	0.1998	0.6685	4.0192×10^{-7}
16	0.19998	0.66683	5.0831×10^{-12}

The computation was terminated after 16 iterations, since each partial derivative became less than or equal to 10^{-5} in magnitude. Notice that this problem required less than one tenth as many iterations as optimal steepest descent for a comparable solution (cf. Example 3.2.3). The computational effort per iteration is slightly greater than with optimal steepest descent.

Variable-metric algorithm

The variable-metric algorithm is another sophisticated gradient technique, originally devised by Davidon [1959] to minimize a quadratic function with no more than n steps. The method was later modified and improved upon by Fletcher and Powell [1964]. This later version of the algorithm is an extremely powerful gradient method for extremizing any unconstrained, continuous, and differentiable objective function. We shall present the Fletcher and Powell version of the algorithm and discuss its advantages and disadvantages.

Like the method of conjugate gradients, the variable-metric algorithm is designed to extremize the function

$$y = \tfrac{1}{2}(\mathbf{X} - \mathbf{X}_o)'\mathbf{A}(\mathbf{X} - \mathbf{X}_o) \tag{3.2.58}$$

by conducting a sequence of one-dimensional searches. These searches begin at some arbitrary point $\mathbf{X}_0$ and proceed to locate a succession of improved points in accordance with

$$\mathbf{X}_{i+1} = \mathbf{X}_i + \alpha_i \mathbf{P}_i, \qquad i = 0, 1, 2, \ldots, \tag{3.2.59}$$

where α_i is some positive constant.

Now let us recall a few significant relationships that relate to the minimization of a quadratic form. First, each $\mathbf{X}_{i+1}$ represents the position of an extremum along $\mathbf{P}_i$. Hence the gradient of the objective function will be orthogonal to $\mathbf{P}_i$ at $\mathbf{X}_{i+1}$; i.e.,

$$\nabla y'_{i+1}\mathbf{P}_i = \mathbf{P}'_i \nabla y_{i+1} = 0, \tag{3.2.60}$$

where

$$\nabla y_i = \mathbf{A}(\mathbf{X}_i - \mathbf{X}_o). \tag{3.2.61}$$

[The validity of Equation (3.2.60) is, of course, not restricted to a quadratic form.] Also, we have established that

$$\nabla y_{i+1} = \nabla y_i + \alpha_i \mathbf{A}\mathbf{P}_i. \tag{3.2.62}$$

Finally, we have shown that an extremization algorithm will be quadratically convergent, providing the $\mathbf{P}_i$ are $\mathbf{A}$-conjugate; i.e.,

$$\mathbf{P}'_i \mathbf{A} \mathbf{P}_j = 0, \qquad i \neq j. \tag{3.2.63}$$

Any algorithm that satisfies Equation (3.2.63) will, apart from numerical roundoff errors, minimize a quadratic form with no more than n one-dimensional searches in the $\mathbf{P}_i$ directions, $i = 0, 1, \ldots, n - 1$.

Thus far the description of the variable metric algorithm parallels that of the method of conjugate gradients. The algorithms differ in the manner in which the search vectors $\mathbf{P}_1, \mathbf{P}_2, \ldots, \mathbf{P}_{n-1}$ are chosen.

In the variable metric algorithm the search vectors are chosen as

$$\mathbf{P}_i = -\mathbf{H}_i \nabla y_i, \qquad i = 0, 1, \ldots, n - 1, \tag{3.2.64}$$

where $\mathbf{H}_i$ is a symmetric, positive-definite* $n \times n$ matrix that must be specified for $i = 0$ and determined recursively for $i = 1, 2, \ldots, n - 1$. ($\mathbf{H}_i$ must be negative-definite for a maximization problem.) Fletcher and Powell show that this particular choice of the search vectors $\mathbf{P}_i$ will satisfy Equation (3.2.63), and therefore guarantee quadratic convergence, providing $\mathbf{H}_i$ is positive

---

* See Appendix B for definition of positive-definite and negative-definite matrices.

(negative) definite. Moreover, Fletcher and Powell show that the particular manner in which the $\mathbf{H}_i$ are chosen (which will be discussed below) assures that $\mathbf{H}_i$ will be symmetric and positive (negative) definite, providing $\mathbf{H}_{i-1}$ is symmetric and positive (negative) definite. Hence the algorithm will be quadratically convergent as long as $\mathbf{H}_0$ is chosen to be symmetric and positive (negative) definite. For lack of a better choice, $\mathbf{H}_0$ can be set equal to the (negative) identity matrix; i.e.,

$$\mathbf{H}_0 = \mathbf{I} \tag{3.2.65}$$

for a minimization problem and

$$\mathbf{H}_0 = -\mathbf{I} \tag{3.2.66}$$

for a maximization.

Once $\mathbf{H}_0$ is specified, the subsequent $\mathbf{H}_i$ are determined as follows:

$$\mathbf{H}_{i+1} = \mathbf{H}_i + \mathbf{B}_i + \mathbf{C}_i, \tag{3.2.67}$$

where $\mathbf{B}_i$ and $\mathbf{C}_i$ are also symmetric $n \times n$ matrices that are determined at each step of the computation. These matrices are defined in such a manner that

$$\mathbf{H}_n = \mathbf{A}^{-1}. \tag{3.2.68}$$

To relate Equations (3.2.67) and (3.2.68), let us apply Equation (3.2.67) recursively to obtain

$$\mathbf{H}_{i+1} = \mathbf{H}_0 + \sum_{j=0}^{i} \mathbf{B}_j + \sum_{j=0}^{i} \mathbf{C}_j. \tag{3.2.69}$$

In particular, let $n = i + 1$, so that

$$\mathbf{H}_n = \mathbf{H}_0 + \sum_{j=0}^{n-1} \mathbf{B}_j + \sum_{j=0}^{n-1} \mathbf{C}_j. \tag{3.2.70}$$

Now let us choose $\mathbf{B}_i$ and $\mathbf{C}_i$ such that

$$\sum_{i=0}^{n-1} \mathbf{B}_i = \mathbf{A}^{-1} \tag{3.2.71}$$

and

$$\sum_{i=0}^{n-1} \mathbf{C}_i = -\mathbf{H}_0, \tag{3.2.72}$$

so that Equation (3.2.70) thus reduces to Equation (3.2.68).

The individual $\mathbf{B}_i$ can be determined through the use of Equation (3.2.63). Let $\mathbf{T}$ be an $n \times n$ matrix whose columns consist of the vectors $\mathbf{P}_i$, $i = 0, 1, \ldots, n-1$. The matrix product $\mathbf{T}'\mathbf{A}\mathbf{T}$ will result in a *diagonal* $n \times n$ matrix, which we shall refer to as the matrix $\mathbf{D}$. This must be so because of Equation (3.2.63); i.e., the off-diagonal terms of $\mathbf{D}$ are given by

$$d_{ij} = \mathbf{P}_i'\mathbf{A}\mathbf{P}_j = \mathbf{P}_j'\mathbf{A}\mathbf{P}_i = 0. \tag{3.2.73}$$

Writing

$$\mathbf{T}'\mathbf{A}\mathbf{T} = \mathbf{D}, \tag{3.2.74}$$

we can invert the above expression to obtain

$$\mathbf{A} = (\mathbf{T}')^{-1}\mathbf{D}\mathbf{T}^{-1}, \tag{3.2.75}$$

which, from the properties of the inverse of a matrix product, can be written

$$\mathbf{A} = (\mathbf{T}\mathbf{D}^{-1}\mathbf{T}')^{-1}. \tag{3.2.76}$$

Inverting this expression,

$$\begin{aligned} \mathbf{A}^{-1} &= \mathbf{T}\mathbf{D}^{-1}\mathbf{T}' \\ \sum_{j=0}^{n-1} \mathbf{B}_j &= \mathbf{T}\mathbf{D}^{-1}\mathbf{T}'. \end{aligned} \tag{3.2.77}$$

Let us now make use of a result from linear algebra—that $\mathbf{D}^{-1}$ is also a diagonal matrix whose nonzero terms are equal to $1/d_{ii}$. From this result one can show that Equation (3.2.77) can be written

$$\sum_{j=0}^{n-1} \mathbf{B}_j = \sum_{j=0}^{n-1} \frac{1}{d_{jj}} \mathbf{P}_j\mathbf{P}_j' \tag{3.2.78}$$

where $\mathbf{P}_j\mathbf{P}_j'$ represents a matrix $\mathbf{M}$ whose elements m_{ik} are given by

$$m_{ik} = p_{i,j}p_{k,j}, \tag{3.2.79}$$

and $p_{i,j}$ denotes the ith element of the vector $\mathbf{P}_j$.

Finally, recall that

$$d_{jj} = \mathbf{P}_j'\mathbf{A}\mathbf{P}_j. \tag{3.2.80}$$

Thus,

$$\sum_{j=0}^{n-1} \mathbf{B}_j = \frac{\mathbf{P}_j\mathbf{P}_j'}{\mathbf{P}_j'\mathbf{A}\mathbf{P}_j}. \tag{3.2.81}$$

Combining this result with Equation (3.2.62), we have

$$\sum_{j=0}^{n-1} \mathbf{B}_j = \sum_{j=0}^{n-1} \frac{\alpha_j \mathbf{P}_j \mathbf{P}'_j}{\mathbf{P}'_j(\nabla y_{j+1} - \nabla y_j)}. \tag{3.2.82}$$

If we now equate corresponding terms in the summations for each j, we have

$$\mathbf{B}_i = \frac{\alpha_i \mathbf{P}_i \mathbf{P}'_i}{\mathbf{P}'_i(\nabla y_{i+1} - \nabla y_i)}, \qquad i = 0, 1, \ldots, n - 1, \tag{3.2.83}$$

which is our final result.

Let us now turn our attention to the choice of $\mathbf{C}_i$. Consider the expression $\mathbf{H}_{i+1}\mathbf{A}\mathbf{P}_i$. From Equation (3.2.67) we can write

$$\mathbf{H}_{i+1}\mathbf{A}\mathbf{P}_i = \mathbf{H}_i\mathbf{A}\mathbf{P}_i + \mathbf{B}_i\mathbf{A}\mathbf{P}_i + \mathbf{C}_i\mathbf{A}\mathbf{P}_i. \tag{3.2.84}$$

Combining this expression with Equations (3.2.62) and (3.2.83) yields

$$\begin{aligned} \mathbf{H}_{i+1}\mathbf{A}\mathbf{P}_i &= \mathbf{H}_i\mathbf{A}\mathbf{P}_i + \frac{\mathbf{P}_i\mathbf{P}'_i\mathbf{A}\mathbf{P}_i}{\mathbf{P}'_i\mathbf{A}\mathbf{P}_i} + \mathbf{C}_i\mathbf{A}\mathbf{P}_i \\ &= \mathbf{H}_i\mathbf{A}\mathbf{P}_i + \mathbf{P}_i + \mathbf{C}_i\mathbf{A}\mathbf{P}_i. \end{aligned} \tag{3.2.85}$$

Observe that the expression for $\mathbf{B}_i$ that we have just used is based upon the condition expressed by Equation (3.2.71), which in turn implies that Equation (3.2.72) must be satisfied. Hence any expression for $\mathbf{C}_i$ that satisfies Equation (3.2.85) will also satisfy Equation (3.2.72).

Suppose that we base our choice of $\mathbf{C}_i$ on the condition that

$$\mathbf{H}_{i+1}\mathbf{A}\mathbf{P}_i = \mathbf{P}_i, \tag{3.2.86}$$

which implies that $\mathbf{P}_i$ is an eigenvector of $\mathbf{H}_{i+1}\mathbf{A}$ with a corresponding eigenvalue of unity. (Fletcher and Powell make use of this fact to show that the algorithm is quadratically convergent.)

If Equation (3.2.86) holds, then it follows immediately that

$$\mathbf{C}_i\mathbf{A}\mathbf{P}_i = -\mathbf{H}_i\mathbf{A}\mathbf{P}_i. \tag{3.2.87}$$

This expression can be written

$$\mathbf{C}_i\mathbf{A}\mathbf{P}_i = -\mathbf{H}_i\mathbf{A}\mathbf{P}_i \frac{\mathbf{P}'_i\mathbf{A}'\mathbf{H}_i\mathbf{A}\mathbf{P}_i}{\mathbf{P}'_i\mathbf{A}'\mathbf{H}_i\mathbf{A}\mathbf{P}_i}, \tag{3.2.88}$$

since we have simply multiplied the numerator and denominator of the right-hand side of Equation (3.2.88) by the same scalar quantity.

One way Equation (3.2.88) can be satisfied is to let

$$\mathbf{C}_i = -\frac{\mathbf{H}_i\mathbf{A}\mathbf{P}_i\mathbf{P}'_i\mathbf{A}'\mathbf{H}'_i}{\mathbf{P}'_i\mathbf{A}'\mathbf{H}_i\mathbf{A}\mathbf{P}_i} \tag{3.2.89}$$

where we have made use of the symmetry of $\mathbf{H}_i$. Combining Equations (3.2.62) and (3.2.89) yields the desired result:

$$\mathbf{C}_i = -\frac{\mathbf{H}_i(\nabla y_{i+1} - \nabla y_i)(\nabla y_{i+1} - \nabla y_i)'\mathbf{H}'_i}{(\nabla y_{i+1} - \nabla y_i)'\mathbf{H}_i(\nabla y_{i+1} - \nabla y_i)}, \tag{3.2.90}$$

where the matrix multiplication in the numerator of the above expression is as described after Equation (3.2.78).

To summarize the algorithm, choose $\mathbf{X}_0$ arbitrarily and evaluate ∇y_0 at that point. Choose some positive-(or negative-)definite $\mathbf{H}_0$ and evaluate $\mathbf{P}_0$ in accordance with Equation (3.2.64). The next search point is then established with Equation (3.2.59), a new gradient vector is determined, and the next $\mathbf{H}_i$ and $\mathbf{P}_i$ are obtained from Equations (3.2.83), (3.2.90), (3.2.67), and (3.2.64). The procedure thus continues recursively until some convergence criterion has been satisfied. [Fletcher and Powell suggest checking the magnitude of the individual components of the latest $\mathbf{P}_i$, and also the magnitude of $(\mathbf{P}'_i\mathbf{P}_i)^{1/2}$ as stopping criterion.] Aside from roundoff errors, the algorithm will extremize a quadratic objective function with no more than n searches in the $\mathbf{P}_i$ directions, $i = 0, 1, \ldots, n - 1$. Some care is required in the implementation of the algorithm, however, since it is highly susceptible to the accumulation of roundoff errors.

Notice from Equation (3.2.61) that the vector $-\mathbf{A}^{-1}\nabla y_i$ is equal to the difference between a given point $\mathbf{X}_i$ and the desired solution $\mathbf{X}_o$. This difference, which can be thought of as an error vector, is called a *metric*. The present algorithm does not evaluate the metric $-\mathbf{A}^{-1}\nabla y_i$ at each step, but instead approximates the metric with the search vector $-\mathbf{H}_i\nabla y_i$. Thus the algorithm is named variable-metric search.

Other gradient methods

Finally, the variable-metric algorithm is but one of a class of similar minimization algorithms that share the common feature

$$\mathbf{X}_{i+1} = \mathbf{X}_i - \alpha_i\mathbf{H}_i\nabla y_i. \tag{3.2.91}$$

This class of algorithms, known as *quasi-Newton methods*, is discussed by Broyden [1967], Zeleznik [1968], and Pearson [1969].

A different gradient technique, which makes use of the geometric properties of quadratic objective functions, is that of Shah, Buehler, and Kempthorne [1964]. This method, known as *partan* (parallel tangents) alternates steepest-descent moves with certain acceleration moves. It can be shown that a class of quadratic objective functions (concentric hyperellipsoids) can be minimized with this method in no more than $2n$ search moves. Although this represents a distinct improvement over the method of steepest descent, partan is generally less efficient than the quasi-Newton methods.

EXAMPLE 3.2.6

Resolve Example 3.2.1 using the variable-metric algorithm, with $x_{10} = 1$ and $x_{20} = 1$ as an initial point.

As before, we have

$$\mathbf{X}_0 = \begin{bmatrix} 1 \\ 1 \end{bmatrix} \quad \text{and} \quad \nabla y_0 = -\begin{bmatrix} 4 \\ 72 \end{bmatrix}.$$

Selecting $\mathbf{H}_0 = \mathbf{I}$, we have, from Equation (3.2.64),

$$\mathbf{P}_0 = -\mathbf{I}\nabla y_0 = -\nabla y_0 = \begin{bmatrix} 4 \\ 72 \end{bmatrix}.$$

Thus we can determine $\alpha_0 = 0.0557$,

$$\mathbf{X}_1 = \begin{bmatrix} 1.223 \\ 5.011 \end{bmatrix}, \qquad \nabla y_1 = \begin{bmatrix} -3.554 \\ 0.197 \end{bmatrix},$$

and $y_1 = 3.1594$, as in Example 3.2.4.

Using Equations (3.2.83) and (3.2.90), the matrices $\mathbf{B}_0$ and $\mathbf{C}_0$ are determined as

$$\mathbf{B}_0 = \frac{0.0557}{4(0.446) + 72(72.197)} \begin{bmatrix} 4^2 & 4(72) \\ 72(4) & 72^2 \end{bmatrix} = \begin{bmatrix} 0.00017 & 0.00308 \\ 0.00308 & 0.05553 \end{bmatrix}$$

$$\mathbf{C}_0 = -\frac{1}{(0.446)^2 + (72.197)^2} \begin{bmatrix} (0.446)^2 & (0.446)(72.197) \\ (72.197)(0.446) & (72.197)^2 \end{bmatrix}$$

$$= -\begin{bmatrix} 0.00004 & 0.00618 \\ 0.00618 & 0.99996 \end{bmatrix}.$$

From Equation (3.2.67), $\mathbf{H}_1$ can be written

$$\mathbf{H}_1 = \begin{bmatrix} 1 & 0 \\ 0 & 1 \end{bmatrix} + \begin{bmatrix} 0.00017 & 0.00308 \\ 0.00308 & 0.05553 \end{bmatrix} - \begin{bmatrix} 0.00004 & 0.00618 \\ 0.00618 & 0.99996 \end{bmatrix}$$

$$= \begin{bmatrix} 1.00013 & -0.00310 \\ -0.00310 & 0.05557 \end{bmatrix}.$$

Again utilizing Equation (3.2.64),

$$\mathbf{P}_1 = -\mathbf{H}_1\nabla y_1 = \begin{bmatrix} -1.00013 & 0.00310 \\ 0.00310 & -0.05557 \end{bmatrix}\begin{bmatrix} -3.554 \\ 0.197 \end{bmatrix} = \begin{bmatrix} 3.555 \\ -0.022 \end{bmatrix}.$$

Writing

$$\mathbf{X}_2 = \begin{bmatrix} 1.223 \\ 5.011 \end{bmatrix} + \alpha_1\begin{bmatrix} 3.555 \\ -0.022 \end{bmatrix},$$

we can substitute this value of $\mathbf{X}$ into the objective function $y(\mathbf{X})$ and minimize with respect to α_1. This yields a value of $y = 2.020 \times 10^{-5}$ at $\alpha_1 = 0.4999$, which corresponds to

$$\mathbf{X}_2 = \begin{bmatrix} 3.0044 \\ 5.0000 \end{bmatrix}.$$

At this point the computation is finished, the correct solution having been obtained (except for roundoff errors that were incurred). It is instructive to compute $\mathbf{H}_2$, however, since $\mathbf{H}_2$ is expected to equal $\mathbf{A}^{-1}$. Hence, letting $\nabla y_2 = \mathbf{0}$,

$$\mathbf{B}_1 = \frac{0.4999}{(3.555)(3.554) + (0.022)(-0.197)}\begin{bmatrix} (3.555)^2 & (3.555)(-0.022) \\ (-0.022)(3.555) & (-0.022)^2 \end{bmatrix}$$

$$= \begin{bmatrix} 0.4985 & -0.00309 \\ -0.00309 & 0.00002 \end{bmatrix},$$

$$\mathbf{C}_1 = -\frac{1}{12.64}\begin{bmatrix} (-3.555)^2 & (-3.555)(0.022) \\ (0.022)(-3.555) & (0.022)^2 \end{bmatrix} = \begin{bmatrix} -0.99994 & 0.00619 \\ 0.00619 & -0.00004 \end{bmatrix},$$

$$\mathbf{H}_2 = \begin{bmatrix} (1.0013 + 0.4985 - 0.9999) & (-0.0031 - 0.0031 + 0.0062) \\ (-0.0031 - 0.0031 + 0.0062) & (0.05557 + 0.00002 - 0.00004) \end{bmatrix}$$

$$= \begin{bmatrix} 0.4999 & 0 \\ 0 & 0.05555 \end{bmatrix}.$$

It can easily be established that $\mathbf{H}_2 = \mathbf{A}^{-1}$, since

$$\mathbf{A} = \begin{bmatrix} 2 & 0 \\ 0 & 18 \end{bmatrix},$$

and

$$\mathbf{AH}_2 = \mathbf{I} = \mathbf{AA}^{-1},$$

except for roundoff error. Notice also that

$$\mathbf{B}_0 + \mathbf{B}_1 = \mathbf{H}_2 = \mathbf{A}^{-1}$$

and that

$$\mathbf{C}_0 + \mathbf{C}_1 = -\mathbf{I},$$

as expected.

EXAMPLE 3.2.7

Resolve Example 3.2.3 using the variable-metric algorithm. Let $\epsilon = 10^{-5}$ be the largest permissible value of each of the partial derivatives for which the problem will be considered to have converged to an optimum. The partial derivatives will be evaluated numerically by central differences.

Some select results of this problem are shown in Table 3.9.

Table 3.9. RESULTS OF VARIABLE METRIC MINIMIZATION

n	x_{1n}	x_{2n}	y_n
0	0.5000	0.5000	5.7899
1	0.1800	0.7716	7.1772×10^{-3}
2	0.1706	0.7608	3.7259×10^{-3}
4	0.1805	0.7333	1.8108×10^{-3}
6	0.1999	0.6857	1.6970×10^{-4}
8	0.2001	0.6698	4.3031×10^{-6}
10	0.1998	0.6687	5.0601×10^{-7}
12	0.2000	0.6668	3.8858×10^{-12}

The computation was terminated after the twentieth iteration, as the partial derivatives of the objective function were sufficiently small. Although this solution required fewer iterations than the conjugate-gradient method, the computation time was about the same because of the greater amount of computation involved for each iteration.

3.3 Multi-dimensional Optimization by Direct-search Techniques

We have previously examined the use of simple direct-search schemes for one-dimensional optimization. Here we shall examine the use of somewhat more sophisticated direct-search algorithms in multidimensional optimization. All direct-search techniques are hillclimbing methods, which determine the path toward an optimum by evaluating the objective function at several points rather than by calculating derivatives. These techniques are characterized by their simplicity, effectiveness, and applicability to a wide variety of objective functions. Because these methods do not employ derivatives, they can be applied to objective functions that are not differentiable. It is required, however, that the function be continuous.

Pattern search

One direct-search technique which has been found to be particularly effective is that of Hooke and Jeeves [1961], known as *pattern search*. This technique alternates sequences of local exploratory moves with extrapolations (or *pattern moves*, as Hooke and Jeeves refer to them). The basis for the method is the intuitive presumption that a strategy which was successful in the past will be successful in the future (allowing, of course, for the possibility that this presumption may prove to be false).

The algorithm proceeds as follows: An incremental value ϵ_i is assigned to each independent variable x_i. These incremental values control the size of the exploratory moves and they need not be equal for all i. Generally, ϵ_i will represent some small fraction of the distance between the upper and lower bounds of x_i. Once the ϵ_i are selected, however, they remain fixed in magnitude until it becomes necessary at some point in the computation to reduce all the ϵ_i simultaneously.

The computation begins by selecting an initial *exploratory point* $\mathbf{X}_1^*$ within the region of interest, and the objective function is evaluated at $\mathbf{X}_1^*$. Knowing $y(\mathbf{X}_1^*)$, we increase x_1 to $x_1 + \epsilon_1$ and evaluate y at this new point. If y shows improvement as a result of this move [i.e., if

$$y(x_1 + \epsilon_1, x_2, \ldots, x_n) > y(x_1, x_2, \ldots, x_n)$$

for a maximization problem, or if

$$y(x_1 + \epsilon_1, x_2, \ldots, x_n) < y(x_1, x_2, \ldots, x_n)$$

for a minimization], then we let this new point be the next exploratory point, which we shall call $\mathbf{X}_{11}^*$. If y does not improve at $x_1 + \epsilon_1$, we then try $x_1 - \epsilon_1$, and we let this be the new exploratory point $\mathbf{X}_{11}^*$ if y now shows improvement. In the event that y does not show improvement at either $x_1 + \epsilon_1$ or $x_1 - \epsilon_1$, we set $\mathbf{X}_{11}^*$ equal to the initial exploratory point $\mathbf{X}_1^*$.

From $\mathbf{X}_{11}^*$ we proceed to change x_2 by $\pm\epsilon_2$ and thus establish the position of the next exploratory point $\mathbf{X}_{12}^*$ as above. The procedure is repeated, testing changes in one variable at a time, until all the independent variables have been perturbed by their respective incremental distances. At the end of this sequence of exploratory moves, we shall have established some point $\mathbf{X}_{1n}^*$ that will lie in the vicinity of $\mathbf{X}_1^*$. We shall refer to this terminal position as a *base point*, which we shall simply call $\mathbf{X}_1$. Notice that the base point will not coincide with the initial exploratory point unless there has been no improvement in y after all the independent variables have been perturbed by their corresponding incremental values. An exploratory sequence for the maximization of a two-dimensional function is shown in Figure 3.15.

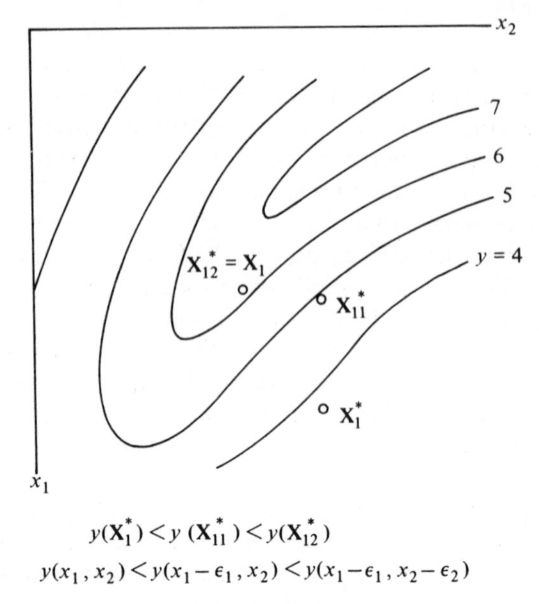

Figure 3.15. Exploratory Search Sequence in Maximization of a Two-Dimensional Function

A pattern move is now made, providing that some improvement has been obtained in the exploratory sequence (i.e., providing that $\mathbf{X}_1 \neq \mathbf{X}_1^*$). This is accomplished by extrapolating from the initial exploratory point $\mathbf{X}_1^*$ through the base point $\mathbf{X}_1$ to establish a new exploratory point $\mathbf{X}_2^*$, where

$$\mathbf{X}_2^* = \mathbf{X}_1^* + 2(\mathbf{X}_1 - \mathbf{X}_1^*). \tag{3.3.1}$$

Another sequence of exploratory moves is then carried out. The best value of the objective function found during this exploratory process is compared with the previous base point. If it is an improvement, it is accepted as the new base point.

The next pattern move is made by extrapolation through the previous two *base points;* i.e.,

$$\mathbf{X}_3^* = \mathbf{X}_1 + 2(\mathbf{X}_2 - \mathbf{X}_1). \tag{3.3.2}$$

Notice that such a pattern move could not have been made after the first exploratory sequence because only one base point had then been established.

The successive alternation of an exploratory sequence with a pattern move is continued as long as the strategy continues to offer improvement in the value of the objective function. Each exploratory sequence begins at an exploratory point $\mathbf{X}_i^*$ and ends at a base point $\mathbf{X}_i$. Each pattern move (except the first) is made in accordance with the expression

$$\mathbf{X}_{i+1}^* = \mathbf{X}_{i-1} + 2(\mathbf{X}_i - \mathbf{X}_{i-1}). \tag{3.3.3}$$

In the event that a pattern move is not successful [i.e., if $y(\mathbf{X}_{i+1}^*)$ does not show improvement over $y(\mathbf{X}_i)$], then $\mathbf{X}_{i+1}^*$ is set equal to $\mathbf{X}_i$ so that a new exploratory sequence is initiated from the previous base point. The successive explore-and-extrapolate strategy is again applied as before, providing that the objective function again continues to improve. The first pattern move is again made in accordance with Equation (3.3.1), and successive pattern moves as expressed by (3.3.3).

If an exploratory sequence is not successful, then all the ϵ_i are reduced by a constant factor and a new exploratory sequence is initiated from the initial exploratory point. If this new exploratory sequence is successful, it is followed by a pattern move; if it is not, then the ϵ_i are again reduced and the exploratory sequence is again initiated. The first pattern move following a repeated exploratory sequence is again made in accordance with Equation (3.3.1).

Once the ϵ_i incremental distances have been decreased, they are maintained at their new values. The entire algorithm terminates, which indicates that an optimum has been found, when improvement cannot be obtained by an exploratory sequence and the values of ϵ_i have been reduced to certain preassigned minimal values.

Although it is not apparent from the above discussion, the pattern moves increase in length as long as the search proceeds in the same direction. To see this, let the vector from an initial exploratory point $\mathbf{X}_i^*$ to the corresponding base point $\mathbf{X}_i$ be represented by $\mathbf{\Delta}_i$. As long as the search strategy continues in the same direction, all the $\mathbf{\Delta}_i$ will be equal; hence let us simply refer to each $\mathbf{\Delta}_i$ as $\mathbf{\Delta}$. From Equation (3.3.1) we see that

$$\mathbf{X}_2^* - \mathbf{X}_1^* = 2\mathbf{\Delta}. \tag{3.3.4}$$

Combining Equations (3.3.2) and (3.3.4) we obtain

$$\mathbf{X}_3^* - \mathbf{X}_1^* = 5\mathbf{\Delta}. \tag{3.3.5}$$

Subtracting Equation (3.3.4) from (3.3.5),

$$\mathbf{X}_3^* - \mathbf{X}_2^* = 3\mathbf{\Delta}. \tag{3.3.6}$$

Hence the distance between the second and third initial exploratory points is greater than the distance between the first and second exploratory points.

Continuing as above, it is easy to show that

$$\mathbf{X}_4^* - \mathbf{X}_3^* = 4\mathbf{\Delta}, \tag{3.3.7}$$

$$\mathbf{X}_5^* - \mathbf{X}_4^* = 5\mathbf{\Delta}, \tag{3.3.8}$$

$$\vdots$$

$$\mathbf{X}_i^* - \mathbf{X}_{i-1}^* = i\mathbf{\Delta}. \tag{3.3.9}$$

Thus the exploratory sequences become increasingly far apart as long as the search proceeds in the same direction. Once a pattern move fails to offer improvement in the objective function, then $\mathbf{X}_{i+1}^*$ becomes equal to $\mathbf{X}_i$, and one can easily see that the algorithm reverts back to

$$\mathbf{X}_{i+1}^* - \mathbf{X}_i^* = \mathbf{\Delta}, \tag{3.3.10}$$

$$\mathbf{X}_{i+2}^* - \mathbf{X}_{i+1}^* = 2\mathbf{\Delta}, \tag{3.3.11}$$

and so on. The progress of the algorithm is shown schematically for the maximization of a two-dimensional function in Figure 3.16.

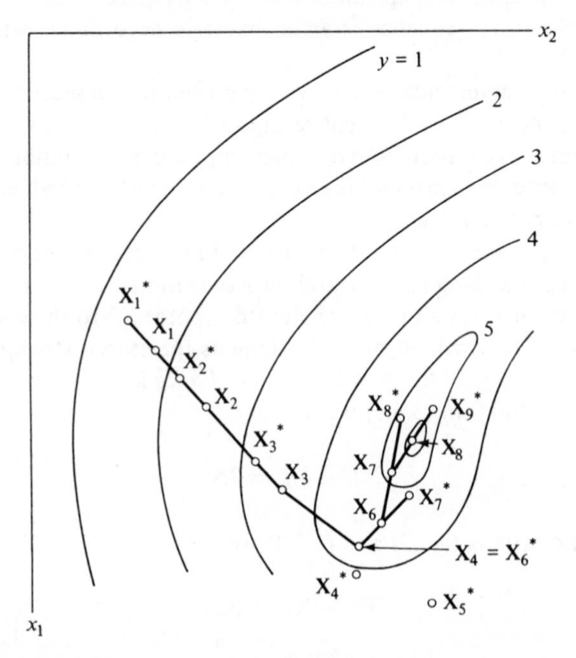

Figure 3.16. Progress of Pattern Search in Maximization of Two-Dimensional Function

An exception to the above set of rules occurs when the independent variables are bounded (as is often the case). Suppose, for example, that as the result of either an exploratory move or a pattern move the variable x_i exceeds its allowable upper bound. In this case the variable is simply set equal to its maximum allowable value, and the algorithm continues. Although it is very simple to test and correct for this condition, it does require that the independent variables be tested continually throughout the course of the computation. Notice the ease with which the bounds are handled, in contrast to the gradient methods which require that bounds be treated as separate constraints.

The algorithm can usually be speeded up (in the sense that the objective function needs to be evaluated fewer times) by "remembering" the direction of improvement for each x_i in the previous exploratory sequence and then applying the same strategy during the current exploration. In other words, if the objective function previously showed improvement by decreasing a given x_i to $x_i - \epsilon_i$, then the current exploratory move should first decrease x_i to $x_i - \epsilon_i$, and then increase x_i to $x_i + \epsilon_i$ only if the move to $x_i - \epsilon_i$ was not successful. As long as the optimization traverses a reasonably well-behaved path in the n-dimensional space, this technique will result in a net increase in computational efficiency. The increased efficiency becomes particularly apparent when each evaluation of the objective function involves lengthy computation.

Hooke and Jeeves suggest that the algorithm can be improved by following every pattern move with an exploratory sequence, without testing the pattern move for success or failure. If the exploratory sequence does not offer improvement over the old base point, then a new exploratory sequence is initiated from the end of the previous exploratory sequence (the old base point). Otherwise, a new pattern move is made by extrapolation through the two successful base points. Alteration of the search procedure in this manner enables the algorithm to follow a curved ridge or valley with greater accuracy.

A further improvement, developed by Weisman, Wood, and Rivlin [1965], can be obtained if one varies the magnitude of each ϵ_i separately, depending upon the history of successes or failures with respect to the exploratory moves in each direction. If the first move in a given direction is a success (i.e., if an improvement results from an independent variable x_i having been perturbed in the same direction as in the previous exploratory sequence), the corresponding value for ϵ_i is doubled in the next exploratory sequence. If the reverse move is successful, the same value for ϵ_i is retained for the next exploratory sequence. If both moves fail, then ϵ_i is halved in the next exploratory search. All the ϵ_i are upper- and lower-bounded, however, so as to maintain control over the size of the exploratory moves. A satisfactory rule of thumb is to let

$$10^{-5}(x_{i_{\max}} - x_{i_{\min}}) \le \epsilon_i \le 10^{-2}(x_{i_{\max}} - x_{i_{\min}}) \tag{3.3.12}$$

when the independent variables are bounded.

The incorporation of this modification into the pattern-search strategy frequently results in a small increase in computational efficiency. The main advantage, however, is greater directional flexibility, which enables the pattern search to perform more effectively on problems that have sharp ridgelike characteristics.

Although the method lacks mathematical elegance, it is a highly efficient optimization procedure, which is becoming increasingly popular for many applications. The simplicity in computer programming furthers its appeal. Hooke and Jeeves suggest its use for curve fitting and solving systems of algebraic equations, where one minimizes a square error criterion, as well as its use in solving functional equations. The technique is particularly well suited to functions exhibiting a straight, sharp ridge or valley.

In concluding the present discussion, it must be pointed out that the pattern search is not foolproof and can, under certain conditions, attain a false optimum. This can be seen in Figure 3.17, which shows the level contours of a two-dimensional function that exhibits a sharp 45° ridge. The point $\mathbf{X}_i^*$ represents the ith initial exploratory point, and the surrounding points represent subsequent exploratory points in the ith sequence. Notice that none of

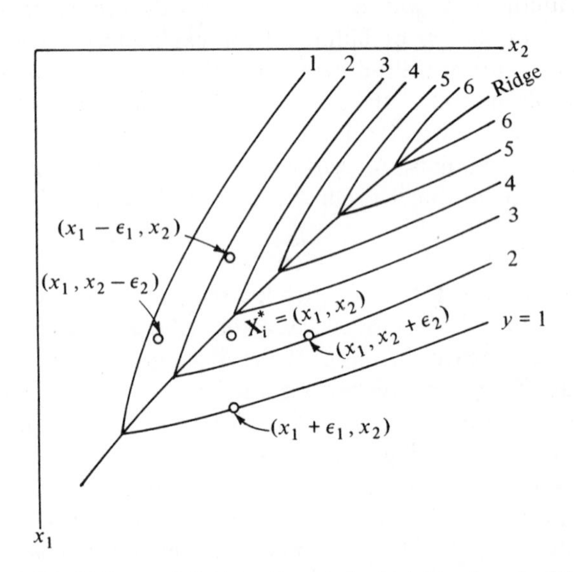

Figure 3.17. Failure of Pattern Search in Neighborhood of Sharp Ridge

the subsequent exploratory points shows improvement in the value of y, although y clearly increases along the ridge. This situation can be corrected by decreasing all the ϵ_i and again carrying out the exploratory moves. If, however, the existing values of ϵ_i are at their smallest allowable values and the point $\mathbf{X}_i^*$ coincides with the previous base point $\mathbf{X}_{i-1}$, then the pattern search terminates, indicating incorrectly that $\mathbf{X}_i^*$ is the approximate location of the maximum value of y. Although situations such as this are relatively uncommon, they can arise on occasion.

EXAMPLE 3.3.1

Resolve Example 3.2.3 using a pattern search with a value of $\epsilon_1 = \epsilon_2 = 0.01$. Each exploratory sequence will terminate when ϵ_i becomes less than or equal to 10^{-5}. Let each ϵ_i be reduced by a factor of 10.

This problem was solved on a digital computer, beginning with $x_{11}^* = x_{21}^* = 0.5000$. Some representative numerical values are shown in Table 3.10.

Table 3.10. RESULTS OF PATTERN SEARCH MINIMIZATION

n	x_{1n}^*	x_{2n}^*	x_{1n}	x_{2n}	y_n
1	0.5000	0.5000	0.4900	0.5100	5.1444
2	0.4900	0.5100	0.4800	0.5200	4.5561
4	0.4700	0.5300	0.4500	0.5500	3.0989
8	0.4000	0.6000	0.3600	0.6400	0.7705
12	0.2900	0.7100	0.2300	0.7700	0.5540×10^{-1}
18	0.1500	0.7900	0.1600	0.7800	0.6613×10^{-2}
24	0.1900	0.7100	0.2000	0.6800	0.7676×10^{-4}
36	0.2000	0.6667	0.2000	0.6665	0.1887×10^{-7}
48	0.2000	0.6668	0.2000	0.6668	0.5051×10^{-9}

The computation ended after the 48th iteration, where each iteration refers to either an exploratory sequence or a pattern move. The machine time required to obtain the solution was somewhat less than that required by the conjugate-gradient or variable-metric algorithms.

Other direct-search techniques

Rosenbrock [1960] has devised a direct-search procedure that consists only of sequences of exploratory searches along sets of orthogonal unit vectors. Essentially, the idea is to search along the axes of a new orthogonal coordinate system that does not coincide with the original set of coordinate axes. A new coordinate system is constructed each time a new exploratory sequence is to begin. This coordinate system has as its origin the previously determined "best" search point. From this point the coordinates are rotated such that the first search vector points in a previously established direction of greatest improvement.

The method is somewhat more complicated than Hooke and Jeeves's pattern search, but it is highly effective in following a curved ridge or valley (cf. Wilde [1964]). Wood [1965] presents a particularly clear description of the algorithm.

The idea of orthogonalized vectors can also be applied to gradient optimization methods. It has been successfully adapted by Pegis, Grey, Vogl, and Rigler (cf. Lavi and Vogl [1966]).

Another effective direct-search technique is the "sequential simplex" method of Spendley, Hext, and Himsworth [1962]. (This is not to be confused with the well known simplex procedure of linear programming, which we will discuss in the next chapter.) In this method the objective function is evaluated at each of the vertices of an n-dimensional polyhedron having $n + 1$ vertices. The search proceeds by moving away from the worst point, through the centroid of the remaining points, thus forming a new polyhedron. The procedure continues until a sufficiently small polyhedron has been developed for which no further improvement is possible.

Like the Rosenbrock procedure, the method of Spendley et al. is not restricted only to moves in the coordinate directions. However, the algorithm does not allow for acceleration moves in the established direction of improvement.

Powell [1964] presents still another direct-search procedure that has been found effective in unconstrained optimization. Powell's method is based upon the properties of quadratic objective functions.

3.4 Multi-dimensional Optimization by Random Search Techniques

In contrast to the sequential searching techniques, which we developed earlier, we now consider a method in which the search points are generated randomly. At first this may seem to be a cumbersome, "brute-force" method for optimizing a function, particularly after we have expounded the virtues of sequential methods. However, random search can be very useful for discontinuous functions and for terminal explorations after using a sequential optimization technique.

It has been argued (Brooks [1958]) that the number of search points required to carry out a random search does not increase with the number of independent variables. Suppose that rather than obtain the exact location of the extremum we are content to locate a value of y within some small fraction of the hyperspace which contains the extremum values of y. For example, suppose that we would be satisfied with any value of y falling within the 1 per cent of the n-dimensional region which contains the largest (or smallest) values of y. We would not know how closely our value of y approached the true extremum. However, it is likely that the difference between the chosen value of the objective function and the true extremum value will decrease as the acceptable fraction of the hyperspace becomes smaller.

Let p be the acceptable fraction of the n-dimensional region within which we wish to estimate the extremum value of y. If the search points are generated randomly, then p will also be the probability that a given search point will fall within the desired subregion. Hence the probability that a given search point will fall *outside* of the desired subregion is $(1 - p)$. In accordance with the rules of joint probabilities, Q, the probability that m search points will fall outside of the desired subregion, is given by

$$Q = (1 - p)^m. \tag{3.4.1}$$

Therefore, P, the probability that *at least one* of the search points falls *within* the desired subregion, becomes

$$P = 1 - Q \tag{3.4.2}$$

or

$$P = 1 - (1 - p)^m. \tag{3.4.3}$$

Solving for m,

$$m = \frac{\log(1 - P)}{\log(1 - p)}. \tag{3.4.4}$$

Hence, by specifying the size of the acceptable subregion p and the probability of at least one successsful search point P, we can determine the required number of search points from Equation (3.4.4). For example, if $p = 0.01$ and $P = 0.99$, then it is necessary to evaluate y at 461 search points. Notice that this result is not a function of n, the number of independent variables, although the amount of computation required to generate one random search point is proportional to this number.

The above argument is, however, misleading. A small fraction of the search region can still entail deceptively large errors in the independent variables when the dimensionality is large. The ratio of the range of uncertainty in a variable value to its original range is a more usual criterion for the effectiveness of a search procedure. When this criterion is used, random search is not found to be particularly efficient.

For example, notice that a reduction of a one-dimentional region to 10 per cent of its original size requires a corresponding reduction of the range of the independent variable to 0.100, its original range. To reduce a two-dimensional region to 10 per cent of its original size, however, requires that the range of each independent variable be reduced by only 0.316. This is seen in Figure 3.18, where the large square is 1 unit of area corresponding to $0 \leq x_1 \leq 1$, $0 \leq x_2 \leq 1$, and the shaded square represents 0.10 unit of area corresponding to $0 \leq x_1 \leq 0.316$, $0 \leq x_2 \leq 0.316$. Similarly, we see in Figure 3.19 that a factor-of-10 reduction in a three-dimensional region (corresponding to the shaded region) can be obtained by reducing the bounds on

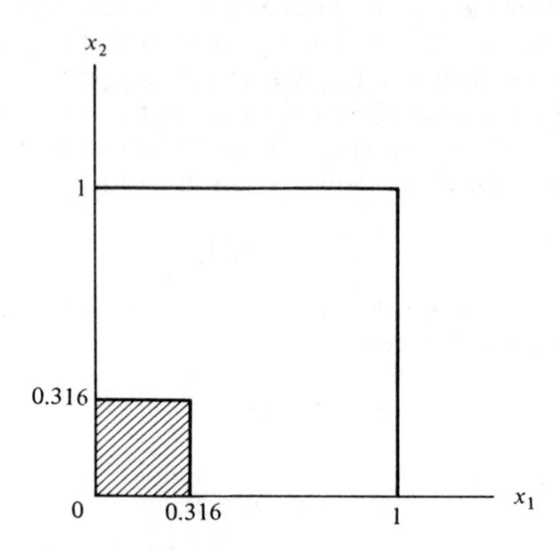

Figure 3.18. Variable Range Which Reduces Two-Dimensional Region to 10% of Original Size

the individual variables by only 0.464. This effect continues to become more pronounced as the number of independent variables increases. Thus if we wish to solve a problem of high dimensionality by random-search techniques, we must require the final search region to be an extremely small fraction (p) of the original search region to obtain a meaningful set of bounds on the independent variables. This requires a large number of observations. Let us define F as the fraction of the original range and assume that F is to be constant for all variables. For small values of F, it may be shown that m, the number of observations, is given by

$$m = -\frac{1}{F^n} \log(1 - P) \tag{3.4.5}$$

where n is the number of independent variables and P is the probability that an observation will have been made in the desired small hypercube. For example, if $n = 10$ and $F = 0.2$, we need 6.75×10^6 observations for P to be 0.5.

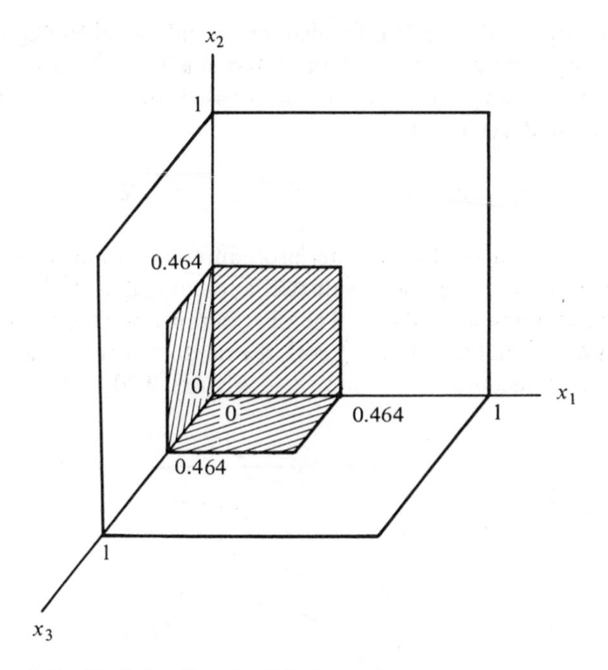

Figure 3.19. Variable Range Which Reduces Three-Dimensional Region to 10% of Original Size

Simple random search

To carry out a random search we must have either a table of random numbers or, preferably, a digital computer with a routine available for generating random numbers. It is important that the random numbers generated be bounded so that they do not exceed the allowable ranges of the search-point coordinates. We then proceed simply to choose random values for the components $x_1, x_2, \ldots, x_n$ of a given point, and then evaluate y at that point. The value of y obtained is then compared with the previously calculated extremum value. If the new value of y represents an improvement over the previously calculated extremum, then the new value of y becomes the extremum; otherwise, the previously calculated extremum is left unchanged. A new search point is then generated randomly, y is again calculated and compared with the previous extremum, and so on. The procedure terminates when the required number of search points has been generated.

Stratified random search

A variation of the random-search technique that assures a more uniform sampling of the hyperspace is the *stratified-random-search* method. In this method the region of interest is divided into m individual subregions of equal volume. If $x_{i_{max}}$ and $x_{i_{min}}$ are the largest and smallest allowable values of x_i within a given subregion, then a random value for x_i can be generated by choosing some θ_i such that

$$x_i = x_{i_{min}} + \theta_i(x_{i_{max}} - x_{i_{min}}), \qquad 0 \leq \theta_i \leq 1, \tag{3.4.6}$$

where θ_i is selected randomly. This procedure is continued for all i within each subregion, and y is then evaluated and compared with the previously determined extremum as above. The distributions of random search points in a two-dimensional region using both the random-search and the stratified-random-search techniques are shown in Figures 3.20 and 3.21.

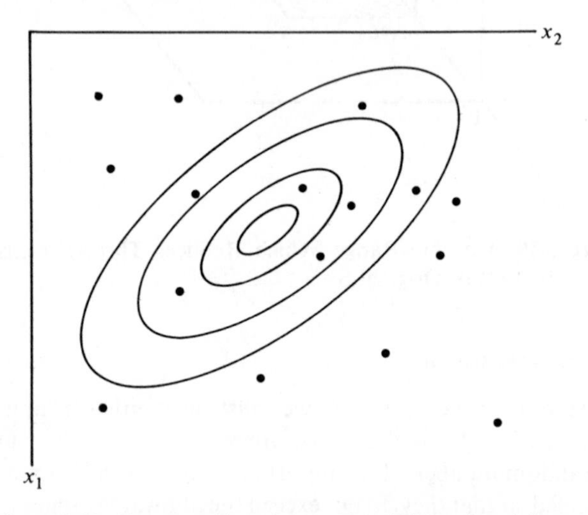

Figure 3.20. Random Search in a Two-Dimensional System

Brooks [1959] has found the stratified random search to be somewhat more effective than the pure random-search technique when the total number of search points is relatively small. As the number of search points increases, however, the two methods become equally effective, so that the additional complexity of the stratified random search is not justifiable under these circumstances.

Adaptive random search

Another variation of the random-search technique, which is usually more efficient than a simple random search, is a *pseudo random* or *adaptive random search.* The technique generates subsequent search points in the vicin-

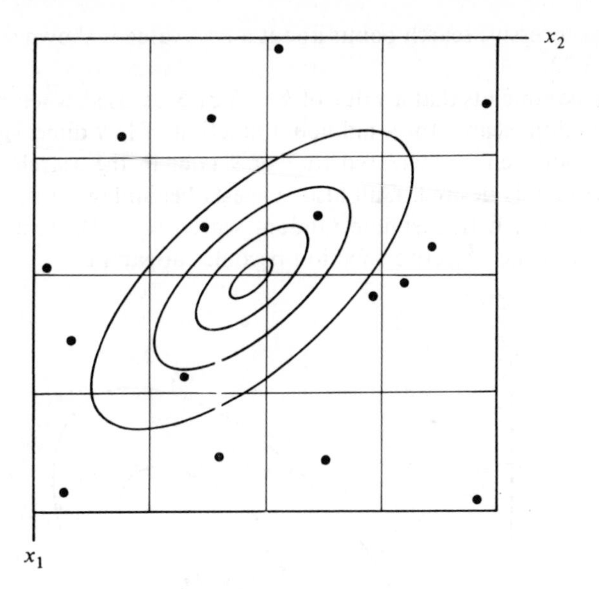

Figure 3.21. Stratified Random Search in a Two-Dimensional System

ity of a "best" search point in such a manner that subsequent search points which are close to the "best" point are more likely to be chosen than those points farther away. Thus the selection of the search points is governed by a *probability density* $f(\mathbf{X})$, which is centered about the search point yielding the best known value of y. When a new search point is found that yields a better value for y than the previous best search point, then the new point becomes the best search point and the probability density becomes centered about it for the determination of subsequent search points. The procedure can be continued as long as desired.

There are many methods that one can devise to select search points in accordance with some probability density. Gall [1966] suggests the use of the expression

$$x_i = x_i^* + (x_{i_{\max}} - x_{i_{\min}})(2\theta_i - 1)^k \tag{3.4.7}$$

as a particularly simple and effective method for generating search-point components about the best search-point component x_i^*. In this expression k is an odd integer and θ_i is a random number between 0 and 1. [Notice that Equation (3.4.7) reduces to a simple random search when $k = 1$.] Care must be taken in using Equation (3.4.7) so that the value of x_i selected does not exceed the allowable range; i.e., $x_{i_{\min}} \leq x_i \leq x_{i_{\max}}$. Values of x_i that do not fall within this range should be discarded. The use of Equation (3.4.7) in generating successive search points about a best value is shown schematically in Figure 3.22.

Gall recommends that a value of $k = 3$ or 5 be used when initiating an adaptive random search for a unimodal function of low dimensionality. The value of k can then be increased to 7 or 9 later in the search, particularly if high accuracy is desired. Gall also suggests beginning a search of a more complex function with a simple random search ($k = 1$), with k increasing as the shape of the objective function becomes apparent.

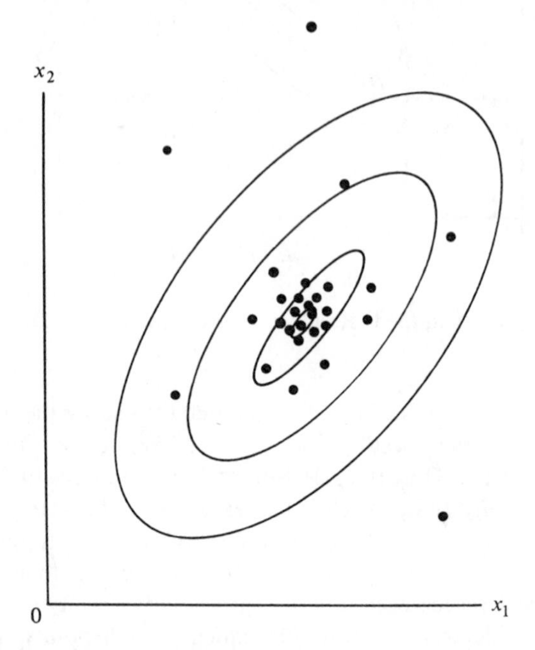

Figure 3.22. Adaptive Random Search in a Two-Dimensional System

An alternative procedure is suggested by Karnopp [1963], in which a simple random search over the entire region is followed by another simple random search in a small subregion surrounding the most current "best" search point. The subregion changes its location as new search points are found that yield more favorable values for y than the previous best values. The method is less elegant than that developed by Gall, but it is efficient and easy to implement.

We shall not dwell upon the random-search idea or the various ways it can be implemented. Before leaving the subject, however, let us momentarily pause and reflect upon some of the unique features of a random search. We have already commented on the fact that a random search is a simultaneous search, as opposed to the sequential methods discussed earlier. Since it is not necessary to determine partial derivatives or make moves in particular directions within the search region, the objective function need not be continuous. Moreover, the objective function need not be unimodal; hence random-search techniques offer one approach to the very nasty problem of optimizing an objective function that may have several local optima. In this context it may be desirable to begin by exploring a function with random-search techniques, switching to sequential methods once the approximate shape of the surface defined by the objective function becomes known.

Providing one has access to a random number generator, random-search techniques are very easy to program and implement. Difficulties created by numerical roundoff problems are usually not present. Because of the simplicity of random-search techniques, a large number of search points can be generated and tested very quickly and easily. Hence the drawbacks of high dimensionality are somewehat less severe than they first appeared. Finally, the adaptive random search can be used very effectively to carry out a terminal exploration in the neighborhood of a tentative extremum.

EXAMPLE 3.4.1

Resolve Example 3.2.3 using a random search and a stratified random search, with the components of the randomly generated search points being bounded between 0 and 1.

This problem was solved on a digital computer. Two thousand search points were generated for each method. For the stratified random search the region ($0 \leq x_i \leq 1$, $0 \leq x_2 \leq 1$) was divided into 400 equally spaced subintervals, and five search points were generated randomly within each subinterval. Shown in Tables 3.11 and 3.12 are the best (smallest) computed values of y, the index number

Table 3.11. RANDOM SEARCH RESULTS

y	m	x_1	x_2
1.0180×10^{-3}	1866	0.0018	0.0092
1.0467×10^{-3}	1091	0.0019	0.0093
2.7923×10^{-3}	1454	0.0028	0.0140
4.2960×10^{-3}	1845	0.0038	0.0192
5.5300×10^{-3}	1466	0.0055	0.0276
5.8718×10^{-3}	754	0.0045	0.0225
5.8727×10^{-3}	941	0.0045	0.0225
6.9547×10^{-3}	1180	0.1558	0.7791
6.9715×10^{-3}	316	0.1551	0.7754
7.1974×10^{-3}	1956	0.1545	0.7726
7.7890×10^{-3}	28	0.1572	0.7858
7.8742×10^{-3}	306	0.1537	0.7685

Table 3.12. STRATIFIED RANDOM SEARCH RESULTS

y	m	x_1	x_2
1.0167×10^{-4}	370	0.1966	0.6832
1.2335×10^{-4}	366	0.1981	0.6903
1.2950×10^{-4}	363	0.1994	0.6468
1.5531×10^{-4}	469	0.2027	0.6633
4.9556×10^{-4}	229	0.1171	0.2854
5.5802×10^{-4}	359	0.1954	0.5772
5.6424×10^{-4}	240	0.1407	0.3535
5.9466×10^{-4}	228	0.1167	0.2834
9.6596×10^{-4}	362	0.1960	0.6299
1.0182×10^{-3}	237	0.1450	0.3750
1.0248×10^{-3}	350	0.1768	0.4839
1.8925×10^{-3}	231	0.1231	0.3155

of the calculation (m ran consecutively from 1 to 2000), and the corresponding values of x_1 and x_2.

The stratified random search comes much closer to the correct answer of $x_1 = \frac{1}{5}$, $x_2 = \frac{2}{3}$ for this particular problem. Notice that neither answer is particularly accurate, however, compared with the answers obtained using the sequential techniques described earlier. Furthermore, the pure random search favors the trivial solution for this particular problem of $y = 0$ at $x_1 = 0$, $x_2 = 0$.

EXAMPLE 3.4.2

Resolve the last example using an adaptive random search with an initial value of $k = 3$, changing to $k = 7$ once y becomes less than 10^{-2}. Let the search initially be conducted about the point $x_1 = 0.5000$, $x_2 = 0.5000$.

Table 3.13. ADAPTIVE RANDOM SEARCH RESULTS

y	m	x_1	x_2
1.1177	1	0.5029	0.8673
0.7914	2	0.4293	0.8664
0.2932	10	0.2844	0.8526
0.2842	23	0.3138	0.7545
0.2230	35	0.2985	0.7228
0.2707×10^{-1}	50	0.1695	0.6570
0.4531×10^{-3}	65	0.2035	0.6194
0.1110×10^{-5}	88	0.2001	0.6643
0.6659×10^{-6}	122	0.2001	0.6649
0.3903×10^{-6}	179	0.2001	0.6653
0.6488×10^{-7}	251	0.2001	0.6662
0.1782×10^{-7}	354	0.2000	0.6665
0.9397×10^{-8}	455	0.2000	0.6668

The increase in efficiency of the adaptive random search, compared with the previous search techniques, can be seen from Table 3.13. The programming considerations are only slightly more complex than the simpler random methods, making the adaptive random search clearly more desirable.

3.5 Concluding Remarks

In concluding our exposition of various search techniques it is natural that we inquire as to which technique is best suited to which problem. Although hard and fixed rules do not exist, we can establish some general guidelines.

The Fibonacci and golden-ratio search techniques are the most efficient one-dimensional methods, although they are somewhat tricky to program for a computer. If one anticipates conducting many one-dimensional searches (as is often the case when conducting a sequential multidimensional search), then the involved programming is worthwhile. On the other hand, a dichotomous search or some other heuristic search technique that is easily programmed may turn out to be a more practical means of obtaining an occasional quick answer.

Of the gradient methods presented, the variable-metric algorithm seems to have gained the greatest acceptance. Conjugate gradients is also an efficient technique worthy of consideration. The method of steepest ascent, however, cannot be recommended as a useful tool, in spite of its relative simplicity. Finally, there are several direct-search algorithms, including pattern search, that appear to compete effectively with the best gradient techniques. Unfortunately, there does not seem to be any one "best" method that can be recommended, as a particular method might work very well on one problem and show up poorly on another. With this in mind the reader is cautioned about forming generalizations from the sample problems discussed earlier.

Random-search techniques are very useful for extremizing functions that are discontinuous. In addition, they offer a practical approach to the initial exploration of a function that may be multimodal, and they provide a simple means of distinguishing between an extremum, a saddle point, and a ridge. Their use in combination with sequential methods is often highly effective.

Finally, the reader is reminded that the methods developed in this chapter are appropriate only for unconstrained optimization problems—an idealized situation that is frequently unattainable. The presence of constraints generally introduces a great deal of additional complexity into an optimization problem. In Chapter 4 we shall be concerned exclusively with the optimization of linear objective functions with linear constraints. We shall then turn to the more general problem of optimizing a nonlinear objective function with nonlinear constraints in Chapter 5.

REFERENCES

Beckman, F. S., "The Solution of Linear Equations by the Conjugate Gradient Method," in *Mathematical Methods for Digital Computers*, A. Ralston and H. S. Wilf, eds., Wiley, New York, 1960.

Brooks, S. H., "A Discussion of Random Methods for Seeking Maxima," *Operations Res.*, *6* (1958), 2.

———, "A Comparison of Maximum Seeking Methods," *Operations Res.*, *2* (1959), 4.

Broyden, G. G., "Quasi-Newton Methods and Their Application to Function Minimization," *Mathematics of Computation*, *21* (1967).

Crockett, J. B., and H. Chernoff, "Gradient Methods of Maximization," *Pacific J. Math.*, *5* (1955).

Davidon, W. C., "Variable Metric Method for Minimization," *A.E.C. Research and Development*, *ANL-5990* (1959).

Fletcher, R., and M. J. D. Powell, "A Rapidly Convergent Descent Method for Minimization," *Computer J.*, *6* (1963–1964).

———, and C. M. Reeves, "Function Minimization by Conjugate Gradients," *Computer J. 7* (1964).

Gall, D. A., "A Practical Multifactor Optimization Criterion," in *Recent Advances in Optimization Techniques*, A. Lavi and T. P. Vogl, eds., Wiley, New York, 1966.

Hestenes, M. R., and E. Steifel, "Methods of Conjugate Gradients for Solving Linear Systems," *Nat. Bur. Std. Report*, No. 1659 (1952).

Hooke, R., and T. A. Jeeves, "Direct Search Solution of Numerical and Statistical Problems," *J. Assoc. Comp. Mach.*, *8* (1961), 2.

Karnopp, D. C., "Random Search Techniques for Optimization Problems," *Automatica*, *1* (1963).

Kiefer, J., "Sequential Minimax Search for a Maximum," *Proc. Amer. Math. Soc.*, *4* (1953).

Leitmann, G., *Optimization Techniques—With Applications to Aerospace Systems*, Academic Press, New York, 1962.

Pearson, J. D., "Variable Metric Methods of Minimization," *Computer J.*, *12* (1969).

Pegis, R. J., D. S. Grey, T. P. Vogl, and A. K. Rigler, "The Generalized Orthonormal Optimization Program and Its Applications," in *Recent Advances in Optimization Techniques*, A. Lavi and T. P. Vogl, eds., Wiley, New York, 1966.

Powell, M. J. D., "An Iterative Method for Finding Stationary Values of a Function of Several Variables Without Calculating Derivatives," *Computer J.*, *7* (1964).

Rosenbrock, H. H., "An Automatic Method for Finding the Greatest or Least Value of a Function," *Computer J.*, *3* (1960).

Shah, B. V., R. J. Buehler, and O. Kempthorne, "Some Algorithms for Minimizing a Function of Several Variables," *J. Soc. Ind. Appl. Math.*, *12* (1964), 1.

Spendley, W.. G. R. Hext, and F. R. Himsworth, "Sequential Applications of Simplex Designs in Optimization and Evolutionary Operation," *Technometrics*, *4* (1962).

Weisman, J., C. Wood, and L. Rivlin, "Optimal Design of Chemical Process Systems," *Chem. Eng. Progr. Symp. Ser*,. *61*, No. 55 (1965), 50.

Wilde, D. J., *Optimum Seeking Methods*, Prentice-Hall, Englewood Cliffs, N.J., 1964.

———, and C. S. Beightler, *Foundations of Optimization*, Prentice-Hall, Englewood Cliffs, N.J., 1967.

Wood, C. F., "Review of Design Optimization Techniques," *IEEE Trans. Sys. Sci. Cybernetics*, *SSC-1* (1965), 1.

Zeleznik, F. J., "Quasi-Newton Methods for Nonlinear Equations," *J. Assoc. Comp. Mach. 15* (1968).

PROBLEMS

3.1. Given the following one-dimensional optimization problems:

(a) Minimize $y = (x - 1)^2$, $0 \leq x \leq 3$.
(b) Maximize $y = x \sin x$, $0 \leq x \leq \pi$.
(c) Minimize $y = e^{-0.01x^2} \cos(0.5x)$, $0 \leq x \leq 10$.

Solve each of these problems as accurately as possible with

(i) A six-point dichotomous search.
(ii) A six-point Fibonacci search.
(iii) A six-point golden-ratio search.

Assume that $\epsilon = 0.10$ for each set of calculations.

3.2. Resolve problem 3.1(b) using a dichotomous search and a golden-ratio search. In each case let $\epsilon = 0.10$, and utilize as many search points as possible without losing the distinction between adjacent values of $y(x)$.

3.3. Determine a *global* maximum of the function

$$y = \sin(x) \sin(2.8x) \sin(27x), \quad 0 \leq x \leq \pi.$$

(This problem will require considerable computational effort, as it has many local maxima.)

3.4. Use a Fibonacci search to maximize the function $y = \sin x$, $0 \leq x \leq 2$, with the restriction that x can only take on values which are multiples of 0.10; i.e., $x = 0, 0.10, 0.20, \ldots, 1.90, 2.00$. Assume that $\epsilon = 0$ for this problem.

3.5. Determine the minimum of the function

$$y = 2(x_1 - 2)^2 + 3(x_2 + 1)^2 + (x_3 - 3)^2$$

along the line connecting the points (3, −4, 2) and (2, 1, −3). Solve this problem by means of

(a) Differential calculus, substituting the equation for the line into the objective function.

(b) A one-dimensional search technique along the line between the two given end points.

(Note that the line can be expressed as $\mathbf{X} = \beta\mathbf{X}_0 + (1 - \beta)\mathbf{X}_1$, $0 \leq \beta \leq 1$, where $\mathbf{X}_0$ and $\mathbf{X}_1$ are the given end points.)

3.6. The equivalence of Equations (3.1.18) and (3.1.19) hinges about the identity

$$F_{n-1}F_{n-2} - F_nF_{n-3} = (-1)^n.$$

Prove by induction that the above identity is valid. (Show first that the identity is valid for the special cases of $n = 3$ and $n = 4$. Then show that the identity is valid for $n = k$, providing the identity holds for $n = k - 1$.)

3.7. Draw the contours of the function

$$y = x_1^2 + 4x_2^2$$

for $y = 3$ and $y = 8$. Derive equations for the tangent line and the gradient through the point (2, 1). Draw these lines over the previously drawn contours. If one starts at the point (2, 1), in what direction should one search in order that y will decrease as rapidly as possible? How far along this search line should one proceed? Show graphically.

3.8. Consider the unconstrained function

$$y = 10x_1 - 2x_1^2 - 4x_2 + 5 \log x_2.$$

(a) Choose (1, 2) as your staring point and locate a relative maximum using the method of steepest ascent with $\Delta\tau = 0.075$.

(b) Repeat (a), starting from the same location, using the method of optimal steepest ascent. Use $\Delta\tau = 1.0$.

Which of the two procedures requires less computational effort?

3.9. Write a computer program to find the minimum of an unconstrained, multidimensional function. Use one or more of the following methods:

(a) Optimal steepest descent

(b) Conjugate gradients

(c) Variable-metric search

(d) Pattern search

(e) Adaptive random search

Include a provision for satisfying variable bounds. Use a golden-ratio search wherever a one-dimensional search is required.

Restate problem 3.8 as a minimization problem and solve it with the program you have written.

3.10. Given the following unconstrained, multidimensional optimization problems:

(a) Minimize $y = x_1^2 + 4x_2^2$.

(b) Minimize $y = 3(x_1 - 1)^2 + (x_2 + 2)^2 + 5(x_3 - 4)^2$.

(c) Minimize $y = 100(x_2 - x_1^2)^2 + (1 - x_1)^2$.

This well-known test problem is sometimes referred to as Rosenbrock's "banana" because of the shape of its contours (cf. Rosenbrock [1960]).

(d) Minimize $y = 100[(x_3 - 10\theta)^2 + (r - 1)^2] + x_3^2$,
where $r = (x_1^2 + x_2^2)^{1/2}$ and

$$\theta = \begin{cases} \dfrac{1}{2\pi}\tan^{-1}\left(\dfrac{x_2}{x_1}\right) & \text{if } x_1 > 0 \\ \dfrac{1}{4} & \text{if } x_1 = 0 \\ \dfrac{1}{2} + \dfrac{1}{2\pi}\tan^{-1}\left(\dfrac{x_2}{x_1}\right) & \text{if } x_1 < 0. \end{cases}$$

(Cf. Flecher and Powell [1963–1964].)

(e) Maximize $y = 10 - [(x_1 + 10x_2)^2 + 5(x_3 - x_4)^2 + (x_2 - 2x_3)^4 + 10(x_1 - x_4)^4]$ (cf. Fletcher and Powell [1963–1964]).

Solve each of these problems using one or more of the following techniques:

(i) Optimal steepest descent
(ii) Conjugate gradients
(iii) Variable-metric search
(iv) Pattern search
(v) Adaptive random search

(Note that problems (c), (d), and (e) will require considerable computational effort. If these problems are assigned to a group of students, then a particular starting point should be specified for each problem.)

3.11. Solve problem 3.10(c) using a

(a) Random search
(b) Stratified random search
(c) Adaptive random search

Compare the number of search points required to obtain an acceptable solution using each of the above random-search techniques.

3.12. Solve, using an appropriate computational technique: Minimize, with respect to (x_1, x_2, x_3, x_4):

$$y = \sum_{i=1}^{20} [u_i(x_1, x_2, t) - u_i^*]^2 + \sum_{i=1}^{20} [v_i(x_3, x_4, t) - v_i^*]^2$$

where u_i and v_i are defined by the differential equations

$$\frac{du}{dt} = -x_1 uve^{-x_2}, \qquad 0 \leq t \leq 20,\ u(0) = 1.10,$$

$$\frac{dv}{dt} = -x_3 uve^{-x_4}, \qquad 0 \leq t \leq 20,\ v(0) = 1.80,$$

and u_i^* and v_i^* are linearly interpolated values obtained from the following table:

t	u^*	t	v^*
0.0	1.10	0.0	1.80
0.6	0.94	0.6	1.71
0.7	0.82	1.2	1.56
1.0	0.67	1.5	1.34
2.2	0.68	2.3	1.21
2.8	0.56	3.8	1.18
5.4	0.43	5.4	1.00
7.1	0.39	8.0	0.92
9.4	0.20	10.1	0.81
11.7	0.15	10.8	0.69
18.8	0.07	14.0	0.60
30.0	0.00	18.2	0.55
		26.0	0.50

The terms in the summations are evaluated at $t_1 = 0.5,\ t_2 = 1.5,\ t_3 = 2.5,\ \ldots,\ t_{20} = 19.5$.

3.13. Wilde [*A.I.Ch.E. Journal* 9, 186 (1963)] has suggested a search method based on contour tangents. In this procedure the contour tangent is obtained by constructing a perpendicular to the gradient at some starting point. The region to one side of the tangent is then eliminated as unfavorable. A new search point is then selected in the center of the remaining region. The contour tangent is again obtained and the region to one side of this new tangent is discarded as unfavorable. The procedure continues in this manner until the optimum is located within the desired precision.

How does one determine which of the regions separated by the contour tangent should be discarded? What properties must the objective function have for this search procedure to succeed?

3.14. A particular multivariable objective function $y(x_1, x_2 \ldots x_n)$ is meaningful only when all the x_i are integers. Devise a modified pattern-search scheme that could be used to minimize such a function.

3.15. Show that problem 3.10(c) can be simplified considerably by the following change of variables:

$$u = (1 - x_1)$$
$$v = 10(x_2 - x_1)^2.$$

Show that minimization of the new function, $y(u, v)$, yields the same results as are obtained by minimizing the original function, $y(x_1, x_2)$.

3.16. If $y(\mathbf{X})$ is a continuous and differentiable function with a stationary point at $\mathbf{X}_0$, show that $\mathbf{X}_0$ can be located approximately by the expression

$$\mathbf{X}_0 - \mathbf{X} = \mathbf{P} = -\mathbf{A}^{-1}\nabla y(\mathbf{X}),$$

where $\mathbf{A}$ is the Hessian matrix, evaluated at point $\mathbf{X}_0$. This is known as *Newton's method.* Under what conditions is the above expression exact? Under what conditions can $\mathbf{X}_0$ be guaranteed to be a minimizing point? Maximizing point? From a computational point of view, how effective would you expect this method to be? What are its drawbacks?

3.17. If in problem 3.16 the inverse of the Hessian matrix is approximated by a matrix H, obtained using only first-derivative information, then the following iterative scheme can be constructed:

$$\mathbf{X}_{i+1} - \mathbf{X}_i = \mathbf{P}_i = -\mathbf{H}_i\nabla y(\mathbf{X}_i).$$

This is known as a *quasi-Newton method.* How can this method be implemented on a computer? What must be said about the approximating matrices in order that the method yield a minimizing point? Maximizing point? What does one obtain if $\mathbf{H}_i$ is taken to be the identity matrix $\mathbf{I}$? Show an analogy between the quasi-Newton method suggested above and the well-known *Newton–Raphson method* for solving the nonlinear algebraic equation $f(x) = 0$.

3.18. Construct a quadratic function whose minimum point, $\mathbf{X}_0$, is also a solution to the system of equations

$$\begin{aligned} x_1 + 2x_2 + x_3 &= 3, \\ 2x_1 + x_2 - x_3 &= 3, \\ x_1 - x_2 + 3x_3 &= 10. \end{aligned}$$

Show that the two problems are equivalent.

3.19. Show that a set of n-dimensional vectors $\mathbf{P}_0, \mathbf{P}_1, \ldots, \mathbf{P}_{n-1}$ are linearly independent if they are $\mathbf{A}$-conjugate [cf. Equation (3.2.32)].

3.20. Show that the search vectors defined by Equation (3.2.34) are $\mathbf{A}$-conjugate.

3.21. Show that the search vectors defined by Equation (3.2.64) are $\mathbf{A}$-conjugate.

3.22. Using the approximation

$$\log(1 + x) = x - \tfrac{1}{2}x^2 + \tfrac{1}{3}x^3 - \tfrac{1}{4}x^4 + \ldots, \qquad -1 < x < 1,$$

show that Equation (3.4.5) is valid.

LINEAR PROGRAMMING

In Chapter 3 we considered several different approaches to the optimization of continuously varying, unconstrained functions. We now turn our attention to a more difficult class of problems: the optimization of a continuous objective function, subject to one or more continuous constraining conditions. Our interest in this chapter will be restricted to problems that are entirely linear. That is, we shall consider the optimization of a linear, algebraic objective function, subject to a number of linear, algebraic constraints. We shall require that the independent variables be nonnegative, but otherwise they will vary continuously. Optimization problems that satisfy all these conditions are known as *linear programming* problems.

Linear programming is an important cornerstone in the aggregate of subjects known as optimization theory. There are two reasons for this. First, a great many physical problems can be adequately represented by means of linear mathematical models. Indeed, linear programming models are frequently applied to problems that range from the highly specialized, such as fitting response surfaces to detailed scientific data, to the very broad, such as the analysis and control of a nationa economy. In the areas of economics, engineering, and the physical sciences, many diverse kinds of problems (such as production scheduling, portfolio analysis, contract awarding, per-

4

sonnel assignment, resource allocation, and inventory control) have been successfully solved with linear programming. Thus, from an applications standpoint, linear programming is very far-reaching.

Second, the simplicity of linear programming models has made it possible to analyze such problems thoroughly. This has resulted in highly efficient computational algorithms. Global optima, if they exist, are always found in a finite number of steps. Unforseen difficulties, such as unbounded optima or inconsistent constraints, are recognized and identified as such. Post-optimal analyses (as, for example, the variation in the optimal solution that corresponds to a change in certain of the problem parameters) can be carried out easily and quickly. With all these features easily implemented on a digital computer, it is well within the realm of practicality to solve linear programming problems that contain several thousand independent variables and several thousand constraints. In fact, problems of this size are solved routinely in many industries.

The fundamental concepts of linear programming are presented in this chapter. Much of the material is concerned with the basic computational algorithm, known as the *simplex algorithm*, for solving linear programming problems. This method was first developed in 1947 by George B. Dantzig and his co-workers, and has since received extensive refinement by Dantzig and others. This is followed by a description of post-optimal analysis. The chapter concludes with a discussion of duality and some of its implications.

4.1 Statement of the Linear Programming Problem

The objective of the linear programming problem is to maximize or minimize the linear function

$$y = c_1x_1 + c_2x_2 + \cdots + c_nx_n,^* \tag{4.1.1}$$

subject to the linear constraints

$$\begin{aligned}
&a_{11}x_1 + a_{12}x_2 + \cdots && + a_{1n}x_n \leq b_1 \\
&\vdots \\
&a_{k1}x_1 + a_{k2}x_2 + \cdots && + a_{kn}x_n \leq b_k \\
&a_{(k+1)1}x_1 + a_{(k+1)2}x_2 + \cdots && + a_{(k+1)n}x_n \geq b_{(k+1)} \\
&\vdots \\
&a_{l1}x_1 + a_{l2}x_2 + \cdots && + a_{ln}x_n \geq b_l \\
&a_{(l+1)1}x_1 + a_{(l+1)2}x_2 + \cdots && + a_{(l+1)n}x_n = b_{(l+1)} \\
&\vdots \\
&a_{m1}x_1 + a_{m2}x_2 + \cdots && + a_{mn}x_n = b_m
\end{aligned} \tag{4.1.2}$$

and the condition that $x_i \geq 0$, $i = 1, 2, \ldots, n$. We require that $(m - l)$, the number of equality constraints, be less than n; otherwise the values of x_i would be either uniquely determined or overspecified.

Slack and surplus variables

The problem can be formulated in an alternative manner if we desire. The equality constraints are a system of $(m - l)$ equations for $(m - l)$ of the variables in terms of the remaining $(n - m + l)$ variables. Substituting into the objective function and the inequality constraints, we would then have a

* The c_i in this equation are generally referred to as cost coefficients since they often represent unit costs.

linear programming problem of lower dimensionality than the original problem, with the $(n - m + l)$ variables restricted only by inequality constraints. Although the solution to this problem may require less work than the original problem, we have gained this simplicity at the expense of solving a system of simultaneous algebraic equations. Thus this procedure may not be worth the trouble, particularly if the number of equality constraints is large.

We shall specify that the values of b_j, $j = 1, 2, \ldots, m$, be nonnegative. This does not decrease the generality of the problem, because we can always multiply the jth constraint by -1 if b_j is originally negative. However, multiplication by -1 will reverse the inequality (i.e., from $\leq$ to $\geq$, or from $\geq$ to $\leq$) if the jth constraint is expressed as an inequality.

The inequalities can be converted into equalities by introducing slack and surplus variables $x_i \geq 0$ for $i = n + 1, n + 2, \ldots, n + l$. The linear programming problem can then be written: Extremize

$$y = c_1x_1 + c_2x_2 + \cdots + c_nx_n + c_{n+1}x_{n+1} + \cdots + c_{n+l}x_{n+l}, \quad (4.1.3)$$

subject to the constraints

$$\begin{aligned}
&a_{11}x_1 + a_{12}x_2 + \cdots \qquad + a_{1n}x_n + x_{n+1} = b_1,\\
&\quad\vdots\\
&a_{k1}x_1 + a_{k2}x_2 + \cdots \qquad + a_{kn}x_n + x_{n+k} = b_k,\\
&a_{(k+1)1}x_1 + a_{(k+1)2}x_2 + \cdots + a_{(k+1)}x_n - x_{n+k+1} = b_{(k+1)}, \qquad (4.1.4)\\
&\quad\vdots\\
&a_{l1}x_1 + a_{l2}x_2 + \cdots \qquad + a_{ln}x_n - x_{n+l} = b_l,\\
&a_{(l+1)1}x_1 + a_{(l+1)2}x_2 + \cdots \quad + a_{(l+1)n}x_n = b_{(l+1)},\\
&\quad\vdots\\
&a_{m1}x_1 + a_{m2}x_2 + \cdots \qquad + a_{mn}x_n = b_m,
\end{aligned}$$

and the conditions that $c_i = 0$, $i = n + 1, \ldots, n + l$, and $x_i \geq 0$, $i = 1, 2, \ldots, n + l$. Notice that the introduction of the slack and surplus variables has not altered the objective function, since the cost coefficients of these variables have been set equal to zero. Also, observe that m, the number of constraining equations, is always less than $n + l$, the number of independent variables, providing $(m - l)$ was originally less than n; hence we have an underspecified system of simultaneous linear algebraic equations, which allows the objective function to take on an infinity of values.

Vector-matrix formulation

We can reformultate the linear programming problem in matrix notation as follows: Let

$$\mathbf{X} = \begin{bmatrix} x_1 \\ x_2 \\ \cdot \\ \cdot \\ \cdot \\ x_n \\ \cdot \\ \cdot \\ \cdot \\ x_{n+l} \end{bmatrix}, \qquad \mathbf{C} = \begin{bmatrix} c_1 \\ c_2 \\ \cdot \\ \cdot \\ \cdot \\ c_n \\ \cdot \\ \cdot \\ \cdot \\ c_{n+l} \end{bmatrix}, \qquad \mathbf{B} = \begin{bmatrix} b_1 \\ b_2 \\ \cdot \\ \cdot \\ \cdot \\ b_m \end{bmatrix},$$

$$\mathbf{A} = \begin{bmatrix} a_{11} & a_{12} & \cdots & a_{1n} & 1 & \cdots & 0 & & \cdots & 0 \\ \cdot \\ \cdot \\ \cdot \\ a_{k1} & a_{k2} & \cdots & a_{kn} & 0 & \cdots & 1 & 0 & \cdots & 0 \\ a_{(k+1)1} & a_{(k+1)2} & \cdots & a_{(k+1)n} & 0 & \cdots & 0 & -1 & \cdots & 0 \\ \cdot \\ \cdot \\ \cdot \\ a_{l1} & a_{l2} & \cdots & a_{ln} & 0 & \cdots & 0 & 0 & \cdots & -1 \\ a_{(l+1)1} & a_{(l+1)2} & \cdots & a_{(l+1)n} & 0 & \cdots & & & \cdots & 0 \\ \cdot \\ \cdot \\ \cdot \\ a_{m1} & a_{m2} & \cdots & a_{mn} & 0 & \cdots & & & \cdots & 0 \end{bmatrix}.$$

(4.1.5)

The linear programming problem becomes: Extremize

(4.1.6) $$y = \mathbf{C}'\mathbf{X},$$

subject to

(4.1.7) $$\mathbf{AX} = \mathbf{B}, \qquad \mathbf{X} \geq \mathbf{0}.$$

Any value of **X** that satisfies Equation (4.1.7) is called a *feasible solution.* In addition, those feasible solutions which extriemize Equation (4.1.6) are called *maximum* or *minimum feasible solutions.* The objective of linear programming is the determination of the maximum or minimum feasible solution to a given problem.

Decomposition of the **A** *matrix*

Finally, it is convenient to decompose the $m \times (n + l)$ matrix **A** into the vectors

$$\mathbf{P}_1 = \begin{bmatrix} a_{11} \\ a_{21} \\ \cdot \\ \cdot \\ \cdot \\ a_{m1} \end{bmatrix}, \quad \mathbf{P}_2 = \begin{bmatrix} a_{12} \\ a_{22} \\ \cdot \\ \cdot \\ \cdot \\ a_{m2} \end{bmatrix}, \quad \ldots, \quad \mathbf{P}_n = \begin{bmatrix} a_{1n} \\ a_{2n} \\ \cdot \\ \cdot \\ \cdot \\ a_{mn} \end{bmatrix},$$

$$\mathbf{P}_{n+1} = \begin{bmatrix} 1 \\ 0 \\ \cdot \\ \cdot \\ \cdot \\ \cdot \\ \cdot \\ \cdot \\ \cdot \\ 0 \end{bmatrix}, \quad \mathbf{P}_{n+2} = \begin{bmatrix} 0 \\ 1 \\ \cdot \\ \cdot \\ \cdot \\ \cdot \\ \cdot \\ \cdot \\ \cdot \\ 0 \end{bmatrix}, \quad \ldots, \quad \mathbf{P}_{n+l} = \begin{bmatrix} 0 \\ 0 \\ \cdot \\ \cdot \\ \cdot \\ -1 \\ \cdot \\ \cdot \\ \cdot \\ 0 \end{bmatrix}, \qquad (4.1.8)$$

and to let $\mathbf{P}_0 = \mathbf{B}$. We can then rewrite our problem statement in vector form as: Extremize

$$y = \mathbf{C}'\mathbf{X}, \qquad (4.1.9)$$

subject to the constraints

$$x_1\mathbf{P}_1 + x_2\mathbf{P}_2 + \cdots + x_N\mathbf{P}_N = \mathbf{P}_0, \qquad N = (n + l), \qquad (4.1.10)$$

and the nonnegativity requirements

$$\mathbf{X} \geq \mathbf{0}. \qquad (4.1.11)$$

The advantages of this last formulation of the linear programming problem will become apparent shortly. In our subsequent discussion, however, we shall let n represent the total number of variables, including slack variables. Thus the term N appearing in Equation (4.1.10) will hereafter be replaced by n.

4.2 Characteristics of Linear Objective Functions and Linear Inequality Constraints

To gain insight into the linear programming problem, let us examine the characteristics of some simple linear objective functions. We shall in particular be concerned with the location of extremum values of the objective function when the independent variables are nonnegative and possibly subject to inequality constraints.

Let us begin by considering the simple expression $y = cx$, $c > 0$, which is shown in Figure 4.1 for $0 \leq x \leq b$. This is, of course, a straight line whose minimum occurs at $x = 0$ and whose maximum is at $x = b$. Notice that the function takes on its extremum values at the *boundaries* of the interval in which it is defined and not at any interior point. We cannot locate the extrema using calculus, because we would have the condition $dy/dx = c = 0$, which is inconsistent with the earlier assertion that $c > 0$. The failure of the calculus to extremize this simple function is actually quite logical when we recall that calculus applies only to functions that exhibit an extremum *within* the region in which they are defined.

As a second example consider the plane represented by the expression $y = y_0 - c_1x_1 - c_2x_2$, as shown in Figure 4.2. We will restrict our attention to positive values of the constants y_0, c_1, and c_2 and to nonnegative values of the variables y, x_1, and x_2. These restrictions cause the function to be defined only in the shaded triangle of the x_1x_2 plane in Figure 4.2. Again we see that the linear objective function takes on its extremum values on the

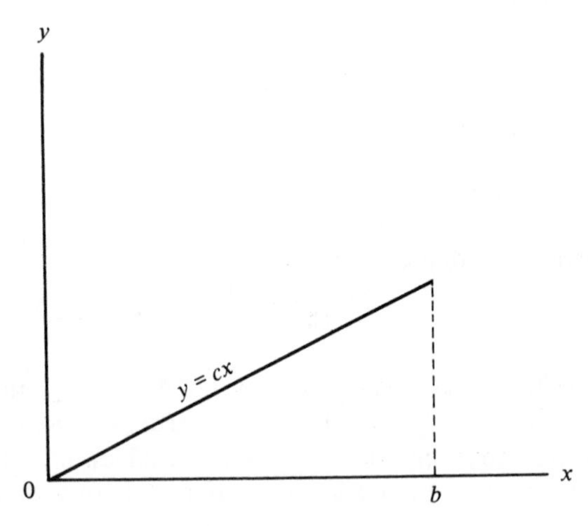

Figure 4.1. Linear Objective Function of One Variable

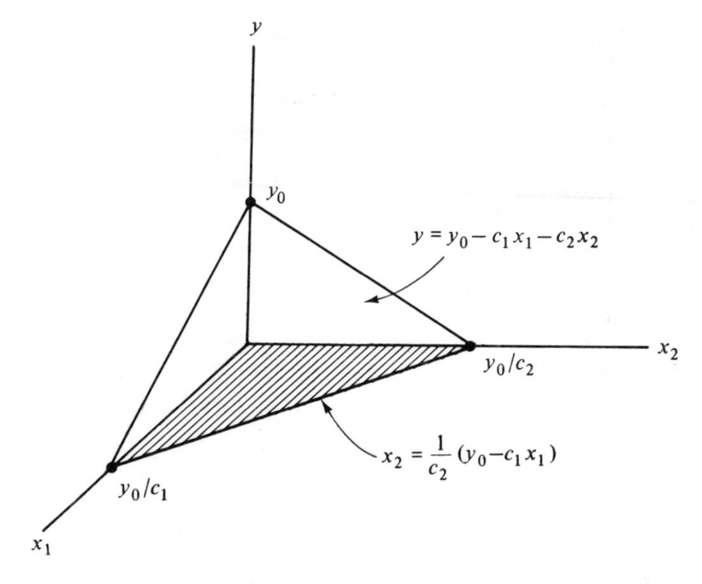

Figure 4.2. Linear Objective Function of Two Variables

region boundaries. Clearly, y takes on its maximum value y_0 at $x_1 = x_2 = 0$, which forms one vertex of the triangle. The function equals zero and is hence minimized on the line representing the intersection of the objective plane and the x_1x_2 plane, the equation of this line being $x_2 = (y_0 - c_1x_1)/c_2$. Once again we see that the given objective function cannot be extremized by the classical use of calculus because the function does not take on extremal values within the region of interest.

Finally, consider maximizing the linear function $y = x_1 + x_2$; x_1, $x_2 \geq 0$, subject to the linear inequality constraints $x_1 + 2x_2 \leq 6$ and $2x_1 + x_2 \leq 8$. The nonnegative values of x_1 and x_2 that satisfy both of the constraints fall within the shaded polygon shown in Figure 4.3. Thus we see that the constraints form a part of the region boundary. (We reached this same conclusion in our discussion of Lagrange multipliers and inequality constraints in Chapter 2.) Notice that the region has the interesting property that no part of the boundary is hidden from the view of an observer placed anywhere within the region. Stated differently, we see that if we connect any two points on the boundary with a straight line, then all points on that line lie within the region. The collection of points contained in such a region is called a *convex set*, and the region itself is known as a *convex polyhedron*.

When the objective function is plotted in the x_1x_2 plane for various values of y, we obtain the family of parallel lines shown in Figure 4.4. Notice that only those line segments which lie within or on the convex polygon satisfy the constraints and the nonnegativity requirements of x_1 and x_2.

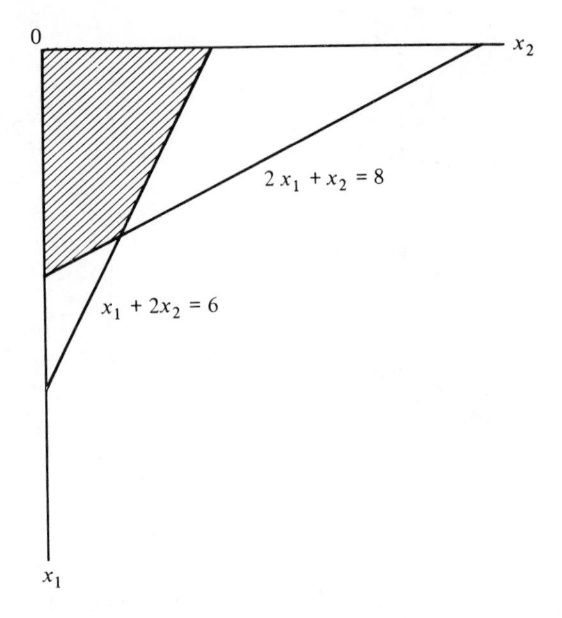

Figure 4.3. Region Satisfying Two Linear Constraints

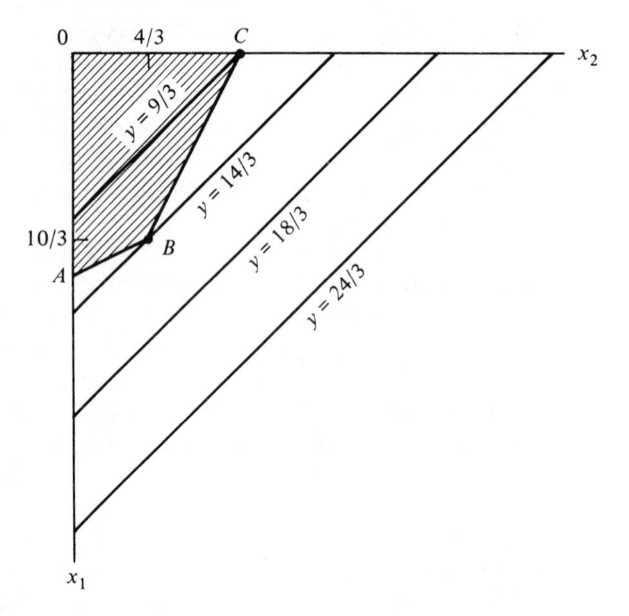

Figure 4.4. Graphical Solution of Two-Dimensional Linear Program

Any point (x_1, x_2) lying within or on the convex polygon is therefore a feasible solution to the linear programming problem. Recall that any feasible solution which maximizes the objective function is called a maximum feasible solution (cf. Section 4.1). The maximum feasible solution to the example shown in Figure 4.4 is clearly the vertex B of the convex polyhedron; i.e., the constrained maximum of the given objective function occurs along the line $x_1 + x_2 = \frac{14}{3}$ at the point $x_1 = \frac{10}{3}$, $x_2 = \frac{4}{3}$.

It is significant that the maximum feasible solution to our problem occurred at a vertex of the convex polygon. We shall see that the constraints and the nonnegativity conditions of a general n-dimensional linear programming problem always form a convex polyhedron. Furthermore, the optimal feasible solution always occurs at a vertex representing the intersection of two or more hyperplanes, providing the optimal feasible solution is finite and unique.

Not all linear programming problems admit to a finite and unique solution. For example, consider the problem of again maximizing the function $y = x_1 + x_2$; $x_1 \geq 0$, $x_2 \geq 0$, subject to the constraints $x_1 + 2x_2 \geq 6$ and $2x_1 + x_2 \geq 8$. The inequality constraints are plotted in the same positions in the x_1x_2 plane as in Figure 4.3; now, however, the inequalities are reversed, so that the convex polygon, which is partly bounded by sides CDE in Figure 4.5, is infinite. Consequently, values for x_1 and x_2 can be infinite and still be feasible solutions. Therefore, the maximum feasible solution to this problem is not finite. On the other hand, the same problem has a *minimum* feasible

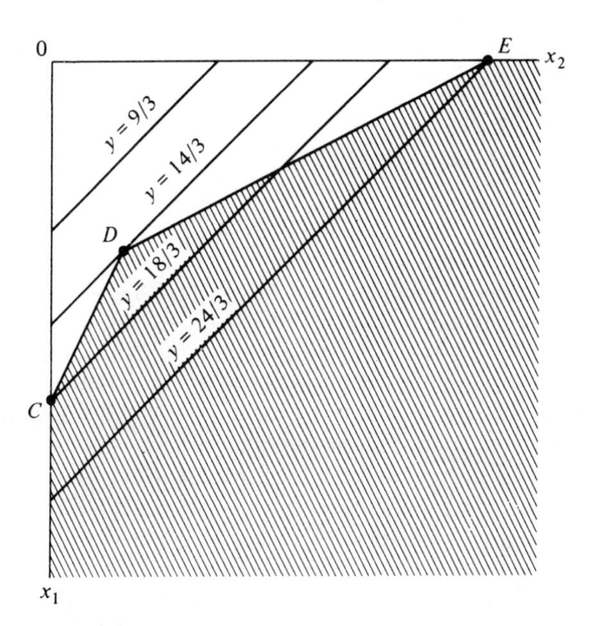

Figure 4.5. Two-Dimensional Linear Program with Unbounded Solution

solution at point D in Figure 4.5, i.e., along the line $y = \frac{14}{3}$ at $x_1 = \frac{10}{3}$ and $x_2 = \frac{4}{3}$. Notice that the minimum feasible solution to this problem is the same as the maximum feasible solution to the problem shown in Figure 4.4.

Suppose now that we wish to maximize the function $y = 2x_1 + 4x_2$; $x_1, x_2 \geq 0$, subject to the constraints $x_1 + 2x_2 \leq 6$ and $2x_1 + x_2 \leq 8$. This problem is represented in Figure 4.6. Observe that the line which represents the objective function is parallel to the constraint $x_1 + 2x_2 = 6$. Clearly, the objective function takes on its constrained maximum when $y = 12$. Although the maximum allowable value for y is finite, however, there are an infinite number of finite maximum feasible solutions that satisfy the problem, viz., all points lying along line segment BC.

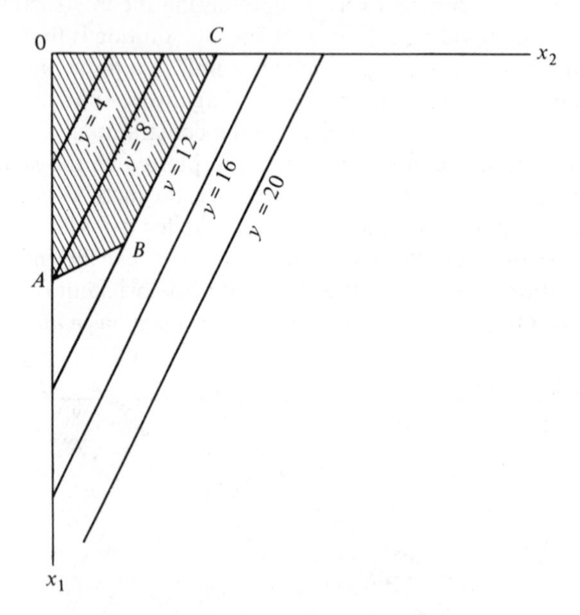

Figure 4.6. Linear Program where Objective Function Lines are Parallel to Constraint

Most linear programming models that represent realistic, physical optimization problems admit to a finite optimal value for the objective function (though there may be an infinite number of optimal policies, as pointed out above). It may turn out, however, that a problem has an unbounded optimal value for the objective function, particularly if the number of degrees of freedom is large. This may result from a mathematical model which, from a *physical* point of view, is incorrect. Moreover, it sometimes happens that a particular problem does not have any feasible region. This is either a consequence of some difficulty with the mathematical representation of a valid problem, or else the problem actually cannot be solved because of two or more conflicting conditions. Unexpected results of this nature should not be dismissed without careful study, as this is one of the few indications the analyst may have of an error in the model formulation.

4.3 Properties of Convex Sets

In this section we shall lay some of the algebraic groundwork for our later development of the simplex algorithm. We shall also analyze more critically some characteristics of feasible solutions of the linear programming problem.

Convex sets

Suppose that we choose any two points $\mathbf{X}_1$ and $\mathbf{X}_2$ which are members of a collection of points contained in a closed region $\mathbf{R}$. If these points are connected by a straight line and all points lying on this straight line are also contained in $\mathbf{R}$, then the collection of points is known as a *convex set.* In other words, let $\mathbf{X}_1$ and $\mathbf{X}_2$ be any two points contained in the closed region $\mathbf{R}$. If all points

$$\mathbf{X} = \beta\mathbf{X}_1 + (1 - \beta)\mathbf{X}_2, \qquad 0 < \beta < 1, \tag{4.3.1}$$

lie within $\mathbf{R}$ for all $\mathbf{X}_1$ and $\mathbf{X}_2$, then the collection of points within $\mathbf{R}$ is a convex set. Figures 4.7 and 4.8 show examples of convex and nonconvex sets, respectively. The shaded regions in Figures 4.2–4.6 are all examples of convex sets.

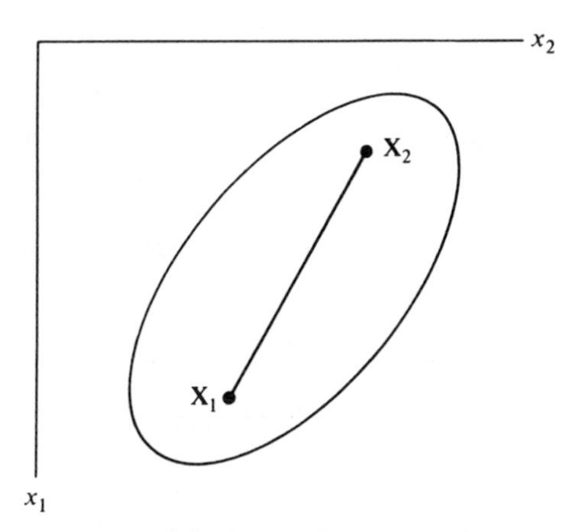

Figure 4.7. A Convex Set

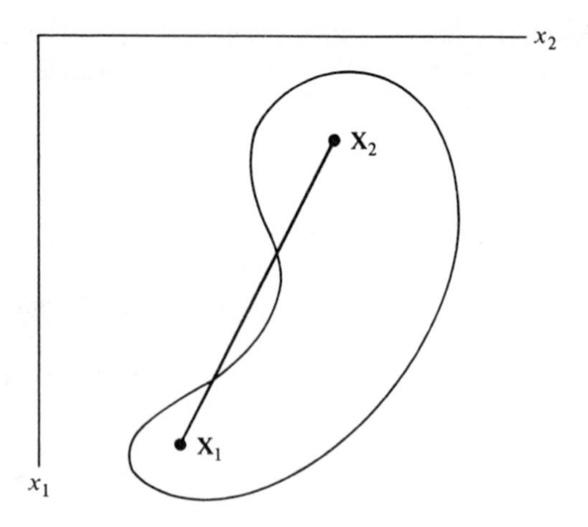

Figure 4.8. A Non-Convex Set

Suppose now that some point **X** of a convex set *cannot* be expressed as a linear combination of two other points, as given by Equation (4.3.1). Such a point is known as an *extreme point*. The vertices of the polygons in Figures 4.2–4.6 are extreme points.

A point **X** that is a combination of points $\mathbf{X}_1, \mathbf{X}_2, \ldots, \mathbf{X}_k$ in a convex set

(4.3.2) $$\mathbf{X} = \sum_{i=1}^{k} \mu_i \mathbf{X}_i, \qquad \mu_i \geq 0, \qquad i = 1, 2, \ldots, k,$$

where

(4.3.3) $$\sum_{i=1}^{k} \mu_i = 1,$$

is known as a *convex combination*. Such a point obviously is also a member of the convex set. The region containing all convex combinations of a finite number of points $\mathbf{X}_1, \mathbf{X}_2, \ldots, \mathbf{X}_k$ is known as a *convex polyhedron*. This definition is more formally correct than that presented in the last section, although it may be somewhat less appealing from an intuitive viewpoint.

Hyperplanes

Consider now an equation for a hyperplane in an n-dimensional space; e.g.,

(4.3.4) $$d_1 x_1 + d_2 x_2 + \cdots + d_n x_n = b$$

or, in vector notation,

$$\mathbf{D}'\mathbf{X} = b, \tag{4.3.5}$$

where

$$\mathbf{D} = \begin{bmatrix} d_1 \\ d_2 \\ \cdot \\ \cdot \\ \cdot \\ d_n \end{bmatrix}. \tag{4.3.6}$$

Choose two points $\mathbf{X}_1$ and $\mathbf{X}_2$ that are on the hyperplane; i.e., $\mathbf{D}'\mathbf{X}_1 = b$ and $\mathbf{D}'\mathbf{X}_2 = b$. Let us form the convex combination

$$\mathbf{X} = \beta\mathbf{X}_1 + (1 - \beta)\mathbf{X}_2, \qquad 0 < \beta < 1. \tag{4.3.7}$$

We can see that $\mathbf{X}$ is also a point on the hyperplane, since

$$\begin{aligned} \mathbf{D}'\mathbf{X} &= \mathbf{D}'\beta\mathbf{X}_1 + \mathbf{D}'(1 - \beta)\mathbf{X}_2 \\ &= \beta\mathbf{D}'\mathbf{X}_1 + (1 - \beta)\mathbf{D}'\mathbf{X}_2 \\ &= \beta b + (1 - \beta)b \\ &= b. \end{aligned} \tag{4.3.8}$$

Now consider a point $\mathbf{X}_1$ that lies on the hyperplane, i.e., $\mathbf{D}'\mathbf{X}_1 = b$; and a point $\mathbf{X}_2$ that lies in the half-space *above* the hyperplane, i.e., $\mathbf{D}'\mathbf{X}_2 > b$. Again forming a convex combination as given by Equation (4.3.7), we see that

$$\begin{aligned} \mathbf{D}'\mathbf{X} &= \beta\mathbf{D}'\mathbf{X}_1 + (1 - \beta)\mathbf{D}'\mathbf{X}_2 \\ &> \beta b + (1 - \beta)b \\ &> b. \end{aligned} \tag{4.3.9}$$

Hence we see that point $\mathbf{X}$ also lies above the hyperplane. We conclude, then, that *the half-space above or below a hyperplane is a convex set.* Moreover, it is easily established that *the space enclosed by the intersection of two or more hyperplanes is a convex set, as is the intersection of two or more half-planes.*

Feasible solutions

Now recall that the hyperplane represented by Equation (4.3.4) is equivalent to an equality constraint in the linear programming problem [cf., for example, Equation (4.1.2)]. The set of equality constraints represents the intersection of a set of hyperplanes and is therefore a convex set. Furthermore, a linear programming problem that contains both equality and inequality constraints produces a convex set, providing the half-spaces that correspond to the inequalities all intersect. We see, then, that a properly posed linear programming problem will give rise to feasible solutions which are contained within a closed convex set. Moreover, this convex set will be a convex polyhedron, providing the number of constraints is finite.

Let us now form the convex combination of feasible solutions $\mathbf{X}_1, \mathbf{X}_2, \ldots, \mathbf{X}_k$:

$$\mathbf{X} = \mu_1\mathbf{X}_1 + \mu_2\mathbf{X}_2 + \cdots + \mu_k\mathbf{X}_k, \qquad \mu_i \geq 0, \qquad i = 1, 2, \ldots, k, \tag{4.3.10}$$

and

$$\mu_1 + \mu_2 + \cdots + \mu_k = 1. \tag{4.3.11}$$

Let us form the product $\mathbf{AX}$. We can write the product as

$$\mathbf{AX} = \mu_1\mathbf{AX}_1 + \mu_2\mathbf{AX}_2 + \cdots + \mu_k\mathbf{AX}_k \tag{4.3.12}$$

$$= \mu_1\mathbf{B} + \mu_2\mathbf{B} + \cdots + \mu_k\mathbf{B} \tag{4.3.13}$$

$$= (\mu_1 + \mu_2 + \cdots + \mu_k)\mathbf{B} \tag{4.3.14}$$

$$= \mathbf{B}. \tag{4.3.15}$$

Hence a convex combination of feasible solutions is also a feasible solution, from which we conclude that *the collection of all feasible solutions to the linear programming problem is a convex set.*

Optimal feasible solutions

Now let $\mathbf{X}_1, \mathbf{X}_2, \ldots, \mathbf{X}_k$ be the *extreme* points of a convex set, and let $\mathbf{X}^*$ be the *minimum feasible solution.* Suppose that $\mathbf{X}^*$ is *not* an extreme point. We can then write $\mathbf{X}^*$ as a convex combination

$$\mathbf{X}^* = \sum_{i=1}^{k} \mu_i\mathbf{X}_i, \qquad \mu_i \geq 0, \qquad i = 1, 2, \ldots, k, \tag{4.3.16}$$

and

$$\sum_{i=1}^{k} \mu_i = 1, \tag{4.3.17}$$

and the objective function can be written

$$\mathbf{C}'\mathbf{X}^* = \mathbf{C}' \sum_{i=1}^{k} \mu_i\mathbf{X}_i = \sum_{i=1}^{k} \mu_i\mathbf{C}'\mathbf{X}_i. \tag{4.3.18}$$

Let us now replace each of the $\mathbf{X}_i$ in Equation (4.3.18) with the particular $\mathbf{X}_j$ that yields the minimum value of $\mathbf{C}'\mathbf{X}_i$. We can hence write

$$\mathbf{C}'\mathbf{X}^* \geq \sum_{i=1}^{k} \mu_i \mathbf{C}'\mathbf{X}_j = \mathbf{C}'\mathbf{X}_j \sum_{i=1}^{k} \mu_i = \mathbf{C}'\mathbf{X}_j. \tag{4.3.19}$$

Since we defined $\mathbf{X}^*$ to be the *minimum* feasible solution, however, we require that $\mathbf{C}'\mathbf{X}^* \leq \mathbf{C}'\mathbf{X}_j$. Therefore, Equation (4.3.19) can only be satisfied if

$$\mathbf{C}'\mathbf{X}^* = \mathbf{C}'\mathbf{X}_j. \tag{4.3.20}$$

We see that $\mathbf{X}_j$, which is an extreme point of the convex set, is a minimum feasible solution. The same type of reasoning can be applied to a maximum feasible solution, from which we conclude that *the feasible solutions that extremize the given objective function are extreme points of the convex set.* Recall that we have observed this in the two-dimensional problem discussed in the last section and represented in Figure 4.4.

Suppose now that $\mathbf{X}_1, \mathbf{X}_2, \ldots, \mathbf{X}_k$ are all feasible solutions that extremize the objective function; i.e., $\mathbf{C}'\mathbf{X}_1 = \mathbf{C}'\mathbf{X}_2 = \cdots = \mathbf{C}'\mathbf{X}_k$. Let us form the convex combination

$$\mathbf{X} = \sum_{i=1}^{k} \mu_i \mathbf{X}_i, \qquad \mu_i \geq 0, \qquad i = 1, 2, \ldots, k, \tag{4.3.21}$$

and

$$\sum_{i=1}^{k} \mu_i = 1. \tag{4.3.22}$$

We can write the objective function as

$$\mathbf{C}'\mathbf{X} = \sum_{i=1}^{k} \mu_i \mathbf{C}'\mathbf{X}_i. \tag{4.3.23}$$

Since all the $\mathbf{C}'\mathbf{X}_i$ on the right-hand side of Equation (4.3.23) are equal, however, we can replace each of these terms with any term, say $\mathbf{C}'\mathbf{X}_j$, so that

$$\mathbf{C}'\mathbf{X} = \mathbf{C}'\mathbf{X}_j \sum_{i=1}^{k} \mu_i = \mathbf{C}'\mathbf{X}_j. \tag{4.3.24}$$

Thus we see that *if the objective function is extremized at more than one point, then the objective function takes on the same extremum value at any convex combination of the extremum points.* We have encountered such a situation in the example problem of the last section, which is represented in Figure 4.6.

Now let us consider the expression

$$x_1\mathbf{P}_1 + x_2\mathbf{P}_2 + \cdots + x_k\mathbf{P}_k = \mathbf{P}_0 \tag{4.3.25}$$

where $x_i \geq 0$, $i = 1, 2, \ldots, k$; the vectors $\mathbf{P}_1, \mathbf{P}_2, \ldots, \mathbf{P}_k$ are linearly independent; and k is the number of linearly independent vectors required to represent the given vector $\mathbf{P}_0$, $k \leq m$.† Recall that this is a way to express the constraints of a linear programming problem [cf. Equation (4.1.10)]. We can form an n-dimensional vector $\mathbf{X}$ from the coefficients in Equation (4.3.25) as

$$\mathbf{X} = \begin{bmatrix} x_1 \\ x_2 \\ \cdot \\ \cdot \\ \cdot \\ x_k \\ 0 \\ \cdot \\ \cdot \\ \cdot \\ 0 \end{bmatrix}. \tag{4.3.26}$$

Let $\mathbf{X}$ be a convex combination of the feasible solutions $\mathbf{X}_1$ and $\mathbf{X}_2$; i.e.,

$$\mathbf{X} = \beta\mathbf{X}_1 + (1 - \beta)\mathbf{X}_2, \qquad 0 < \beta < 1. \tag{4.3.27}$$

Since $\mathbf{X}_1$ and $\mathbf{X}_2$ are feasible solutions, we have $x_{i1} \geq 0$ and $x_{i2} \geq 0$, $i = 1, 2, \ldots, n$. However, we have defined $\mathbf{X}$ in such a manner that $x_i = 0$ for $i = k + 1, k + 2, \ldots, n$. From Equation (4.3.27) we see that x_{i1} and x_{i2} must also equal zero for $i = k + 1$, $k + 2, \ldots, n$. *Hence the number of nonzero components in any feasible solution vector cannot exceed* m, *the number of rows in the constraint matrix.*

We can write the constraints in terms of $\mathbf{X}_1$ and $\mathbf{X}_2$ as

$$\mathbf{A}\mathbf{X}_1 = \mathbf{B}, \tag{4.3.28}$$

$$\mathbf{A}\mathbf{X}_2 = \mathbf{B}, \tag{4.3.29}$$

which in vector notation becomes

$$x_{11}\mathbf{P}_1 + x_{21}\mathbf{P}_2 + \cdots + x_{k1}\mathbf{P}_k = \mathbf{P}_0, \tag{4.3.30}$$

$$x_{12}\mathbf{P}_1 + x_{22}\mathbf{P}_2 + \cdots + x_{k2}\mathbf{P}_k = \mathbf{P}_0. \tag{4.3.31}$$

† A set of vectors $\mathbf{Y}_1, \mathbf{Y}_2, \ldots, \mathbf{Y}_k$ are said to be *linearly independent* if and only if the expression $\alpha_1\mathbf{Y}_1 + \alpha_2\mathbf{Y}_2 + \cdots + \alpha_k\mathbf{Y}_k = 0$ only when $\alpha_1 = \alpha_2 = \cdots = \alpha_k = 0$. This implies that no single vector $\mathbf{Y}_i$ is a multiple of some linear combination of the remaining $k - 1$ vectors.

Subtracting Equation (4.3.31) from (4.3.30) gives

$$(x_{11} - x_{12})\mathbf{P}_1 + (x_{21} - x_{22})\mathbf{P}_2 + \cdots + (x_{k1} - x_{k2})\mathbf{P}_k = \mathbf{0}. \qquad (4.3.32)$$

Since the vectors $\mathbf{P}_1, \mathbf{P}_2, \ldots, \mathbf{P}_k$ are linearly independent, however, we require that $(x_{i1} - x_{i2}) = 0$, $i = 1, 2, \ldots, k$, so that $\mathbf{X}_1 = \mathbf{X}_2 = \mathbf{X}$. Since $\mathbf{X}$ cannot be expressed as a convex combination of two *distinct* feasible solutions, we conclude that *the n-dimensional vector whose nonzero elements are the coefficients of the expression*

$$x_1\mathbf{P}_1 + x_2\mathbf{P}_2 + \cdots + x_k\mathbf{P}_k = \mathbf{P}_0, \qquad x_i \geq 0, \qquad i = 1, 2, \ldots, k,$$

and k $\leq$ m, *is an extreme point of the convex set of all feasible solutions, providing* $\mathbf{P}_1, \mathbf{P}_2, \ldots, \mathbf{P}_k$ *are linearly independent.*

Suppose now that $\mathbf{X}$ is an extreme point of the convex set of all feasible solutions and that $x_i \geq 0$, $i = 1, 2, \ldots, k$, and $x_i = 0$, $i = k + 1, k + 2, \ldots, n$. Consider the expression

$$x_1\mathbf{P}_1 + x_2\mathbf{P}_2 + \cdots + x_k\mathbf{P}_k = \mathbf{P}_0. \qquad (4.3.33)$$

We shall for the time being assume that $\mathbf{P}_1, \mathbf{P}_2, \ldots, \mathbf{P}_k$ are linearly *dependent.* Hence we can write

$$\mu_1\mathbf{P}_1 + \mu_2\mathbf{P}_2 + \cdots + \mu_k\mathbf{P}_k = \mathbf{0}, \qquad (4.3.34)$$

where at least one $\mu_i \neq 0$. However, we can also write

$$x_1\mathbf{P}_1 + x_2\mathbf{P}_2 + \cdots + x_k\mathbf{P}_k = \mathbf{P}_0. \qquad (4.3.35)$$

Let us now multiply Equation (4.3.34) by some $\epsilon > 0$ and add and subtract this result from Equation (4.3.35). This results in

$$(x_1 + \epsilon\mu_1)\mathbf{P}_1 + (x_2 + \epsilon\mu_2)\mathbf{P}_2 + \cdots + (x_k + \epsilon\mu_k)\mathbf{P}_k = \mathbf{P}_0, \qquad (4.3.36)$$

$$(x_1 - \epsilon\mu_1)\mathbf{P}_1 + (x_2 - \epsilon\mu_2)\mathbf{P}_2 + \cdots + (x_k - \epsilon\mu_k)\mathbf{P}_k = \mathbf{P}_0. \qquad (4.3.37)$$

We shall choose ϵ sufficiently small so that all the coefficients in Equation (4.3.37) are nonnegative. Thus X′ and X″ are both feasible solutions, where

$$\mathbf{X}' = \begin{bmatrix} x_1 + \epsilon\mu_1 \\ x_2 + \epsilon\mu_2 \\ \cdot \\ \cdot \\ \cdot \\ x_k + \epsilon\mu_k \end{bmatrix}, \qquad \mathbf{X}'' = \begin{bmatrix} x_1 - \epsilon\mu_1 \\ x_2 - \epsilon\mu_2 \\ \cdot \\ \cdot \\ \cdot \\ x_k - \epsilon\mu_k \end{bmatrix}, \qquad (4.3.38)$$

and the given vector **X** can be expressed as the convex combination

$$\mathbf{X} = \tfrac{1}{2}(\mathbf{X}' + \mathbf{X}''). \tag{4.3.39}$$

However, Equation (4.3.39) violates our original supposition that **X** is an extreme point. Therefore, the vectors $\mathbf{P}_1, \mathbf{P}_2, \ldots, \mathbf{P}_k$ cannot be linearly dependent. Moreover, we require that $k \leq m$, since the vectors $\mathbf{P}_1, \mathbf{P}_2, \ldots, \mathbf{P}_k$ are m-dimensional, as is $\mathbf{P}_0$, and the vector $\mathbf{P}_0$ cannot be expressed in terms of more than m linearly independent vectors. Hence the number of nonzero members of the extreme point **X** cannot exceed m, the number of constraints. To summarize, *the vectors* $\mathbf{P}_1, \mathbf{P}_2, \ldots, \mathbf{P}_k$ *in the expression* $x_1\mathbf{P}_1 + x_2\mathbf{P}_2 + \cdots + x_k\mathbf{P}_k = \mathbf{P}_0$ *are linearly independent, providing* **X** *is an extreme point of the convex set of all feasible solutions. Furthermore, not more than* m *members of* **X** *can be greater than zero.*

From the foregoing analysis we see that the solution to a linear programming problem is an extreme point of a convex set, providing that the solution exists. Moreover, the solution vector will have no more than m linearly independent vectors associated with it, where m is the total number of constraints. Therefore, we seek a systematic method which will successively choose adjacent extreme points in such a manner that the objective function monotonically increases or decreases. The simplex algorithm, which we shall develop in the next section, is an efficient method for carrying out such a sequence of moves.

4.4 The Simplex Algorithm

In the last section, we saw that the optimal feasible solution to the linear programming problem is an extreme point of the convex set of all feasible solutions. Furthermore, we established that **X** is an extreme point of the convex set of all feasible solutions if and only if there are no more than m linearly independent vectors $\mathbf{P}_1, \mathbf{P}_2, \ldots, \mathbf{P}_k$, such that

$$x_1\mathbf{P}_1 + x_2\mathbf{P}_2 + \cdots + x_k\mathbf{P}_k = \mathbf{P}_0, \qquad k \leq m, \tag{4.4.1}$$

where the coefficients in Equation (4.4.1) are the positive components of the n-dimensional vector **X**, the remaining $(n - k)$ components being zero.

We seek the particular subset of k linearly independent vectors, $k \leq m$, out of the set of n vectors $\mathbf{P}_1, \mathbf{P}_2, \ldots, \mathbf{P}_n$ resulting from the constraints of the linear programming problem [cf. Equation (4.1.10)], whose associated values of x_i will cause the objective function

$$y = \mathbf{C}'\mathbf{X} \tag{4.4.2}$$

to be extremized subject to the given constraints. Hence we desire an algorithm that will allow us to search systematically from one set of k linearly independent vectors to another until the associated value of $\mathbf{X}$ which will extremize Equation (4.4.2) is found.

Let us assume that the objective function is to be minimized. (Recall that we can always convert a maximization problem into an equivalent minimization problem.) Furthermore, we shall for the time being assert that every extreme point $\mathbf{X}$ has exactly m linearly independent vectors associated with it. We shall discuss the case where $k < m$ in a later section.

Changing a basis vector

Suppose that we have a set of m linearly independent vectors $\mathbf{P}_1, \mathbf{P}_2, \ldots, \mathbf{P}_m$ such that

$$x_{10}\mathbf{P}_1 + x_{20}\mathbf{P}_2 + \cdots + x_{m0}\mathbf{P}_m = \mathbf{P}_0, \tag{4.4.3}$$

where the x_{i0} are the positive components of a given extreme point $\mathbf{X}_0$. We can express any other vector, say $\mathbf{P}_j$, of the original set of n vectors as a linear combination of the given m vectors:

$$\mathbf{P}_j = x_{1j}\mathbf{P}_1 + x_{2j}\mathbf{P}_2 + \cdots x_{mj}\mathbf{P}_m, \tag{4.4.4}$$

where x_{ij} is the ith component of $\mathbf{P}_j$, and at least one $x_{ij} \neq 0$. In other words, the given set of m linearly independent vectors forms a *basis* for the m-dimensional vector space.

Let us now replace one of the basis vectors in Equation (4.4.3), say $\mathbf{P}_l$, with $\mathbf{P}_j$. Solving Equation (4.4.4) for $\mathbf{P}_l$, with $x_{lj} \neq 0$,

$$\mathbf{P}_l = \frac{1}{x_{lj}}[\mathbf{P}_j - x_{1j}\mathbf{P}_1 - x_{2j}\mathbf{P}_2 - \cdots - x_{(l-1)j}\mathbf{P}_{(l-1)} - x_{(l+1)j}\mathbf{P}_{(l+1)} - \cdots - x_{mj}\mathbf{P}_m]. \tag{4.4.5}$$

Substituting into Equation (4.4.3) gives

$$x_{10}\mathbf{P}_1 + x_{20}\mathbf{P}_2 + \cdots + x_{(l-1)0}\mathbf{P}_{(l-1)} + x_{(l+1)0}\mathbf{P}_{(l+1)} + \cdots + x_{m0}\mathbf{P}_m + \frac{x_{l0}}{x_{lj}}[\mathbf{P}_j - x_{1j}\mathbf{P}_1 - x_{2j}\mathbf{P}_2 - \cdots - x_{(l-1)j}\mathbf{P}_{(l-1)} - x_{(l+1)j}\mathbf{P}_{(l+1)} - \cdots - x_{mj}\mathbf{P}_m] = \mathbf{P}_0. \tag{4.4.6}$$

Rearranging Equation (4.4.6), we have

$$\left(x_{10} - x_{1j}\frac{x_{l0}}{x_{lj}}\right)\mathbf{P}_1 + \left(x_{20} - x_{2j}\frac{x_{l0}}{x_{lj}}\right)\mathbf{P}_2 + \cdots$$
$$+ \left[x_{(l-1)0} - x_{(l-1)j}\frac{x_{l0}}{x_{lj}}\right]\mathbf{P}_{(l-1)} + \left[x_{(l+1)0} - x_{(l+1)j}\frac{x_{l0}}{x_{lj}}\right]\mathbf{P}_{(l+1)} + \cdots$$
$$+ \left(x_{m0} - x_{mj}\frac{x_{l0}}{x_{lj}}\right)\mathbf{P}_m + \frac{x_{l0}}{x_{lj}}\mathbf{P}_j = \mathbf{P}_0.$$

(4.4.7)

Equation (4.4.7) again expresses the vector $\mathbf{P}_0$ in terms of m vectors, but the original vector $\mathbf{P}_l$ has now been replaced by $\mathbf{P}_j$. Also, the coefficients are different. We can easily see that the new vectors are linearly independent, for if they were not linearly independent, we could write

$$\alpha_1\mathbf{P}_1 + \alpha_2\mathbf{P}_2 + \cdots + \alpha_{(l-1)}\mathbf{P}_{(l-1)} + \alpha_{(l+1)}\mathbf{P}_{(l+1)} + \cdots + \alpha_m\mathbf{P}_m + \alpha_j\mathbf{P}_j = \mathbf{0}$$

(4.4.8)

where at least one of the $\alpha_i \neq 0$. Solving Equation (4.4.8) for $\mathbf{P}_j$, where $\alpha_j \neq 0$, we have

$$\mathbf{P}_j = \frac{1}{\alpha_j}[\alpha_1\mathbf{P}_1 + \alpha_2\mathbf{P}_2 + \cdots + \alpha_{(l-1)}\mathbf{P}_{(l-1)} + \alpha_{(l+1)}\mathbf{P}_{(l+1)} + \cdots + \alpha_m\mathbf{P}_m].$$

(4.4.9)

Subtracting this result from Equation (4.4.4) gives

$$\left(x_{1j} - \frac{\alpha_1}{\alpha_j}\right)\mathbf{P}_1 + \left(x_{2j} - \frac{\alpha_2}{\alpha_j}\right)\mathbf{P}_2 + \cdots + \left[x_{(l-1)j} - \frac{\alpha_{(l-1)}}{\alpha_j}\right]\mathbf{P}_{(l-1)}$$
$$+ x_{lj}\mathbf{P}_l + \left[x_{(l+1)j} - \frac{\alpha_{(l+1)}}{\alpha_j}\right]\mathbf{P}_{(l+1)} + \cdots + \left(x_{mj} - \frac{\alpha_m}{\alpha_j}\right)\mathbf{P}_m = \mathbf{0}.$$

(4.4.10)

However, the vectors $\mathbf{P}_1, \mathbf{P}_2, \ldots, \mathbf{P}_m$ appearing in Equation (4.4.10) were defined as being linearly independent, and we asserted that $x_{lj} \neq 0$. Hence we have a contradiction, and our supposition that the vectors appearing in Equation (4.4.7) are linearly dependent is incorrect. These vectors therefore form a new basis in the m-dimensional vector space.

Since we now have a new set of m linearly independent vectors, the coefficients appearing in Equation (4.4.7) are the elements of a new extreme point of the set of all feasible solutions, providing each of these coefficients is positive; i.e.,

$$\left(x_{i0} - x_{ij}\frac{x_{l0}}{x_{lj}}\right) > 0, \qquad i = 1, 2, \ldots, (l-1), (l+1), \ldots, m, \tag{4.4.11}$$

and

$$\frac{x_{l0}}{x_{lj}} > 0. \tag{4.4.12}$$

Notice that Equations (4.4.11) and (4.4.12) are written as strict inequalities, because we require that $\mathbf{P}_0$ be expressed in terms of exactly m linearly independent vectors.

We require that $x_{lj} > 0$ to satisfy Equation (4.4.12). If $x_{ij} \leq 0$ for $i \neq l$, then all the coefficients expressed by Equation (4.4.11) will always be positive. If $x_{ij} > 0$ for $i \neq l$, we can still satisfy Equation (4.4.11), providing

$$\frac{x_{l0}}{x_{lj}} < \frac{x_{i0}}{x_{ij}}. \tag{4.4.13}$$

Hence *the vector* $\mathbf{P}_l$*, which is eliminated from the original basis, is not arbitrary, but is that particular vector which yields a minimum value for* x_{i0}/x_{ij} *in accordance with Equations* (4.4.12) *and* (4.4.13).

To summarize, we have replaced one of the vectors in our original basis with one of the $(n - m)$ vectors originally not in the basis. Moreover, the coefficients of the expression for $\mathbf{P}_0$ in terms of the new basis are the nonzero components of a new extreme point of the convex set of all feasible solutions.

Changing the objective function

Let us now examine the effect of such a vector substitution on the objective function. Suppose that the function

$$x_{10}c_1 + x_{20}c_2 + \cdots + x_{m0}c_m = y_0 \tag{4.4.14}$$

is the objective function corresponding to Equation (4.4.3), and suppose that, by means of Equation (4.4.4), we express a vector $\mathbf{P}_j$ in terms of the original basis vectors. Then the function

$$x_{1j}c_1 + x_{2j}c_2 + \cdots + x_{mj}c_m = y_j \tag{4.4.15}$$

may be considered to be the equivalent cost of a unit of j. Let us eliminate one of the coefficients, say c_l, from Equations (4.4.14) and (4.4.15). Solution of Equation (4.4.15) for c_l and substitution into Equation (4.4.14) yields, after some rearrangement,

$$\left(x_{10} - x_{1j}\frac{x_{l0}}{x_{lj}}\right)c_1 + \left(x_{20} - x_{2j}\frac{x_{l0}}{x_{lj}}\right)c_2 + \cdots$$
$$+ \left[x_{(l-1)0} - x_{(l-1)j}\frac{x_{l0}}{x_{lj}}\right]c_{(l-1)} + \left[x_{(l+1)0} - x_{(l+1)j}\frac{x_{l0}}{x_{lj}}\right]c_{(l+1)} + \cdots$$
$$+ \left(x_{m0} - x_{mj}\frac{x_{l0}}{x_{lj}}\right)c_m = y_0 - \frac{x_{l0}}{x_{lj}}y_j.$$

(4.4.16)

If we add the quantity $x_{l0}c_j/x_{lj}$ to both sides of Equation (4.4.16), we obtain

$$\left(x_{10} - x_{1j}\frac{x_{l0}}{x_{lj}}\right)c_1 + \left(x_{20} - x_{2j}\frac{x_{l0}}{x_{lj}}\right)c_2 + \cdots$$
$$+ \left[x_{(l-1)0} - x_{(l-1)j}\frac{x_{l0}}{x_{lj}}\right)c_{(l-1)} + \left[x_{(l+1)0} - x_{(l+1)j}\frac{x_{l0}}{x_{lj}}\right]c_{(l+1)} + \cdots$$
$$+ \left(x_{m0} - x_{mj}\frac{x_{l0}}{x_{lj}}\right)c_m + \frac{x_{l0}}{x_{lj}}c_j = y_0 - \frac{x_{l0}}{x_{lj}}(y_j - c_j).$$

(4.4.17)

Notice that Equation (4.4.17) is the expression for the objective function which corresponds to the extreme point associated with the new basis vectors [see Equation (4.4.7)].

We have already seen that all the coefficients in Equation (4.4.17) must be positive if they are to be the components of an extreme point of our convex set. For these conditions Equation (4.4.17) shows us that *the new value of the objective function, obtained by replacing one of the vectors in the original basis, has a value less than the original objective function, providing* $(y_j - c_j) > 0$ *and* $x_{lj} > 0$. We can continue to choose new vectors $\mathbf{P}_j$ and enter them into the basis as long as $(y_j - c_j) > 0$ for each j. In this manner we successively decrease the value of the objective function. The procedure can be continued until we reach a situation where $(y_j - c_j) \leq 0$ for all remaining j; at this point we will have minimized our objective function, subject to the given constraints and nonnegativity conditions, and therefore solved our problem.

Multiple solutions

Once the optimum has been obtained, it is not unusual that $(y_j - c_j) = 0$ for one or more $\mathbf{P}_j$ not in the final basis. If such a vector is entered into the basis, replacing some vector $\mathbf{P}_l$, then clearly the value of the objective function will be unchanged, although we shall have found a different extreme point. Thus *we have a criterion for recognizing the existence of multiple solu-*

tion points; i.e., $(y_j - c_j) = 0$ *for one or more nonbasis vectors at an optimum.* All the multiple solution points can be found by entering the appropriate vectors $\mathbf{P}_j$ into the basis. Hadley [1962] suggests a systematic procedure for carrying out the computation.

Unbounded solutions

In our earlier discussion of Equation (4.4.12) we saw that the coefficients in Equation (4.4.7), which also appear in Equation (4.4.17), can all be positive only if $x_{ij} > 0$ for at least one i in Equation (4.4.4); the term x_{lj} in Equation (4.4.17) is that particular $x_{ij} > 0$ which satisfies Equations (4.4.12) and (4.4.13). If some of the other $x_{ij} > 0$, then the choice of x_{lj} in accordance with Equation (4.4.13) assures us that all the coefficients in Equations (4.4.7) and (4.4.17) will be positive. Furthermore, these coefficients will *always* be positive if none of the remaining $x_{ij} > 0$, $i \neq l$. Let us now investigate the result of a change in the basis if *all* the $x_{ij} \leq 0$ in Equation (4.4.4).

Suppose that we multiply Equation (4.4.4) by some positive constant θ and subtract the result from Equation (4.4.3). This gives us

$$(x_{10} - \theta x_{1j})\mathbf{P}_1 + (x_{20} - \theta x_{2j})\mathbf{P}_2 + \cdots + (x_{m0} - \theta x_{mj})\mathbf{P}_m + \theta \mathbf{P}_j = \mathbf{P}_0. \tag{4.4.18}$$

All the coefficients in the above expression will be positive if all the $x_{ij} \leq 0$, $i = 1, 2, \ldots, m$. Notice that we now have expressed the vector $\mathbf{P}_0$ in terms of the $m + 1$ vectors $\mathbf{P}_1, \mathbf{P}_2, \ldots, \mathbf{P}_m$ and $\mathbf{P}_j$. Hence for any $\theta > 0$ the vector $\mathbf{X}$ associated with Equation (4.4.18) has $m + 1$ positive components and therefore is not an extreme point of the convex set of all feasible solutions.

In a similar fashion let us multiply Equation (4.4.15) by some $\theta > 0$ and subtract the result from Equation (4.4.14). We obtain

$$(x_{10} - \theta x_{1j})c_1 + (x_{20} - \theta x_{2j})c_2 + \cdots + (x_{m0} - \theta x_{mj})c_m = y_0 - \theta y_j. \tag{4.4.19}$$

Adding the term θc_j to each side of Equation (4.4.19), we have

$$\begin{aligned}(x_{10} - \theta x_{1j})c_1 &+ (x_{20} - \theta x_{2j})c_2 + \cdots \\ &+ (x_{m0} - \theta x_{mj})c_m + \theta c_j = y_0 - \theta(y_j - c_j).\end{aligned} \tag{4.4.20}$$

If we again assert that $x_{ij} \leq 0$, $i = 1, 2, \ldots, m$, we see that the coefficients of Equation (4.4.20) are positive for any $\theta > 0$, and if $(y_j - c_j) > 0$ the new value of the objective function can be made arbitrarily small by allowing θ to become sufficiently large. In other words, since θ can be arbitrarily large when $x_{ij} \leq 0$, $i = 1, 2, \ldots, m$, the value of the objective function that solves the linear programming problem is infinite. Thus we see that *the re-*

peated application of the simplex algorithm indicates the presence of an unbounded solution to the linear programming minimization problem if, at some stage in the computation, all $x_{ij} \leq 0$, $i = 1, 2, \ldots, m$, *and* $(y_j - c_j) > 0$ *for some nonbasis vector* $\mathbf{P}_j$.

Summary of computational procedure

We can summarize our development of the simplex algorithm as follows: We wish to replace some vector $\mathbf{P}_l$ in the basis by some other vector $\mathbf{P}_j$ that is not in the basis. The vector $\mathbf{P}_j$ which enters the basis is selected as that vector corresponding to the largest positive value* of $(y_j - c_j)$, providing at least one $x_{ij} > 0$, $i = 1, 2, \ldots, m$ [cf. Equations (4.4.4) and (4.4.15)]. If, however, there exists some vector $\mathbf{P}_q$ not originally in the basis for which $(y_q - c_q) > 0$ but $x_{iq} \leq 0$ for all i, then the solution to the linear programming problem is unbounded and the computation ceases. The vector $\mathbf{P}_l$ which is eliminated from the basis (providing the solution is not unbounded) is that vector for which $x_{lj} > 0$ and $x_{l0}/x_{lj} < x_{i0}/x_{ij}$, $x_{ij} > 0$ and $x_{ij} \neq x_{lj}$ [cf. Equation (4.4.7)]. Each application of the simplex algorithm results in a new extreme point of the convex set of all feasible solutions. The procedure is continued until one cannot find any vectors $\mathbf{P}_j$ outside the basis with $(y_j - c_j) > 0$. When all $(y_j - c_j) \leq 0$, then the coefficients that appear in the expression for $\mathbf{P}_0$ [cf. Equation (4.4.3)] are the nonzero components of the minimum feasible solution vector, and the problem is then solved. If at the optimum $(y_j - c_j) = 0$ for one or more $\mathbf{P}_j$ not in the basis, then alternative solution vectors can be found by entering such $\mathbf{P}_j$ into the basis.

Unit basis vectors

Thus far we have said nothing about the vectors which appear in the basis except that there be no more than m vectors and that they be linearly independent. Suppose, however, that we choose a set of m *unit* vectors as a basis. That is, let

$$(4.4.21) \qquad \mathbf{P}_1 = \begin{bmatrix} 1 \\ 0 \\ \cdot \\ \cdot \\ \cdot \\ 0 \end{bmatrix}, \qquad \mathbf{P}_2 = \begin{bmatrix} 0 \\ 1 \\ \cdot \\ \cdot \\ \cdot \\ 0 \end{bmatrix}, \ldots, \qquad \mathbf{P}_m = \begin{bmatrix} 0 \\ 0 \\ \cdot \\ \cdot \\ \cdot \\ 1 \end{bmatrix}.$$

---

* From Equation (4.4.17) we see that the choice of the vector $\mathbf{P}_j$ to enter the basis should be that vector corresponding to the maximum value of $(x_{l0}/x_{lj})(y_j - c_j)$. Since the computation of all values of this quantity may be tedious, however, we choose the vector corresponding simply to $\max(y_j - c_j)$.

It is easy to show that these vectors are linearly independent. Moreover, if we express some vector not in the basis, say $\mathbf{P}_j$, in terms of the unit basis vectors, then

$$\mathbf{P}_j = x_{1j}\mathbf{P}_1 + x_{2j}\mathbf{P}_2 + \cdots + x_{mj}\mathbf{P}_m, \tag{4.4.22}$$

which can be written as

$$\mathbf{P}_j = x_{1j}\begin{bmatrix} 1 \\ 0 \\ \cdot \\ \cdot \\ \cdot \\ 0 \end{bmatrix} + x_{2j}\begin{bmatrix} 0 \\ 1 \\ \cdot \\ \cdot \\ \cdot \\ 0 \end{bmatrix} + \cdots + x_{mj}\begin{bmatrix} 0 \\ 0 \\ \cdot \\ \cdot \\ \cdot \\ 1 \end{bmatrix}. \tag{4.4.23}$$

However, $\mathbf{P}_j$ can be expressed in terms of unit vectors only if

$$\mathbf{P}_j = \begin{bmatrix} p_{1j} \\ p_{2j} \\ \cdot \\ \cdot \\ \cdot \\ p_{mj} \end{bmatrix} = p_{1j}\begin{bmatrix} 1 \\ 0 \\ \cdot \\ \cdot \\ \cdot \\ 0 \end{bmatrix} + p_{2j}\begin{bmatrix} 0 \\ 1 \\ \cdot \\ \cdot \\ \cdot \\ 0 \end{bmatrix} + \cdots + p_{mj}\begin{bmatrix} 0 \\ 0 \\ \cdot \\ \cdot \\ \cdot \\ 1 \end{bmatrix}. \tag{4.4.24}$$

Thus we see that $x_{1j} = p_{1j}, x_{2j} = p_{2j}, \ldots, x_{mj} = p_{mj}$. In particular, we can write

$$\begin{aligned} \mathbf{P}_0 &= x_{10}\mathbf{P}_1 + x_{20}\mathbf{P}_2 + \cdots + x_{m0}\mathbf{P}_m \\ &= p_{10}\mathbf{P}_1 + p_{20}\mathbf{P}_2 + \cdots + p_{m0}\mathbf{P}_m. \end{aligned} \tag{4.4.25}$$

In other words, *the components of the vector* $\mathbf{P}_0$ *are equal to the nonzero components of the extreme point* $\mathbf{X}_0$, *providing the basis consists of* m *linearly independent unit vectors.* We shall see certain computational advantages to the use of unit basis vectors in the next example.

Finally, we could have developed the simplex algorithm from the alternative viewpoint of maximizing an objective function if we had wished. All the arguments would have been identical to those developed previously, except that the criterion for selecting a new vector would have been the following: Select the vector $\mathbf{P}_j$ corresponding to the largest *negative* value of $(y_j - c_j)$, i.e., the minimum value of $(y_j - c_j) < 0$, with at least one $x_{ij} > 0$. The computation would cease when all $(y_j - c_j) \geq 0$ or when some $(y_j - c_j) < 0$ with all $x_{ij} \leq 0$. Other than these changes, the computational scheme for obtaining a maximum feasible solution is the same as that for a minimum feasible solution.

EXAMPLE 4.4.1

As an example of how the simplex algorithm is applied to a given problem, let us again consider one of the linear programming problems that we solved graphically in Section 4.2; i.e., maximize the function

$$y = x_1 + x_2,$$

subject to the constraints

$$x_1 + 2x_2 \leq 6$$
$$2x_1 + x_2 \leq 8$$

and the nonnegativity conditions $x_1, x_2 \geq 0$. (The reader is again referred to Figure 4.4 for a graphical representation of this problem.)

Adding the slack variables x_3 and x_4 to the constraints in order to obtain strict equality constraints, the problem becomes: Maximize

$$y = x_1 + x_2,$$

subject to

$$x_1 + 2x_2 + x_3 \qquad = 6,$$
$$2x_1 + x_2 \qquad + x_4 = 8,$$

and $x_i \geq 0$, $i = 1, 2, 3, 4$. In vector notation, the problem can be stated as: Maximize

$$y = x_1 + x_2,$$

subject to

$$x_1\mathbf{P}_1 + x_2\mathbf{P}_2 + x_3\mathbf{P}_3 + x_4\mathbf{P}_4 = \mathbf{P}_0$$

and $x_i \geq 0$, $i = 1, 2, 3, 4$, where

$$\mathbf{P}_1 = \begin{bmatrix} 1 \\ 2 \end{bmatrix}, \quad \mathbf{P}_2 = \begin{bmatrix} 2 \\ 1 \end{bmatrix}, \quad \mathbf{P}_3 = \begin{bmatrix} 1 \\ 0 \end{bmatrix}, \quad \mathbf{P}_4 = \begin{bmatrix} 0 \\ 1 \end{bmatrix}, \quad \mathbf{P}_0 = \begin{bmatrix} 6 \\ 8 \end{bmatrix}.$$

Table 4.1. TABULAR REPRESENTATION OF INITIAL BASIS—EXAMPLE 4.4.1

	c		1	1	0	0
Basis		$\mathbf{P}_0$	$\mathbf{P}_1$	$\mathbf{P}_2$	$\mathbf{P}_3$	$\mathbf{P}_4$
$\mathbf{P}_3$	0	6	1	2	1	0
$\mathbf{P}_4$	0	8	2	1	0	1
	$(y_j - c_j)$	0	−1	−1	0	0

It is convenient to show each iteration of the computational procedure in tabular form, as in Table 4.1. Here we have chosen the unit vectors $\mathbf{P}_3$ and $\mathbf{P}_4$ as our initial basis, since these vectors are clearly linearly independent. Each column labeled $\mathbf{P}_0, \mathbf{P}_1, \ldots, \mathbf{P}_4$ contains the components of the corresponding vector. The column labeled c contains the cost coefficients corresponding to the vectors in the basis, and the row labeled c contains the cost coefficients assoicated with the vectors in the row beneath.

Notice that we have chosen a set of *unit* vectors as the initial basis.* Therefore, the components of any nonbasis vector $\mathbf{P}_j$ are equal to the coefficients x_{ij} of the basis vectors. In this example

$$\mathbf{P}_0 = \begin{bmatrix} 6 \\ 8 \end{bmatrix} = x_{30}\mathbf{P}_3 + x_{40}\mathbf{P}_4 = 6\begin{bmatrix} 1 \\ 0 \end{bmatrix} + 8\begin{bmatrix} 0 \\ 1 \end{bmatrix},$$

$$\mathbf{P}_1 = \begin{bmatrix} 1 \\ 2 \end{bmatrix} = x_{31}\mathbf{P}_3 + x_{41}\mathbf{P}_4 = 1\begin{bmatrix} 1 \\ 0 \end{bmatrix} + 2\begin{bmatrix} 0 \\ 1 \end{bmatrix},$$

and

$$\mathbf{P}_2 = \begin{bmatrix} 2 \\ 1 \end{bmatrix} = x_{32}\mathbf{P}_3 + x_{42}\mathbf{P}_4 = 2\begin{bmatrix} 1 \\ 0 \end{bmatrix} + 1\begin{bmatrix} 0 \\ 1 \end{bmatrix}.$$

Since we know the x_{ij} for all the nonbasis vectors, we can easily compute the values of y_j and $(y_j - c_j)$ for these vectors. Thus

$$\begin{aligned} y_0 &= c_3x_{30} + c_4x_{40} \\ &= (0)(6) + (0)(8) = 0, \\ y_1 - c_1 &= c_3x_{31} + c_4x_{41} - c_1 \\ &= (0)(1) + (0)(2) - 1 = -1, \\ y_2 - c_2 &= c_3x_{32} + c_4x_{42} - c_2 \\ &= (0)(2) + (0)(1) - 1 = -1. \end{aligned}$$

In Table 4.1, since the basis vectors are *unit* vectors, any y_j can be obtained by summing the product of an entry in the $\mathbf{P}_j$ column and a corresponding (by row) entry in the basis cost column, for all rows. Also the value of c_j is given at the top of the $\mathbf{P}_j$ column. Hence it is very easy to obtain y_0 and $(y_j - c_j)$ directly from the table. The quantities are entered in the bottom row of Table 4.1. Notice that $(y_j - c_j)$ is set equal to zero for the vectors which are in the basis. Also notice that c_0 is undefined, so that the last entry in the $\mathbf{P}_0$ column is simply y_0, the prevailing value of the objective function.

We are now ready to select a new vector for the basis. Since the objective function is to be maximized, we seek the largest negative value for $(y_j - c_j)$. In this case both $(y_1 - c_1)$ and $(y_2 - c_2)$ equal -1, so we shall arbitrarily decide to pick

* The appearance of unit vectors results from the introduction of slack variables into this particular problem. Not all linear programming problems involve unit vectors. As we shall see in the next section, however, it is always possible to introduce a set of unit vectors artificially.

the value of $(y_j - c_j)$ having the lowest value of j in case of a tie. Hence we choose $\mathbf{P}_1$ as the new vector to enter the basis.

The vector to be removed will be that vector yielding a minimum positive value for x_{i0}/x_{i1}. The values of x_{30} and x_{40} are 6 and 8, and the values of x_{31} and x_{41} are 1 and 2; hence the minimum value of x_{i0}/x_{i1} is $\min(x_{30}/x_{31}, x_{40}/x_{41}) = \min(\frac{6}{1}, \frac{8}{2}) = 4$, and the vector corresponding to $i = 4$, i.e., $\mathbf{P}_4$, will be removed from the basis.

We have established that we will replace the vector $\mathbf{P}_4$ in the basis with vector $\mathbf{P}_1$. This could be carried out by means of equation (4.4.7); i.e., we could express vectors $\mathbf{P}_0$, $\mathbf{P}_2$, and $\mathbf{P}_4$ in terms of $\mathbf{P}_1$ and $\mathbf{P}_3$ using the coefficients given in equation (4.4.7). However, it is convenient to retain *unit* vectors as members of the basis so that the components of any vector $\mathbf{P}_j$ which is not in the basis remain equal to the coefficients x_{ij}. In other words, we wish to replace $\mathbf{P}_4$ by $\mathbf{P}_1$ and then transform $\mathbf{P}_1$ to the unit vector (0, 1). This can be directly accomplished by performing a Gauss–Jordan elimination (cf. Appendix B) on Table 4.1. The result of the elimination is shown in Table 4.2.

Table 4.2. SECOND BASIS—EXAMPLE 4.4.1

	c		1	1	0	0
Basis		$\mathbf{P}_0$	$\mathbf{P}_1$	$\mathbf{P}_2$	$\mathbf{P}_3$	$\mathbf{P}_4$
$\mathbf{P}_3$	0	2	0	$\frac{3}{2}$	1	$-\frac{1}{2}$
$\mathbf{P}_1$	1	4	1	$\frac{1}{2}$	0	$\frac{1}{2}$
$(y_j - c_j)$		4	0	$-\frac{1}{2}$	0	$\frac{1}{2}$

To obtain Table 4.2, we first divide the $\mathbf{P}_4$ row in Table 4.1 by 2; the result is entered in the $\mathbf{P}_1$ row in Table 4.2. Notice that this causes the second component of $\mathbf{P}_1$ to become 1. We now return to Table 4.1, multiply the $\mathbf{P}_4$ row by $-\frac{1}{2}$, and add the resultant to the $\mathbf{P}_3$ row. The results of this calculation are entered in the $\mathbf{P}_3$ row of Table 4.2. Notice that the new basis vector $\mathbf{P}_1$ now has components (0, 1), as desired.

From Table 4.2 we see that $\mathbf{P}_2$ is the new vector to enter the basis. We also observe that $\min(x_{30}/x_{32}, x_{10}/x_{12}) = \min(\frac{4}{3}, 8) = \frac{4}{3}$, so that $\mathbf{P}_3$ is removed from the basis. Therefore, $\mathbf{P}_2$ must be transformed to (1, 0) by Gauss–Jordan elimination, the result of which is shown in Table 4.3.

We now see that Table 4.3 does not contain negative entries for $(y_j - c_j)$, which indicates that the problem is solved. The optimum value of the objective function is given in the last row of column $\mathbf{P}_0$; i.e., $y = \frac{14}{3}$. The maximum feasible solution has components of $x_1 = \frac{10}{3}$, $x_2 = \frac{4}{3}$; these values are read in the first two rows of column $\mathbf{P}_0$. The result agrees, of course, with the graphical solution shown in Figure 4.4.

Table 4.3. FINAL BASIS—EXAMPLE 4.4.1

	c		1	1	0	0
Basis		$\mathbf{P}_0$	$\mathbf{P}_1$	$\mathbf{P}_2$	$\mathbf{P}_3$	$\mathbf{P}_4$
$\mathbf{P}_2$	1	$\frac{4}{3}$	0	1	$\frac{2}{3}$	$-\frac{1}{3}$
$\mathbf{P}_1$	1	$\frac{10}{3}$	1	0	$-\frac{1}{3}$	$\frac{2}{3}$
$(y_j - c_j)$		$\frac{14}{3}$	0	0	$\frac{1}{3}$	$\frac{1}{3}$

4.5 Selecting the Initial Basis

The introduction of slack variables into the last example made it possible for us to select an initial basis consisting entirely of unit vectors with positive coefficients x_{i0}. Thus we were assured that our initial basis was linearly independent, and we had the additional computational advantage that the components of the vector $\mathbf{P}_0$ were equal to the coefficients x_{i0}. This latter condition facilitated the use of the tabular entries in obtaining a solution to our problem.

Unfortunately, not all linear programming problems lend themselves to the introduction of m slack variables. Many problems involve either equality constraints, or inequality constraints that become equality constraints upon subtraction of a surplus variable. Thus it is not always possible to select an initial basis of m unit vectors having positive coefficients from the given set of constraints.

Artificial vectors and artificial cost coefficients

Because of the advantages in choosing unit vectors having positive coefficients as an initial basis, we ask if it is possible to introduce a set of unit vectors into the problem artificially if unit vectors do not enter naturally. One way this can be done is as follows: Suppose that our initial problem is to minimize

$$y = c_1x_1 + c_2x_2 + \cdots + c_nx_n, \tag{4.5.1}$$

subject to the constraints

$$\begin{aligned} a_{11}x_1 + a_{12}x_2 + \cdots + a_{1n}x_n &= b_1, \\ a_{21}x_1 + a_{22}x_2 + \cdots + a_{2n}x_n &= b_2, \\ &\vdots \\ a_{m1}x_1 + a_{m2}x_2 + \cdots + a_{mn}x_n &= b_m, \end{aligned} \tag{4.5.2}$$

and $x_i \geq 0$, $i = 1, 2, \ldots, n$. Equations (4.5.2) may represent either equality constraints or a set of inequality constraints of the form $a_{i1}x_1 + a_{i2}x_2 + \cdots + a_{in}x_n \geq b_i$, which were converted to equality constraints through the introduction of surplus variables.

We now augment the above system by the introduction of m *artificial* variables $x_{n+1}, x_{n+2}, \ldots, x_{n+m}$ and an *artificial* cost coefficient μ. Our problem now becomes: Minimize

$$y' = c_1x_1 + c_2x_2 + \cdots + c_nx_n + \mu(x_{n+1} + x_{n+2} + \cdots + x_{n+m}), \tag{4.5.3}$$

subject to

$$\begin{aligned}
a_{11}x_1 + a_{12}x_2 + \cdots + a_{1n}x_n + x_{n+1} \qquad\qquad\qquad &= b_1, \\
a_{21}x_1 + a_{22}x_2 + \cdots + a_{2n}x_n \qquad + x_{n+2} \qquad\quad &= b_2, \\
\vdots \qquad\qquad\qquad\qquad\qquad\qquad & \\
a_{m1}x_1 + a_{m2}x_2 + \cdots + a_{mn}x_n \qquad\qquad\qquad + x_{n+m} &= b_m,
\end{aligned} \tag{4.5.4}$$

and $x_i \geq 0$, $i = 1, 2, \ldots, n, n + 1, \ldots, n + m$. In vector notation Equations (4.5.4) become

$$x_1\mathbf{P}_1 + x_2\mathbf{P}_2 + \cdots + x_n\mathbf{P}_n + x_{n+1}\mathbf{P}_{n+1} + \cdots + x_{n+m}\mathbf{P}_{n+m} = \mathbf{P}_0, \tag{4.5.5}$$

where $\mathbf{P}_{n+1}, \mathbf{P}_{n+2}, \ldots, \mathbf{P}_{n+m}$ are a set of artificial, linearly independent unit vectors that can serve as an initial basis.

The artificial cost coefficient μ in Equation (4.5.3) is an arbitrarily large number that is taken to be positive for a minimization problem and negative for a maximization problem. Thus the presence of one or more artificial variables in the objective function must cause the value of the objective function to be far removed from its optimum (recall that the artificial variables are restricted to being nonnegative). Furthermore, the simplex algorithm will favor the removal of the artificial vectors from the basis (and the corresponding removal of the artificial variables from the objective function), providing that μ is chosen sufficiently large in magnitude. If all the artificial vectors can be removed from the basis and an optimal feasible solution of the augmented problem can be obtained, then this optimal feasible solution will be the optimal feasible solution of the original problem.

We can now see a strategy evolving from the discussion. Let us apply the simplex algorithm to the augmented problem with the idea of first removing the artificial variables from the basis, thus attaining feasibility, and then continuing until an optimal feasible solution has been found. Once an artificial vector leaves the basis it will never be allowed to reenter (the simplex algorithm imposes no such restriction on the vectors that are a part of the original problem). In this manner we attempt to force the problem into a feasible region as soon as possible, and then continue on our merry way.

The success of this strategy cannot, however, be guaranteed. Whether the artificial vectors can be removed from the basis will depend largely on the characteristics of the original problem. If a problem does not have a finite optimal feasible solution, then the simplex algorithm applied to the augmented problem will reach a terminal condition (no further improvement in the objective function will be possible) with one or more artificial vectors still in the basis.

A more subtle difficulty can result from a poorly posed problem in which the constraints are incompatible or are not independent. Under these conditions it is possible that the simplex algorithm applied to the augmented problem will result in an optimal feasible solution which contains one or more artificial vectors, but with the corresponding artificial variables equal to zero. The attainment of such a situation obviously requires that the original problem formulation be closely scrutinized.

The two-phase method

When using a digital computer to solve linear-programming problems that require the addition of artificial variables, it is most convenient to assign a numerical value to μ and to combine the real and artificial parts of $(y_j - c_j)$ into one numerical value. The success of the method is sensitive to the particular choice of μ. If μ is chosen to be too small in magnitude, then the problem may terminate falsely without removing all the artificial vectors from the basis. On the other hand, too large a choice of $|\mu|$ can cause the real part of the objective function to be lost (because of truncation) when added to the artificial part. Now the artificial variables can generally be removed, but the resulting feasible solutions may be considerably in error. As a rule of thumb, the magnitude of μ can often be chosen such that it is about 100 times as great as the largest natural cost coefficient.

To circumvent the computational difficulties that can arise when using artificial variables and an arbitrary artificial cost coefficient, a variation of the artificial-variable technique, known as the *two-phase method*, is used. Phase I of the two-phase method sets all the *real* cost coefficients equal to zero, and $|\mu|$ is allowed to equal one. The simplex algorithm is then applied systematically until either all the artificial vectors are removed from the basis and the objective function becomes equal to zero, or else there is an indication of an infeasible solution. In the latter case the objective function will be greater than zero, but it will not be possible to add to the basis a new vector whose corresponding value of $(y_j - c_j)$ is greater than zero.

In Phase II all the *artificial* cost coefficients are set equal to zero and the natural cost coefficients are allowed to assume their correct values. The simplex algorithm then continues where Phase I stopped [with the final values of $(y_j - c_j)$ from Phase I altered to correct for the new cost coefficients] until an optimum feasible solution is reached. A more complete discussion of the two-phase method as applied to the simplex algorithm is given by Hadley [1962].

The solution of a simple problem through the introduction of artificial variables is shown in the following example.

EXAMPLE 4.5.1

Let us minimize the function

$$y = x_1 + x_2 + x_3,$$

subject to the constraints

$$x_1 - x_2 + 2x_3 = 5,$$
$$2x_1 + 3x_2 - x_3 = 4,$$

and the nonnegativity conditions $x_1, x_2, x_3 \geq 0$.

Adding the artificial variables x_4 and x_5, the problem becomes: Minimize

$$y = x_1 + x_2 + x_3 + \mu(x_4 + x_5),$$

subject to

$$x_1\mathbf{P}_1 + x_2\mathbf{P}_2 + x_3\mathbf{P}_3 + x_4\mathbf{P}_4 + x_5\mathbf{P}_5 = \mathbf{P}_0,$$

where

$$\mathbf{P}_1 = \begin{bmatrix} 1 \\ 2 \end{bmatrix}, \quad \mathbf{P}_2 = \begin{bmatrix} -1 \\ 3 \end{bmatrix}, \quad \mathbf{P}_3 = \begin{bmatrix} 2 \\ -1 \end{bmatrix}, \quad \mathbf{P}_4 = \begin{bmatrix} 1 \\ 0 \end{bmatrix},$$
$$\mathbf{P}_5 = \begin{bmatrix} 0 \\ 1 \end{bmatrix}, \quad \mathbf{P}_0 = \begin{bmatrix} 5 \\ 4 \end{bmatrix},$$

and $x_i \geq 0$, $i = 1, 2, \ldots, 5$.

We now represent the problem by Table 4.4, which is similar in form to the tables in Example 4.4.1. There are, however, certain differences: we now have two rows for the value of $(y_j - c_j)$, which contain, respectively, the portion of $(y_j - c_j)$ obtained from the original problem, and the remaining portion which results from the introduction of the artificial variables. Also, the table has sufficient columns for entering both the vectors defined by the original problem and the artificial vectors, which constitute the initial basis.

The computation proceeds by introducing into the basis the vector containing the largest positive coefficient of μ, i.e., the vector having the largest artificial value

Table 4.4. INITIAL BASIS—EXAMPLE 4.5.1

	c		1	1	1	μ	μ
Basis		$\mathbf{P}_0$	$\mathbf{P}_1$	$\mathbf{P}_2$	$\mathbf{P}_3$	$\mathbf{P}_4$	$\mathbf{P}_5$
$\mathbf{P}_4$	μ	5	1	-1	2	1	0
$\mathbf{P}_5$	μ	4	2	3	-1	0	1
$(y_j - c_j)$		0	-1	-1	-1		
$\mu \sum x_{i0}$		9μ	3μ	2μ	μ		

of $(y_j - c_j)$, which happens to be $\mathbf{P}_1$ in this case. Notice that the values of $(y_j - c_j)$ corresponding to the artificial vectors are set equal to zero, so that an artificial vector which is removed from the basis can never reenter the basis.

The vector which leaves the basis is, as usual, that vector for which x_{l0}/x_{lj} is a minimum, $x_{lj} > 0$, and $x_{l0}/x_{lj} > 0$. For this example we seek $\min(x_{40}/x_{41}, x_{50}/x_{51}) = \min(5, 2) = 2$; hence $\mathbf{P}_5$ is replaced by $\mathbf{P}_1$ in the new basis, as shown in Table 4.5.

Table 4.5. SECOND BASIS—EXAMPLE 4.5.1

	c		1	1	1	μ
Basis		$\mathbf{P}_0$	$\mathbf{P}_1$	$\mathbf{P}_2$	$\mathbf{P}_3$	$\mathbf{P}_4$
$\mathbf{P}_4$	μ	3	0	$-\frac{5}{2}$	$\frac{5}{2}$	1
$\mathbf{P}_1$	1	2	1	$\frac{3}{2}$	$-\frac{1}{2}$	0
$(y_j - c_j)$		2	0	$\frac{1}{2}$	$-\frac{3}{2}$	
$\mu \sum x_{i0}$		3μ	0	$-\frac{5\mu}{2}$	$\frac{5\mu}{2}$	

Again applying the simplex algorithm to Table 4.5, we replace the artificial vector $\mathbf{P}_4$ with $\mathbf{P}_3$, as shown in Table 4.6. Once the artificial variables have been removed from the basis, the simplex algorithm is continued as before until either all $(y_j - c_j) \leq 0$, which indicates that the optimum feasible solution has been found, or until $(y_j - c_j) > 0$ but all $x_{ij} \leq 0$ for some vector $\mathbf{P}_j$, which indicates the presence of an unbounded solution to the original problem. In this example we see that the optimum feasible solution has been found as soon as the artificial vectors are removed from the basis, because all the $(y_j - c_j) \leq 0$ in Table 4.6. The minimum value of y is therefore $\frac{19}{5}$, and it occurs at $x_1 = \frac{13}{5}$, $x_2 = 0$, $x_3 = \frac{6}{5}$.

Table 4.6. FINAL BASIS—EXAMPLE 4.5.1

	c		1	1	1
Basis		$\mathbf{P}_0$	$\mathbf{P}_1$	$\mathbf{P}_2$	$\mathbf{P}_3$
$\mathbf{P}_3$	1	$\frac{6}{5}$	0	-1	1
$\mathbf{P}_1$	1	$\frac{13}{5}$	1	1	0
	$(y_j - c_j)$	$\frac{19}{5}$	0	-1	0
	$\mu \sum x_{i0}$	0	0	0	0

4.6 Degeneracy

In our previous discussion of the simplex algorithm we have assumed that the vector $\mathbf{P}_0$ can be expressed in terms of exactly m linearly independent vectors $\mathbf{P}_1, \mathbf{P}_2, \ldots, \mathbf{P}_m$ in accordance with Equation (4.4.3). We have seen in Section 4.3, however, that any vector whose nonzero elements are the coefficients of the expression

$$x_{10}\mathbf{P}_1 + x_{20}\mathbf{P}_2 + \cdots + x_{k0}\mathbf{P}_k = \mathbf{P}_0, \tag{4.6.1}$$

$x_{i0} \geq 0$ and $k \leq m$, is a feasible solution to the linear programming problem, providing $\mathbf{P}_1, \mathbf{P}_2, \ldots, \mathbf{P}_k$ are linearly independent. Thus it is not necessary that the number of linearly independent vectors in the basis equal the number of constraints. Let us investigate the conditions under which k becomes less than m.

Suppose that we begin our elimination using the simplex algorithm, with $\mathbf{P}_0$ expressed in terms of m linearly independent vectors. At some point in the computation, however, suppose that two values of i exist which yield the same minimum, positive value for x_{i0}/x_{ij}. Let us refer to the two vectors in the basis for which the quantity $x_{i0}/x_{ij} > 0$ is minimized as $\mathbf{P}_\alpha$ and $\mathbf{P}_\beta$; hence $0 < x_{\alpha 0}/x_{\alpha j} = x_{\beta 0}/x_{\beta j} < x_{i0}/x_{ij}$, $i \neq \alpha, \beta$. Replacing the basis vector $\mathbf{P}_\alpha$ with $\mathbf{P}_j$ and performing the usual vector transformation, we obtain, for the new value of $x_{\beta 0}$,

$$\begin{aligned} x_{\beta 0}^{(1)} &= x_{\beta 0}^{(0)} - x_{\alpha 0}^{(0)} \frac{x_{\beta j}}{x_{\alpha j}} \\ &= x_{\beta 0}^{(0)} - x_{\beta j} \frac{x_{\alpha 0}^{(0)}}{x_{\alpha j}} \\ &= x_{\beta 0}^{(0)} - x_{\beta j} \frac{x_{\beta 0}^{(0)}}{x_{\beta j}} \\ &= x_{\beta 0}^{(0)} - x_{\beta 0}^{(0)} \\ &= 0. \end{aligned} \tag{4.6.2}$$

Since the new value of $x_{\beta 0} = 0$, the vector $\mathbf{P}_\beta$ will not appear in the basis, and the new vector $\mathbf{P}_0$ will be expressed as

$$\mathbf{P}_0 = x_{10}^{(1)}\mathbf{P}_1 + x_{20}^{(1)}\mathbf{P}_2 + \cdots + x_{\alpha 0}^{(1)}\mathbf{P}_\alpha + 0\mathbf{P}_\beta + \cdots + x_{m0}^{(1)}\mathbf{P}_m, \qquad (4.6.3)$$

which is simply another way of writing Equation (4.6.1) for $k = m - 1$. Hence $\mathbf{P}_0$ is now expressed in terms of $m - 1$ linearly independent vectors. The solution vector $\mathbf{X}$, whose $m - 1$ nonzero components are the coefficients of the vectors $\mathbf{P}_1, \mathbf{P}_2, \ldots, \mathbf{P}_k$ in Equation (4.6.1), is called a *degenerate* solution since $k < m$.

One characteristic of a degenerate solution is that there may be no improvement in the value of the objective function in the succeeding solution. This lack of improvement could go on for several successive steps, even though the vector substitution continues at each step. A possibility which can arise as a result of such a lack of improvement in y is that some basis may eventually repeat itself. If this were to happen, then the continued application of the simplex algorithm would result in a cycling of the bases, which could continue indefinitely without any change in the value of the objective function. Under these circumstances it would not be possible to obtain an optimum solution through the use of the simplex algorithm.

Fortunately, several techniques have been developed for preventing cycling when solving degenerate linear programming problems. These techniques are based upon application of the simplex algorithm, but they entail additional computational devices which ensure that cycling will not occur (cf. Hadley [1962] for a thorough discussion of two such techniques). Thus it is possible to continue using the simplex algorithm on problems that yield degenerate solutions, and, for the price of some extra computation, one can be confident that the computation will not cycle.

In practical problems, however, cycling is never encountered, although degeneracy is not at all uncommon. The use of specialized computational procedures to avoid degeneracy therefore becomes unnecessary, although the existence of such procedures has considerable theoretical significance. In a practical sense, any degenerate problem possessing a finite optimum feasible solution can still be solved by normal application of the simplex algorithm. If a finite optimum feasible solution does not exist, the simplex algorithm will still indicate this, as before.

It is interesting that only two examples of degenerate problems which cycle are reported in the literature (cf. Beale [1955] and Hoffman [1953]). Both of these examples are artificial in the sense that they were deliberately constructed with great care so as to experience cycling.

Having mentioned the characteristics of degenerate solutions and having discussed the existence of special degeneracy procedures, we shall not pursue the topic of degeneracy any further. The interested reader is referred to the books by Dantzig [1963] and Hadley [1962].

An example of a degenerate problem solved by the simplex procedure is presented next.

EXAMPLE 4.6.1

Maximize the function

$$y = 2x_1 + x_2 + 3x_3,$$

subject to

$$x_1 + 3x_2 + 2x_3 \leq 6,$$
$$2x_1 - x_2 + x_3 \leq 3,$$

and

$$x_1, x_2, x_3 \geq 0.$$

The constraints can be transformed into equalities by the addition of the slack variables $x_4 \geq 0$ and $x_5 \geq 0$. Table 4.7 shows the linear programming problem resulting from the addition of the slack variables.

Table 4.7. INITIAL BASIS—EXAMPLE 4.6.1

	c		2	1	3	0	0
Basis		$\mathbf{P}_0$	$\mathbf{P}_1$	$\mathbf{P}_2$	$\mathbf{P}_3$	$\mathbf{P}_4$	$\mathbf{P}_5$
$\mathbf{P}_4$	0	6	1	3	2	1	0
$\mathbf{P}_5$	0	3	2	-1	1	0	1
$(y_j - c_j)$		0	-2	-1	-3	0	0

Since we wish to maximize the objective function, we enter into the basis the vector having the largest negative value of $(y_j - c_j)$, which from Table 4.7 is seen to be $\mathbf{P}_3$. The vector which leaves the basis is that vector having the minimum value of x_{i0}/x_{ij}, $x_{ij} > 0$. For this problem we seek the minimum of $(x_{40}/x_{43}, x_{50}/x_{53})$ $= \min(3, 3)$. Since we have a tie, we choose $\mathbf{P}_4$, the vector having the lowest index, as the vector to be replaced.

The result of replacing the vector $\mathbf{P}_4$ with the vector $\mathbf{P}_3$ is shown in Table 4.8. Notice that x_{50} is equal to zero, so that we now have a degenerate solution. Con-

Table 4.8. SECOND BASIS—EXAMPLE 4.6.1

	c		2	1	3	0	0
Basis		$\mathbf{P}_0$	$\mathbf{P}_1$	$\mathbf{P}_2$	$\mathbf{P}_3$	$\mathbf{P}_4$	$\mathbf{P}_5$
$\mathbf{P}_3$	3	3	$\frac{1}{2}$	$\frac{3}{2}$	1	$\frac{1}{2}$	0
$\mathbf{P}_5$	0	0	$\frac{3}{2}$	$-\frac{5}{2}$	0	$-\frac{1}{2}$	1
$(y_i - c_j)$		9	$-\frac{1}{2}$	$\frac{7}{2}$	0	$\frac{3}{2}$	0

tinuing as before, we see that $\mathbf{P}_1$ now enters the basis, replacing $\mathbf{P}_5$. Table 4.9 shows the result of this operation.

In Table 4.9 we see that all the $(y_j - c_j) \geq 0$, which indicates that the optimum feasible solution has been found. Thus $y_{\max} = 9$ when $x_1 = x_2 = 0$, $x_3 = 3$. Notice that this value of y is unchanged from the previous step shown in Table 4.8. In going from Table 4.8 to Table 4.9, however, we have obtained the value of x_1 corresponding to the optimum feasible solution. This was not known previously.

Notice that we have obtained a solution to this problem using the rules of the simplex algorithm which we developed for the case where $k = m$; i.e., the number of linearly independent vectors equals the number of constraints. Thus we were able to obtain an optimum feasible solution to this problem by simply ignoring the degenerate condition that arose in the course of the computation.

Table 4.9. FINAL BASIS—EXAMPLE 4.6.1

	c		2	1	3	0	0
Basis		$\mathbf{P}_0$	$\mathbf{P}_1$	$\mathbf{P}_2$	$\mathbf{P}_3$	$\mathbf{P}_4$	$\mathbf{P}_5$
$\mathbf{P}_3$	3	3	0	$\frac{7}{3}$	1	$\frac{2}{3}$	$-\frac{1}{3}$
$\mathbf{P}_1$	2	0	1	$-\frac{5}{3}$	0	$-\frac{1}{3}$	$\frac{2}{3}$
$(y_j - c_j)$		9	0	$\frac{8}{3}$	0	$\frac{4}{3}$	$\frac{1}{3}$

4.7 The Revised Simplex Algorithm

Another technique often used to solve linear programming problems is the *revised simplex algorithm*. This technique makes use of the inverse of the $m \times m$ matrix containing the current basis vectors at each stage of the computation.

Suppose that $\mathbf{P}_1, \mathbf{P}_2, \ldots, \mathbf{P}_m$ form a set of basis vectors, and that x_1, $x_2, \ldots, x_m$ are the corresponding independent variables. (Recall that all the remaining independent variables will equal zero.) We can then write

$$\mathbf{P}_0 = x_1\mathbf{P}_1 + x_2\mathbf{P}_2 + \cdots + x_m\mathbf{P}_m \tag{4.7.1}$$

or

$$\mathbf{P}_0 = \mathbf{B}\mathbf{X}_B, \tag{4.7.2}$$

where $\mathbf{X}_B$ is an m-dimensional vector whose components are $x_1, x_2, \ldots,$ x_m, and $\mathbf{B}$ is an $m \times m$ matrix whose columns are the basis vectors $\mathbf{P}_1, \mathbf{P}_2,$ $\ldots, \mathbf{P}_m$. If Equation (4.7.2) is solved for $\mathbf{X}_B$, we obtain

$$\mathbf{X}_B = \mathbf{B}^{-1}\mathbf{P}_0. \tag{4.7.3}$$

The corresponding objective function can be expressed as

$$y = \mathbf{C}_B'\mathbf{X}_B = \mathbf{C}_B'\mathbf{B}^{-1}\mathbf{P}_0, \tag{4.7.4}$$

where $\mathbf{C}_B$ is an m-dimensional vector whose components are the cost coefficients corresponding to the variables $x_1, x_2, \ldots, x_m$. Furthermore, an expression analogous to Equation (4.7.3) can be written as

$$\mathbf{X}_j = \mathbf{B}^{-1}\mathbf{P}_j, \tag{4.7.5}$$

where $\mathbf{P}_j$ is an m-dimensional, nonbasis vector, and $\mathbf{X}_j$ contains the coefficients needed to express $\mathbf{P}_j$ in terms of the basis vectors.

Equations (4.7.3), (4.7.4), and (4.7.5) are fundamental to the development of the revised simplex algorithm. The computational procedure is similar to the basic simplex algorithm, except that the operations are performed on the $(m + 1) \times (m + 1)$ matrix $\mathbf{D}^{-1}$, where

$$\mathbf{D}^{-1} = \begin{bmatrix} 1 & \mathbf{C}_B'\mathbf{B}^{-1} \\ \mathbf{0} & \mathbf{B}^{-1} \end{bmatrix}. \tag{4.7.6}$$

Rather than calculate $\mathbf{D}^{-1}$ directly each time the basis is changed, it is possible to calculate a new inverse from an old inverse by means of the expression

$$\mathbf{D}_i^{-1} = \mathbf{E}_i\mathbf{D}_{i-1}^{-1}, \tag{4.7.7}$$

where $\mathbf{D}_i^{-1}$ and $\mathbf{D}_{i-1}^{-1}$ are the new and old inverse matrices, respectively, and $\mathbf{E}_i$ is an $(m + 1) \times (m + 1)$ matrix that is calculated as a part of the ith vector transformation procedure. It can be shown that $\mathbf{D}_i^{-1}$ can be written as

$$\mathbf{D}_i^{-1} = \mathbf{E}_i\mathbf{E}_{i-1} \cdots \mathbf{E}_1. \tag{4.7.8}$$

This is known as the *product form of the inverse*.

The revised simplex algorithm is generally used when solving linear programming problems on a digital computer. The reason for this is two-fold: first, the revised simplex algorithm involves fewer arithmetic calculations than the basic simplex algorithm, and, second, it requires that less information be stored. Thus it is possible to solve larger problems, in less time, than with the basic simplex algorithm.

Most computer codes utilize the product form of the inverse. Thus Equation (4.7.7) is generally used to determine a new inverse corresponding to each new set of basis vectors. To minimize the cumulative buildup of numerical errors, however, $\mathbf{D}_i^{-1}$ is periodically calculated by direct matrix inversion rather than by Equation (4.7.7).

We shall not discuss the revised simplex algorithm any further. The interested reader is referred to Gass [1969] or Hadley [1962].

4.8 Post-optimal Analysis

Once an optimum has been obtained for a linear programming problem, it is natural to ask how the optimal solution would be affected by changes in the model parameters. Of particular interest are the results of changing one or more coefficients in the objective function or the right-hand side of the constraint set. For example, one might wish to estimate the importance of error in certain of these coefficients by changing their values and examining the consequences of these changes on the optimal solution. Another reason for altering the model parameters is to determine by how much certain costs or requirements must be changed in order to alter the optimal policy (as determined by the particular choice of basis vectors). Information concerning the sensitivity of the optimum to changes in these parameters is often as important as the determination of the optimum itself. Therefore, we shall discuss some of the aspects of post-optimal analysis in the paragraphs below.

The most obvious method for carrying out such parametric studies is, of course, simply to change the parameters in the desired manner and then resolve the problem. For large problems, however, this can be prohibitively expensive. Moreover, this procedure involves a great deal of unnecessary computation, for, as we shall see below, the information we seek can be obtained from a given optimal solution with very little additional computational effort.

Variations in the cost coefficients

For the time being we shall confine our attention to variations in the coefficients that appear in the objective function. Suppose that we have just obtained the optimal solution to a minimization problem, and, for simplicity, we shall assume that the solution is finite, unique, and nondegenerate. We ask first what will happen if some cost coefficient corresponding to a basis vector is changed from c_i to $(c_i + \delta_i)$, assuming that the modification in c_i is not so large as to cause a change in the basis. (Note that δ_i can be either positive or negative.) The optimal value of the objective function now becomes

$$y_{\min} = c_1x_1 + c_2x_2 + \cdots + (c_i + \delta_i)x_i + \cdots + c_nx_n, \qquad (4.8.1)$$

where only m of the independent variables (including the variable x_i) are nonzero. Clearly, the change in the optimal value of the objective function corresponding to the increase in c_i is given by the product $\delta_i x_i$. Hence we can write

$$\frac{\partial y_{\min}}{\partial \delta_i} = x_i, \qquad x_i > 0. \qquad (4.8.2)$$

Thus we see that the activity level, x_i, of a basis vector determines the change in the objective function resulting from a change in the corresponding coefficient, c_i.

Now let us inquire as to the effect of changing some cost coefficient c_j that corresponds to a nonbasis vector. Since $x_j = 0$, we see that a change in c_j will have no effect unless the change in c_j is large enough to bring $\mathbf{P}_j$ into the basis. We know that $(y_j - c_j) < 0$ at the optimum. If c_j were to take on some different value, say c_j^*, such that $(y_j - c_j^*) = 0$, then $\mathbf{P}_j$ would enter the basis and the optimal policy would change. (The vector to leave the basis would be determined in the usual manner.) Hence

$$c_j^* = y_j \tag{4.8.3}$$

is the critical value of c_j that brings about a change in the optimal policy. If we denote the original value of c_j as c_j^o, then the quantity

$$\delta_j = c_j^* - c_j^o \tag{4.8.4}$$

is the minimum amount by which c_j must be changed in order that $\mathbf{P}_j$ enter the basis. This quantity is sometimes referred to as the *reduced cost*.

So far we have been concerned with changing only one cost coefficient at a time, with all other cost coefficients held at their original values. If we wish, however, we can alter several cost coefficients at once, and then see what effect these changes might have on the optimum. Let $\boldsymbol{\delta}$ be an n-dimensional vector whose nonzero elements are the desired changes in the individual cost coefficients; i.e.,

$$\boldsymbol{\delta} = \begin{bmatrix} \delta_1 \\ \delta_2 \\ \cdot \\ \cdot \\ \cdot \\ \delta_n \end{bmatrix}. \tag{4.8.5}$$

Note that the nonzero δ_i can correspond either to vectors in the optimal basis or to nonbasis vectors. Also, the δ_i are not restricted in sign.

We can express the altered set of cost coefficients as

$$\mathbf{C} = \mathbf{C}^o + \theta\boldsymbol{\delta}, \tag{4.8.6}$$

where $\mathbf{C}^o$ is the cost vector of the original problem and θ is a nonnegative scalar. Now, given an optimal solution and given some particular vector $\boldsymbol{\delta}$, we would like to determine by how much θ can be increased (from zero) before the optimal basis is changed.

Recall once again that the criterion for a unique minimum is that $(y_j - c_j) < 0$ for all nonbasis vectors. When the cost vector is altered in accordance with Equation (4.8.6), then $(y_j - c_j)$ can be written as

$$(y_j - c_j) = (y_j^o - c_j^o) + \theta(\boldsymbol{\delta}_B'\mathbf{P}_j - \delta_j), \tag{4.8.7}$$

where y_j^o is the minimum value of y_j obtained with the original coefficients, and $\boldsymbol{\delta}_B$ is an m-dimensional vector whose components are those δ_i which correspond to the optimal basis vectors. Now suppose that there exists some value of θ, say θ_j^*, such that $(y_j - c_j) = 0$. [Note that this condition can be obtained only for those $\mathbf{P}_j$ which yield $(\boldsymbol{\delta}_B'\mathbf{P}_j - \delta_j) > 0$, since $(y_j^o - c_j^o) < 0$ for all nonbasis vectors when degeneracy is not present.] Thus we obtain

$$\theta_j^* = -\frac{(y_j^o - c_j^o)}{(\boldsymbol{\delta}_B'\mathbf{P}_j - \delta_j)}. \tag{4.8.8}$$

If we evaluate θ_j^* for all nonbasis vectors $\mathbf{P}_j$ (including the nonbasis vectors which result from the addition of slack and surplus variables but excluding those vectors which correspond to artificial variables), then the smallest $\theta_j^* > 0$ is the largest value that θ can take on without change in the basis. Hence we can write

$$\theta^* = \min_j \left\{ -\frac{(y_j^o - c_j^o)}{(\boldsymbol{\delta}_B'\mathbf{P}_j - \delta_j)} \right\}, \tag{4.8.9}$$

where only those $(\boldsymbol{\delta}_B'\mathbf{P}_j - \delta_j) > 0$ are considered in the above expression. If none of the $(\boldsymbol{\delta}_B'\mathbf{P}_j - \delta_j)$ are positive, then θ can be increased indefinitely without bringing about a change in the basis.

If some $\theta^* > 0$ can be found for a particular problem, then the corresponding vector $\mathbf{P}_j$ can be entered into the basis, and θ can again increase until another θ^* is found, and so on. Eventually a point will be reached where θ can be increased indefinitely without any further change in the basis. The entire procedure, beginning with $\theta = 0$ and allowing θ to increase through all basis changes, is known as *parameterization* or *parametric linear programming.*

A particularly common application of parametric programming is to choose the $\boldsymbol{\delta}$ vector such that all its components are zero except for any one component δ_i, which corresponds to some optimal basis vector. This nonzero component is given a value of either $\delta_i = 1$ or $\delta_i = -1$. By finding the values of θ^* that cause some nonbasis vector $\mathbf{P}_j$ to enter the basis, replacing $\mathbf{P}_i$, we can establish a range of values over which c_i can be varied without bringing about a change in the original optimal basis. This procedure is called *ranging* or *sensitivity analysis.*

Variations in the right-hand side

Now let us turn our attention to variations in the right-hand side of the constraint set $\mathbf{AX} = \mathbf{P}_0$. Suppose that some component of the vector $\mathbf{P}_0$, say p_{j0}, is altered by an amount ρ_j; i.e.,

$$p_{j0} = p_{j0}^o + \rho_j, \tag{4.8.10}$$

where p_{j0}^o is the original value of p_{j0}†. All the other components of $\mathbf{P}_0$ will be held at their original values. We wish to determine the change in the optimal value of the objective function resulting from the alteration of p_{j0}, providing that ρ_j is sufficiently small in magnitude that the optimal basis is unchanged.

Recall, from Section 4.7, that the optimal solution vector $\mathbf{X}_B$ can be expressed as

$$\mathbf{X}_B = \mathbf{B}^{-1}\mathbf{P}_0, \tag{4.8.11}$$

where $\mathbf{X}_B$ is an m-dimensional vector that contains only the nonzero values of the independent variables. If $\mathbf{X}_B^o$ is the optimal solution vector obtained with the original right-hand side, then the change in the optimal solution vector can be written as

$$\Delta\mathbf{X}_B = (\mathbf{X}_B - \mathbf{X}_B^o) = \begin{bmatrix} b_{11}^{-1} & b_{12}^{-1} & \cdots & b_{1j}^{-1} & \cdots & b_{1m}^{-1} \\ b_{21}^{-1} & b_{22}^{-1} & \cdots & b_{2j}^{-1} & \cdots & b_{2m}^{-1} \\ \cdot & & & & & \cdot \\ \cdot & & & & & \cdot \\ \cdot & & & & & \cdot \\ b_{i1}^{-1} & b_{i2}^{-1} & \cdots & b_{ij}^{-1} & \cdots & b_{im}^{-1} \\ \cdot & & & & & \cdot \\ \cdot & & & & & \cdot \\ \cdot & & & & & \cdot \\ b_{m1}^{-1} & b_{m2}^{-1} & \cdots & b_{mj}^{-1} & \cdots & b_{mm}^{-1} \end{bmatrix} \begin{bmatrix} 0 \\ 0 \\ \cdot \\ \cdot \\ \cdot \\ \rho_j \\ \cdot \\ \cdot \\ \cdot \\ 0 \end{bmatrix}, \tag{4.8.12}$$

and the change in any one independent variable is simply

$$\Delta x_i = b_{ij}^{-1}\rho_j. \tag{4.8.13}$$

Thus we can write

$$x_i = x_i^o + b_{ij}^{-1}\rho_j. \tag{4.8.14}$$

The evaluation of the objective function is now straightforward:

$$y_{\min} = \sum_{i=1}^{m} c_i x_i = \sum_{i=1}^{m} c_i x_i^o + \sum_{i=1}^{m} c_i b_{ij}^{-1}\rho_j. \tag{4.8.15}$$

---

† Recall that the vector $\mathbf{P}_0$ is actually the vector $\mathbf{B}$ in the constraint set $\mathbf{AX} = \mathbf{B}$ (cf. Section 4.1). This should not be confused with the *matrix* $\mathbf{B}$, introduced in the last section, which is an $m \times m$ matrix whose columns are the basis vectors $\mathbf{P}_{i\cdot}$, $i = 1, 2, \ldots, m$. In this section we shall continue to use the symbols $\mathbf{P}_0$ for the right-hand side of the constraint set, and $\mathbf{B}$ for the basis matrix.

The change in the objective function corresponding to a unit change in the right-hand side of the jth constraint is, then,

$$\frac{\partial y_{\min}}{\partial \rho_j} = \sum_{i=1}^{m} c_i b_{ij}^{-1}. \tag{4.8.16}$$

This quantity is known as a *shadow price* or a *marginal value* (cf. footnote, page 59). The shadow prices are easy to obtain when the revised simplex algorithm is used, since the inverse of the basis matrix is readily available.

Suppose that we now alter several of the terms on the right-hand side at once. We can express these changes in terms of a new right-hand vector, given by

$$\mathbf{P}_0 = \mathbf{P}_0^o + \phi\boldsymbol{\rho}, \qquad \phi \geq 0, \tag{4.8.17}$$

where $\mathbf{P}_0^o$ is the original right-hand vector and $\boldsymbol{\rho}$ contains the changes in the various right-hand terms. Notice that the components of $\boldsymbol{\rho}$ are unrestricted in sign.

Let us now see how large ϕ can become before the set of optimal basis vectors will change. If we combine Equations (4.8.11) and (4.8.17), we obtain

$$\begin{aligned} \mathbf{X}_B &= \mathbf{B}^{-1}\mathbf{P}_0^o + \phi\mathbf{B}^{-1}\boldsymbol{\rho} \\ &= \mathbf{X}_B^o + \phi\mathbf{B}^{-1}\boldsymbol{\rho}. \end{aligned} \tag{4.8.18}$$

We wish to avoid letting ϕ increase by so much that one or more components of $\mathbf{X}_B$ will become negative. If each of the components of the vector $\mathbf{B}^{-1}\boldsymbol{\rho}$ is ≥ 0, then the elements of $\mathbf{X}_B$ will remain nonnegative for any $\phi > 0$; hence ϕ can be increased without limit. Observe that any such change in ϕ will alter both the policy vector $\mathbf{X}_B$ and the optimal value of the objective function, $y_{\min}$; only the *choice of vectors* in the optimal basis will remain unchanged.

If one or more components of $\mathbf{B}^{-1}\boldsymbol{\rho}$ are < 0, however, then ϕ will have an upper bound beyond which $\mathbf{X}_B$ will contain one or more negative elements. Let this limiting value of ϕ be denoted by ϕ^*. Hence, from Equation (4.8.18), we have

$$\phi^* = \min_i \left\{\frac{-x_i^o}{\psi_i}\right\}, \qquad \psi_i < 0, \tag{4.8.19}$$

where

$$\boldsymbol{\psi} = \begin{bmatrix} \psi_1 \\ \psi_2 \\ \cdot \\ \cdot \\ \cdot \\ \psi_m \end{bmatrix} = \mathbf{B}^{-1}\boldsymbol{\rho}. \tag{4.8.20}$$

Notice that Equation (4.8.19) includes only those ψ_i which are negative.

If some limiting ϕ^* can be obtained for a given problem, then the basis can be changed accordingly and ϕ can again be allowed to increase. The vector which leaves the basis will be that $\mathbf{P}_i$ which corresponds to the limiting ψ_i in Equation (4.8.19). The vector that enters the basis, $\mathbf{P}_j$, is determined by first forming the vectors

$$\mathbf{X}_j = \begin{bmatrix} x_{1j} \\ x_{2j} \\ \cdot \\ \cdot \\ \cdot \\ x_{mj} \end{bmatrix} = \mathbf{B}^{-1}\mathbf{P}_j \tag{4.8.21}$$

for all nonbasis vectors, and then finding that $\mathbf{P}_j$ for which $x_{ij} < 0$ and the quantity

$$q_j = \frac{(y_j - c_j)}{x_{ij}} \tag{4.8.22}$$

is a minimum.* Note that q_j will be > 0, since all $(y_j - c_j) < 0$ for the nonbasis vectors in a nondegenerate minimal solution.

The procedure of increasing ϕ, finding a ϕ^* (if one exists), and then changing the basis accordingly can be continued through several basis changes if desired. With any given problem a point will eventually be reached where either ϕ can increase indefinitely without any further change in the vectors, or else any further increase in ϕ will result in an infeasible solution. [This latter condition is obtained if all x_{ij} that could enter into Equation (4.8.22) are ≥ 0.] This procedure, which parallels our earlier treatment of the cost vector, is also known as *parameterization* or *parametric linear programming.*

To perform a *sensitivity analysis,* or *ranging* study, of the right-hand side, we can proceed by considering a special case of the parameterization algorithm, as we did with the cost vector. Let $(m - 1)$ components of the m-dimensional vector $\boldsymbol{\rho}$ be zero, and let the remaining element, ρ_j, take on a value of either $+1$ or -1. The value for ϕ^* that causes $\mathbf{P}_j$ to leave the basis is then obtained using the above procedure. Two values for ϕ^* are sought —one for $\rho_j = +1$, the other for $\rho_j = -1$. These values represent the upper and lower bounds of the variation in the right-hand side of the jth constraint. The term ϕ cannot exceed these bounds without causing a change in the optimal basis vectors.

---

* The criterion expressed by Equation (4.8.22) is obtained from the *dual simplex algorithm,* which is described in Chapter 6.

Application to maximization problems

The conditions that we have developed are easily applied to post-optimal analyses of maximization problems as well as minimization problems. The foregoing development applies to maximization problems, providing one recognizes the following modifications:

1. Equations (4.8.8) and (4.8.9) apply only to those $\mathbf{P}_j$ for which $(\boldsymbol{\delta}'_B\mathbf{P}_j - \delta_j) < 0$.
2. Equation (4.8.22) is replaced with

$$q_j = -\frac{(y_j - c_j)}{x_{ij}}. \tag{4.8.23}$$

We still seek that particular $\mathbf{P}_j$ for which $x_{ij} < 0$ and $q_j > 0$ is a minimum.

Variations in the A *matrix*

Post-optimal analyses need not be confined to the cost vector or the right-hand side of the constraint set, although these are the most common and usually the most important items of interest. It is also possible to study the effects of varying one or more of the coefficients a_{ij} in the matrix $\mathbf{A}$ (the "technological coefficients"). The rationale is similar to that given above. Because the detailed treatment is more involved, however, we shall not consider this problem in this book. The reader is referred to the books of Gass [1969] and Orchard-Hays [1968] for more information on this subject.

EXAMPLE 4.8.1

Examine the effects of varying one or more coefficients in the objective function for the problem presented in Example 4.5.1.

The problem is: Minimize

$$y = x_1 + x_2 + x_3,$$

subject to

$$\begin{aligned} x_1 - x_2 + 2x_3 &= 5, \\ 2x_1 + 3x_2 \quad - x_3 &= 4, \\ x_1, x_2, x_3 &\geq 0. \end{aligned}$$

The optimal solution is $y_{\min} = \frac{19}{5}$, at $x_1 = \frac{13}{5}$, $x_2 = 0$, $x_3 = \frac{6}{5}$. The steps in obtaining the solution are shown in Tables 4.4–4.6.

Consider first the cost coefficients corresponding to the basis vectors. If c_1 is increased by one unit (i.e., if c_1 is changed from 1 to 2) and the optimal basis does not change, then, from Equation (4.8.2),

$$\frac{\partial y_{\min}}{\partial \delta_1} = x_1 = \tfrac{13}{5}$$

and

$$y_{\min} = \tfrac{19}{5} + \tfrac{13}{5} = \tfrac{32}{5}.$$

Similarly, if c_1 is held at its original value but c_3 is increased by one unit, again assuming that the optimal basis does not change, then

$$\frac{\partial y_{\min}}{\partial \delta_3} = x_3 = \tfrac{6}{5}$$

and

$$y_{\min} = \tfrac{19}{5} + \tfrac{6}{5} = 5.$$

Thus we see that the minimum value of the objective function is more sensitive to changes in c_1 than to changes in c_3.

We now inquire as to the change required in c_2 in order that $\mathbf{P}_2$ enter the basis. It is easy to establish that $(y_2 - c_2) = -1$ at the optimum. Thus, from Equation (4.8.3), the value of c_2 required in order that $\mathbf{P}_2$ enter the basis is

$$c_2^* = y_2 = (c_2 - 1) = (1 - 1) = 0.$$

The change in c_2 is given by

$$\delta_2 = (c_2^* - c_2) = (0 - 1) = -1.$$

Thus, the reduced cost associated with the nonbasis vector $\mathbf{P}_2$ is -1.

Now let us undertake a ranging study of the cost coefficients.

1. Let $\delta_1 = 1$ and $\delta_2 = \delta_3 = 0$.

$$\boldsymbol{\delta}_B = \begin{bmatrix} \delta_3 \\ \delta_1 \end{bmatrix} = \begin{bmatrix} 0 \\ 1 \end{bmatrix}.$$

(Notice that the optimal solution vector, $\mathbf{X}_B$, is expressed as

$$\mathbf{X}_B = \begin{bmatrix} x_3 \\ x_1 \end{bmatrix}$$

in Table 4.6. The ordering of the components in $\boldsymbol{\delta}_B$ must conform to the ordering of the corresponding components in $\mathbf{X}_B$.) The only nonbasis vector in this problem is $\mathbf{P}_2$. Thus, from Equation (4.8.8),

$$\boldsymbol{\delta}_B'\mathbf{P}_2 - \delta_2 = [0 \quad 1]\begin{bmatrix} -1 \\ 1 \end{bmatrix} - 0 = 1,$$

and

$$\theta^* = -\frac{y_2 - c_2}{\boldsymbol{\delta}_B'\mathbf{P}_2 - \delta_2} = 1.$$

Hence c_1 can be increased up to one unit without a change in the optimal policy.

2. Let $\delta_1 = -1$ and $\delta_2 = \delta_3 = 0$.

$$\boldsymbol{\delta}_B = \begin{bmatrix} 0 \\ -1 \end{bmatrix}$$

and

$$\boldsymbol{\delta}' \mathbf{P}_2 - \delta_2 = -1.$$

Since $\boldsymbol{\delta}'_B\mathbf{P}_2 - \delta_2 < 0$, we see that c_1 can be decreased indefinitely without a change in the optimal policy.

3. Let $\delta_3 = 1$, $\delta_1 = \delta_2 = 0$.

$$\boldsymbol{\delta}_B = \begin{bmatrix} 1 \\ 0 \end{bmatrix}$$

and

$$\boldsymbol{\delta}'_B\mathbf{P}_2 - \delta_2 = -1.$$

Therefore, c_3 can be increased indefinitely without changing the optimal policy.

4. We now let $\delta_3 = -1$, $\delta_1 = \delta_2 = 0$.

$$\boldsymbol{\delta}_B = \begin{bmatrix} -1 \\ 0 \end{bmatrix},$$

$$\boldsymbol{\delta}'_B\mathbf{P}_2 - \delta_2 = 1,$$

and, from Equation (4.8.8),

$$\theta^* = 1.$$

Thus c_3 cannot be reduced by more than one unit without bringing about a change in the optimal policy.

EXAMPLE 4.8.2

Examine the effect of varying the right-hand side of the constraint set for the problem presented in Example 4.8.1.

To carry out these post-optimal calculations, we must know $\mathbf{B}^{-1}$, the inverse of the optimal basis matrix. Observe that the columns of the optimal basis matrix are the *original* (untransformed) vectors which comprise the optimal basis. Hence the matrix $\mathbf{B}$ can be thought of as an operator that transforms the *optimal* solution vector $\mathbf{X}_B$ into the *original* (untransformed) right-hand side vector $\mathbf{P}_0$, in accordance with the expression

$$\mathbf{P}_0 = \mathbf{B}\mathbf{X}_B.$$

For this problem, we can write the above expression as

$$\begin{bmatrix} p_{10} \\ p_{20} \end{bmatrix} = \begin{bmatrix} p_{13} & p_{11} \\ p_{23} & p_{21} \end{bmatrix} \begin{bmatrix} x_3 \\ x_1 \end{bmatrix}$$

or

$$\begin{bmatrix} 5 \\ 4 \end{bmatrix} = \begin{bmatrix} 2 & 1 \\ -1 & 2 \end{bmatrix} \begin{bmatrix} \frac{6}{5} \\ \frac{13}{5} \end{bmatrix}.$$

The columns of the matrix **B** are simply the original vectors $\mathbf{P}_3$, $\mathbf{P}_1$.

The above matrix is easily inverted to yield

$$\mathbf{B}^{-1} = \begin{bmatrix} b_{31}^{-1} & b_{32}^{-1} \\ b_{11}^{-1} & b_{12}^{-1} \end{bmatrix} = \begin{bmatrix} \frac{2}{5} & -\frac{1}{5} \\ \frac{1}{5} & \frac{2}{5} \end{bmatrix}.$$

The inversion can be accomplished by any one of several well-known techniques. (Some matrix inversion techniques are presented in Appendix B.)

We can easily calculate the shadow prices once $\mathbf{B}^{-1}$ has been determined. If we increase the right-hand side of the first constraint by one unit, Equation (4.8.16) yields

$$\begin{aligned} \frac{\partial y_{\min}}{\partial p_1} &= c_3 b_{31}^{-1} + c_1 b_{11}^{-1} \\ &= (1)(\tfrac{2}{5}) + (1)(\tfrac{1}{5}) = \tfrac{3}{5}. \end{aligned}$$

Hence $y_{\min} = \frac{22}{5}$ if p_{10} is increased from 5 to 6 and the choice of optimal basis vectors does not change. Note, however, that the optimal policy does change, even though the choice of basis vectors may remain the same. From Equation (4.8.14) we have

$$\begin{aligned} x_3 &= x_3^o + b_{31}^{-1} p_1 \\ &= \tfrac{6}{5} + (\tfrac{2}{5})(1) = \tfrac{8}{5} \end{aligned}$$

and

$$\begin{aligned} x_1 &= x_1^o + b_{11}^{-1} p_1 \\ &= \tfrac{13}{5} + (\tfrac{1}{5})(1) = \tfrac{14}{5}. \end{aligned}$$

Similarly, we can increase the right-hand side of the second constraint, assuming that the first constraint has been restored to its original form and that the optimal basis does not change. This yields

$$\begin{aligned} \frac{\partial y_{\min}}{\partial p_2} &= c_3 b_{32}^{-1} + c_1 b_{12}^{-1} \\ &= (1)(-\tfrac{1}{5}) + (1)(\tfrac{2}{5}) = \tfrac{1}{5} \end{aligned}$$

and

$$y_{\min} = \tfrac{19}{5} + \tfrac{1}{5} = 4.$$

Also,

$$\begin{aligned} x_3 &= x_3^o + b_{32}^{-1} p_2 \\ &= \tfrac{6}{5} + (-\tfrac{1}{5})(1) = 1 \end{aligned}$$

and

$$x_1 = x_1^o + b_{12}^{-1}\rho_2$$
$$= \tfrac{13}{5} + (\tfrac{2}{5})(1) = 3.$$

We see that the minimum value of the objective function is more sensitive to a change in p_{10} than to a like change in p_{20}.

Now consider a parametric study of the right-hand side of the constraint set. Let

$$\boldsymbol{\rho} = \begin{bmatrix} \rho_1 \\ \rho_2 \end{bmatrix} = \begin{bmatrix} -2 \\ 3 \end{bmatrix}.$$

From Equation (4.8.20), we have

$$\boldsymbol{\psi} = \begin{bmatrix} \psi_3 \\ \psi_1 \end{bmatrix} = \begin{bmatrix} \frac{2}{5} & -\frac{1}{5} \\ \frac{1}{5} & \frac{2}{5} \end{bmatrix} \begin{bmatrix} -2 \\ 3 \end{bmatrix} = \begin{bmatrix} -\frac{7}{5} \\ \frac{4}{5} \end{bmatrix}.$$

Since $\psi_3 < 0$, we see that $\mathbf{P}_3$ will leave the basis for $\phi > \phi^*$, where

$$\phi^* = -\left(\frac{x_3^o}{\psi_3}\right) = \frac{6/5}{7/5} = \frac{6}{7}$$

from Equation (4.8.19).

The only vector that can enter the optimal basis, replacing $\mathbf{P}_3$, is $\mathbf{P}_2$, and this is possible only if x_{32}, as determined by Equation (4.8.21), is negative. Calculating $\mathbf{X}_2$ from Equation (4.8.21),

$$\mathbf{X}_2 = \begin{bmatrix} x_{32} \\ x_{12} \end{bmatrix} = \begin{bmatrix} \frac{2}{5} & -\frac{1}{5} \\ \frac{1}{5} & \frac{2}{5} \end{bmatrix} \begin{bmatrix} -1 \\ 3 \end{bmatrix} = \begin{bmatrix} -1 \\ 1 \end{bmatrix}.$$

Hence $x_{32} < 0$, and $\mathbf{P}_2$ enters the basis, replacing $\mathbf{P}_3$.

The new basis matrix is

$$\mathbf{B} = [\mathbf{P}_2, \mathbf{P}_1] = \begin{bmatrix} p_{12} & p_{11} \\ p_{22} & p_{21} \end{bmatrix} = \begin{bmatrix} -1 & 1 \\ 3 & 2 \end{bmatrix},$$

whose inverse is

$$\mathbf{B}^{-1} = \begin{bmatrix} b_{21}^{-1} & b_{22}^{-1} \\ b_{11}^{-1} & b_{12}^{-1} \end{bmatrix} = \begin{bmatrix} -\frac{2}{5} & \frac{1}{5} \\ \frac{3}{5} & \frac{1}{5} \end{bmatrix}.$$

When $\phi = \phi^*$, the optimal solution vector can be expressed in terms of either the old basis matrix, which we now call $\mathbf{B}_0$, or the new basis matrix $\mathbf{B}$. In terms of the old basis,

$$\mathbf{X}_B = \mathbf{B}_0^{-1}\mathbf{P}_0 + \phi^*\mathbf{B}_0^{-1}\rho$$
$$= \mathbf{X}_B^o + \phi^*\boldsymbol{\psi};$$
$$\begin{bmatrix} x_3 \\ x_1 \end{bmatrix} = \begin{bmatrix} \frac{6}{5} \\ \frac{13}{5} \end{bmatrix} + \frac{6}{7}\begin{bmatrix} -\frac{7}{5} \\ \frac{4}{5} \end{bmatrix} = \begin{bmatrix} 0 \\ \frac{23}{7} \end{bmatrix}.$$

If the new basis is employed,

$$\mathbf{X}_B = \mathbf{B}^{-1}\mathbf{P}_0 + \phi^*\mathbf{B}^{-1}\boldsymbol{\rho}$$

$$\begin{bmatrix} x_2 \\ x_1 \end{bmatrix} = \begin{bmatrix} -\frac{2}{5} & \frac{1}{5} \\ \frac{3}{5} & \frac{1}{5} \end{bmatrix}\begin{bmatrix} 5 \\ 4 \end{bmatrix} + \frac{6}{7}\begin{bmatrix} -\frac{2}{5} & \frac{1}{5} \\ \frac{3}{5} & \frac{1}{5} \end{bmatrix}\begin{bmatrix} -2 \\ 3 \end{bmatrix} = \begin{bmatrix} 0 \\ \frac{23}{7} \end{bmatrix}.$$

Thus the optimal solution, for $\phi = \phi^*$, is $x_1 = \frac{23}{7}$, $x_2 = x_3 = 0$, $y_{\min} = \frac{23}{7}$.

Now let $\phi > \phi^*$. We can express $\mathbf{X}_B$ as

$$\mathbf{X}_B = \mathbf{X}_B^o + \phi^*\mathbf{B}^{-1}\boldsymbol{\rho} + (\phi - \phi^*)\mathbf{B}^{-1}\boldsymbol{\rho},$$

or

$$\begin{bmatrix} x_2 \\ x_1 \end{bmatrix} = \begin{bmatrix} 0 \\ \frac{23}{7} \end{bmatrix} + (\phi - \phi^*)\begin{bmatrix} \frac{7}{5} \\ -\frac{3}{5} \end{bmatrix}.$$

Since $\psi_1 < 0$, we see that $\mathbf{P}_1$ will leave the basis at some value of $\phi > \phi^*$, say ϕ^{**}, providing a nonbasis vector can enter.

The only nonbasis vector in this problem, for $\phi > \phi^*$, is $\mathbf{P}_3$. Therefore, we form the vector $\mathbf{X}_3$ as

$$\mathbf{X}_3 = \mathbf{B}^{-1}\mathbf{P}_3,$$

$$\begin{bmatrix} x_{23} \\ x_{13} \end{bmatrix} = \begin{bmatrix} -\frac{2}{5} & \frac{1}{5} \\ \frac{3}{5} & \frac{1}{5} \end{bmatrix}\begin{bmatrix} 2 \\ -1 \end{bmatrix} = \begin{bmatrix} -1 \\ 1 \end{bmatrix}.$$

Since x_{13} is not negative, we see that $\mathbf{P}_3$ cannot enter the basis. We thus conclude that a feasible solution does not exist for $\phi > \phi^{**}$.

The numerical value for ϕ^{**} is easily obtained from Equation (4.8.19) as follows:

$$(\phi^{**} - \phi^*) = -\left(\frac{x_1^o}{\psi_1}\right) = \frac{23/7}{3/5} = \frac{115}{21},$$

$$\phi^{**} = \frac{6}{7} + \frac{115}{21} = \frac{19}{3}.$$

The optimal solution vector, at $\phi = \phi^{**}$, is

$$\begin{bmatrix} x_2 \\ x_1 \end{bmatrix} = \begin{bmatrix} 0 \\ \frac{23}{7} \end{bmatrix} + \frac{115}{21}\begin{bmatrix} \frac{7}{5} \\ -\frac{3}{5} \end{bmatrix} = \begin{bmatrix} \frac{23}{3} \\ 0 \end{bmatrix}.$$

Thus when $\phi = \frac{19}{7}$, $x_1 = x_3 = 0$, $x_2 = \frac{23}{3}$, and $y_{\min} = \frac{23}{3}$.

4.9 Duality

Recall that the concept of duality was introduced in Section 2.4. We observed that under appropriate conditions the maximum with respect to $\mathbf{X}$ of the Lagrangian $z(\mathbf{X}, \boldsymbol{\lambda})$, subject to the given constraints, is equal to the minimum with respect to $\boldsymbol{\lambda}$ of $z(\mathbf{X}, \boldsymbol{\lambda})$, subject to a different set of constraints. This second problem was referred to as the *dual*. For linear programming problems, the concept of duality can be further exploited to considerable advantage, as we shall see below.

In this section, we shall see that any linear programming problem of a given type (e.g., a minimization problem) corresponds to another linear programming problem of opposite type (a maximization problem). The number of independent variables of the given, or *primal*, problem equals the number of constraints of the corresponding *dual* problem, and the number of constraints of the primal equals the number of independent variables of the dual. Moreover, the value of the objective function that corresponds to the optimal feasible solution of the primal equals the value of the objective function that represents the optimal feasible solution of the dual.

Primal-dual properties

We shall begin by proving several interesting properties of dual problems, including those stated above. The practical application of these properties to the solution of linear programming problems will then be discussed.

We shall first show that the basic properties of the dual directly follow from the Kuhn–Tucker conditions which we presented in Section 2.7. Let us consider the problem of maximizing

$$y = c_1x_1 + c_2x_2 + \cdots + c_nx_n \tag{4.9.1}$$

subject to m constraints of the form

$$\begin{aligned} a_{11}x_1 + a_{12}x_2 + \cdots + a_{1n}x_n &\leq b_1, \\ &\vdots \\ a_{m1}x_1 + a_{m2}x_2 + \cdots + a_{mn}x_n &\leq b_m, \end{aligned} \tag{4.9.2}$$

and $x_j \geq 0$, $j = 1, 2, \ldots, n$. We shall also assume that there is a global maximum y_o at $(\mathbf{X}_o, \boldsymbol{\lambda}_o)$. Since the objective function and constraints are linear, we can think of them as concave or convex. The Kuhn–Tucker conditions are then necessary and sufficient for $y(\mathbf{X})$ to take on its global maximum. Substituting Equations (4.9.1) and (4.9.2) into Equation (2.7.12) yields

$$\left.\frac{\partial z}{\partial x_j}\right|_{(\mathbf{X}_o, \boldsymbol{\lambda}_o)} = c_j + \sum_{i=1}^{m} \lambda_{io}a_{ij} \leq 0. \tag{4.9.3}$$

Substituting into Equation (2.7.13) and summing over j results in

$$\sum_{j=1}^{n} x_{jo}\left.\frac{\partial z}{\partial x_j}\right|_{(\mathbf{X}_o, \boldsymbol{\lambda}_o)} = \sum_{j=1}^{n} x_{jo}\left[c_j + \sum_{i=1}^{m} \lambda_{io}a_{ij}\right] = 0, \tag{4.9.4}$$

while substitution into (2.7.18) and summation over i yields

$$\sum_{i=1}^{m} \lambda_{io} \frac{\partial z}{\partial \lambda_i}\bigg|_{(\mathbf{x}_o, \boldsymbol{\lambda}_o)} = \sum_{i=1}^{m} \lambda_{io}\left[-b_i + \sum_{j=1}^{n} a_{ij}x_{jo}\right] = 0. \tag{4.9.5}$$

For the constraint inequalities shown, $\lambda_{io} \leq 0$ for $i = 1, \ldots, m$. We can rewrite Equation (4.9.4) in the form

$$\sum_{j=1}^{n} x_{jo}c_j = -\sum_{i=1}^{m} \lambda_{io}\left(\sum_{j=1}^{n} a_{ij}x_{jo}\right). \tag{4.9.6}$$

If any constraint is active, we observe that the corresponding equality sign will hold in Equation (4.9.2). When a constraint is inactive, $\lambda_{io} = 0$. Hence we can substitute b_i for $\sum_{j=1}^{n} a_{ij}x_{jo}$ in Equation (4.9.6). We then have

$$\sum_{j=1}^{n} x_{jo}c_j = -\sum_{i=1}^{m} \lambda_{io}b_i. \tag{4.9.7}$$

We have thus shown that at the optimum we have an equivalent objective function. We now examine this new function at nonoptimum values of $(\boldsymbol{\lambda})$.

Let us multiply our constraint equations by λ_i. Since the λ_i are negative, we have

$$\lambda_i \sum_{j=1}^{n} a_{ij}x_j \geq \lambda_i b_i, \qquad i = 1, \ldots, m. \tag{4.9.8}$$

Changing signs and summing over all i, we obtain

$$-\sum_{i=1}^{m} \lambda_i b_i \geq -\sum_{i=1}^{m} \lambda_i\left(\sum_{j=1}^{n} a_{ij}x_j\right) = -\sum_{j=1}^{n}\left(\sum_{i=1}^{m} \lambda_i a_{ij}\right)x_j. \tag{4.9.9}$$

If Equation (4.9.3) is rewritten as

$$-\sum_{i=1}^{m} \lambda_i a_{ij} \geq c_j, \qquad j = 1, 2, \ldots, n, \tag{4.9.10}$$

then, combining with Equation (4.9.9),

$$-\sum_{i=1}^{m} \lambda_i b_i \geq \sum_{j=1}^{n} c_j x_j. \tag{4.9.11}$$

Therefore, if we choose

$$-\sum_{i=1}^{m} \lambda_i b_i$$

as our objective function, subject to the constraints of Equation (4.9.10), any vector $\boldsymbol{\lambda} < \mathbf{0}$ (other than $\boldsymbol{\lambda}_o$) will produce a value $y \geq y_o$.

The minus sign in front of the left side of Equation (4.9.11) may be eliminated by letting $w_j = -\lambda_j$; hence $w_{jo} = -\lambda_{jo}$, or

$$\mathbf{w}_o = -\boldsymbol{\lambda}_o. \tag{4.9.12}$$

Thus we have as our equivalent (dual) problem: Minimize

$$y = w_1 b_1 + w_2 b_2 + \cdots + w_m b_m, \tag{4.9.13}$$

subject to

$$w_1 a_{1j} + w_2 a_{2j} + \cdots + w_m a_{mj} \geq c_j, \qquad j = 1, 2, \ldots, n, \tag{4.9.14}$$

with $w_i \geq 0$, $i = 1, \ldots, m$.

Let us summarize our results. If the *primal* problem is: Maximize

$$y_p = \mathbf{C'X}, \tag{4.9.15}$$

subject to

$$\mathbf{AX} \leq \mathbf{P} \tag{4.9.16}$$

and

$$\mathbf{X} \geq \mathbf{0}, \tag{4.9.17}$$

the corresponding *dual* is: Minimize

$$y_d = \mathbf{P'W}, \tag{4.9.18}$$

subject to

$$\mathbf{A'W} \geq \mathbf{C} \tag{4.919}$$

and

$$\mathbf{W} \geq \mathbf{0}. \tag{4.9.20}$$

In the primal, **A** is an $m \times n$ matrix, **C** and **X** are n-dimensional vectors, and **P** is an m-dimensional vector. In the dual, **A′** is an $n \times m$ matrix (the transpose of **A**) and **W** is an m-dimensional solution vector. Note that the components of **W** are the negatives of the Lagrange multipliers. We have shown that *if the vectors* $\mathbf{X}_o$ *and* $\mathbf{W}_o$ *are optimal feasible solutions to the primal and dual problems, respectively, the corresponding objective fucntions are equal.*

Furthermore, since we have shown for the dual that

$$y_d \geq y_o \tag{4.9.21}$$

and for the primal

$$y_p \leq y_o, \tag{4.9.22}$$

the primal and dual objective functions can only be equal at the optimum. We therefore also conclude that *the vectors* $\mathbf{X}_o$ *and* $\mathbf{W}_o$ *are optimal feasible solutions to the primal and dual problems, respectively, if the corresponding objective functions are equal.*

The dual problem was derived on the assumption that the primal had a finite optimal solution. Since the Kuhn–Tucker conditions are necessary and sufficient for an optimum to exist in this case, there exists a $\boldsymbol{\lambda}_o$ that solves the dual. We thus may state: *The dual has an optimal feasible solution if the primal has an optimal feasible solution.*

Suppose now that primal has an unbounded solution; i.e., $\max y_p = \infty$. From Equations (4.9.21) and (4.9.22) we see that $y_d \geq y_p$. Hence we require that $\min y_d = \infty$, which implies that in general $y_d \geq \infty$. Since this is clearly impossible, we conclude that *the dual has no optimal solution if the optimal solution to the primal is unbounded.*

Let us now rewrite the dual problem as follows: Maximize

$$y_d^* = -\mathbf{P}'\mathbf{W}, \tag{4.9.23}$$

subject to

$$-\mathbf{A}'\mathbf{W} \leq -\mathbf{C} \tag{4.9.24}$$

and $\mathbf{W} \geq \mathbf{0}$. Since we can in general write $\max\{y(\mathbf{X})\} = -\min\{-y(\mathbf{X})\}$, we see that $\max(y_d^*) = -\min(-y_d)$. Hence we have recast the dual problem into a maximization problem equivalent to the original dual except for the sign of the objective function. Since Equations (4.9.23) and (4.9.24) now represent a maximization problem subject to ($\leq$) constraints, we can write the dual of this problem as before: Minimize

$$y_p^* = -\mathbf{C}'\mathbf{X}, \tag{4.9.25}$$

subject to

$$-\mathbf{AX} \geq -\mathbf{P}, \qquad \mathbf{X} \geq 0, \tag{4.9.26}$$

or, again allowing for a sign reversal in the objective function, maximize

$$y_p = \mathbf{C}'\mathbf{X}, \tag{4.9.27}$$

subject to

$$\mathbf{AX} \leq \mathbf{P} \tag{4.9.28}$$

and $\mathbf{X} \geq \mathbf{0}$. This is simply our original primal problem. Hence we conclude that *the dual of the dual is the primal.*

We could have considered Equations (4.9.18)–(4.9.20) as the primal problem and Equations (4.9.15)–(4.9.17) as the dual. The result would have been the same. In this case the primal would have been a minimization problem with ($\geq$) constraints, however, and the dual would have been a maximization problem with ($\leq$) constraints. Since we have the same symmetric relationship between primal and dual in either case, it does not matter which problem is chosen as the primal.

Use of the dual

From the foregoing discussion we can now see a practical application of duality theory in solving linear programming problems. Suppose that the original problem, as stated in the form of Equations (4.9.15)–(4.9.17), has a small number of independent variables but a large number of constraints. Recall that the number of constraints determines the number of vectors in the basis, and hence the maximum number of nonzero components in the solution vector. By solving the dual rather than the primal, we would be concerned with a problem having a small number of constraints and a large number of independent variables. The solution to the dual would therefore require less computation for each step (i.e., each vector transformation) than the primal. Moreover, the total number of steps required to obtain an optimal solution would be fewer with the dual than with the primal, since the number of required steps tends to be proportional to the number of constraints.

If the basic simplex algorithm is used to solve the dual problem, it can be shown that the nonzero components of the optimal primal solution vector appear in the bottom row [i.e., the $(y_j - c_j)$ row] of the table, under the columns which correspond to slack vectors.

EXAMPLE 4.9.1

Minimize the function

$$y_p = x_1 + 2x_2,$$

subject to

$$\begin{aligned} x_1 + x_2 &\geq 3, \\ 2x_1 + x_2 &\geq 2, \\ x_1 + 3x_2 &\geq 4, \\ 3x_1 - x_2 &\geq 1, \end{aligned}$$

and

$$x_1, x_2 \geq 0.$$

The dual of this problem is: Maximize

$$y_d = 3w_1 + 2w_2 + 4w_3 + w_4,$$

subject to

$$\begin{aligned} w_1 + 2w_2 + w_3 + 3w_4 &\leq 1, \\ w_1 + w_2 + 3w_3 - w_4 &\leq 2, \end{aligned}$$

and

$$w_1, w_2, w_3, w_4 \geq 0.$$

Adding two slack variables to the dual constraints and solving by means of the simplex algorithm, we obtain the final tabulation shown in Table 4.10. The maximum value of the dual objective function appears in the last row of the $\mathbf{P}_0$ column as $y_d = \frac{7}{2}$; the corresponding components of the solution vector also appear under $\mathbf{P}_0$ as $w_1 = \frac{1}{2}$ and $w_3 = \frac{1}{2}$, which implies that $w_2 = w_4 = 0$. Since this is an optimal feasible solution of a dual problem, we know that it is equal to the optimal feasible solution of the primal. Hence the minimum value of the primal objective function is $y_p = \frac{7}{2}$. The components of the primal solution vector appear in the last row of the $\mathbf{P}_5$ and $\mathbf{P}_6$ columns as $x_1 = \frac{5}{2}$ and $x_2 = \frac{1}{2}$.

Table 4.10. SOLUTION OF EXAMPLE 4.9.1

	c		3	2	4	1	0	0
Basis		$\mathbf{P}_0$	$\mathbf{P}_1$	$\mathbf{P}_2$	$\mathbf{P}_3$	$\mathbf{P}_4$	$\mathbf{P}_5$	$\mathbf{P}_6$
$\mathbf{P}_1$	3	$\frac{1}{2}$	1	$\frac{5}{2}$	0	5	$\frac{3}{2}$	$-\frac{1}{2}$
$\mathbf{P}_3$	4	$\frac{1}{2}$	0	$\frac{1}{2}$	1	-2	$-\frac{1}{2}$	$\frac{1}{2}$
$(y_j - c_j)$		$\frac{7}{2}$	0	$\frac{15}{2}$	0	6	$\frac{5}{2}$	$\frac{1}{2}$

Only three steps were required to solve this problem, and each step involved only a two-dimensional vector space. The reader can verify the minimal computational effort required to solve this problem with the dual by solving the primal problem directly.

4.10 Unrestricted Dual Variables

Most linear programming problems of practical interest are not stated in the particular form of Equations (4.9.1) and (4.9.2). Fortunately, however, the properties of the dual problem are independent of the particular form of the primal problem statement, so that any primal problem can be solved in terms of its dual if desired. To see this more clearly, let us investigate the applicability of duality theory to a more general primal problem. Consider the problem of minimizing the function

$$y_p = c_1x_1 + \cdots + c_jx_j + \cdots + c_nx_n, \tag{4.10.1}$$

subject to

$$
\begin{aligned}
&a_{11}x_1 + a_{12}x_2 + \cdots && + a_{1n}x_n \leq b_1,\\
&\vdots\\
&a_{k1}x_1 + a_{k2}x_2 + \cdots && + a_{kn}x_n \leq b_k,\\
&a_{(k+1)1}x_1 + a_{(k+1)2}x_2 + \cdots && + a_{(k+1)n}x_n \geq b_{(k+1)},\\
&\vdots\\
&a_{l1}x_1 + a_{l2}x_2 + \cdots && + a_{ln}x_n \geq b_l,\\
&a_{(l+1)}x_1 + a_{(l+1)2}x_2 + \cdots && + a_{(l+1)n}x_n = b_{(l+1)},\\
&\vdots\\
&a_{m1}x_1 + a_{m2}x_2 + \cdots && + a_{mn}x_n = b_m,
\end{aligned}
\tag{4.10.2}
$$

and $x_i \geq 0$, $i = 1, 2, \ldots, n$. Recall that the b_i are assumed to be non-negative; however, the cost coefficients are unrestricted in sign.

Now let us return to the Kuhn–Tucker conditions that we stated in Equations (4.9.3)–(4.9.5). These remain valid, but we must now rewrite our restrictions on the sign of λ_{io} in accordance with the rules of Section 2.7. We thus require that $\lambda_{io} \leq 0$, $i = 1, \ldots, k$; $\lambda_{io} \geq 0$, $i = k + 1, \ldots, l$; and λ_{io} unrestricted as to sign, $i = l + 1, \ldots, m$. All the arguments of Section 4.9. apply and the final results of Equations (4.9.15)–(4.9.20) hold. Since the dual variables are the negatives of the Lagrange multipliers, we see that the dual variables w_i must be restricted in sign such that $w_{io} \geq 0$, $i = 1, \ldots, k$; $w_{io} \leq 0$, $i = k + 1, \ldots, l$; and w_{io} unrestricted in sign, $i = l + 1, \ldots, m$. Note that the dual variables of the ($\geq$) inequalities are opposite in sign from those of the ($\leq$) inequalities. We may reverse any such inequality constraint by simply multiplying through by -1. We thus see that *a reversal in one of the primal inequality constraints changes the sign of the corresponding dual variable. Furthermore, the presence of an equality constraint in the primal causes the corresponding dual variable to be unrestricted in sign.*

Suppose that we now start out with the dual problem and consider it to be the primal. The dual of the given problem is, of course, the original problem, For each ($\leq$) inequality in the given problem we have $w_{io} \geq 0$ in the dual (i.e., $x_{io} \geq 0$ in the original), whereas for each ($\geq$) inequality in the given problem we have $w_{io} \leq 0$. Each equality in the given problem corresponds to a w_{io} that is unrestricted in sign. We may therefore conclude that *reversal of one of the dual inequality constraints changes the sign of the cor-*

responding primal variable. Moreover, *the dual constraint that corresponds to an unrestricted primal variable is an equality.*

The practicality of the above relationships can be seen in the next example.

EXAMPLE 4.10.1

Minimize the function

$$y_p = x_1 - 2x_2$$

subject to

$$x_1 + x_2 \geq 3,$$
$$x_1 + 3x_2 \geq 4,$$
$$3x_1 + x_2 \leq 8,$$

and

$$x_1, x_2 \geq 0.$$

The dual of this problem is: Maximize

$$y_d = 3w_1 + 4w_2 + 8w_3'$$

subject to

$$w_1 + w_2 + 3w_3' \leq 1,$$
$$w_1 + 3w_2 + w_3' \leq -2,$$

and

$$w_1, w_2 \geq 0, \qquad w_3' \leq 0.$$

Notice that $w_3' \leq 0$ because of the reversal in the third primal constraint. If we let $w_3 = -w_3'$, however, the dual problem can be written: Maximize

$$y_d = 3w_1 + 4w_2 - 8w_3$$

subject to

$$w_1 + w_2 - 3w_3 \leq 1,$$
$$-w_1 - 3w_2 + w_3 \geq 2,$$

and

$$w_1, w_2, w_3 \geq 0.$$

Adding a slack, a surplus, and an artificial variable and solving the dual by means of the simplex algorithm, we obtain the final tabulation given in Table 4.11. The maximum value of the dual objective function is $y_d = -16$, which occurs at $w_1 = w_2 = 0$, $w_3 = 2$ (or $w_3' = -2$). The minimum value of the primal objective function is $y_p = -16$, and the components of the corresponding solution vector are $x_1 = 0$, $x_2 = 8$, as seen in the last row of the $\mathbf{P}_4$ and $\mathbf{P}_5$ columns.

Table 4.11. SOLUTION OF EXAMPLE 4.10.1

	c		3	4	−8	0	0
Basis		$\mathbf{P}_0$	$\mathbf{P}_1$	$\mathbf{P}_2$	$\mathbf{P}_3$	$\mathbf{P}_4$	$\mathbf{P}_5$
$\mathbf{P}_4$	0	7	−2	−8	0	1	−3
$\mathbf{P}_3$	−8	2	−1	−3	1	0	−1
$(y_j - c_j)$		−16	5	20	0	0	8

REFERENCES

Beale, E. M., "Cycling in the Dual Simplex Algorithm," *Naval Res. Logistics Quart.*, *2*, No. 4 (1955).

Dantzig, G. B., *Linear Programming and Extensions*, Princeton University Press, Princeton, N.J., 1963.

Gass, S. I., *Linear Programming*, 3rd ed., McGraw-Hill, New York, 1969.

Hadley, G., *Linear Programming*, Addison-Wesley, Reading, Mass., 1962.

Hoffman, A. J., "Cycling in the Simplex Algorithm," *National Bureau of Standards Report No. 2974*, December, 1953.

Orchard-Hays, W., *Advanced Linear Programming Computing Techniques*, McGraw-Hill, New York, 1968.

PROBLEMS

4.1. Find the extreme points of the convex polyhedron defined by the inequalities

$$\begin{aligned} 3x_1 + 4x_2 + 6x_3 &\leq 24, \\ 10x_1 + 5x_2 - 6x_3 &\leq 30, \\ x_2 &\leq 4, \\ x_1, x_2, x_3 &\geq 0. \end{aligned}$$

Express the point $\mathbf{X} = \begin{bmatrix} 1 \\ 1 \\ 2 \end{bmatrix}$ as a linear combination of the extreme points.

Plot the convex polyhedron and the given point $\mathbf{X}$.

4.2. Show that the intersection of k n-dimensional hyperplanes, $k < n$, is a convex set.

4.3. Determine whether or not the following vectors are linearly independent:

(a) $\mathbf{P}_1 = \begin{bmatrix} 6 \\ 0 \\ 8 \\ 3 \end{bmatrix}, \quad \mathbf{P}_2 = \begin{bmatrix} 2 \\ -1 \\ 0 \\ 3 \end{bmatrix}, \quad \mathbf{P}_3 = \begin{bmatrix} 0 \\ 4 \\ 7 \\ -2 \end{bmatrix}, \quad \mathbf{P}_4 = \begin{bmatrix} 3 \\ -5 \\ -4 \\ 0 \end{bmatrix}$

(b) $\mathbf{P}_1 = \begin{bmatrix} 3 \\ 0 \\ 0 \\ 7 \end{bmatrix}, \quad \mathbf{P}_2 = \begin{bmatrix} 0 \\ 4 \\ 1 \\ 5 \end{bmatrix}, \quad \mathbf{P}_3 = \begin{bmatrix} 2 \\ 1 \\ 0 \\ 3 \end{bmatrix}, \quad \mathbf{P}_4 = \begin{bmatrix} 7 \\ -2 \\ -1 \\ 8 \end{bmatrix}$

(c) $\mathbf{P}_1 = \begin{bmatrix} 1 \\ 0 \\ 0 \\ 0 \end{bmatrix}, \quad \mathbf{P}_2 = \begin{bmatrix} 0 \\ 1 \\ 0 \\ 0 \end{bmatrix}, \quad \mathbf{P}_3 = \begin{bmatrix} 0 \\ 0 \\ 1 \\ 0 \end{bmatrix}, \quad \mathbf{P}_4 = \begin{bmatrix} 0 \\ 0 \\ 0 \\ 1 \end{bmatrix}$

4.4. A well-known theorem in linear algebra states that every basis in an m-dimensional vector space must contain the same number of vectors. Utilize this theorem to show that not more than m vectors of a set of n m-dimensional vectors, $m < n$, can be linearly independent. (*Hint:* Choose some particular basis for which the above statement is easily shown to be true.)

4.5. Given the accompanying table, which contains five three-dimensional vectors, where $\mathbf{P}_2$, $\mathbf{P}_4$, and $\mathbf{P}_5$ are the basis vectors, use the Gauss–Jordan elimination procedure to transform $\mathbf{P}_1$ into a unit vector, replacing $\mathbf{P}_2$. Express the nonbasis vectors in terms of the basis vectors for both the new and the old bases.

	$\mathbf{P}_1$	$\mathbf{P}_2$	$\mathbf{P}_3$	$\mathbf{P}_4$	$\mathbf{P}_5$
$i = 1$	-2	0	3	0	1
2	3	1	-1	0	0
3	0	0	4	1	0

4.6. Solve the following linear-programming problems graphically:
(a) Maximize $y = x_1 + 2x_2$, subject to

$$\begin{aligned} 2x_1 - x_2 &\leq 12, \\ -x_1 + x_2 &\leq 3, \\ 2x_1 + 3x_2 &\leq 24, \\ x_1, x_2 &\geq 0. \end{aligned}$$

(b) Maximize $y = x_1 + 1.5x_2$, subject to the constraints in (a).

(c) Maximize $y = x_1 - x_2$, subject to the constraints in (a).
(d) Minimize $y = x_1 + x_2$, subject to

$$\begin{aligned} 2x_1 - x_2 &\geq 6, \\ x_1 - 2x_2 &\geq 8, \\ x_1, x_2 &\geq 0. \end{aligned}$$

(e) Maximize $y = x_1 + x_2$, subject to the constraints in (d).
(f) Minimize $y = 3x_1 + 2x_2$, subject to

$$\begin{aligned} x_1 + x_2 &\leq 2, \\ 2x_1 - x_2 &\geq 6, \\ x_1, x_2 &\geq 0. \end{aligned}$$

4.7. Convert the inequality constraints in problem 4.6(a) to equality constraints by adding slack variables. Express the constraint set as

$$\mathbf{P}_0 = x_1\mathbf{P}_1 + x_2\mathbf{P}_2 + \cdots + x_5\mathbf{P}_5.$$

Choose every possible combination of basis vectors from the set $\mathbf{P}_1, \mathbf{P}_2, \ldots, \mathbf{P}_5$ and determine the corresponding x_i for each choice. Do all the combinations result in a feasible solution to the original problem? What is the optimal feasible solution?

4.8. Solve problems 4.6(a) through 4.6(f) by means of the simplex algorithm, adding slack, surplus, and artificial variables wherever necessary. Identify the type of solution (e.g., unique solution, multiple solutions, unbounded solution, infeasible solution, etc.) from the information generated in the simplex tables. Compare the solutions with those obtained graphically.

4.9. Solve the following problems analytically, using the simplex algorithm:
(a) Maximize $y = 4x_1 + 2x_2 + 3x_3$, subject to

$$\begin{aligned} 3x_1 + 4x_2 + 6x_3 &\leq 24, \\ 10x_1 + 5x_2 - 6x_3 &\leq 30, \\ x_2 &\leq 4, \\ x_1, x_2, x_3 &\geq 0. \end{aligned}$$

(b) Minimize $y = x_1 + x_2 - 3x_3$, subject to the constraints in (a).
(c) Maximize $y = x_1 + 5x_2 + 2x_3$, subject to

$$\begin{aligned} 4x_1 + x_2 + 3x_3 &\leq 6, \\ x_1 + 4x_2 + 2x_3 &\geq 8, \\ 3x_1 + 2x_2 + 6x_3 &\leq 12, \\ x_1, x_2, x_3 &\geq 0. \end{aligned}$$

(d) Minimize $y = x_1 + 2x_2 + 2x_3 - x_4$, subject to

$$\begin{aligned} 2x_1 - x_2 + 3x_3 + x_4 &\leq 10, \\ x_1 + x_2 - x_3 + 2x_4 &\geq 1, \\ -x_1 \qquad - x_3 + x_4 &\leq 12, \\ 3x_1 + 2x_2 \qquad + x_4 &= 5, \\ x_1, x_2, x_3, x_4 &\geq 0. \end{aligned}$$

(e) Maximize $y = 2x_1 + x_2 + 3x_3 + 2x_4$, subject to

$$\begin{aligned} x_1 + 2x_2 + x_3 + 2x_4 &\leq 12, \\ 3x_1 + 4x_2 + 2x_3 + 5x_4 &\geq 20, \\ x_1, x_2, x_3, x_4 &\geq 0. \end{aligned}$$

(f) Solve problem 4.9(e) as a minimization problem.
(g) Maximize $y = 2x_1 + x_2 + 3x_3 + 2x_4$, subject to

$$\begin{aligned} x_1 - 2x_2 + x_3 - 2x_4 &\leq 12, \\ 3x_1 + 4x_2 - 2x_3 + 5x_4 &\geq 20, \\ x_1, x_2, x_3, x_4 &\geq 0. \end{aligned}$$

(h) Maximize $y = 4x_1 + 5x_2 - x_3 - 2x_4$, subject to

$$\begin{aligned} 4x_1 - x_2 + 2x_3 - x_4 &= 6, \\ x_1 + 2x_2 - x_3 - x_4 &= 3, \\ -3x_1 + 3x_2 - 2x_3 + x_4 &\leq 6, \\ x_1, x_2, x_3, x_4 &\geq 0. \end{aligned}$$

4.10. Solve problem 1.8 by means of the simplex algorithm.

4.11. Solve the following linear programming problems, carrying two significant figures:

(a) Maximize $y = x_1 + x_2 + x_3$, subject to

$$\begin{aligned} 0.9785x_1 + 1.0023x_2 + 1.0310x_3 &\leq 4, \\ 2.7633x_1 + 2.4228x_2 + 2.4530x_3 &\leq 10, \\ 13.4052x_1 + 12.3488x_2 + 12.1087x_3 &\leq 50, \\ x_1, x_2, x_3 &\geq 0. \end{aligned}$$

(b) Maximize $y = 10x_1 + x_2 - 10x_3$, subject to

$$\begin{aligned} 1.1183x_1 + 0.00817x_2 + 0.9370x_3 &\leq 1, \\ 0.9816x_1 + 0.01259x_2 + 1.1504x_3 &\leq 1, \\ 0.9608x_1 + 0.00810x_2 + 0.9217x_3 &\leq 1, \\ x_1, x_2, x_3 &\geq 0. \end{aligned}$$

Resolve these problems using a desk calculator or (preferably) a digital computer. Is the additional accuracy necessary to obtain a valid solution? Do you suppose that numerical accuracy becomes more significant in problems of much higher dimensionality? Why?

4.12. Given the linear programming problem: Maximize $y = ax_1 + bx_2$, subject to

$$x_2 + 2x_1 \leq 6$$
$$2x_1 + x_2 \leq 8,$$
$$x_1, x_2 \geq 0.$$

Solve graphically for the following values of (a, b):

(i) (0, 1) (iv) (1, 0.5)
(ii) (0.2, 1) (v) (1, 0.2)
(iii) (0.5, 1) (vi) (1, 0)

Geometrically, what is the effect of changing the ratio a/b? (Note that $a = 6$, $b = 1$ corresponds to Example 4.4.1, shown in Figure 4.4.)

4.13. Given the linear programming problem: Maximize $y = x_1 + x_2$, subject to

$$x_1 + 2x_2 \leq a,$$
$$2x_1 + x_2 \leq b,$$
$$x_1, x_2 \geq 0,$$

solve graphically for the following values of (a, b):

(i) (1, 4) (iv) (4, 2)
(ii) (2, 4) (v) (4, 1)
(iii) (4, 4) (vi) (4, 0)

What is the effect of changing the value of either a or b? (Note that $a = 6$, $b = 8$ corresponds to Example 4.4.1 shown in Figures 4.3 and 4.4.)

4.14. Beginning with the optimal solution of problem 4.9(a), perform the following post-optimal analysis:

(a) Determine the reduced costs.
(b) Determine whether or not the optimal policy is affected by an increase in the coefficient of x_3 in the objective function. In what manner is the optimal basis altered?
(c) Determine the shadow prices.
(d) Repeat the above calculations for problem 4.9(b).

4.15. Beginning with the optimal solution of problem 4.9(e), perform the following post-optimal analysis:

(a) Determine the reduced costs.
(b) Determine what changes in the basis, and hence in the optimal policy, result from the following alteration in the coefficient vector:

$$\boldsymbol{\delta} = \begin{bmatrix} 1 \\ 2 \\ -1 \\ 1 \end{bmatrix}.$$

(c) What is the effect of altering the right-hand side of the constraint set by

$$\boldsymbol{\rho} = \begin{bmatrix} -1 \\ 2 \end{bmatrix}.$$

(d) Repeat the above calculations for problem 4.9(f).

4.16. Obtain the optimal solution to the following problems by solving the corresponding dual problems:
(a) Problem 4.9(a)
(b) Problem 4.9(c)
(c) Problem 4.9(g)
(d) Problem 4.9(h)
(*Hint:* A variable w_i that is unrestricted in sign can be expressed as the difference of two nonnegative variables; i.e., $w_i = w_i' - w_i''$, where $w_i', w_i'' \geq 0$.)

4.17. The milling department of a large manufacturing plant is required to process k different kinds of products in a given period of time. The department has available l milling machines. These machines differ in age and efficiency, so that certain of the machines may be better suited to a given kind of product than other machines. The following information is known (or can be estimated with reasonable accuracy): number of units of each product to be processed, total time available for each machine, time required to process one unit of a given product type on any particular machine, and the corresponding processing cost.

Derive a linear programming model to determine a minimum-cost processing policy (i.e., to determine how many products of each type shall be allocated to each machine). Assume that each product need not be processed by more than one machine.

4.18. Solve problem 4.17 for the data shown in the accompanying table.

Job No.	*No. Units Per Job*		*Procesing Cost ($/unit) and Processing Time (hr/unit)*			
			Machine No.			
			1	*2*	*3*	*4*
1	40	$	2.00	1.80	2.20	2.50
		hr	0.80	1.10	0.95	1.25
2	10	$	4.65	5.40	5.25	6.50
		hr	0.75	0.90	0.95	1.00
3	50	$	1.50	1.75	1.75	1.90
		hr	0.70	0.95	1.10	1.20
4	25	$	3.20	3.50	3.00	3.75
		hr	1.05	1.00	0.95	1.10
5	35	$	1.00	1.20	1.35	1.20
		hr	0.85	0.95	1.05	0.90
6	15	$	2.25	2.00	2.50	2.25
		hr	1.25	1.25	1.50	1.35

Each machine is available up to the following maximum number of hours:

	Machine No.			
	1	*2*	*3*	*4*
Maximum available hours	50	35	45	60

How is the optimal policy affected by a 50 per cent increase in the processing cost of job 3? What would be the optimal policy if machine 3 were to break down unexpectedly at the start of the time period? What would happen if job 5 was suddenly canceled and job 4 was increased to 50 units?

4.19. An automobile manufacturer has k assembly plants scattered geographically around the country, and l marketing regions (each marketing region being a collection point for a major metropolitan area). Each assembly plant has a maximum capacity of a_i units per month, and each marketing region has a known demand of b_j units per month. The cost of transporting an automobile from the ith assembly plant to the jth marketing region is given by c_{ij}.

Derive a linear programming model to determine how many automobiles should be assembled at each plant and where they should be sent.
(*Note:* Problems of this type, where the constraint matrix consists entirely of zeros and ones and each independent variable appears only twice, are known as *transportation problems.* Because of its particular structure, the transportation problem can be solved by special algorithms more easily than by the simplex procedure, although the simplex or revised simplex algorithms will yield correct results. We shall say more about transportation problems in Chapter 6, where we shall present a specialized algorithm for solving such problems.)

4.20. Solve problem 4.19 for the following set of data:

		Transportation Cost, \$/automobile							
Assembly Plant	*Capacity, Units/mo.*	*Marketing Region*							
		1	*2*	*3*	*4*	*5*	*6*	*7*	*8*
1	12,000	72	85	43	27	118	86	57	63
2	10,000	176	45	153	187	22	66	125	85
3	8,000	45	22	53	78	110	36	85	70
4	6,000	120	100	65	48	37	84	21	39
5	14,000	27	36	18	125	53	22	110	27
Market demand, units/mo.		6800	5400	4800	8500	7200	5700	6100	4500

4.21. A large oil refinery produces a stream of hydrocarbon base fuel from each of four different processing units. Each processing unit has a capacity of Q_i

barrels per day. A portion of each base fuel is fed to a central blending station where the base fuels are mixed into three grades of gasoline. The remainder of each base fuel is sold "as is" at the refinery. Storage facilities are available at the blending station for temporary storage of the blended gasolines, if desired.

Let N_1, N_2, N_3, and N_4 be the octane ratings of the respective base fuels, and let R_1, R_2, and R_3 be the minimum octane requirements of the three grades of gasoline that must be blended on a given day to satisfy customer demands, D_1, D_2, and D_3. Let x_{ij} be the quantity of the ith base fuel used to blend the jth gasoline. Let P_1, P_2, and P_3 be the profit per gallon that is realized from the sale of each gasoline, and let S_1, S_2, S_3, and S_4 be the profit derived from direct sale of the base fuels.

Determine how much of each base fuel should be blended into gasoline and how much should be sold "as is" in order to maximize profit. The following numerical data may be used:

$$\begin{array}{llll}
Q_1 = 13{,}000 \text{ bbl/day} & & N_1 = 82 \text{ octane} \\
Q_2 = 7{,}000 \text{ bbl/day} & & N_2 = 95 \text{ octane} \\
Q_3 = 25{,}000 \text{ bbl/day} & & N_3 = 102 \text{ octane} \\
Q_4 = 15{,}000 \text{ bbl/day} & & N_4 = 107 \text{ octane} \\
D_1 = 13{,}000 \text{ bbl/day} & & R_1 = 93 \text{ octane} \\
D_2 = 25{,}000 \text{ bbl/day} & & R_2 = 97 \text{ octane} \\
D_3 = 18{,}000 \text{ bbl/day} & & R_3 = 102 \text{ octane} \\
S_1 = 45¢\text{/bbl} \\
S_2 = 52¢\text{/bbl} \\
S_3 = 57¢\text{/bbl} \\
S_4 = 60¢\text{/bbl} \\
P_1 = 1\tfrac{3}{4}¢\text{/gal} \\
P_2 = 2\tfrac{1}{2}¢\text{/gal} \\
P_3 = 4¢\text{/gal}
\end{array}$$

Note that 42 gal = 1 bbl.

Assume that the octane number of each gasoline can be expressed as the weighted average of the octane numbers which correspond to the constituent base fuels, the weighting factors being the fractions of each base fuel in each gasoline.

How would the optimal profit and the optimal policy be affected if the selling prices of the gasolines were changed to 2¢/gal, $2\frac{1}{2}$¢/gal (as before), and $3\frac{1}{2}$¢/gal, respectively? What would be the consequences of increasing D_3 by 2000 bbl/day while simultaneously decreasing D_2 by 7000 bbl/day?

4.22. A financial analyst has been asked by a client to invest a sizeable sum of money in a portfolio of common stocks. The overall objective is to maximize the growth of capital. However, the portfolio is not to exceed a prescribed degree of risk. Also, the portfolio must provide at least enough income to pay for taxes and other expenses.

The analyst has available to him information on a large number of companies, broken down into various industry groups. The growth potential of any company in a given industry group is measured on an integer scale ranging from 1 to 5; i.e., a rating of 1 indicates minimum likelihood of capital appreciation, whereas a rating of 5 indicates the highest likelihood. Similar scales are assigned as measures of risk and of yield (income).

The particular portfolio in question is to contain several industry groups. Each industry group is to represent no more than 20 per cent of the entire portfolio. Each individual company is to represent no more than 10 per cent of the portfolio. Derive an appropriate linear programming model to solve this problem.

Suppose that the problem is altered slightly by requiring that the stock of each individual company in the portfolio represent at least 5 per cent but no more than 10 per cent of the entire portfolio. Would this complicate the model? How might such a problem be solved? Explain.

4.23. Solve problem 4.22 for the data shown in the accompanying tables. Assume that the yield factor must be at least 2.2, and the risk factor cannot exceed 3.7. Let G represent the growth factor, R the risk factor, and Y the yield factor. Assume that the portfolio risk factor and the portfolio yield factor are weighted averages of the corresponding factors of the constituent stocks.

	Group 1					*Group 2*				*Group 3*						*Group 4*			
	Company					*Company*				*Company*						*Company*			
	A	*B*	*C*	*D*	*E*	*A*	*B*	*C*	*D*	*A*	*B*	*C*	*D*	*E*	*F*	*A*	*B*	*C*	*D*
G	4	3	2	5	3	3	3	1	2	4	5	2	3	4	3	3	2	3	1
R	3	5	1	4	1	4	3	2	1	2	3	3	3	5	3	1	2	3	4
Y	1	2	4	1	4	3	3	5	5	3	2	3	5	1	3	3	3	2	1

	Group 5					*Group 6*						*Group 7*				
	Company					*Company*						*Company*				
	A	*B*	*C*	*D*	*E*	*A*	*B*	*C*	*D*	*E*	*F*	*A*	*B*	*C*	*D*	*E*
G	4	3	1	4	3	5	4	5	4	3	3	3	1	2	3	2
R	3	2	1	2	3	3	3	2	4	4	2	5	2	5	3	3
Y	3	3	5	3	4	3	2	1	2	1	5	3	4	2	1	4

4.24. A distributor maintains an inventory of some particular commodity in a warehouse whose capacity is 50,000 units. The supply and demand for this commodity are seasonal. Therefore, the distributor will purchase excess quantities of the commodity when the purchase cost is low, store the excess, and sell it at some later time when the selling price is high. The demand, purchase cost, storage cost, and selling price vary on a monthly basis as shown in the accompanying table.

Month	*Demand (units)*	*Purchase Cost ($/unit)*	*Monthly Storage Cost ($/unit)*	*Selling Price ($/unit)*
Jan.	20,000	20	1.4	40
Feb.	18,000	16	1.6	36
Mar.	15,000	13	1.8	30
Apr.	11,000	10	1.6	24
May	9,000	12	1.4	20
June	7,000	15	1.2	17
July	6,000	18	1.0	15
Aug.	7,500	22	0.8	18
Sept.	10,000	26	0.6	22
Oct.	13,000	30	0.8	27
Nov.	16,000	28	1.0	32
Dec.	19,000	24	1.2	37

If the total inventory is 30,000 units on Jan. 1, determine an optimal buy–sell strategy that will maximize the yearly profit. Would the optimal policy be significantly different if the storage costs were negligible?

NONLINEAR PROGRAMMING

We have seen that constrained optimization problems which consist entirely of linear relationships can be solved by the powerful techniques of linear programming. Solutions can be obtained with relative ease on a digital computer, and a global optimum will always be obtained (providing, of course, that a finite optimum exists for a given problem). Thus there is a tendency for the analyst to express a given problem in terms of a linear model if it is at all possible to do so.

Many optimization problems contain a few relatively insignificant nonlinearities. These problems can be adequately represented by a linear model and solved by means of linear programming. There are, however, many other optimization problems that are highly nonlinear. Such problems are particularly prevalent in the areas of engineering design and control. These problems cannot be represented by linear models without destroying many of the salient features that characterize the problem. Therefore, there is considerable interest in the development of optimization techniques that can accommodate models containing significant nonlinearities. We shall discuss several such techniques in this chapter.

If all the constraints are equalities, we may form the Lagrangian function and find the vectors $(\mathbf{X}, \boldsymbol{\lambda})$ that produce a saddle point, as discussed in

5

Chapter 2. Furthermore, we can transform any inequality constraint into an equality by introducing an appropriate slack or surplus variable. Thus we can form the Lagrangian even if some constraints are not equalities initially. Numerical solution techniques based upon this classical procedure have been proposed but have performed poorly. Not only does this method substantially increase the number of variables by introducing the Lagrange multipliers, but we must locate a saddle point rather than an extremum. Considerably greater success can usually be obtained by representing the nonlinear programming problem in a form such that the techniques developed for linear problems or unconstrained nonlinear problems are applicable. We shall examine several of these methods and then look at a procedure specifically developed for polynomial optimization that makes use of the Lagrangian.

5.1 Convex and Concave Programming

Before examining the available nonlinear programming techniques, it is desirable to consider the circumstances under which we may be certain that our problem does not contain local optima. From our previous discussion of optimization fundamentals, we know that a minimization problem may contain separate local minima unless the objective function is convex (cf. Sections 1.8, 2.6, and 2.7). This situation may be further aggravated when constraints are present. The first requirement is therefore a convex objective function in the region of interest.

For $y(\mathbf{X})$ to be convex, it must be defined on a convex set. When $(\mathbf{X})$ is unconstrained, we encounter no difficulties since the set of all real numbers is convex. In the presence of constraints, we must be able to distinguish a convex set. To do so, we make use of the theorem which states that if $y(\mathbf{X})$ is convex, the set

$$R = \{(\mathbf{X}) \mid y(\mathbf{X}) \leq k\} \tag{5.1.1}$$

is convex for all positive (k).* We may restrict the definition of $y(\mathbf{X})$ to the set R and still retain the property of convexity. We now make use of the theorem which states that the intersection of any number of convex sets is convex. Hence if $y(\mathbf{X})$, which was originally convex over the set of real numbers, is defined only over the constraint set

$$g_i(\mathbf{X}) \leq 0 \qquad i = 1, \ldots, m, \tag{5.1.2}$$

where each $g_i(\mathbf{X})$ is convex, then $y(\mathbf{X})$ is still convex. We should then conclude that such a function would not have separate local minima.

The problem of minimizing a convex function over a convex constraint set is called a *convex programming problem*. We see, from our previous discussion, that any local minimum of a convex programming problem is a global minimum (Figure 5.1a). If the global minimum exists at a number of points, the set of such points will be convex (e.g., on a line or hyperplane). No separated local minima will exist. It will be recalled that linear functions are both convex or concave. Hence any inequality constrained linear programming problem is a convex programming problem. When the constraints are not convex, we may introduce local minima as shown in Figure 5.1b.

In a *concave programming problem* we: Maximize

$$y(\mathbf{X}) \tag{5.1.3}$$

subject to

$$g_i(\mathbf{X}) \geq 0, \qquad i = 1, \ldots, m, \tag{5.1.4}$$

under the conditions that $y(\mathbf{X})$ and the $g_i(\mathbf{X})$ are concave and defined over a convex set. Any such problem can be converted to an equivalent minimiza-

* The notation $R = \{(\mathbf{X}) \mid y(\mathbf{X}) \leq k\}$ means that R is that set of vectors $(\mathbf{X})$, which satisfies the inequality $y(\mathbf{X}) \leq k$.

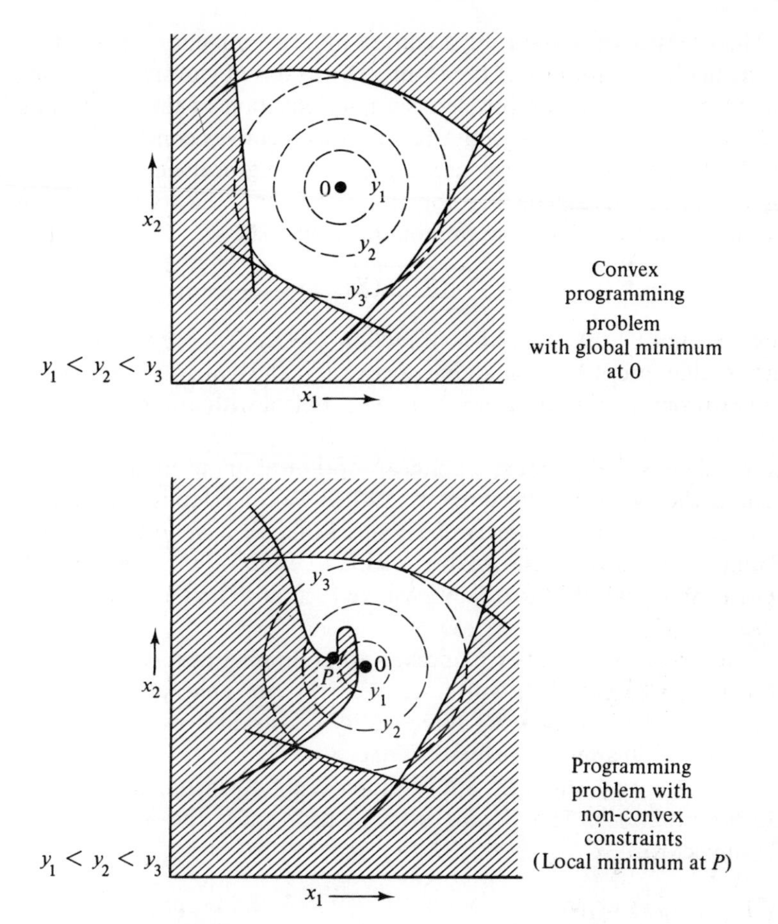

Figure 5.1. Appearance of Local Minimum in Presence of Non-Convex Constraints

tion problem through multiplication by -1. Thus we have: Minimize

$$-y(\mathbf{X}) \tag{5.1.5}$$

subject to

$$-g_i(\mathbf{X}) \leq 0, \qquad i = 1, \ldots, m. \tag{5.1.6}$$

Since the negative of a concave function is convex, this is a convex programming problem with a unique value for the global minimum, the value of the minimum being the negative of the maximum of the original problem. Therefore, we conclude that in any concave programming problem any local maximum is a global maximum.

The problem of minimizing a convex objective function subject to concave inequality constraints of the form $g_i(\mathbf{X}) \geq 0$ is obviously transformable to a convex programming problem by multiplication of the constraints by -1. Similarly, a concave objective function with convex constraints $g_i(\mathbf{X}) \leq 0$ can be transformed into a concave programming problem. However, the pressence of nonlinear equality constraints prevents the transformation of a problem into a convex programming problem. The set

$$R = \{(\mathbf{X}) \mid g_i(\mathbf{X}) = k\} \tag{5.1.7}$$

is convex only if $g_i(\mathbf{X})$ is a linear function of $(\mathbf{X})$. Thus the general linear programming problem, containing inequality and equality constraints, is a convex programming problem, but any problem with a nonlinear equality constraint is not.

We shall see that most nonlinear optimization techniques can only determine the location of local optima. When we can establish that the problem is convex or concave, we can be assured that we have located the global optimum. In most real cases we are unable to establish the convexity of the system and we shall therefore lack positive assurance that a global optimum has been obtained.

When we have a convex (concave) programming problem, the concept of duality can be useful. It will be recalled that the concept of duality was first introduced in Section 2.4 for equality constrained extremization. We later (Section 4.9) applied the duality concept to inequality constrained linear programs. We may similarly apply the concept to inequality constrained nonlinear programs. Thus if our original problem is the minimization of $y(\mathbf{X})$, subject to

$$g_i(x_1, x_2, \ldots, x_n) \leq 0, \qquad i = 1, 2, \ldots, m, \tag{5.1.8}$$

we can rewrite this primal problem as the minimization of the Lagrangian $z(\mathbf{X}, \boldsymbol{\lambda})$ with respect to $\mathbf{X}$, subject to the constraints of (5.1.8). When the Kuhn–Tucker theorem holds, we can write the dual as the maximization of $z(\mathbf{X}, \boldsymbol{\lambda})$ with respect to $\boldsymbol{\lambda}$, subject to the constraints

$$\frac{\partial z}{\partial x_j}(\mathbf{X}, \boldsymbol{\lambda}) = \frac{\partial y}{\partial x_j} + \sum_{i=1}^{m} \lambda_i \frac{\partial g_i}{\partial x_j} \leq 0 \qquad j = 1, \ldots, n \tag{5.1.9}$$

[cf. Equation (2.7.12)]. The Kuhn–Tucker theorem tells us that if $(\mathbf{X}_0, \boldsymbol{\lambda}_0)$ solves the primal problem, the quantities together solve the dual problem. When $y(\mathbf{X})$ and $g_i(\mathbf{X})$ are convex (concave), the Kuhn–Tucker conditions hold, and the saddle point $(\mathbf{X}_0, \boldsymbol{\lambda}_0)$ is a global saddle point (cf. Section 2.7). Hence if our original program ming problem is convex (concave), we can locate its global optimum by solving its dual. We shall make use of this concept in our discussion of geometric programming.

5.2 Linearization

Linear approximations

In the real world there are few problems that are entirely linear. However, the availability of powerful linear programming procedures provides a major incentive for approximating problems in this form. Hence one of the better means for solving nonlinear programming problems remains replacement of the nonlinear relationships by linear approximations and solution of the resulting linear programming problem. When the nonlinearities are small, the result may be a close approximation to the correct solution. As we have noted previously, however, this simple procedure is not adequate as the nonlinearities become significant.

There are some instances when it is possible to obtain a good linear approximation for a problem with significant nonlinearities. Let us consider the case of a so-called "pessimistic" economy. Assume that the profit y from a particular activity u increases monotonically with u but at an ever-decreasing rate. This is a very common situation. For example, u may be the total plant capital investment, which yields a decreasing profit from each incremental amount invested, as shown in Figure 5.2.

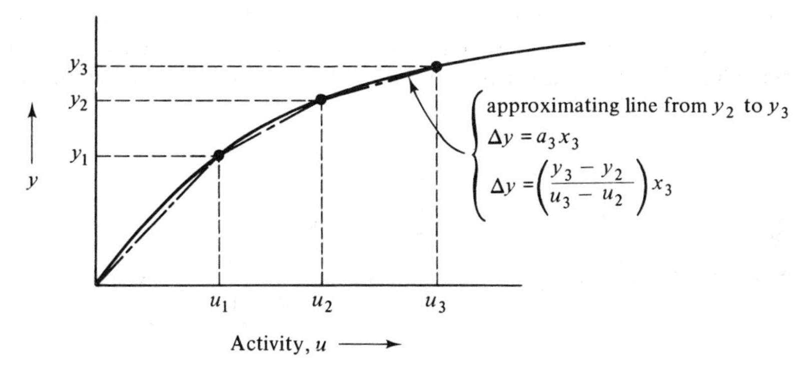

Figure 5.2. Linearization of Function with Diminishing Marginal Returns

To use the linear programming approach, yet preserve the nonlinearity of the profit function, we approximate the profit curve by a series of linear segments. Thus we make the approximation (cf. Figure 5.2) that the profit y is given by

$$y = a_1 x_1 + a_2 x_2 + a_3 x_3 + a_4 x_4 + \dots, \tag{5.2.1}$$

where the total investment u is

(5.2.2) $$u = x_1 + x_2 + x_3 + \dots.$$

Since the linear approximations hold only over narrow ranges, we must require that

$$x_1 \leq u_1$$

and

(5.2.3) $$x_j \leq u_j - u_{j-1}, \qquad j = 2, 3, \dots$$

We may approximate the curve to any desired accuracy by increasing the number of segments selected. If we proceed to maximize the profit subject to some limit on u, we are assured that unless x_j is at its maximum value, the optimum value for x_{j+1} will be zero. Since $a_j > a_{j+1}$, a unit of x_j increases the objective function more than a unit of x_{j+1}. The variable x_{j+1} does not appear in the solution unless it is not possible to add more x_j. Thus if u had a value between u_2 and u_3, x_1 and x_2 would be at their maximum values, x_3 would be between zero and its maximum, and x_4 would be zero.

EXAMPLE 5.2.1

The use of this technique may be illustrated through the gasoline-blending operation shown in Figure 5.3. Suppose that tetraethyl lead (TEL) is being added to a mixture of two streams to raise the octane number of the product. The incremental effect of TEL is independent of the amounts of the two feed materials and

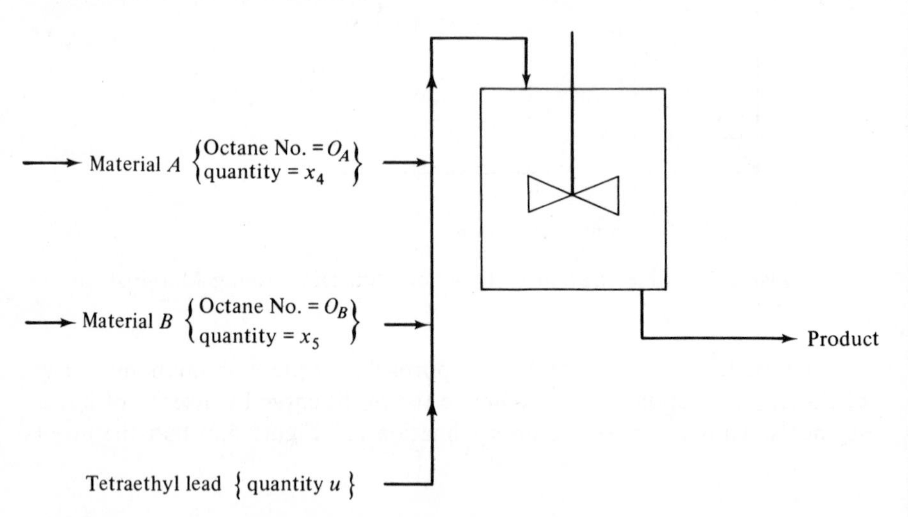

Figure 5.3. Gasoline Blending Operation

depends only on the amount of TEL present. We shall assume that the octane number, y, varies with the quantity of TEL, u, as shown in Figure 5.2 (each incremental addition of TEL produces a decreasing change in octane number). The octane number of the untreated gasoline is a linear function of the untreated composition. The minimum allowable octane number of the product is T_1 and the maximum allowable is T_2. We are to determine the quantities of materials A and B to be used to produce a barrel of product at minimum cost.

We first define a set of constants a_i that approximate the octane number improvement per unit TEL for several ranges. Thus, from Figure 5.2, we have

$$a_1 = \frac{y_1}{u_1}, \qquad a_2 = \frac{y_2 - y_1}{u_2 - u_1}, \qquad a_3 = \frac{y_3 - y_2}{u_3 - u_2}. \tag{5.2.4}$$

If we follow our previous notation, the total amount of TEL is given by

$$u = x_1 + x_2 + x_3 \tag{5.2.5}$$

and the total octane improvement it produces is

$$a_1x_1 + a_2x_2 + a_3x_3. \tag{5.2.6}$$

Our constraints on the minimum and maximum octane numbers of the product then become

$$a_1x_1 + a_2x_2 + a_3x_3 + O_Ax_4 + O_Bx_5 \geq T_1, \tag{5.2.7}$$

$$a_1x_1 + a_2x_2 + a_3x_3 + O_Ax_4 + O_Bx_5 \leq T_2, \tag{5.2.8}$$

where O_A and O_B are the octane numbers of materials A and B. We also require that, for each barrel of product, the sum of feed A (represented by x_4) and B (represented by x_5) be given by

$$x_4 + x_5 = 1. \tag{5.2.9}$$

The final constraints are those which restrict the x_j to the proper range:

$$x_1 \leq u_1, \tag{5.2.10}$$

$$x_2 \leq u_2 - u_1, \tag{5.2.11}$$

$$x_3 \leq u_3 - u_2, \tag{5.2.12}$$

$$x_j \geq 0, \qquad j = 1, 2, \ldots, 5. \tag{5.2.13}$$

The objective function is then: Minimize

$$c_1(x_1 + x_2 + x_3) + c_2x_4 + c_3x_5. \tag{5.2.14}$$

Since the costs of x_1, x_2, and x_3 are identical, it is less expensive (we achieve the maximum increase in the octane number with the minimum addition of TEL) to use as much x_1 as possible before using x_2, and as much x_2 as possible before using x_3. We can therefore be assured that x_2 will not appear in the solution until x_1 is at its maximum, and x_3 will not appear until x_2 is at its maximum value. A contiguous set of line segments will thus be in the optimum solution. A solution to the linear approximating problem is obtained readily from the simplex method. It should be noted, however, that by linearizing the problem we have increased the number of independent variables. In large problems this increase in dimensionality can be significant.

Quadratic programming

Some of the first attempts to extend linear programming techniques were directed at problems in which the only nonlinearities were quadratic terms. If we write the Lagrangian for an equality constrained problem, we have

$$z = y(\mathbf{X}) + \sum_i \lambda_i g_i(\mathbf{X}), \tag{5.2.15}$$

and we obtain the constraints to be satisfied by taking the partial derivatives with respect to each of the x_j. If the nonlinear terms appear only in $y(\mathbf{X})$ and the highest-order terms are quadratic, then these partial derivatives will be linear in the x_j. It would seem that there ought to be some way in which we can take advantage of this linearity. Quadratic programming does just this.

Quadratic programming, as normally defined, requires not only that the nonlinearities be quadratic but that they be confined to the objective function. With this restriction our objective function and constraints become: Maximize

$$y = \sum_{j=1}^{n} c_j x_j + \sum_{k=1}^{n} \sum_{j=1}^{n} d_{kj} x_k x_j, \tag{5.2.16}$$

subject to

$$g_i(\mathbf{X}) = a_{i1}x_1 + a_{i2}x_2 + \cdots + a_{in}x_n - b_i = 0, \qquad i = 1, \ldots, m, \tag{5.2.17}$$

where the c_j, a_{ij}, and d_{jk} are constants and $d_{kj} = d_{jk}$. We also require that $x_j \geq 0$, $j = 1, 2, \ldots, n$. The condition for the optimum may be expressed in terms of Lagrange multipliers in accordance with the Kuhn–Tucker conditions. At the optimum, we require [cf. Equation (2.7.12)] that there be a set of x_{j0} and λ_{i0} such that for each x_j

$$\left[\frac{\partial y(\mathbf{X}_0)}{\partial x_j} + \sum_{i=1}^{m} \lambda_{i0} \frac{\partial g_i(\mathbf{X}_0)}{\partial x_j}\right] \leq 0, \qquad j = 1, \ldots, n. \tag{5.2.18}$$

For the problem at hand, (5.2.18) may be written

$$\left[c_j + 2\sum_{k=1}^{n} x_{k0}d_{kj} + \sum_{i=1}^{m} \lambda_{i0}a_{ij}\right] \leq 0, \qquad j = 1, \ldots, n. \tag{5.2.19}$$

Note that the term inside the brackets is zero whenever x_{j0} is nonzero and also that it is linear in both x_j and λ_i. We shall make use of this linearity so that the simplex method can be applied.

We now introduce a set of slack variables s_j such that

$$\left[c_j + 2\sum_{k=1}^{n} x_k d_{kj} + \sum_{i=1}^{m} \lambda_i a_{ij}\right] + s_j = 0, \qquad j = 1, \ldots, n. \tag{5.2.20}$$

If $x_j = 0$, the term inside the brackets of (5.2.20) is nonzero, and s_j will be nonzero. When x_j is nonzero, the bracketed terms will be zero, and s_j will be zero [see Equation (2.7.13)]. This condition may be written as

$$x_{j0}s_j = 0, \qquad j = 1, \ldots, n. \tag{5.2.21}$$

The optimum solution is then one which satisfies the equation sets (5.2.17) and (5.2.20) while maintaining $x_j s_j$ equal to zero. This latter requirement is nonlinear, and we cannot impose it directly.

The quadratic programming scheme due to Wolfe [1959] makes use of the simplex method to solve the problem. Wolfe suggests that a two-step procedure be followed. The first step is to obtain a feasible solution to the constraint set (5.2.17). This may be accomplished by means of phase I of the simplex procedure. Let us refer to this feasible point as $\mathbf{X}_f$. (Note that $\mathbf{X}_f$ will contain no more than m nonzero components, x_{fj}).

The second step in the procedure is to introduce a set of non-negative artificial variables, u_j. These are added to equation set (5.2.20), and the objective function is set equal to the negative of the sum of these artificial variables. Our revised problem is then: Maximize

$$-\sum_j u_j, \tag{5.2.22}$$

subject to

$$g_i(\mathbf{X}) = a_{i1}x_1 + a_{i2}x_2 + \cdots + a_{in}x_n - b_i = 0, \qquad i = 1, \ldots, m, \tag{5.2.23}$$

and

$$c_j + 2\sum_{k=1}^{n} x_k d_{kj} + \sum_{i=1}^{m} \lambda_i a_{ij} + s_j + k_j u_j = 0, \qquad j = 1, \ldots, n, \tag{5.2.24}$$

where

$$k_j = +1 \quad \text{if} \left[c_j + 2 \sum_f d_{kj} x_k \right] \leq 0, \tag{5.2.25}$$
$$k_j = -1 \quad \text{if} \left[c_j + 2 \sum_f d_{kj} x_k \right] > 0,$$

and $\sum_f$ refers to a summation which includes only those $x_{fj} > 0$ which satisfy Equation (5.2.23). Once the values for k_j have been established we can set $u_j = |c_j + 2 \sum_f x_{fk} d_{kj}|$, $s_j = 0$ and $\lambda_i = 0$. This provides an initial feasible solution to the full problem.

The optimal solution can be found by means of the simplex procedure provided that each of the λ_i (which are unrestricted in sign) is expressed as $\lambda_i = \mu_i - \nu_i$, $\mu_i, \nu_i \geq 0$, and provided that $x_j s_j = 0$ throughout the procedure. This latter condition is maintained by modifying the simplex procedure so that s_j is not allowed to enter the basis if $x_j > 0$ and x_j is not allowed to enter if $s_j > 0$. By this scheme, the computations deal only with linear equations; yet the one nonlinear condition is maintained.

If y is strictly concave, a global optimum is obtained. If y is concave but not strictly concave (or strictly convex for a minimization problem), the procedure is generally satisfactory, but, in some instances, convergence may not be achieved. When this occurs, the problem can usually be eliminated by slightly perturbing the diagonal elements of the d_{kj} matrix. Quadratic programming may be unsatisfactory, however, for nonconcave (nonconvex) problems.

EXAMPLE 5.2.2.

As an illustration of the quadratic programming scheme, consider the maximization of

$$y = 9x_2 + x_1^2, \tag{5.2.26}$$

subject to

$$x_1 + 2x_2 - 10 = 0 \tag{5.2.27}$$

and the usual nonnegativity restrictions. In terms of our previous nomenclature, we have

$$\begin{aligned} c_1 &= 0, & c_2 &= 9, & & \\ d_{11} &= 1, & d_{12} &= 0, & d_{22} &= 0, \\ a_{11} &= 1, & a_{12} &= 2, & b_1 &= 10. \end{aligned}$$

Our revised program is then: Maximize

$$-u_1 - u_2, \tag{5.2.28}$$

subject to

$$x_1 + 2x_2 - 10 = 0, \tag{5.2.29}$$

$$0 + 2[(x_1)(1) + (x_2)(0)] - (\lambda_1)(1) + s_1 + k_1u_1 = 0, \tag{5.2.30}$$

$$9 + 2[(x_1)(0) + (x_2)(0)] - (\lambda_1)(2) + s_2 + k_2u_2 = 0. \tag{5.2.31}$$

We can see that $x_1 = 10$, $x_2 = 0$ provides a feasible solution for Equation (5.2.29). From Equation (5.2.25) we have that $k_1 = k_2 = -1$. We now set $u_1 = +20$, $u_2 = +9$, $s_1 = 0$, $s_2 = 0$, $\lambda_1 = 0$, and have available a basic feasible solution to the problem. We solve our problem by means of the simplex method, simply observing the additional restrictions that s_1 and x_1 may not both be in the basis, and also that s_2 and x_2 may not both be in the basis.

Although a great deal of attention has been devoted to quadratic programming in the past, it is now considered of much less importance. The requirement that the nonlinearities be quadratic and confined to the objective function restricts application of quadratic programming to a narrow class of problems. In general, nonlinearities in the constraints will be of equal importance with those of the objective function. It is unfortunate that quadratic programming in its original form cannot handle nonlinear constraints. When we consider approximation programming, we shall indicate how a modified quadratic programming approach can be applied when the constraints are nonlinear.

Separable programming

The desire to avoid the restriction of quadratic programming has lead to a search for more general procedures. We saw earlier, when we considered the gasoline-blending problem, that under some conditions it is possible to approximate a function by a series of linear segments. In the gasoline-blending problem this was easy to do, since the nonlinearities involved only a single variable. We could follow a similar approximation procedure for any function of the form

$$\phi = f_1(x_1) + f_2(x_2) + \ldots + f_n(x_n). \tag{5.2.32}$$

Although we now have nonlinearities in several variables, the nonlinearities are independent, and each $f_j(x_j)$ can be separately approximated. Functions that are in the form $\sum_{j=1}^{n} f_j(x_j)$ are called *separable* functions. If all the nonlinear functions in a problem are of this form, we can linearize the problem and obtain a solution by a slight modification of the simplex method.

Let us consider such a separable problem, which we shall write as: Minimize

$$y = \sum_{j=1}^{n} f_j(x_j), \tag{5.2.33}$$

subject to

$$\sum_{j=1}^{n} g_{ij}(x_j)\begin{Bmatrix}\geq\\=\\\leq\end{Bmatrix} b_i, \qquad i = 1, 2, \ldots, m, \tag{5.2.34}$$

$$0 \leq x_j \leq a_j, \qquad j = 1, 2, \ldots, n, \tag{5.2.35}$$

where the a_j, are positive constants. Now let us assume that some $f_j(x_j)$ has the appearance shown in Figure 5.4. The function is not monotonic and has local maxima. We can approximate the function to any desired accuracy by dividing the interval considered into an appropriate number of units and assuming a linear approximation to hold within each unit.

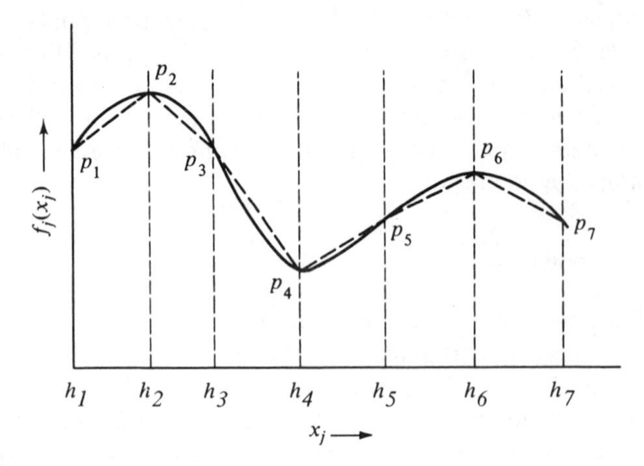

Figure 5.4. Approximation of Non-Convex Function by Linear Segments

Let us consider two adjacent points p_3 and p_4, having the coordinates (h_3, k_3) and (h_4, k_4). We can represent our approximating function $\bar{f}_j(x_j)$ between points 3 and 4 by a linear combination of k_3 and k_4; i.e.,

$$\begin{aligned} x_j &= \alpha_3 h_3 + \alpha_4 h_4 \\ \bar{f}_j(x_j) &= \alpha_3 k_3 + \alpha_4 k_4, \end{aligned} \tag{5.2.36}$$

where

$$0 \leq \alpha_j \leq 1 \quad \text{and} \quad \alpha_3 + \alpha_4 = 1.$$

More generally, we can represent the approximating function over the entire range of interest, say s_j segments, by

$$\bar{f}_j(x_j) = \sum_{l=1}^{s_j} \alpha_{lj} k_{lj}, \tag{5.2.37}$$

where

$$\sum_{l=1}^{s_j} \alpha_{lj} = 1 \tag{5.2.38}$$

$$\alpha_{lj} \geq 0 \qquad \text{for all } l \text{ and } j. \tag{5.2.39}$$

We encounter one difficulty with this notation. We have not eliminated the possibility that two nonadjacent α (e.g., α_{1j} and α_{4j}) be nonzero. Such a situation would be meaningless, as would the situation where, for the same index j, more than two α_{lj} are nonzero. We must therefore impose the additional restriction that for a given index j no more than two α_{lj} are in the solution at any one time and that these two are adjacent.

The technique used to linearize one of the components of the objective function may obviously be used for each $f_j(x_j)$. Our objective function is then approximated by: Minimize

$$\sum_{j=1}^{n} \sum_{l=1}^{s_j} \alpha_{lj} k_{lj}. \tag{5.2.40}$$

Since we have assumed that each of the constraints is separable, these may also be represented in the same manner. If we retain the division of each variable range into the s_j segments used in the approximation of our objective function, we can also express the constraints in terms of α_{lj}. We thus have the approximations

$$\sum_{j=1}^{n} \sum_{l=1}^{s_j} \alpha_{lj} k_{ilj} \begin{Bmatrix} \geq \\ = \\ \leq \end{Bmatrix} b_i, \qquad \text{for } i = 1, 2, \ldots, m, \tag{5.2.41}$$

in place of the original constraints. Since the k_{lj} and k_{ilj} are constants, we have a linear programming problem with the α_{lj} as variables. This linearity is retained when we add the required additional restrictions of (5.2.38) and (5.2.39). We can readily solve this problem by the simplex method. The simplex solution obtained would, however, not meet the restriction that for each j only two adjacent α may be in the basis. This difficulty can easily be resolved by applying a restriction on those variables allowed to enter the basis. If α_{lj} is the only variable from set j in the basis, we eliminate from consideration all other variables from set j except $\alpha_{(l+1)j}$ and $\alpha_{(l-1)j}$. If two variables from set j are already in the basis, no variable from set j is admitted unless it is (a) adjacent to one of those already in the basis and (b) the variable eliminated is such that only two adjacent α from set j remain in the basis. The other rules for selecting variables to be admitted and eliminated from the basis remain unchanged.

The modified simplex procedure will, of course, terminate in a finite number of steps. However, if the objective and constraint functions are not convex, it is quite likely that the procedure will terminate at a local optimum. We have no way of knowing how close the solution is to the global optimum. We do obtain a globally optimum solution in a convex programming problem. Indeed, in such a situation we need not apply the restricted basis entry techniques since the optimum solution will have no inadmissable combination of variables.

EXAMPLE 5.2.3

A simple illustration may clarify the procedure. Let us consider the convex programming problem: Minimize

(5.2.42)
$$y = \left[\frac{1}{x_1 + 1}\right] + x_2^3,$$

subject to

(5.2.43)
$$x_1^2 - x_2^3 \leq 5.0.$$

This simple problem may be solved graphically to obtain $x_1 = 2.24$ and $x_2 = 0$. We shall therefore assume that we are told that both x_1 and x_2 will have values between 0 and 3. We shall arbitrarily divide the range into four intervals and evaluate the $f_j(x_j)$ and the $g_{1j}(x_j)$ for x_j equal to 0.5, 1.0, 2.0, and 3.0. The values of $f_1(x_1)$, $f_2(x_2)$, $g_{11}(x_1)$, and $g_{12}(x_2)$ are tabulated in Table 5.1. Note that we need not have chosen the same intervals for x_1 and x_2 but did so simply for convenience.

Table 5.1. VALUES OF SEPARABLE FUNCTIONS OF ILLUSTRATIVE PROBLEM AT INTERVAL LIMITS

x_i	$f_1 = 1/(x_1 + 1)$	$g_{11} = x_1^2$	x_2	$f_2 = g_{12} = x_2^3$
0	1.00	0.00	0	0.00
0.5	0.667	0.250	0.5	0.125
1.0	0.500	1.00	1.0	1.00
2.0	0.333	4.00	2.0	8.00
3.0	0.250	9.00	3.0	27.0

The functional values in Table 5.1 provide the constants needed to linearize the objective function and constraints. We shall assign the linear variables α_1 through α_5 to represent x_1, the linear variables α_6 through α_{10} to represent x_2, and α_{11} as the slack factor. Our linearized problem is then: Minimize

$$\begin{aligned} y = 1.0\alpha_1 + 0.667\alpha_2 + 0.500\alpha_3 + 0.333\alpha_4 + 0.250\alpha_5 \\ + (0)\alpha_6 + 0.125\alpha_7 + 1.00\alpha_8 + 8.00\alpha_9 + 27.0\alpha_{10}, \end{aligned} \tag{5.2.44}$$

subject to

$$\begin{aligned} 0\alpha_1 + 0.250\alpha_2 + \alpha_3 + 4.0\alpha_4 + 9.0\alpha_5 - (0)\alpha_6 \\ - 0.125\alpha_7 - 1.00\alpha_8 - 8.00\alpha_9 - 27.0\alpha_{10} + \alpha_{11} = 5.0, \end{aligned} \tag{5.2.45}$$

$$\alpha_1 + \alpha_2 + \alpha_3 + \alpha_4 + \alpha_5 = 1.0, \tag{5.2.46}$$

$$\alpha_6 + \alpha_7 + \alpha_8 + \alpha_9 + \alpha_{10} = 1.0. \tag{5.2.47}$$

The constraint matrix contains a unit matrix, and therefore we may take this as our initial basis: $\mathbf{P}_1$, $\mathbf{P}_6$, and $\mathbf{P}_{11}$.

The initial tableau is shown in Table 5.2. The largest value of $(y_j - c_j)$ corresponds to vector $\mathbf{P}_5$, which we therefore introduce into the basis. Since the minimum value of x_{i0}/x_{ij} corresponds to vector $\mathbf{P}_{11}$, we eliminate $\mathbf{P}_{11}$. As our program is convex, we need not apply any basis-entry restriction. Had this not been the case, we could not have admitted $\mathbf{P}_5$ to the basis. We would have entered $\mathbf{P}_2$, and our progress towards a minimum would have been slower.

The second tableau, shown in Table 5.3, shows that $\mathbf{P}_4$ should be admitted to the basis. From the minimum of the x_{i0}/x_{ij} we find that $\mathbf{P}_1$ should be eliminated and obtain the tableau shown in Table 5.4. Since all the $(y_j - c_j) \leq 0$, we have obtained the optimum solution ($\alpha_5 = 0.2$, $\alpha_4 = 0.8$, $\alpha_6 = 1.0$) to our linearized problem. We readily translate these results in terms of our original variables by linearly interpolating between the variable values:

$$x_1 = 0.2(3) + 0.8(2) = 2.2, \tag{5.2.48}$$

$$x_2 = 1.0(0) = 0. \tag{5.2.49}$$

This result compares quite closely to the exact solution.

Separable programming can be a highly useful tool in those situations in which it is applicable. Not only must we be able to express all functions as separable functions, but we must be able to specify a range for each variable. However, the area of applicability can be considerably expanded by simple variable transformainons. For example, a product term such as $x_1^{a_1}x_2^{a_2}x_3^{a_3}$ can be replaced by new variable x_4 if we add the additional nonlinear constraint

$$\ln x_4 = a_1 \ln x_1 + a_2 \ln x_2 + a_3 \ln x_3. \tag{5.2.50}$$

The chief disadvantage of separable programming is the large number of variables introduced by the linear approximations. This is compounded by any additional variables added because of variable transformations. Prob-

Table 5.2. INITIAL TABLEAU OF SEPARABLE PROGRAMMING PROBLEM

Basis	c → ↓		1.0	0.667	0.500	0.333	0.250	0.0	0.125	1.00	8.00	27.00	0.0
		$\mathbf{P}_0$	$\mathbf{P}_1$	$\mathbf{P}_2$	$\mathbf{P}_3$	$\mathbf{P}_4$	$\mathbf{P}_5$	$\mathbf{P}_6$	$\mathbf{P}_7$	$\mathbf{P}_8$	$\mathbf{P}_9$	$\mathbf{P}_{10}$	$\mathbf{P}_{11}$
$\mathbf{P}_{11}$	0	5.0	0	0.250	1	4.00	9.00	0	−0.125	−1.00	−8.00	−27.0	1
$\mathbf{P}_1$	1.0	1.0	1	1	1	1	1	0	0	0	0	0	0
$\mathbf{P}_6$	0	1.0	0	0	0	0	0	1	1	1	1	1	0
$(y_j - c_j)$		1.0	0	0.333	0.500	0.663	0.750	0	−0.125	−1.0	−8.00	−27.0	0

Table 5.3. SECOND TABLEAU OF SEPARABLE PROGRAMMING PROBLEM

Basis	c → ↓		1.0	0.667	0.500	0.333	0.250	0.0	0.125	1.00	8.00	27.00	0.0
		$\mathbf{P}_0$	$\mathbf{P}_1$	$\mathbf{P}_2$	$\mathbf{P}_3$	$\mathbf{P}_4$	$\mathbf{P}_5$	$\mathbf{P}_6$	$\mathbf{P}_7$	$\mathbf{P}_8$	$\mathbf{P}_9$	$\mathbf{P}_{10}$	$\mathbf{P}_{11}$
$\mathbf{P}_5$	0.25	0.555	0	0.028	0.111	0.444	1	0	−0.0139	−0.111	−0.889	−3.00	0.111
$\mathbf{P}_1$	1.0	0.455	1	0.972	0.889	0.556	0	0	0.0139	0.111	0.889	3.00	−0.111
$\mathbf{P}_6$	0	1.0	0	0	0	0	0	1	1	1	1	1	0
$(y_j - c_j)$		0.579	0	0.312	0.417	0.434	0	0	−0.1354	−0.917	−7.33	−24.75	−0.072

Table 5.4. FINAL TABLEAU OF SEPARABLE PROGRAMMING PROBLEM

Basis	c → ↓		1.0	0.667	0.500	0.333	0.250	0.0	0.125	1.00	8.00	27.00	0.0
		$\mathbf{P}_0$	$\mathbf{P}_1$	$\mathbf{P}_2$	$\mathbf{P}_3$	$\mathbf{P}_4$	$\mathbf{P}_5$	$\mathbf{P}_6$	$\mathbf{P}_7$	$\mathbf{P}_8$	$\mathbf{P}_9$	$\mathbf{P}_{10}$	$\mathbf{P}_{11}$
$\mathbf{P}_5$	0.25	0.20	−8.0	−0.747	−0.609	0	1	0	−0.025	−0.199	−1.59	−5.38	−0.199
$\mathbf{P}_4$	0.333	0.80	1.795	1.745	1.60	1	0	0	0.025	0.199	1.59	5.38	−0.199
$\mathbf{P}_6$	0	1.0	0	0	0	0	0	1	1	1	1	1	0
$(y_j - c_j)$		0.317	−0.62	−0.274	−0.124	0	0	0	−0.123	−0.984	−7.87	−25.35	−0.061

lems that appeared quite small originally can be of significant size after the linear approximations are made. If most of the variables enter the problem linearly, the problem may remain of reasonable size, since such variables need not be expressed in terms of α_{lj}.

In those cases in which the number of variables makes the problem intractable, it is possible to deal with the problem by the method of *decomposition*. We observe that the solution to our approximating problem will contain at most $2n$ of the α_{lj}. It is thus possible to generate feasible solutions with a restricted number of linear segments. The decomposition method endeavors to solve a set of small subproblems that are separate except for certain linking variables. The procedure is certain to be successful only if the problem is convex. However, it is found to work in many practical cases. We shall describe the method of decomposition in Chapter 6.

Although direct application of separable programming to nonconvex functions only guarantees a local optimum, it is possible to combine this technique with the requirement that certain additional variables be integers, and locate the global optimum. This requires the solution of a series of linear programming problems. We shall also indicate how this may be done in Chapter 6. This technique is significant, since it is one of the few ways we have of finding global optima to problems that are not convex or concave.

Cutting plane procedures

None of the procedures previously considered is readily applicable to the general nonlinear programming problem. Quadratic programming is limited to a narrow class of problems. Separable programming, in conjunction with variable transformations, has wider applicability, but, as we have just observed, the large number of variables introduced limits the usefulness of this technique.

Let us consider other ways in which we can linearize our problem. First, we can always linearize our objective function. If the objective to be minimized contains some nonlinear function $f(\mathbf{X})$, i.e.,

$$y = \sum_{j=1}^{n} c_j x_j + f(\mathbf{X}), \tag{5.2.51}$$

we can always write the objective function as: Minimize

$$y = \sum_{j=1}^{n} c_j x_j + x_{n+1}, \tag{5.2.52}$$

providing we introduce the restraint

$$x_{n+1} \geq f(\mathbf{X}). \tag{5.2.53}$$

Hence, minimization of (5.2.52) will yield the minimum of (5.2.51).

We now must devise a means for linearizing the constraints. We know that we can always approximate a nonlinear function by a Taylor's series. If we are satisfied with only a crude approximation over some region, we can retain only first-order terms. We thus could replace a nonlinear constraint $g_i(\mathbf{X}) \leq 0$ by the linear relationship

$$g_i(\mathbf{X}^o) + \nabla g_i(\mathbf{X}, \mathbf{X}^o)(\mathbf{X} - \mathbf{X}^o) \leq 0, \tag{5.2.54}$$

where $\mathbf{X}^o$ is some arbitrary point in whose neighborhood we wish the approximation to be useful, and $\nabla g_i(\mathbf{X}, \mathbf{X}^o)$ represents the gradient at point $\mathbf{X}^o$.

Kelley [1960] has devised an algorithm that makes use of this method of constraint approximation when the objective function is linear. For simplicity, let us consider the case in which we have several linear constraints

$$\mathbf{AX} \leq \mathbf{B} \tag{5.2.55}$$

but only a single nonlinear constraint

$$g(\mathbf{X}) \leq 0. \tag{5.2.56}$$

Also, let us assume that $g(\mathbf{X})$ is convex. Let us choose our starting point $\mathbf{X}^o$ such that it satisfies the linear constraints but does not satisfy the nonlinear constraint. Now, using the first-order Taylor's series approximation of our constraint given in (5.2.54), let us define the hyperplane $G(\mathbf{X}, \mathbf{X}^o)$ such that

$$G(\mathbf{X}, \mathbf{X}^o) = g(\mathbf{X}^o) + \nabla g(\mathbf{X}, \mathbf{X}^o)(\mathbf{X} - \mathbf{X}^o). \tag{5.2.57}$$

If the constraint

$$G(\mathbf{X}, \mathbf{X}^o) \leq 0 \tag{5.2.58}$$

is imposed, we find that every point that satisfies the inequality of (5.2.56) also satisfies (5.2.58). We can see that this follows immediately from our assumption that $g(\mathbf{X})$ is convex. This is illustrated graphically in Figure 5.5 for the case in which $g(\mathbf{X})$ is a function of a single variable. The convexity property tells us that the value of $G(x)$ can never exceed the value of $g(x, x_1)$ for a given x (see Fig. 5.5). Therefore, in the illustrative case, the value of x at which $G(x) = 0$ is greater than or equal to the value of x at which $g(x) = 0$. The constraint $G(x) \leq 0$ is then looser than $g(x) \leq 0$ and will allow some values of x that are not feasible. However, any values of x excluded by $G(x) \leq 0$ must also be excluded by $g(x) \leq 0$. The tangent plane $G(x) \leq 0$ thus excludes or "cuts" away a large infeasible area, i.e., all that area in which the constraint $G(x) \leq 0$ is not met. This property is true not only in this particular example but also in the more general multidimensional case.

By imposing a sequence of the cutting planes we obtain a convergent

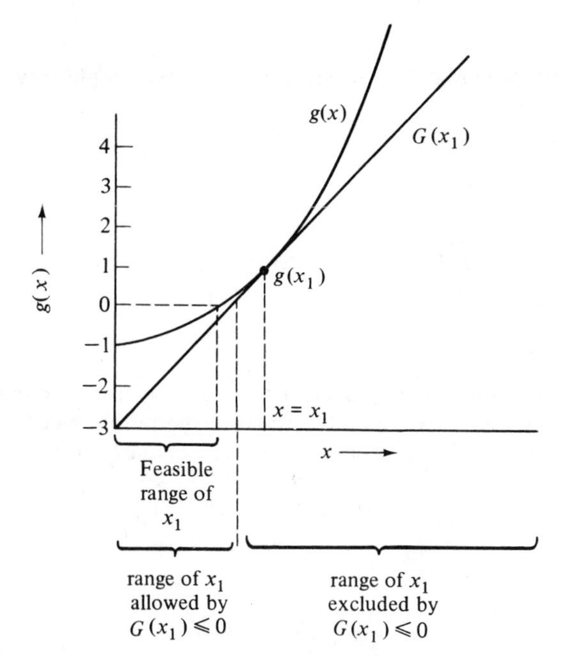

Figure 5.5. Cutting Plane with Single Variable Constraint

process for locating the optimum to our original problem. We first solve the linear programming problem defined by: Minimize

$$y = \sum_{j=1}^{n} c_j x_j \tag{5.2.59}$$

and the linear constraints of (5.2.55), while ignoring our nonlinear constraint. If the minimum found meets the nonlinear constraint, the problem is solved. Otherwise, we then let $(\mathbf{X}_1^o)$ be the minimum point just determined and then add to our constraint set

$$G_1(\mathbf{X}, \mathbf{X}_1^o) \leq 0. \tag{5.2.60}$$

The linear programming problem, with the original linear constraints and the additional constraint, is resolved. The new minimum, $\mathbf{X}_2^o$, is checked against our nonlinear constraint, and if, as is most likely, the constraint is not satisfied, we add still another linear constraint,

$$G_2(\mathbf{X}, \mathbf{X}_2^o) \leq 0, \tag{5.2.61}$$

and resolve the problem. The process continues, adding a linear constraint after each linear programming solution, until the nonlinear constraint is satisfied or becomes smaller than some preassigned tolerance.

EXAMPLE 5.2.4

To illustrate the procedure, we shall consider the example provided by Kelley: Minimize

$$y = x_1 - x_2, \tag{5.2.62}$$

subject to

$$g(\mathbf{X}) = 3x_1^2 - 2x_1x_2 + x_2^2 - 1 \leq 0, \tag{5.2.63}$$

$$-2 \leq x_1 \leq 2, \tag{5.2.64}$$

$$-2 \leq x_2 \leq 2. \tag{5.2.65}$$

As illustrated in Figure 5.6, $g(\mathbf{X})$ is an ellipse and the constrained minimum occurs at $\mathbf{X} = (0, 1)$ with $y = -1$. Initially, we solve the problem defined by (5.2.62), (5.2.64) and (5.2.65) via linear programming* and obtain $\mathbf{X}_1^o = (-2, 2)$. By evalua-

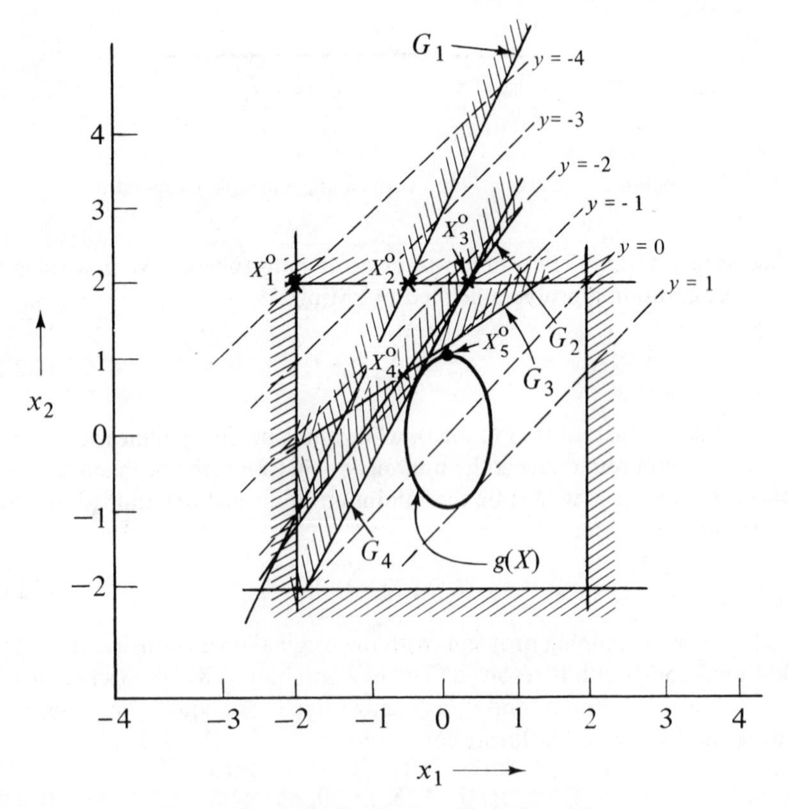

Figure 5.6. Kelley's Cutting Plane Procedure

* The actual linear programming problem is solved with respect to transformed nonnegative variables. In the example, however, the solution can be obtained graphically.

tion of the gradient at that point, we obtain G_1, our first additional constraint:

$$G_1(\mathbf{X}, \mathbf{X}_1^o) = -16x_1 + 8x_2 - 25 \leq 0. \tag{5.2.66}$$

We then solve the linear programming problem: Minimize

$$y = x_1 - x_2, \tag{5.2.67}$$

subject to

$$-2 \leq x_1 \leq 2, \tag{5.2.68}$$

$$-2 \leq x_2 \leq 2, \tag{5.2.69}$$

$$-16x_1 + 8x_2 \leq 25. \tag{5.2.70}$$

Solution of this problem yields $\mathbf{X}_2^o = (-0.5625, 2.0)$ as the minimum and provides the new constraint $G_2(-7.375x_1 + 5.125x_2 \leq 8.20)$. The procedure is repeated successively, adding constraints $G_3(-2.33x_1 + 3.44x_2 \leq 4.12)$ and $G_4(-4.85x_1 + 2.73x_2 \leq 3.43)$. Observe that point $\mathbf{X}_5^o$, which is then obtained, is quite close to the true minimum.

The example chosen is, of course, highly artificial. Normally there will be a number of nonlinear constraints. It is then not obvious how to obtain the gradient needed to establish $G(\mathbf{X}, \mathbf{X}^o)$. A simple technique is to define

$$G(\mathbf{X}) = \max_i\,[g_i(\mathbf{X})]. \tag{5.2.71}$$

The successive solution of problems with increasing numbers of constraints is undertaken until $\max_i\,[g_i(\mathbf{X})]$ is below a specified tolerance.

When Kelley's procedure is applicable, we are able to turn the solution of a nonlinear programming problem into the sequential solution of a series of linear programming problems in which, at each step, we increase the number of rows in the matrix. We may solve each new linear programming problem by starting with the basis furnished by slack and artificial variables and again proceeding through the entire solution. This is, however, a relatively inefficient way of conducting the computation.

It would be desirable to be able to use the solution to the preceding problem as a starting point for a solution of the new problem. We find that we encounter some difficulty in doing this. Consider the situation in which we have added our first constraint G_1. We modify the original constraint matrix so that the simplex procedure can be used (replacing each negative variable by the difference of two positive variables). We then augment the problem by the addition of constraint G_1. After addition of a slack variable to create an equality and use of the Gauss–Jordan elimination procedure to place zeros appropriately, we would find a negative value in the $\mathbf{P}_0$ column. This negative value arises since our initial solution is not feasible with respect to G_1. We would also find that all $(y_j - c_j)$ are zero or negative. We seem to have no way to choose a new vector to enter the basis. We can surmount this difficulty by considering the dual of our linear program.

Although our original solution to the augmented primal problem is infeasible, the dual problem is feasible but not optimal. When we take the transpose of the coefficient matrix to establish the dual, we need not be concerned about negative values in the $\mathbf{P}_0$ column. These are allowable, since the dual problem places no restriction on the sign of the variables. Since the optimality criterion is not met, we can determine the new vector to enter the basis. Each *constraint* we add to the primal means an additional *variable* in the dual. Thus, when the dual is used, we really increase the number of columns in the matrix at each step instead of increasing the number of rows.*

In our discussion of the cutting-plane procedures, we assumed that the constraint set was convex. When this is so, the procedure will converge to a global optimum. The cutting-plane procedure can still be used when this is not the case, providing the constraints are differentiable. However, when the convexity requirement is not met, we have no assurance the procedure will converge. Furthermore, if it does converge, the optimum found is not necessarily global. As we noted earlier, this latter difficulty occurs in almost all nonlinear problems. Unless the problem is convex, or we can transform it into a convex problem, we cannot be certain that the optimum we obtain is anything more than a local optimum. The major drawback to the cutting-plane method for nonconvex problems is that oscillations may occur and prevent convergence.

Approximation programming

In the cutting-plane method we added one linearized constraint at a time and successively eliminated portions of the infeasible region. Suppose that we were to extend this idea and simultaneously approximate the objective function and all constraints by their tangent planes at a given starting point. We would then have a linear programming problem that we could solve. For example, let us consider Griffith and Stewart's problem [1961] wherein we seek to: Maximize

$$y = 2x_1 + x_2, \tag{5.2.72}$$

subject to

$$x_1^2 + x_2^2 \leq 25, \tag{5.2.73}$$

---

* It is possible to solve the dual problem by operating directly on the primal using the *dual-simplex* algorithm of Lemke [1954]. This algorithm is described in Section 6.3.

$$x_1^2 - x_2^2 \leq 7, \tag{5.2.74}$$

$$x_j \geq 0, \quad j = 1, 2. \tag{5.2.75}$$

Note that this particular example has a linear objective function. The method works equally well, however, when the objective function is not linear.

Let us assume that $x_1 = 1$ and $x_2 = 1$ at our starting point (point 0 of Figure 5.7a). When we write a Taylor's series expansion about point 0, retaining only linear terms for each of the constraints, we obtain the linear inequalities

$$2(x_1 - 1) + 2(x_2 - 1) \leq 23, \tag{5.2.76}$$

$$2(x_1 - 1) - 2(x_2 - 1) \leq 7. \tag{5.2.77}$$

These are shown as lines BC and AC on Figure 5.7a.

If we were to solve the resulting linear program, we would obtain a solution outside the feasible region (at point C). We obviously would have pushed our assumption of linearity too far. Suppose, instead of simply solving the linear programming problem which we generated, that we add additional constraints on how far we would allow the variables to move; e.g.,

$$0 \leq x_1 \leq 2, \tag{5.2.78}$$

$$0 \leq x_2 \leq 2. \tag{5.2.79}$$

The solution obtained ($x_1 = 2, x_1 = 2$) is indicated as solution 1 on Figure 5.7b. It is closer to the true maximum than our starting point and still within the feasible region. We may repeat this process by taking our first solution as a new starting point and obtaining new linear approximations of our constraints. When we again impose the additional constraint that the variables can move no more than one unit in either direction from solution 1, we obtain $x_1 = 3, x_2 = 3$ as the solution (solution 2).

We have now moved appreciably closer to the true maximum. When we now approximate our constraints using solution 2 as the starting point, we obtain

$$6(x_1 - 3) - 6(x_2 - 3) \leq 7, \tag{5.2.80}$$

$$6(x_1 - 3) + 6(x_2 - 3) \leq 7. \tag{5.2.81}$$

We can see from Figure 5.7b that these linear inequalities are now reasonably good approximations of the original nonlinear constraints in the region of their intersection. The solution to our third problem lies at the intersection of these constraints and is quite close to the true maximum. We have thus been able to obtain an approximate solution to our original nonlinear programming problem by solving a series of linear programming problems. The es-

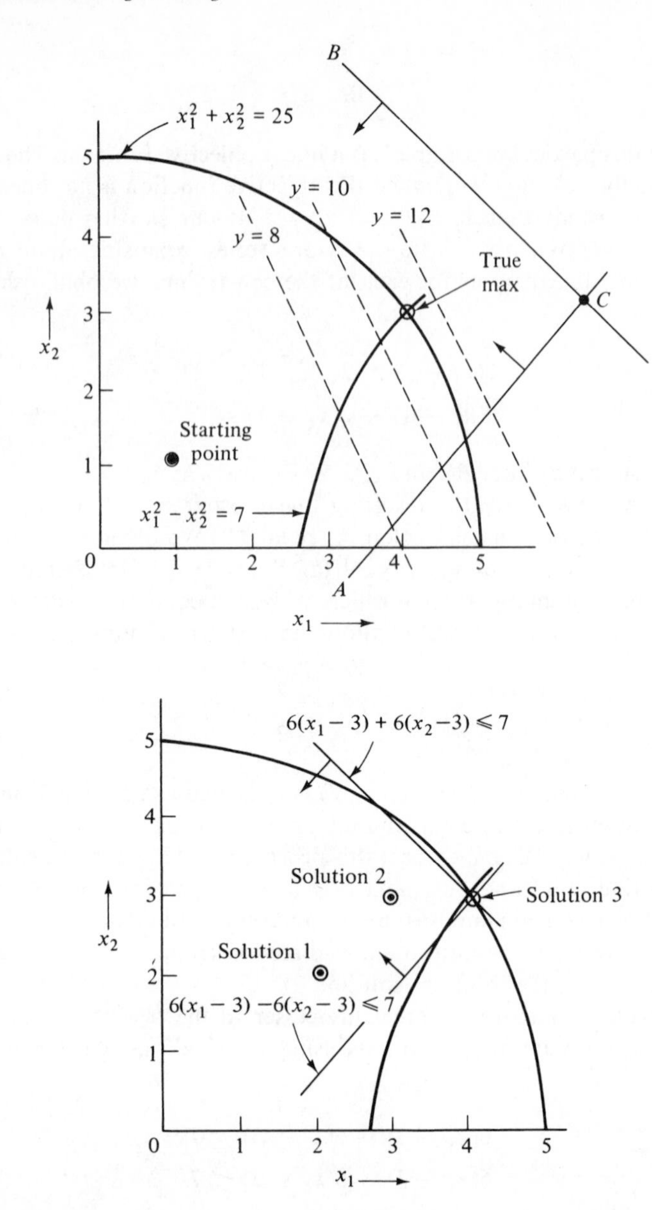

Figure 5.7. Solution of Constrained Maximization Problem by Griffith and Stewart's Approximation Programming

sence of the scheme is the assumption that the linear approximations used are useful only over a narrow range and, therefore, that the values of the independent variables may change by only a small amount at each step.

Griffith and Stewart [1961] have formalized this linear approximation procedure into a well-defined algorithm. Let us formally state our problem as: Maximize

$$y = f(x_1, x_2, \ldots, x_n), \tag{5.2.82}$$

subject to

$$g_i(x_1, x_2, \ldots, x_n)\left\{\begin{matrix} \le \\ = \end{matrix}\right\}b_i, \qquad i = 1, 2, \ldots, m, \tag{5.2.83}$$

$$k_{1j} \le x_j \le k_{2j}, \qquad j = 1, 2, \ldots, n, \tag{5.2.84}$$

where b_i, k_{1j}, and k_{2j} are constants. (Notice that the method applies equally well to equality or inequality constraints.) We then approximate our problem in the region about $(\mathbf{X}^o)$ by: Maximize

$$y = f(x_1^o, x_2^o, \ldots, x_n^o) + \sum_{j=1}^{n} (x_j - x_j^o)\left[\frac{\partial f(x_1^o, x_2^o, \ldots, x_n^o)}{\partial x_j}\right], \tag{5.2.85}$$

subject to

$$g_i = g_i(x_1^o, x_2^o, \ldots, x_n^o) + \sum_{j=1}^{n} (x_j - x_j^o)\left[\frac{\partial g_i(x_1^o, x_2^o, \ldots\ x_n^o)}{\partial x_j}\right]\left\{\begin{matrix} \le \\ = \end{matrix}\right\}b_i,$$

$$i = 1, 2, \ldots, m, \tag{5.2.86}$$

$$k_{1j} \le x_j \le k_{2j}, \qquad j = 1, 2, \ldots\ n. \tag{5.2.87}$$

Since the partial derivatives are taken as constants, c_j and w_{ij}, over a given range, the above may be rewritten as: Maximize

$$y = y^o + \sum_{j=1}^{n} c_j \Delta x_j, \tag{5.2.88}$$

subject to

$$g_i = g_i^o + \sum_{j=1}^{n} w_{ij}\Delta x_j\left\{\begin{matrix} \le \\ = \end{matrix}\right\}b_i, \qquad i = 1, 2, \ldots, m, \tag{5.2.89}$$

$$k_{1j} - x_j^o \le \Delta x_j \le k_{2j} - x_j^o. \tag{5.2.90}$$

We would now have a linear programming problem of the usual form if the (Δx_j) were restricted to be nonnegative. This is easily accomplished by replacing the (Δx_j) by two sets of variables. We let $\Delta^+ x = \Delta x$ when $\Delta x \ge 0$ and $\Delta^- x = -\Delta x$ when $\Delta x \le 0$. We finally have

$$y = y^o + \sum_{j=1}^{n} c_j \Delta^+ x_j - \sum_{j=1}^{n} c_j \Delta^- x_j, \tag{5.2.91}$$

subject to

$$g_i = g_i^o + \sum_{j=1}^{n} w_{ij}\,\Delta^+ x_j - \sum_{j=1}^{n} w_{ij}\,\Delta^- x_j \left\{ \begin{matrix} \leq \\ = \end{matrix} \right\} b_i, \qquad i = 1, 2, \ldots, m.$$

(5.2.92)

The limitations on the size of the Δx_j are imposed by adding the restraints

$$p_j \Delta^+ x_j + q_j \Delta^- x_j \leq k_{3j}, \qquad j = 1, 2, \ldots, n, \tag{5.2.93}$$

where

$$p_j = \max\left[1, \frac{k_{3j}}{k_{2j} - x_j^o}\right], \tag{5.2.94}$$

$$q_j = \max\left[1, \frac{k_{3j}}{x_j^o - k_{1j}}\right]. \tag{5.2.95}$$

This formulation keeps the maximum distance any x_j can move no greater than k_{3j}. Furthermore, if a variable is near an upper or lower limit, the large value of p_j or q_j generated keeps the variable from exceeding the limit while allowing movement away from the limit.

After obtaining a solution to this problem, we obtain a new linear approximation and repeat the process. We have seen in our sample problem that the continued application of this technique leads to the desired optimum. It may be shown that this procedure can be made to converge to any desired accuracy, providing the solution lies on a constraint intersection, the objective function is concave, and the constraints form a convex set.

We thus have again reduced our nonlinear programming problem to a series of linear programming problems, but, in contrast to the cutting-plane method, we have not increased the number of constraints required at each iteration. Also we have not restricted our initial point, and therefore it can be within or outside the constraint set.

Since the size of the allowable Δx_j is purely arbitrary, a step may be taken that is outside the range in which the first partials furnish an adequate description of the behavior of the functions. When the objective function and constraints are evaluated at the new variable values, it may be found that neither the objective function nor constraints have been improved. The next iteration may then move the variable values back toward their original values, causing the solution to oscillate. Such oscillation is particularly likely in the neighborhood of a solution point. Working algorithms that utilize this method contain provisions for forcing convergence by gradually reducing the size of the allowable step. The algorithms are terminated when all Δx_j are below some predetermined limit.

EXAMPLE 5.2.5

Let us again consider the illustrative problem defined by Equations (5.2.72)–(5.2.75) to illustrate our formal method. When we linearize this problem we have

$$y = 2(\Delta x_1) + (\Delta x_2) + 2x_1^o + x_2^o, \tag{5.2.96}$$

$$2x_1^o(\Delta x_1) + 2x_2^o(\Delta x_2) \leq 25 - (x_1^o)^2 - (x_2^o)^2, \tag{5.2.97}$$

$$2x_1^o(\Delta x_1) - 2x_2^o(\Delta x_2) \leq 7 - (x_1^o)^2 + (x_2^o)^2, \tag{5.2.98}$$

$$x_1 \geq 0, \qquad x_2 \geq 0, \tag{5.2.99}$$

where (x_1^o, x_2^o) is the starting point. If we again set $(x_1^o, x_2^o) = (1, 1)$ and require that $|\Delta x_i| \leq 1$ be expressed in terms of non-negative quantities, we obtain

$$y = 2(\Delta^+ x_1) + (\Delta^+ x_2) - 2(\Delta^- x_1) - (\Delta^- x_1) + 3, \tag{5.2.100}$$

$$2(\Delta^+ x_1) + 2(\Delta^+ x_2) - 2(\Delta^- x_1) - 2(\Delta^- x_2) \leq 23, \tag{5.2.101}$$

$$2(\Delta^+ x_1) - 2(\Delta^+ x_2) - 2(\Delta^- x_1) + 2(\Delta^- x_2) \leq 7, \tag{5.2.102}$$

$$\Delta^+ x_1 + \Delta^- x_1 = 1, \qquad \Delta^+ x_2 + \Delta^- x_2 = 1 \tag{5.2.103}$$

and $\Delta^+ x_1, \Delta^- x_1, \Delta^+ x_2, \Delta^- x_2 \geq 0$.

The solution to this problem is then $(\Delta^+ x_1 = 1, \Delta^- x_1 = 0, \Delta^+ x_2 = 1, \Delta^- x_2 = 0)$. In terms of x_1 and x_2, our solution is $(x_1 = 2, x_2 = 2)$, as we saw previously. Our next linear programming problem becomes

$$y = 2(\Delta^+ x_1) + (\Delta^+ x_2) - 2(\Delta^- x_1) - (\Delta^- x_2) + 6, \tag{5.2.104}$$

$$4(\Delta^+ x_1) + 4(\Delta^+ x_2) - 4(\Delta^- x_1) - 4(\Delta^- x_2) \leq 17, \tag{5.2.105}$$

$$4(\Delta^+ x_1) - 4(\Delta^+ x_2) - 4(\Delta^- x_1) + 4(\Delta^- x_2) \leq 7, \tag{5.2.106}$$

which yields $(\Delta^+ x_1 = 1, \Delta^- x_1 = 0, \Delta^+ x_2 = 1, \Delta^- x_2 = 0)$ as the solution. The succession of solutions obtained in terms of the original variables x_1 and x_2 is shown in Table 6.5. The first three solutions are shown in Figure 5.7b along with the linear constraints for the third solution.

Table 5.5. SUCCESSIVE SOLUTIONS OBTAINED FOR EXAMPLE 5.2.5

Iteration	*Linear Programming Solution*
1	2, 2
2	3, 3
3	4, 2.833
4	3.993, 3.025
5	4.000, 3.000

When the convexity and concavity restrictions are not met, the approximation programming method often does converge to a local optimum.

However, in the general case the method may result in oscillation about a local optimum or in divergence. In many areas, such as petroleum-refinery optimization, experience has shown the functions encountered to be well behaved so that convergence is usually obtained. When this is so, use of approximation programming is highly advantageous. Experience in the petroleum industry has indicated approximation programming to be a tool of great power.

In the approximation programming approach, we have assumed that the objective function and all the constraints can be approximated by a Taylor's series in which only the first-order terms are retained. We could have retained both first- and second-order terms for an improved approximation. If the only nonlinearities were in the objective function, we would then have a quadratic programming problem to solve at each iteration. When nonlinearities also appear in the constraints, we cannot use the quadratic programming approach directly. Wilson [1963] and Beale [1967] have suggested that in this case we form the Langrangian

$$z = y(x_1, x_2, \ldots, x_n) + \sum_{i=1}^{m^*} \lambda_i[g_i(x_1, x_2, \ldots, x_n) - b_i], \tag{5.2.107}$$

using only the nonlinear constraints ($i = 1$ to m^*), which are written as equalities. The nonlinear functions are then replaced by their power series approximations to yield

$$\begin{aligned} z \simeq\ & y(x_1^o, x_2^o, \ldots, x_n^o) + \sum_{j=1}^{n} \frac{\partial y}{\partial x_j}(x_1^o, x_2^o, \ldots, x_n^o)(\Delta x_j) \\ & + \sum_{k=1}^{n} \sum_{j=1}^{n} \frac{1}{2} \frac{\partial^2 y(x_1^o, x_2^o, \ldots, x_n^o)}{\partial x_j\, \partial x_k}(\Delta x_j)(\Delta x_k) + \sum_{j=1}^{m^*} \lambda_i g_i(x_1^o, x_2^o, \ldots, x_n^o) \\ & + \sum_{j=1}^{n} \sum_{i=1}^{m^*} \lambda_i \frac{\partial g_i}{\partial x_j}(x_1^o, x_2^o, \ldots, x_n^o)(\Delta x_j) \\ & + \sum_{j=1}^{n} \sum_{k=1}^{n} \sum_{i=1}^{m^*} \frac{\lambda_i}{2} \frac{\partial^2 g_i}{\partial x_j\, \partial x_k}(x_1^o, x_2^o, \ldots, x_n^o)(\Delta x_j)(\Delta x_k). \end{aligned} \tag{5.2.108}$$

If trial values of the λ_i are assumed, the Lagrangian can be minimized (maximized), subject to the linear constraints, using quadratic programming. Once the minimum has been found, a new set of λ_i must be estimated. To do so we make use of the *shadow-price concept*. We recall that the Lagrange multipliers are equivalent to $-\pi_i$, the shadow prices, which at the optimum are given by

$$-\lambda_i = \pi_i = \frac{\partial y}{\partial b_i}(x_1, x_2, \ldots, x_n). \tag{5.2.109}$$

We then use the revised λ_i and reoptimize z. The process is continued until the λ_i from successive estimates converge. The values of the Δx_j obtained are then used to establish a new base point, and the cycle is continued.

The only real justification for retaining second-order terms is that they can allow larger steps to be taken at each iteration. However, each iteration becomes appreciably more complex, and the total computation required is not necessarily reduced. The more complex approach retaining second-order terms has therefore been far less popular than the more straightforward approximation programming of Griffith and Stewart.

5.3 Methods of Feasible Directions

Simple approaches

Just as the availability of rapid linear programming algorithms furnished the incentive for the development of the linearization techniques described in the previous section, the availability of unconstrained optimization methods furnished the incentive for the development of feasible-direction techniques. A feasible-direction method attempts to modify an unconstrained optimization technique so that it will not violate a set of constraints. These methods start at some feasible point and then find a direction along which the objective function can be improved while observing all constraints. After moving a short distance in such a direction, we stop and redetermine the direction in which to move. The procedure is repeated until we cannot improve our objective function. At that time we shall have reached at least a local optimum.

A direction in which a move can be made without violating a constraint is a *feasible direction*. A direction that improves the objective function while violating no constraint is a *useable feasible direction*. Obviously, there can be many ways of establishing criteria for feasible directions, as well as several different minimization (maximization) techniques which can be useful. The pattern search, described in Chapter 3, is particularly well suited to adaptation with these methods.

Suppose that we have located a feasible starting point. By using a univariate search technique we may explore the region around this feasible point. Initially, we would expect to locate new feasible points that have an improved objective function. We may shortly find that the points showing the greatest improvement in the objective function are infeasible, whereas the feasible points show no objective function improvement. If we were to choose a direction between the feasible and infeasible points, we might find a direction that is feasible and in which some objective function improvement occurs. Mugele [1962] has developed an algorithm based upon this idea. A small-step search is used to maximize the objective function until a constraint is encountered. We then move along the constraint by interpolating between feasible and infeasible points. Let us consider Mugele's example in which $y(x_1, x_2)$ is maximized subject to two constraints (cf. Figure 5.8). Assume that we begin at point R. Probes are made by successively increasing and decreasing x_1 and x_2. We locate the improved feasible point Q and accept it. We proceed in the same manner until we reach point M. At M, no probe produces an improved feasible point. We then choose a new point L by interpolating between the high feasible point U and the high nonfeasible point T. A similar interpolation procedure then brings us to point K, which is close to the constrained optimum for the problem.

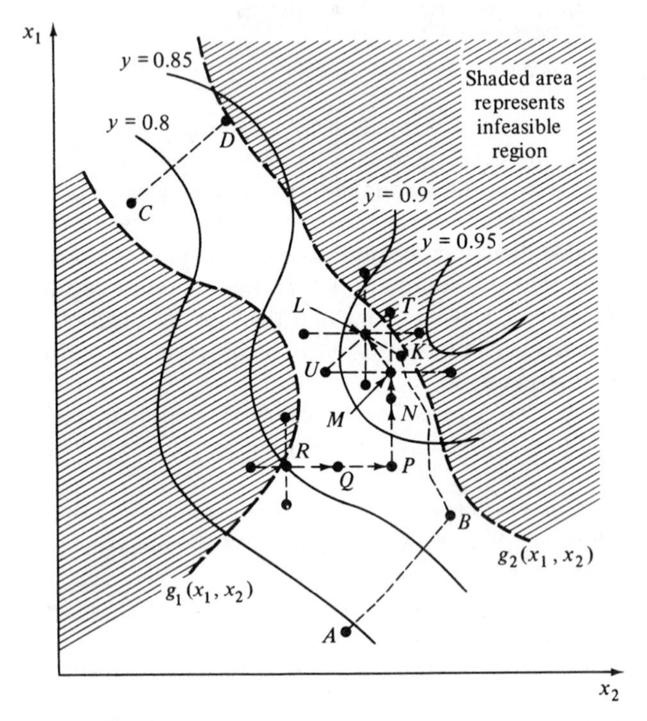

Figure 5.8. Feasible Direction Method of Mugele

If we had started from point A we would have, through a series of steps, approached the constraint $g_2(x_1, x_2)$ at point B, and then the interpolation between feasible and infeasible points would have carried us roughly parallel to the constraints until we reached the vicinity of point K. However, if we had started at point C, we would have contacted $g_2(x_1, x_2)$ at D and would not have found any improved feasible point. Our solution would be only a local minimum. As before, unless our problem is concave (convex), we cannot be assured that we will attain a global minimum.

Mugele's procedure is not as efficient as it might be, since many of the probe evaluations will be infeasible once a constraint has been contacted. If we could define a direction for motion that would avoid probe points in the infeasible area, the efficiency of the search would be improved. Klingman and Himmelblau [1964] have devised a procedure for accomplishing this. Consider the situation shown in Figure 5.9. Probes from point B indicate that we are in contact with the $g_1(x_1, x_2) \geq 0$ constraint. Attempting to maximize $y(x_1, x_2)$ by motion in the direction $\nabla y(x_1, x_2)$ will only lead to movement into the infeasible region. Motion in the direction $\nabla g_1(x_1, x_2)$ will assure feasibility but not necessarily lead to an improvement in the objective function. We define a compromise new feasible direction (**NFD**) such that

$$\mathbf{NFD} = \frac{\nabla y(x_1, x_2)}{\left[\sum_i (\partial y/\partial x_i)^2\right]^{1/2}} + \frac{\nabla g_1(x_1, x_2)}{\left[\sum_i (\partial g_1/\partial x_i)^2\right]^{1/2}}. \tag{5.3.1}$$

Thus **NFD** is the vector sum of the normalized gradients of the contacted constraint and the objective function (cf. Figure 5.9). The search proceeds by small steps in the **NFD** direction until no objective function improvement is obtained or a new constraint is contacted. If a new constraint is contacted, a gradient vector is then added. The revised feasible direction becomes

$$\mathbf{NFD} = \frac{\nabla y(x_1, x_2)}{\left[\sum_i (\partial y/\partial x_i)^2\right]^{1/2}} + \sum_{j=1}^{kc} \frac{\nabla g_i(x_1, x_2)}{\left[\sum_i (\partial g_i/\partial x_i)^2\right]^{1/2}}, \tag{5.3.2}$$

where kc is the number of contacted constraints. Because of the form of this equation, the procedure has been called the *multiple-gradient-summation* technique. The procedure terminates when objective function improvements fall below some prescribed minimum, or multiple constraint contacts show that a local optimum has been reached at the intersection of several constraint hyperplanes.

Best feasible direction

Neither of the previous feasible-direction methods has addressed itself to the question of what is the best feasible direction in which to move. They are content with simply determining a direction that is feasible. Zoutendijk [1964], who first suggested the name for methods of this type, attempted to solve this problem. Consider the problem: Maximize

$$y(\mathbf{X}), \tag{5.3.3}$$

subject to

$$g_i(\mathbf{X}) \leq b_i, \qquad i = 1, \ldots, m. \tag{5.3.4}$$

Assume that we have located a feasible point ($\mathbf{X}^*$) where constraints $g_1(\mathbf{X}), \ldots, g_l(\mathbf{X})$ are binding or nearly so. We wish to take a step in that direction which will maximize the increase, ϵ, in our objective function, yet will not violate any of the l constraints. If we may assume that over a sufficiently small interval we can linearize our constraints and our objective function, the problem can be approximated as a linear programming problem. We may solve for a feasible direction $\mathbf{S}$ such that we: Maximize

$$(\epsilon), \tag{5.3.5}$$

subject to

$$-\nabla y(\mathbf{X}^*)\mathbf{S} + \epsilon \leq 0, \tag{5.3.6}$$

$$\nabla g_i(\mathbf{X}^*)\mathbf{S} + \theta_i \epsilon \leq 0, \qquad i = 1, \ldots, l. \tag{5.3.7}$$

The first restriction (5.3.6) requires that ϵ be no greater than the change in $y(\mathbf{X})$. Hence by maximizing ϵ we maximize the increase in $y(\mathbf{X})$. In the second restriction (5.3.7), θ_i is given a value of zero if $g_i(\mathbf{X})$ is linear and a value of 1 otherwise. For a linear constraint we thus simply require that the change in the constraint not cause infeasibility. In the case of a nonlinear constraint, any appreciable movement directly along the tangent $\nabla g_i(\mathbf{X}^*)$ to the constraint is likely to cause the new solution to be infeasible. Hence we add the term $\theta_i \epsilon$ to force movement away from the constraint boundaries and into the feasible region. After determination of $\mathbf{S}$, we then normalize this vector so that it has a unit length. That is, we find an $\mathbf{S}_n$ such that

$$\mathbf{S}_n'\mathbf{S}_n = 1 \tag{5.3.8}$$

and

$$\mathbf{S}_n = \gamma \mathbf{S}, \tag{5.3.9}$$

where γ is a constant.

Note that ϵ is an independent variable that can assume a negative value. However, if the maximum value of $\epsilon \leq 0$, then no feasible direction exists, and the problem is terminated. The current solution would then be a local optimum. If a positive value of ϵ is found, we then determine a step size α such that $y(\mathbf{X} - \alpha \mathbf{S}_n)$ is maximized subject to the requirement that all constraints $g_i(\mathbf{X} + \alpha \mathbf{S}_n)$ are satisfied. If we are maximizing a concave function subject to concave constraints (minimizing convex function subject to convex constraints), a global optimum is obtained by this procedure.

EXAMPLE 5.3.1

Let us use Zoutendijk's procedure to maximize the simple function

$$y = (x_1 - 4)^2 + (x_2 - 3)^2, \tag{5.3.10}$$

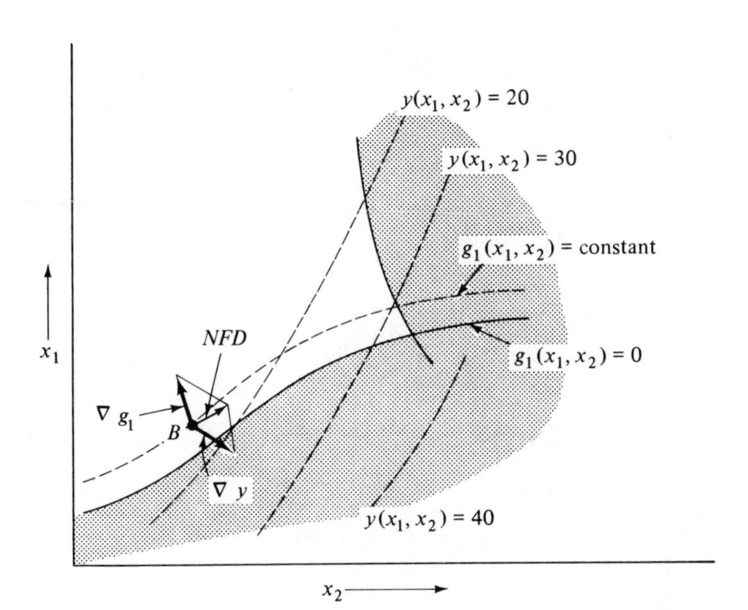

Figure 5.9. Determination of New Feasible Direction (NFD) by Multiple Gradient Summation

subject to the constraints

$$g_1(\mathbf{X}) = x_1 + x_2 \leq 6, \tag{5.3.11}$$

$$g_2(\mathbf{X}) = -x_1 \leq 0, \tag{5.3.12}$$

$$g_3(\mathbf{X}) = -x_2 \leq 0. \tag{5.3.13}$$

Figure 5.10 shows the contour lines of the objective function as well as the constraints. The constrained maximum occurs at $\mathbf{X} = (0, 0)$. Assume that we begin our maximization from $\mathbf{X} = (3, 2)$. At that point no constraint is binding, and we proceed in the direction of the gradient $\nabla y(\mathbf{X})$, where $\nabla y(\mathbf{X}) = (-2, -2)$. By evaluation of the objective function at several points, we find that y continues to increase until the constraint $0 \leq x_2$ is contacted at $\mathbf{X} = (1, 0)$. We redetermine the gradient $\nabla y(\mathbf{X}) = (-6, -6)$, and find $\nabla g_3(\mathbf{X}) = (0, -1)$. We now maximize ϵ, subject to the requirements that

$$-[\nabla y(\mathbf{X})]'\mathbf{S} + \epsilon = -[-6s_1 - 6s_2] + \epsilon \leq 0, \tag{5.3.14}$$

$$[\nabla g_3(\mathbf{X})]'\mathbf{S} + \theta\epsilon = \quad [0 \cdot s_1 - s_2] + 0 \cdot \epsilon \leq 0. \tag{5.3.15}$$

In a real problem we would now determine the values of s_1 and s_2 by solving the above linear programming problem. In this simple problem, we see that $g_3(\mathbf{X})$ requires $s_2 \longrightarrow 0$, and therefore we maximize the ϵ allowed by the first constraint by setting $s_1 = -\infty$. When we normalize $\mathbf{S}$ to a unit length, we obtain $\mathbf{S}_n = (-1, 0)$.

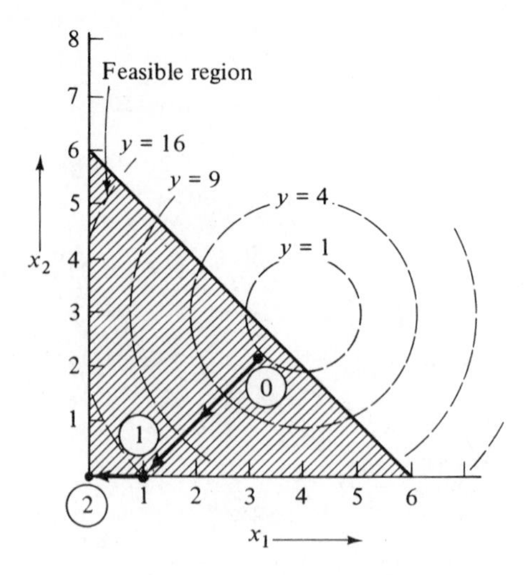

Figure 5.10. Application of Zoutendijk's Feasible Direction Method

If we proceed in the direction $\mathbf{S}_n$ and evaluate y at several points ($\mathbf{X} = \mathbf{X}_{\text{original}} + \alpha\mathbf{S}_n$), we find that the objective function continues to increase in this direction but that we are restrained by $g_2(\mathbf{X})$. (In a more complex problem we might have found that the objective function decreased for the larger α values. We would estimate the location of the maximum by some one-dimensional search procedure.) Our new feasible point is therefore $\mathbf{X} = (0, 0)$. When we again try to maximize ϵ, subject to

$$-[\nabla y(\mathbf{X})]'\mathbf{S} + \epsilon = \quad 8s_1 + 6s_2 + \epsilon \leq 0, \tag{5.3.16}$$

$$[\nabla g_2(\mathbf{X})]'\mathbf{S} + 0 \cdot \epsilon = -s_1 \leq 0, \tag{5.3.17}$$

$$[\nabla g_3(\mathbf{X})]'\mathbf{S} + 0 \cdot \epsilon = -s_2 \leq 0, \tag{5.3.18}$$

we see that since only positive values of s_1 and s_2 are allowed, there is no allowable value of ϵ which will satisfy the first constraint. We have therefore reached the optimum.

Gradient projection

Since Zoutendijk attempts to find some "best" feasible direction at each step, we solve an optimization problem at each move. Such a procedure is very time consuming. Rather than solve a series of optimization problems, Rosen [1960] proposed that the new feasible directions be obtained using the Kuhn–Tucker conditions. His method, called *gradient projection*, was originally developed for linear constraints and works best when that is the case.

We shall begin by consideration of the problem of minimizing $y(\mathbf{X})$, subject to a set of linear inequality constraints

$$g_i(\mathbf{X}) = a_{i1}x_1 + a_{i2}x_2 + \cdots + a_{in}x_n - b_i \leq 0, \qquad i = 1, 2, \ldots, m. \tag{5.3.19}$$

We recognize that the feasible region is a convex polyhedron whose boundaries are the hyperplanes $g_i(\mathbf{X}) = 0$. In the case of a problem with three variables, we could represent this region by a polyhedron such as shown in Figure 5.11. Let us assume that we have located a feasible point, $\mathbf{X}$ (cf. Figure 5.11), within this polyhedron. We improve our objective by taking steps in the direction of $-\nabla y(\mathbf{X})$ until we encounter a constraint (assume this occurs at $\mathbf{X}_2$ of Figure 5.11). At this point we can no longer take a step in the direction of the negative of the gradient, $-\nabla y(\mathbf{X}_2)$, without entering the infeasible region. We can, however, proceed on the bounding hyperplane in a direction aligned with $-\nabla y(\mathbf{X}_2)$. That is, we can proceed along the constraint hyperplane in the direction obtained by *projecting* $-\nabla y(\mathbf{X}_2)$ *on the constraint hyperplane*. We thus will remain within the feasible region. We shall reduce our objective function since some of the components of motion are in the the direction of the negative of the objective function gradient.

We may continue to move in this direction until we reach $(\mathbf{X}_3)$. At that

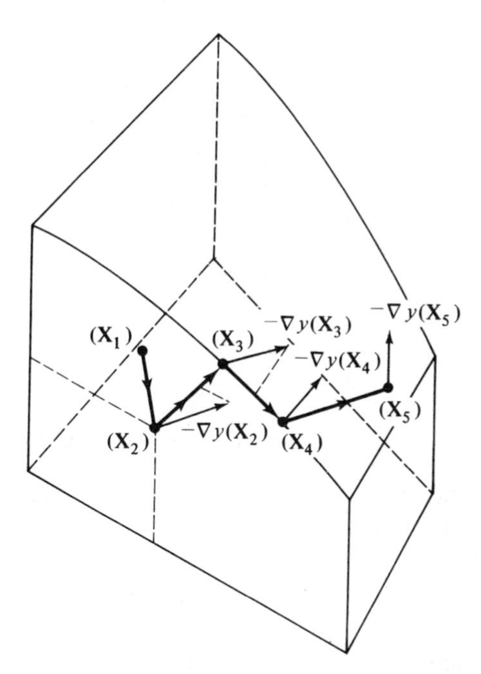

Figure 5.11. Graphical Representation of Gradient Projection Procedure

point (again, cf. Figure 5.11), we encounter two binding constraints. We now move along the intersection of the two hyperplanes in the direction indicated by projecting $-\nabla y(\mathbf{X}_3)$ on the intersection. At $(\mathbf{X}_4)$ a change in direction of the gradient causes us to move across the upper hyperplane. Finally, at $(\mathbf{X}_5)$, the gradient is normal to the constraint surface, and we can no longer reduce our objective function without proceeding into the infeasible region. We have thus reached at least a local optimum.

Before proceeding further, we need to define more exactly what we mean by the projection of a vector in the multidimensional case. The projection $\mathbf{P}$ of vector $\mathbf{Z}$ on the hyperplane $g_i(\mathbf{X}) = 0$ is given by

$$\mathbf{P} = \mathbf{Z} - d \cdot \nabla g_i(\mathbf{X}), \tag{5.3.20}$$

where d is a positive constant determined so that $\mathbf{P}$ is normal to $\nabla g_i(\mathbf{X})$. If $\mathbf{Z} = \nabla y(\mathbf{X})$, then $\mathbf{P}$ is the "gradient projection" on $g_i(\mathbf{X})$. Since the gradient $\nabla g_i(\mathbf{X})$ is itself normal to $g_i(\mathbf{X})$, the only way for $\mathbf{P}$ to be normal to $\nabla g_i(\mathbf{X})$ is for $\mathbf{P}$ to lie in the hyperplane of $g_i(\mathbf{X})$. This is illustrated for a three-variable system in Figure 5.12, where it is apparent that our mathematical definition of the projection as the vector sum of $-\nabla y(\mathbf{X})$ and $-d\nabla g_i(\mathbf{X})$ is consistent with the geometric definition of the projection of $-\nabla y(\mathbf{X})$ on $g_i(\mathbf{X}) = 0$.

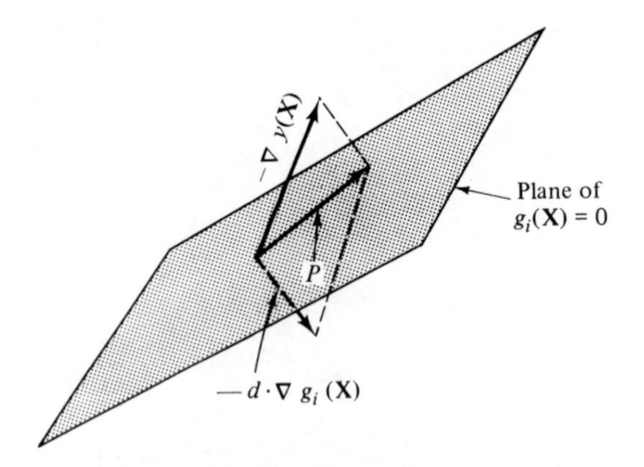

Figure 5.12. Projection of Gradient on Constraint Hyperplane

If we specify that $\mathbf{P}$ be the projection of $-\nabla y(\mathbf{X})$ on the intersection of the constraints $g_1(\mathbf{X}) = 0, \ldots, g_k(\mathbf{X}) = 0$, then

$$\mathbf{P} = -\nabla y(\mathbf{X}) - \sum_{i=1}^{k} d_i \nabla g_i(\mathbf{X}). \tag{5.3.21}$$

The projection $\mathbf{P}$ will now be along the intersection of the constraint hyperplanes 1 through k.

The major computational effort in the gradient projection method is in computing the projections of the gradient vector. Rather than obtain **P** by computing the value of d_i individually so that the orthogonality with $g_i(\mathbf{X})$ is satisfied, Rosen shows that **P** is more effciently obtained by multiplication of the negative of the gradient vector by an $n \times n$ projection matrix (**V**). That is,

$$\mathbf{P} = -\nabla y(\mathbf{X}) \cdot \mathbf{V} \tag{5.3.22}$$

The projection matrix is determined by first writing all the binding constraints, $i = 1, \ldots, k$, in matrix form as

$$(\mathbf{A})(\mathbf{X}) - (\mathbf{B}) = \mathbf{0}. \tag{5.3.23}$$

By making use of this matrix notation, we can write

$$\nabla g_i = [a_{i1}, a_{i2}, \ldots, a_{in}] = \mathbf{A}_i, \tag{5.3.24}$$

where $\mathbf{A}_i$ is the ith row of the $k \times n$ matrix **A**. Hence Equation (5.3.21) becomes

$$\mathbf{P} = -\nabla y - \sum_{i=1}^{k} d_i \mathbf{A}_i \tag{5.3.25}$$

or

$$\mathbf{P} = -\nabla y - \mathbf{DA}, \tag{5.3.26}$$

where ∇y is taken as a row vector and **D** is some k-dimensional row vector, chosen so that **P** will lie in the $(n - k) -$ dimensional hyperplane (the hyperplane defined by the intersection of the k active constraints).

Since **P** must be normal to the gradient of each of the constraints, we can write

$$\mathbf{P}\nabla g' = \mathbf{PA}_i = 0, \qquad i = 1, 2, \ldots, k, \tag{5.3.27}$$

or

$$\mathbf{PA}' = 0, \tag{5.3.28}$$

where $\mathbf{A}'$ is the transpose of **A**. Therefore, we have

$$\mathbf{PA}' = -\nabla y\mathbf{A}' - \mathbf{DAA}'. \tag{5.3.29}$$

Upon solving for **D** ,we obtain

$$\mathbf{DAA}' = -\nabla y\mathbf{A}' \tag{5.3.30}$$

or

$$\mathbf{D} = -\nabla y\mathbf{A}'(\mathbf{AA}')^{-1}. \tag{5.3.31}$$

Substitution of this result into Equation (5.3.26) gives

$$\mathbf{P} = -\nabla y + \nabla y \mathbf{A}'(\mathbf{A}\mathbf{A}')^{-1}\mathbf{A}, \tag{5.3.32}$$

which can be rewritten as

$$\mathbf{P} = -\nabla y[\mathbf{I} - \mathbf{A}'(\mathbf{A}\mathbf{A}')^{-1}\mathbf{A}]. \tag{5.3.33}$$

Equation (5.3.33) can be expressed more simply as

$$\mathbf{P} = -\nabla y \mathbf{V}, \tag{5.3.34}$$

where $\mathbf{V}$ is the *projection matrix* given by

$$\mathbf{V} = \mathbf{I} - \mathbf{A}'(\mathbf{A}\mathbf{A}')^{-1}\mathbf{A}. \tag{5.3.35}$$

The procedure holds, providing the binding constraints are all linearly independent. Rosen shows that the constraints are independent if, and only if, $\mathbf{P} \neq \mathbf{0}$.

We can now succinctly describe the gradient projection algorithm. We start at a feasible point and minimize the objective function by taking steps in the direction of $-\nabla y(\mathbf{X})$ until a constraint is encountered. We then determine which constraint (or constraints) is binding at this point, $(\mathbf{X}_l)$, and determine the projection of $-\nabla y(\mathbf{X}_l)$ on these binding constraints by multiplying $-\nabla y(\mathbf{X}_l)$ by the matrix $(\mathbf{V})$. We shall then have established the direction of the allowable motion but not the magnitude of the step. The step size is determined by evaluating $y(\mathbf{X}_l + \alpha\mathbf{P})$ for several values of α and choosing the α that leads to the greatest change by an interpolation procedure. If the α so obtained leads to a violation of the constraints, a reduced step size must be determined. At location $(\mathbf{X}_{l+1}) = (\mathbf{X}_l + \alpha\mathbf{P})$, any additional constraints or nonnegative restrictions contacted are added to the initial set of binding constraints, and the projection of $-\nabla y(\mathbf{X}_{l+1})$ on this new set is determined. This process continues until the projection $\mathbf{P}$ is zero.

We have yet to relate gradient projection to the Kuhn–Tucker conditions. If we restrict our consideration to those $x_j > 0$ at the optimum, we can write the Kuhn–Tucker conditions as

$$\frac{\partial y}{\partial x_j}(\mathbf{X}_o) + \sum_{i=1}^{m} \lambda_i \frac{\partial g_i}{\partial x_j}(\mathbf{X}_o) = 0, \qquad j = 1, \ldots, p, \tag{5.3.36}$$

where the indices $1, \ldots, p$ indicate the set of nonzero x_j in the solution. In vector form the above becomes

$$\nabla y(\mathbf{X}_o) + \sum_{i=1}^{m} \lambda_i \nabla g_i(\mathbf{X}_o) = \mathbf{0}. \tag{5.3.37}$$

Since $\lambda_i = 0$ for all the nonbinding constraints, $i = k + 1, \ldots, m$, Equation (5.3.37) is unchanged if we write it as

$$\nabla y(\mathbf{X}_o) + \sum_{i=1}^{k} \lambda_i \nabla g_i(\mathbf{X}_o) = \mathbf{0}. \tag{5.3.38}$$

If we multiply through by -1, we have

$$-\nabla y(\mathbf{X}_o) - \sum_{i=1}^{k} d_i \nabla g_i(\mathbf{X}_o) = \mathbf{0}, \tag{5.3.39}$$

where $d_i = \lambda_i$, $i = 1, \ldots, k$. Since the Kuhn–Tucker conditions tell us that λ_i is positive for constraints of the form $g_i(\mathbf{X}) \leq 0$, the d_i are positive quantities. By comparison of (5.3.39) and (5.3.21) it is obvious that at the optimum

$$\mathbf{P} = -\nabla y(\mathbf{X}_o) - \sum_{i=1}^{k} d_i \nabla g_i(\mathbf{X}_o) = \mathbf{0}. \tag{5.3.40}$$

Thus the requirement that the projection of $-\nabla y(\mathbf{X})$ go to zero at the optimum is identical to that imposed by the Kuhn–Tucker conditions. If $\mathbf{P} = \mathbf{0}$ and one $d_i < 0$, we have not yet reached the optimum. The constraint for which $d_i < 0$ must be nonbinding, and hence it must be removed from the set used to define the projection matrix $\mathbf{V}$. A new value of $\mathbf{P}$ would have to be computed and the process continued. Only one constraint at a time may be deleted from the constraint set defining $\mathbf{V}$. Therefore, if more than one $d_i < 0$, one is chosen by some rule, say the most negative, to be eliminated.

It may be shown that the gradient-projection procedure will converge to the global optimum when the programming problem is convex. Difficulties with the procedure can be encountered if the rank of $\mathbf{A}$ is less than the number of its rows. Rosen [1960] shows that this problem can be circumvented in a manner similar to that used for resolving degeneracy in linear programming.

EXAMPLE 5.3.2

As a simple illustration of the gradient-projection technique, let us again use the problem illustrated in Figure 5.10. If we write the problem as a minimization problem, we have: Minimize

$$y = -(x_1 - 4)^2 - (x_2 - 3)^2, \tag{5.3.41}$$

subject to (5.3.11), (5.3.12), and (5.3.13). If we again begin our optimization at $\mathbf{X} = (3, 2)$, we first move in the direction of the gradient until $\mathbf{X} = (1, 0)$, where we encounter constraint $g_3(\mathbf{X})$. We accept $(1, 0)$ as a feasible point and proceed to determine the projection of $-\nabla y(\mathbf{X})$ on $g_3(\mathbf{X})$. As we have only one binding constraint, $\mathbf{A}$ is a row vector and is given by

$$\mathbf{A} = [0, -1]. \tag{5.3.42}$$

We then have

(5.3.43) $$\mathbf{A}' = \begin{bmatrix} 0 \\ -1 \end{bmatrix},$$

(5.3.44) $$(\mathbf{AA}')^{-1} = 1.$$

Hence

(5.3.45) $$[\mathbf{A}'(\mathbf{AA}')^{-1}]\mathbf{A} = \begin{bmatrix} 0 \\ -1 \end{bmatrix} \cdot [0 \quad -1] = \begin{bmatrix} 0 & 0 \\ 0 & 1 \end{bmatrix}$$

and

(5.3.46) $$\mathbf{V}_1 = \mathbf{I} - \begin{bmatrix} 0 & 0 \\ 0 & 1 \end{bmatrix} = \begin{bmatrix} 1 & 0 \\ 0 & 0 \end{bmatrix}.$$

Since

(5.3.47) $$\nabla y(\mathbf{X}) = [6, 6],$$

(5.3.48) $$\mathbf{P}_1 = [-\nabla y(\mathbf{X})](\mathbf{V}_1) = [-6, -6]\begin{bmatrix} 1 & 0 \\ 0 & 0 \end{bmatrix} = [-6, 0].$$

Our projection vector is then along the $g_3(\mathbf{X})$ constraint and is directed to the left. It is identical to the feasible direction determined previously using Zoutendijk's procedure.

We continue in the direction $\mathbf{P}_1$ until we encounter constraint $g_2(\mathbf{X})$ at $(\mathbf{X}) = (0, 0)$. Since both $g_2(\mathbf{X})$ and $g_3(\mathbf{X})$ are binding, our $\mathbf{A}$ matrix becomes

(5.3.49) $$\mathbf{A} = \begin{bmatrix} -1 & 0 \\ 0 & -1 \end{bmatrix}.$$

This leads to

(5.3.50) $$\mathbf{A}'[(\mathbf{AA}')^{-1}]\mathbf{A} = \begin{bmatrix} -1 & 0 \\ 0 & -1 \end{bmatrix}\begin{bmatrix} 1 & 0 \\ 0 & 1 \end{bmatrix}\begin{bmatrix} -1 & 0 \\ 0 & -1 \end{bmatrix} = \begin{bmatrix} 1 & 0 \\ 0 & 1 \end{bmatrix} = \mathbf{I}$$

and a projection

(5.3.51) $$\mathbf{P}_2 = -\nabla y(\mathbf{X})[\mathbf{I} - \mathbf{I}] = \mathbf{0}.$$

Since $\mathbf{P} = \mathbf{0}$, one of the requirements for a stationary point is met. We also require that all the d_i of (5.3.21) be nonnegative.

From Equation (5.3.31) we had

(5.3.52) $$\mathbf{D} = -\nabla y(\mathbf{X})\mathbf{A}'[\mathbf{AA}']^{-1}.$$

Therefore, in our illustrative problem at $\mathbf{X}_2 = (0, 0)$,

$$\mathbf{D}_2 = [-8, 6]\begin{bmatrix} -1 & 0 \\ 0 & -1 \end{bmatrix}\begin{bmatrix} 1 & 0 \\ 0 & 1 \end{bmatrix} = [8, 6]. \tag{5.3.53}$$

Since the components of $\mathbf{D}_2$ are the final values of d_2 and d_3, we meet the nonnegativity requirement. We conclude that $\mathbf{X}_2 = (0, 0)$ is therefore a stationary point.

Although our consideration has been limited to inequality constrained problems, gradient projection may be applied when some or all constraints are equalities. The initial point must satisfy the equality constraints. Since the procedure begins by projecting $-\nabla y(\mathbf{X})$ at the initial point onto the binding constraints, the equality constraints will be part of this set. The equality constraints must remain binding throughout the iterative process; therefore, they remain in the constraint set defining $\mathbf{V}$. It will be recalled that the use of the projective matrix holds, providing the constraints are linearly independent. If they are not independent, they must be reduced to a linearly independent set (cf. Rosen [1960]).

If the constraints to our problem are nonlinear, the gradient is projected on the hyperplanes tangent to the constraints at the point under consideration (cf. Rosen [1961]). The columns of our projection matrix $\mathbf{V}$ consist of the partial derivatives of the constraint functions. Now, however, any movement along the tangent hyperplanes will produce a new point that may be infeasible. A correction procedure is now required to return to the feasible region. This problem is illustrated in Figure 5.13. In Figure 5.13a we have a solution space bounded by linear constraints. If our initial point were 0, we would follow the gradient until constraint AB is met. From point B we would follow the constraints until point C, where the optimum is located. Now consider the system shown in Figure 5.13b, which has an identical objective function but nonlinear constraints. Once the constraint boundaries are reached, motion along the hyperplanes causes us to leave the feasible region. The correction procedure required to return to the constraint boundaries is iterative and tedious. Because of these difficulties, the gradient-projection method is rarely used directly when the constraints are nonlinear.

The gradient-projection method requires, as do all the feasible-direction methods, the location of an initial feasible point. This is not a problem when all the constraints are linear, but may pose great difficulties in the case of nonlinear constraints. If a feasible starting point is not known, there is no simple way of obtaining one. The method does have the advantage of being able to handle equality constraints. Some of the feasible-direction methods (e.g., Mugele's probe and the multiple-gradient-summation technique) can be used only with inequality constraints, whereas Zoutendijk's method can handle equality constraints only if they are linear.

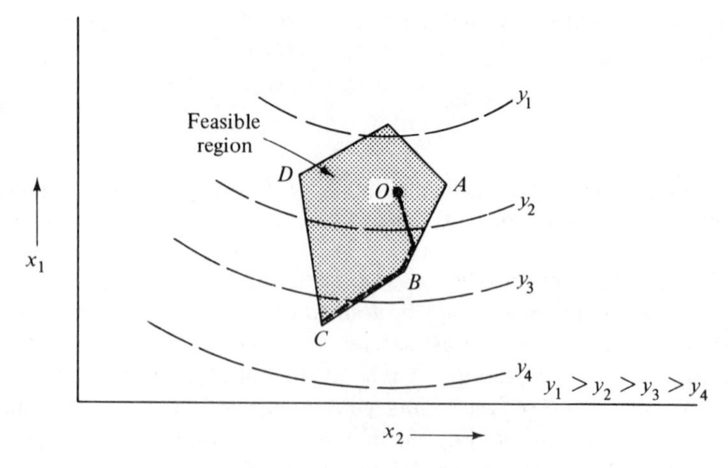

Figure 5.13. (a) Gradient Projection Method in Presence of Linear Constraints

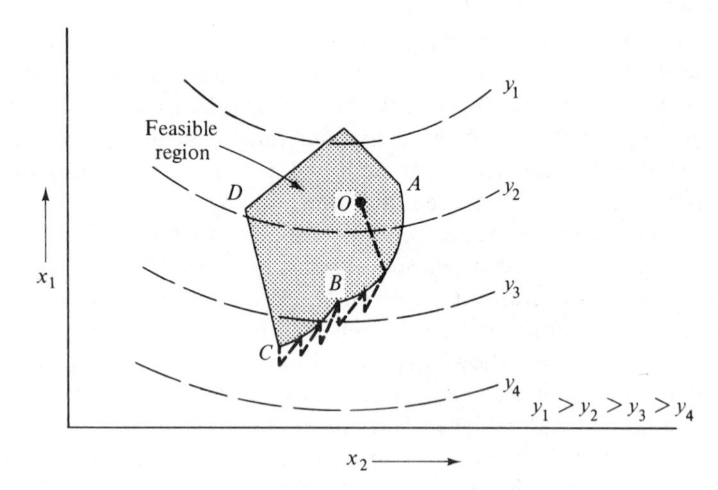

(b) Gradient Projection Method in Presence of Nonlinear Constraints

5.4 Penalty-function Techniques

Basic approach

Each of the general nonlinear programming techniques that we have considered has suffered from some disadvantages. Convergence difficulties can limit the usefulness of linearization techniques. Feasible-direction techniques, as we have just seen, are plagued with the difficulty of determining a feasible starting point. Furthermore, although the development of feasible-direction techniques was motivated by the desire to use unconstrained optimization techniques, the development was not entirely successful in this regard. Each of the feasible-direction techniques required substantial modifications of unconstrained methods. Let us consider how we can reformulate our nonlinear programming problem as an unconstrained problem that does not require a feasible starting point.

Suppose that we wish to minimize

$$y(x) = x^2 - x, \tag{5.4.1}$$

subject to

$$g(x) = x - 2 \geq 0. \tag{5.4.2}$$

As shown in Figure 5.14, the optimum is located at the intersection of the

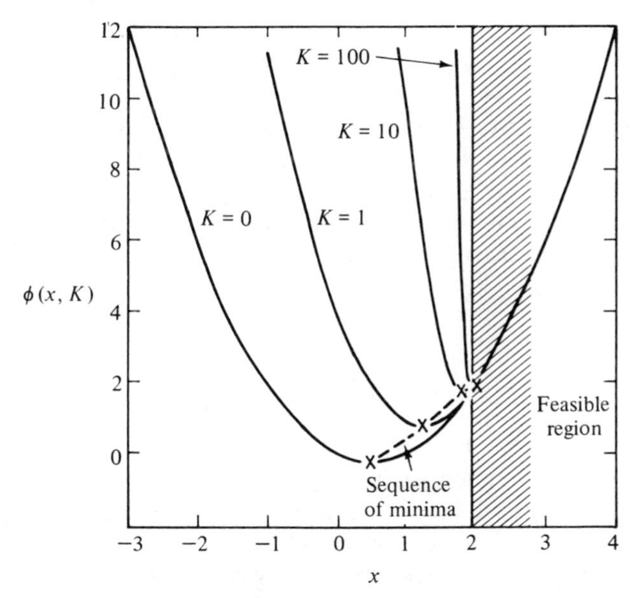

Figure 5.14. Constrained Minimization Using the Penalty Function Technique

constraint and the objective function ($x = 2$). Let us convert this into an unconstrained minimization problem by defining a new objective function $\phi(x)$, which includes some penalty for violating the constraint. Thus, for example,

$$\phi(x) = y(x) + \delta p, \tag{5.4.3}$$

where p is some nonnegative penalty and

$$\delta = 1, \quad \text{if } g(x) \leq 0, \tag{5.4.4}$$

and

$$\delta = 0 \quad \text{otherwise.} \tag{5.4.5}$$

Also, we should like our penalty to reflect how far away we are from the constraint. This is simply accomplished by letting $p = [g(x)]^2$. If our constraint had been of the form $g(x) \leq 0$, we could still use penalties of the same form as Equation (5.4.3), providing we define $\delta = 1$ when the constraint is unsatisfied, and zero otherwise. With this definition of δ, we could also use $|g(x)|\delta$ as the penalty function. With the quadratic penalty function, our objective function becomes

$$\phi(x) = y(x) + \delta[g(x)]^2 \tag{5.4.6}$$

or

$$\phi(x) = x^2 - x + \delta[(x - 2)]^2, \tag{5.4.7}$$

which yields curve A of Figure 5.14. The minimum of this curve is closer to the true optimum than the minimum of the unconstrained original objective function. However, we have not yet put enough emphasis on our constraint penalty. Let us now multiply our constraint penalty by some positive constant K, which is greater than 1. This yields

$$\phi(x) = y(x) + \delta K[g(x)]^2 \tag{5.4.8}$$

or

$$\phi(x) = x^2 - x + \delta K[(x - 2)]^2. \tag{5.4.9}$$

With $K = 0$, our minimum was $x = \frac{1}{2}$. When we had set $K = 1$, as in (5.4.7), the minimum occurred at $x = \frac{5}{4}$ (curve A of Figure 5.14). Now, when we increase K to 10, we find our minimum at $x = \frac{41}{22}$ (curve B of Figure 5.14). If we repeat the procedure using successively higher values of K, we approach the true constrained minimum more closely. We may easily show this by obtaining the derivative analytically in the region where $\delta = 1$, and then setting the derivative equal to zero. We then have

$$x_{\min} = \frac{4K + 1}{2K + 2}. \tag{5.4.10}$$

Thus $x_{\min} \longrightarrow 2$ as $K \longrightarrow \infty$. We may generalize (5.4.8) to include multivariable systems that are inequality constrained by writing

$$\phi(\mathbf{X}) = y(\mathbf{X}) + \delta K[g(\mathbf{X})]^2. \tag{5.4.11}$$

In many problem we are faced with equality constraints. Any such constraint can be written as

$$g(\mathbf{X}) = 0 \tag{5.4.12}$$

by appropriate algebraic rearrangement. We then simply let

$$p = [g(\mathbf{X})]^2. \tag{5.4.13}$$

One reason for choosing the quadratic form of the penalty function is now clear. We observe that by its use both positive and negative deviations from the constraint are seen as penalties. Thus, for a problem in which constraints $1, 2, \ldots, l$ are inequalities and constraints $l + 1, l + 2, \ldots, m$ are equalities, our augmented objective function becomes

$$\phi(\mathbf{X}) = y(\mathbf{X}) + K_j\left\{\sum_{i=1}^{l} \delta_i[g_i(\mathbf{X})]^2 + \sum_{i=l+1}^{m} [g_i(\mathbf{X})]^2\right\}. \tag{5.4.14}$$

Minimization of this function for a series of increasing values of K_j will yield a sequence of solutions that more and more closely satisfies the constraints. When all constraints are satisfied within some preassigned tolerance, the procedure terminates.

When the quadratic form of the penalty function is used, $\phi(\mathbf{X})$ will be continuous and differentiable, providing $y(\mathbf{X})$ and the $g_i(\mathbf{X})$ are continuous and differentiable. With $|g_i(\mathbf{X})|\delta$ as the penalty function for inequality constraints, $\phi(\mathbf{X})$ is not differentiable on the boundaries of the feasible region. We may therefore use any of the gradient or search methods of Chapter 3 for unconstrained optimization, providing we use quadratic penalties. However, if $|g_i(\mathbf{X})|\delta$ is used, we are limited to search techniques that require only function evaluations. Whichever technique is used, it is well to recall that unless the objective function is convex there is no guarantee that the optimization procedure provides anything more than a local optimum.

From consideration of Equation (5.4.4), we might conclude that it would be satisfactory to start the optimization process with a large value of K. For the simple inequality constrained problem illustrated in Figure 5.14, this is possible. By choosing a large value for K we could closely approach the constrained optimum with but a single optimization. However, in most practical problems this is not a desirable approach. Very large values of K introduce sharp valleys (or ridges in a maximization problem), which gradient or search procedures cannot follow. For example, consider a direct-search procedure that uses univariate probes to establish the direction in which to move. As shown in Figure 5.15, in the presence of a sharp enough valley all steps would indicate a positive gradient, and we would erroneously conclude that we were at a minimum. This kind of situation is particularly likely to arise with large K values when nonlinear equality constraints are present. Let line AB of Figure 5.15 represent some equality constraint. If we start deep in the infeasible region, say at C, we would proceed towards a feasible point D. Once having found such a point D the search would not be able to follow the very narrow valley adjacent to the constraint's boundaries, since all univariate probes would fail. If, however, we had first minimized the objective function without consideration of the constraints, we would have located at least a local optimum of the unconstrained objective function. With a series of moderate increases in the value of K, we would expect more gentle valleys at first and that each succeeding optimum would be suffiently close to its predecessor so that it could be located easily. When we have finally increased K to a level that forces compliance with the constraints, we shall have found a constrained minimum without the need to follow the constraint boundaries.

The penalty-function technique is obviously adaptable to maximization problems. If we wish to maximize $y(\mathbf{X})$ subject to a series of constraints, we

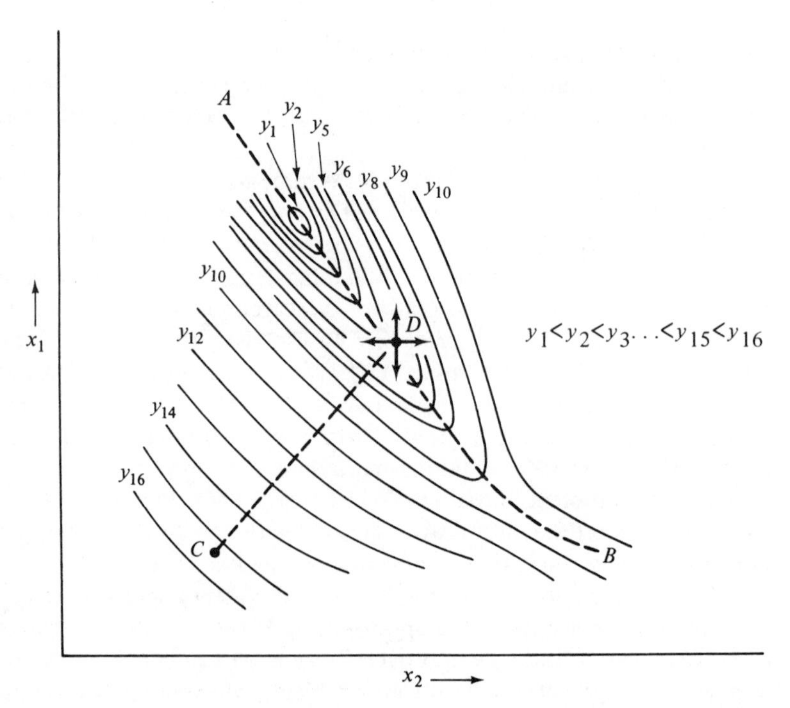

Figure 5.15. Failure of Direct Search in Presence of Sharp Ridge

simply write

$$\phi(\mathbf{X}) = y(\mathbf{X}) - K_j\left\{\sum_{i=1}^{l} \delta_i[g_i(\mathbf{X})]^2 + \sum_{i=l+1}^{m} [g_i(\mathbf{X})]^2\right\}. \tag{5.4.15}$$

In this form the penalties are sometimes referred to as loss functions. Again we conduct our optimization at a succession of increasing values of K_j until the constraints are satisfied to the extent desired.

The use of penalty functions in nonlinear programming was apparently first suggested by Courant and shown to be rigorously correct for functions of two variables by Moser (cf. Courant [1956]). Kelley [1962] subsequently developed a workable algorithm by using a gradient procedure to minimize the augmented objective function. Weisman et al. [1965] developed an alternative algorithm by coupling the penalty-function technique to a direct-search program.

Theoretical foundation

A firm theoretical foundation for the penalty-function procedure applied to multivariable systems was established by Fiacco and McCormick [1964] and by Zangwill [1967]. Zangwill considers the maximization problem under the fairly general conditions where

1. Constraints $1, \ldots, l$ are in the form $g_i(\mathbf{X}) \geq 0$ and constraints $l + 1, \ldots, m$ are equalities.
2. At least one feasible solution exists.
3. $y(\mathbf{X})$ and $g_i(\mathbf{X})$, $i = 1, \ldots, m$, are continuous.
4. There exists an $\epsilon > 0$ such that the region where the inequality constraints are equal to or greater than $-\epsilon$ and the magnitude of the equality constraints equal to or less than ϵ is bounded.
5. There exists a $K_o \geq 0$ such that max $\phi(\mathbf{X}, K)$ = supremum $\phi(\mathbf{X}, K)$ for all $K \geq K_o$.

The first assumptions guarantee that the problem has an optimum solution. The last condition requires further explanation. Let $y_j(\mathbf{X})$ be a series of upper bounds for $f(\mathbf{X})$. Then, if $y_n(\mathbf{X})$ is the lowest possible upper bound for $f(\mathbf{X})$, $y_n(\mathbf{X})$ = supremum $y(\mathbf{X})$. Thus this last condition requires that for any K greater than K_o, the maximum of $\phi(\mathbf{X}, K)$ will be no greater than max $\phi(\mathbf{X}, K_o)$. We must therefore choose loss functions $p[g_i(\mathbf{X})]$ that will ensure that this occurs. Loss functions of the type previously indicated will generally do so.

Under these conditions, it may be shown that *if* $K_1, K_2, K_3, \ldots, K_\infty$ *is an increasing sequence, where* $K_j \geq 0$ *with* $\lim_{j\to\infty} K_j = +\infty$, *then there exists a convergent sequence such that* max $\phi(\mathbf{X}, K_j) \longrightarrow y(\mathbf{X})$ *and the limit of this sequence is an optimum point*. In other words, as K approaches ∞, the penalties are forced to zero since the constraints are satisfied. Thus, under the conditions assumed, when we carry out a series of optimizations at increasing values of K_j, the maxima obtained will converge to at least a local optimum. The method converges even though the objective functions and constraints may not be concave. Providing the general conditions hold, we can be assured that if we start out at any point we can obtain a feasible solution. We are not required to determine a feasible initial point, and the constraints may be violated during the course of the optimization. If we do have a feasible initial point, we can find an improved feasible point if one exists in the region of our initial point.

It may be shown (cf. Zangwill [1967]) that if (a) the objective function and the constraints are concave and (b) all inequality constraints are in the form $g_i(\mathbf{X}) \geq 0$ and (c) we choose $p_i(\mathbf{X})$ so that $\phi(\mathbf{X}, K)$ is concave [penalties of the form $|g_i(\mathbf{X})| \cdot \delta$ meet this requirement when the $g_i(\mathbf{X})$ are concave], then the local unconstrained maximum to which the procedure converges is the global maximum. We should observe that the negative of the convex function is concave. Furthermore, minimizing $\phi(\mathbf{X})$ is equivalent to maximizing $-\phi(\mathbf{X})$. Hence the minimization of a convex function subject to convex inequality constraints is readily convertible to a maximization of a concave function subject to concave constraints. The penalty-function approach thus provides global optima for both convex and concave programming problems.

When all the constraints are of the form

$$g_i(\mathbf{X}) \geq 0 \tag{5.4.16}$$

and the constraints and the unaugmented objective function are concave, only a single maximization problem need by solved. We form the augmented objective function

$$\phi = y(\mathbf{X}) + K \sum_{i=1}^{I} \min[g_i(\mathbf{X}), 0]. \tag{5.4.17}$$

Our penalty terms will be negative unless all constraints are satisfied. Let $(\mathbf{Y})$ represent any feasible point and define the quantity α such that

$$\alpha = \min_i[g_i(\mathbf{Y})] > 0. \tag{5.4.18}$$

Now assume that we have an upper bound y^* for the value of the objective function. We then define a quantity β, where

$$\beta = y^* - y(\mathbf{Y}). \tag{5.4.19}$$

Zangwill then shows that a value of K, such that

$$K \geq \frac{\beta + 1}{\alpha} \tag{5.4.20}$$

used in our augmented objective function (5.4.17) will provide the global optimum.

In many cases we do not have any estimate of an upper bound to our original objective function. Such a bound can be found by simply solving the unconstrained maximization problem: Maximize

$$y(\mathbf{X}), \tag{5.4.21}$$

providing all of the decision variables are bounded. Similarly, if we do not know the location of a feasible point, we may find one by solving the problem: Maximize

$$\sum_{i=1}^{l} \min[g_i(\mathbf{X}), 0]. \tag{5.4.22}$$

Thus, theoretically, we shall have to solve at most three unconstrained maximization problems to solve our original problem. As we have already noted, this procedure can result in a substantial reduction in the number of penalty levels that must be examined, although it is not always desirable to employ the scheme because of computational difficulties.

Numerical techniques

The constraints in any one problem are often concerned with quantities of entirely different magnitudes. In one constraint we may be dealing with values of the magnitude of 10^6 and in another with quantities having a magnitude of 1. In this form, changes in the former constraint will completely mask any changes in the latter. When this situation arises, meaningful results cannot be obtained unless we "scale" all constraints. One method of doing so is to write the penalties as

$$k_i \delta_i [g_i(\mathbf{X})]^2, \tag{5.4.23}$$

where k_i is a scaling factor. To determine the k_i, we establish a single tolerance limit, ϵ, for all constraints. That is, we consider

$$k_i g_i[\mathbf{X}] \leq \epsilon, \qquad i = 1, \ldots, m, \tag{5.4.24}$$

an acceptable approximation of $g_i(\mathbf{X}) \leq 0$. The scaling factors are the set of constants that allow this single tolerance limit to be employed. Keefer and Gottfried [1970] present another method for scaling constraints which has been found to be effective.

It is often found that certain constraints are easier to satisfy then others. Some computational experience indicates that more rapid convergence results if a separate value of K_j is used for each constraint. Thus, for our minimization problem

$$(5.4.25) \qquad \phi = y(\mathbf{X}) + \sum_{i=1}^{l} K_{ji}k_i\delta_i[g_i(\mathbf{X})]^2 + \sum_{i=l+1}^{m} K_{ji}k_i[g_i(\mathbf{X})]^2.$$

After minimization at a given penalty level, each constraint may be checked for feasibility. The values of K_{ji} are increased only for those i where the constraint is not satisfied.

Nothing has been said about how to determine the manner in which the k_i or K_{ji} should be increased after each cycle. Kelley (1962) and Fiacco and McCormick [1968] have made suggestions. However, no firm analytic foundation for this decision has been established, Most algorithms simply increase k_i (or each K_{ji} for violated constraints) by an arbitrary factor C at each cycle. Although a smaller number of steps at each level will be required when C is small, the value selected for C is not crucial since the number of levels to be examined will be increased. Fiacco and McCormick [1968] observe that there is a wide range of values for C over which very little change in computational effort occurs.

Weisman et al. [1965] have suggested that the computational procedures automatically select initial (K_{ji}, k_i) values which scale the constraints and set the sum of the penalty terms as a small fraction of that of the value of $y(\mathbf{X})$. After each minimization the values of K_{ji} are increased by a factor of 8 for those constraints not met within the prescribed tolerance.

The presence of equality constraints means that some of the variables are not independent. The variables that can be considered independent are often referred to as the *decision variables*. In linear programming it is not advantageous to separate the variables into two groups, but, in the use of penalty-function algorithms, it is often desirable. The computation time required increases significantly as the number of variables, whose optimum values must be determined, increases. Thus it is often desirable to conduct the optimization only with respect to the decision variables. When it is possible to solve the equality constraints analytically, it is certainly desirable to do so. It is often worthwhile to solve them numerically even when analytical solutions are not feasible. This approach has been used in determining optimum solutions to some engineering design problems. However, the effort required to solve the equations at each step must be balanced against the reduction in computing time expected by reducing the number of variables. If a highly complex iterative procedure is needed for solution of the equality relationships, minimum computation time may be achieved by treating the equality relationships as constraints. When proper scaling is employed, a penalty-function algorithm is an efficient means for solution of a set of complex nonlinear equations. In some process design problems, this approach has made it possible to obtain an optimum solution in about the same time it previously took to obtain any feasible solution.

Almost any unconstrained minimization (maximization) procedure can be employed with the penalty-function technique. Some of the more successful algorithms have used Fletcher and Powell's [1964] gradient procedure; others have used a minimization procedure based upon Hooke and Jeeves's [1962] pattern search. When only a few of the constraints are nonlinear, it is possible to use gradient projection to advantage. We form an augmented objective function by adding penalties for each nonlinear constraint violated. All linear constraints are retained in their original form. At any given level of K_{ji} we then have a problem with a nonlinear objective function and linear constraints, and this may be efficiently solved by gradient projection. We solve a sequence of such problems at increasing values of K_{ji} until the nonlinear constraints are met. This procedure has an advantage over the usual gradient-projection procedure since it bypasses the difficulties that gradient projection encounters with nonlinear constraints. Also, it does not require that we find an initial point which satisfies the nonlinear constraints; we need only satisfy the linear constraints. We can always locate such a point by use of Phase I of the simplex method.

Exterior- and interior-point algorithms

The techniques we have been discussing are sometimes called *exterior-point* algorithms. If we limit our initial consideration to inequality constrained minimization problems in which the minimum is on a boundary, we recognize that our previous algorithm frequently starts outside (exterior to) the feasible region with an objective function below that of the constrained minimum. We are finally forced to accept a feasible solution with a higher value of the objective function. An alternative, "inside-out" approach to this problem would be the selection of an initial point that is feasible but has an objective function higher than the constrained minimum. A procedure that does this, and approaches the constrained minimum while maintaining feasibility, is called an *interior-point* algorithm.

EXAMPLE 5.4.1

To illustrate an interior-point algorithm, let us return to the problem of finding the minimum for $x^2 - x$, subject to $x - 2 \geq 0$. We now wish to form an augmented objective function that will ensure that our solution will always be feasible. We therefore write

$$\phi(\mathbf{X}) = y(\mathbf{X}) - K \cdot p[g_i(\mathbf{X})], \tag{5.4.26}$$

where $p[g_i(\mathbf{X})]$ is designed to provide a "barrier" preventing our constraint from being violated. This is easily done by letting

$$p[g_i(\mathbf{X})] = \ln[g_i(\mathbf{X})]. \tag{5.4.27}$$

Since the logarithm becomes negative as the boundary of $g_i(\mathbf{X})$ is approached, the minimization procedure forces the acceptance of a feasible point away from the boundary. For the illustrative problem, we have

$$\phi(\mathbf{X}) = x^2 - x - K[\ln (x - 2)]. \tag{5.4.28}$$

At $K = 5$, the minimum of $\phi(\mathbf{X})$ occurs at $x = 3$. When we reduce K to 1, the minimum is at $x = 2.28$, whereas at $K = 0.25$ the minimum occurs at $x = 2.08$. As Figure 5.16 illustrates, the sequence converges toward the constrained minimum at $x = 2$. To prove that this is the case, we determine the minimum analytically by setting the derivative of (5.4.28) equal to zero, and obtain

$$x_{\min} = \frac{5 \pm \sqrt{9 + 8K}}{4}. \tag{5.4.29}$$

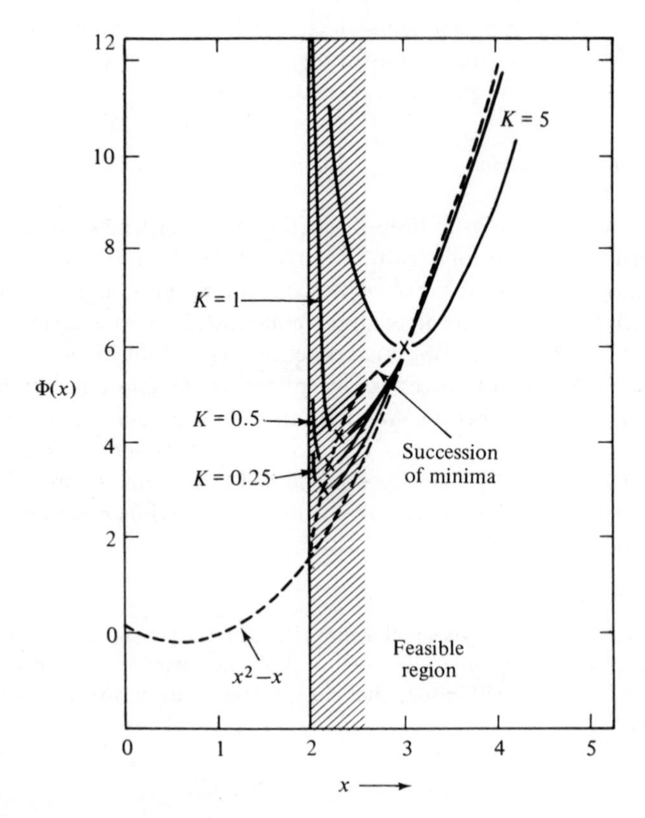

Figure 5.16. Constrained Minimization Using an Interior Point Penalty Function Method

Since x must be positive, only the positive root need be considered:

$$x_{\min} = \frac{5}{4} + \frac{\sqrt{9 + 8K}}{4}. \tag{5.4.30}$$

It is apparent that as $K \longrightarrow 0$, $x_{\min} \longrightarrow 2$.

We may generalize our procedure to handle a number of constraints. To minimize a function subject to a series of inequality constraints of the form

$$g_i(\mathbf{X}) \geq 0, \qquad i = 1, 2, \ldots, m, \tag{5.4.31}$$

we write

$$\phi(\mathbf{X}) = y(\mathbf{X}) - K \sum_i \ln[g_i(\mathbf{X})]. \tag{5.4.32}$$

If we minimize this function for a series of decreasing K's, we obtain a series of minima monotonically approaching the constrained minimum. It should be observed that other forms of the penalty function, such as $+K \sum_{i=1}^{m} 1/[g_i(\mathbf{X})]$, can work equally well.

In simple problems, such as that illustrated in Figure 5.16, it is possible to determine the optimum with any desired degree of precision by optimization of the augmented function at a single value of K. However, just as in the exterior-point methods, this approach is not usually satisfactory for most practical problems. It is seen from Figure 5.16 that at low values of K the barrier function rises very sharply near the constraint boundary, and thus we create a sharp valley adjacent to this boundary. Most unconstrained minimization procedures would quickly find the floor of the valley adjacent to the constraint. However, in many instances they would be unable to follow the floor of the valley along the constraint to find the true minimum. It is thus desirable to conduct the minimization so that when the constraints are finally contacted, no motion along the constraints is required. This we can do by gradually reducing K. The problems encountered here are entirely analogous to those encountered in exterior-point techniques.

The interior-point approach was first suggested by Carroll [1961] and later extended by Fiacco and McCormick [1964]. These latter authors developed an efficient algorithm using the $K \sum_{i=1}^{m} 1/[g_i(\mathbf{X})]$ penalty function. The minimization was carried out by a gradient method that employed approximations based upon Taylor's series expansions through the second-order terms. Fiacco and McCormick showed that the procedure would converge to at least a local minimum under quite general conditions. They also proved that when the objective function is convex, the constraints concave, objective function and constraints twice differentiable, and the penalty function properly chosen, the procedure converges to a global optimum when at least one point satisfying the constraints exists. A thorough discussion of both exterior- and interior-point penalty methods is provided by Fiacco and McCormick [1968].

Duality

It is clear what we mean by convergence to some preassigned tolerance when considering exterior-point methods. There we need only compare the current values of the constraints with a previously assigned tolerance. However, in an interior-point method we are moving within the feasible region toward some unknown improved value of the objective function. If we could bound the possible improvement in the objective function, we could terminate our computations when the possible improvement is small. For a convex programming problem Fiacco and McCormick have established such a bound by considering the dual of the original problem. If our problem is the minimization of $y(\mathbf{X})$, subject to a series of constraints $g_i(\mathbf{X}) \geq 0$, and $y(\mathbf{X})$ is convex and the $g_i(\mathbf{X})$ are concave, we have the required convex programming problem. We can then write the dual of our original (primal) problem as: Maximize

$$z(\mathbf{X}, \boldsymbol{\lambda}) = y(\mathbf{X}) + \sum_{i=1}^{m} \lambda_i g_i(\mathbf{X}) \tag{5.4.33}$$

with respect to $\boldsymbol{\lambda}$, subject to the constraints

$$\frac{\partial z(\mathbf{X}, \boldsymbol{\lambda})}{\partial x_j} = \frac{\partial y(\mathbf{X})}{\partial x_j} + \sum_{i=1}^{m} \lambda_i \frac{\partial g_i(\mathbf{X})}{\partial x_j} \leq 0, \qquad j = 1, 2, \ldots, n. \tag{5.4.34}$$

If we restrict our attention to those x_j having nonzero values at the optimum, the inequalities of (5.4.34) can be written as equalities:

$$\frac{\partial z(\mathbf{X}, \boldsymbol{\lambda})}{\partial x_j} = 0. \tag{5.4.35}$$

Let us compare the dual relationships with those which we obtain from the penalty-function method. If we use $K \sum_{i=1}^{m} 1/[g_i(\mathbf{X})]$ as the penalty function, our augmented objective function is

$$\phi(\mathbf{X}) = y(\mathbf{X}) + \sum_{i=1}^{m} \frac{K}{g_i(\mathbf{X})}. \tag{5.4.36}$$

At the minimum of $\phi(\mathbf{X})$,the condition that

$$\frac{\partial \phi(\mathbf{X})}{\partial x_j} = \frac{\partial y(\mathbf{X})}{\partial x_j} - \sum_{i=1}^{m} \left\{ \frac{K}{[g_i(\mathbf{X})]^2} \right\} \frac{\partial g_i(\mathbf{X})}{\partial x_j} = 0, \qquad j = 1, \ldots, n, \tag{5.4.37}$$

must hold. When we compare (5.4.37) and (5.4.34), we see that they are of the same form, but $-K/[g_i(\mathbf{X})]^2$ replaces λ_i. Since they both hold at the optimum $(\mathbf{X}_o, \boldsymbol{\lambda}_o)$, we conclude that

$$\lambda_{io} = \frac{-K}{[g_i(\mathbf{X}_o)]^2}, \qquad i = 1, \ldots, m. \tag{5.4.38}$$

Thus minimization of the augmented objective function (5.4.36) at any given level of K may be considered to be the evaluation of our dual with a $\boldsymbol{\lambda}$ other than $\boldsymbol{\lambda}_o$. Since the dual problem is maximized with respect to $\boldsymbol{\lambda}$, the current objective function must have a value below that at the optimum

$$z(\mathbf{X}, \boldsymbol{\lambda}) \leq z(\mathbf{X}_o, \boldsymbol{\lambda}_o) \tag{5.4.39}$$

or

$$z(\mathbf{X}, \boldsymbol{\lambda}) = y(\mathbf{X}) - \sum_{i=1}^{m} \frac{K}{[g_i(\mathbf{X})]^2} g_i(\mathbf{X}) \leq z(\mathbf{X}_o, \boldsymbol{\lambda}_o). \tag{5.4.40}$$

Since $z(\mathbf{X}_o, \boldsymbol{\lambda}_o) = y(\mathbf{X}_o)$, we have, simply,

$$y(\mathbf{X}) - \sum_{i=1}^{m} \frac{K}{[g_i(\mathbf{X})]} \leq y(\mathbf{X}_o). \tag{5.4.41}$$

The current value for the dual objective function thus furnishes a lower bound for $y(\mathbf{X}_o)$. The current value of $y(\mathbf{X})$ lies above $y(\mathbf{X}_o)$; hence we consider the necessary convergence to be achieved when $z(\mathbf{X}, \boldsymbol{\lambda})$ has approached $y(\mathbf{X})$ within the desired precision.

We may also use the concept of duality to establish a lower bound for the objective function if the logarithmic form of the penalty function is used. When the penalties are of the form $[-K \sum_{i=1}^{m} \ln g_i(\mathbf{X})]$, we conclude that

$$\lambda_{io} = \frac{-K}{[g_i(\mathbf{X}_o)]}, \tag{5.4.42}$$

and, following our previous reasoning, we obtain

$$y(\mathbf{X}) - mK \leq y(\mathbf{X}_o), \tag{5.4.43}$$

where m is the number of constraints.

Although bounding of the objective function is of less importance in exterior-point methods, duality theory can also be used to establish such bounds for convex (concave) problems at each penalty level.

Comparison of exterior- and interior-point methods

Although both exterior- and interior-point methods have many points of similarity, they represent two different points of view. In an exterior-point procedure, we start from an infeasible point and gradually approach feasibility. While doing so, we move away from the unconstrained optimum of the objective function. In an interior-point procedure we start at a feasible point and gradually improve our objective function while maintaining feasibility. The requirement that we begin at a feasible point and remain within the interior of the feasible inequality-constrained region is the chief difficulty with interior-point methods. In many problems we have no easy way to determine a feasible starting point, and a separate initial computation may be needed. Also, if equality constraints are present, we do not have a feasible inequality constrained region in which to maneuver freely. Thus interior-point methods cannot handle equalities.

We may readily handle equalities by using a "mixed" method in which we use interior-point penalty functions for inequality constraints only. Thus if the first l constraints are inequalities and constraints $(l + 1)$ to m are equalities, our problem becomes: Minimize

$$\phi = y(\mathbf{X}) - K \sum_{i=1}^{l} \ln[g_i(\mathbf{X})] + \frac{1}{K} \sum_{i=l+1}^{m} [g_i(\mathbf{X})]^2. \tag{5.4.44}$$

The function $\phi(\mathbf{X}, K)$ is then minimized for a sequence of monotonically decreasing $K > 0$.

EXAMPLE 5.4.2

To illustrate this approach, consider the example, furnished by Fiacco and McCormick [1968], of: Minimize

$$(\ln x_1) - x_2, \tag{5.4.45}$$

subject to

$$g_1(\mathbf{X}) = x_1 - 1 \geq 0, \tag{5.4.46}$$

$$g_2(\mathbf{X}) = x_1^2 + x_2^2 - 4 = 0. \tag{5.4.47}$$

To solve the problem, we form an augmented objective function $\phi(\mathbf{X})$, where we use the logarithmic penalty function for the inequality constraint and obtain

$$\phi(\mathbf{X}) = (\ln x_1) - x_2 - K \ln(x_1 - 1) + \frac{1}{K}(x_1^2 + x_2^2 - 4)^2. \tag{5.4.48}$$

Minimization of this function yields the succession of values given in Table 5.6. The sequence is seen to be converging to the constrained minimum at $(1, \sqrt{3})$.

Table 5.6. SOLUTIONS OBTAINED TO EXAMPLE 5.4.2 FOR VARIOUS VALUES OF K

K	x_1	x_2	$\phi(X)$
1.0	1.553	1.334	−0.2648
$\frac{1}{4}$	1.159	1.641	−1.0285
$\frac{1}{16}$	1.040	1.711	−1.4693
$\frac{1}{64}$	1.010	1.727	−1.6447
$\frac{1}{256}$	1.002	1.731	−1.7048

Although we have assumed that K is a constant for all inequality constraints, this need not be so. Improved computational efficiencies can often be obtained if separate K's are used for each constraint. Fiacco and McCormick [1968] have suggested that the rule be that the K's be selected so as to be positive and to minimize the magnitude of $\nabla\phi(\mathbf{X})$; i.e., minimize $[\nabla\phi(\mathbf{X})]' \cdot [\nabla\phi(\mathbf{X})]$. A quadratic programming problem would then have to be solved at each penalty level to select the K's.

Both exterior-point and interior-point methods have their partisans. Exterior-point methods pose no difficulty in obtaining a starting point and are easier to program, since they handle equality and inequality constraints in the same manner. Fiacco and McCormick [1968] extensively compare exterior- and interior-point techniques and conclude that minimizing an unconstrained exterior-point function may be harder than minimizing an interior-point unconstrained function. Since both methods rely on the use of one of the previously described unconstrained minimization procedures to find the optimum at a given penalty level, the efficiency of a particular algorithm is just as likely to depend on the unconstrained minimization procedure chosen as on whether an exterior- or interior-point approach is used. An interior-point method may be the only choice if the objective function or one of the constraints is either very badly behaved or just not defined outside the feasible region.

5.5 Geometric Programming

The previous nonlinear programming techniques were all based upon modifications of techniques developed for unconstrained minimization or linear programming. Geometric programming, however, has its roots in the mathematical properties of inequalities. By applying these inequality properties to objective functions and constraints consisting of polynomials with positive coefficients, it is possible to transform the problem into a convex programming problem. This is a significant contribution, since many nonconvex problems are expressed in terms of such polynomials, and our previous techniques would not necessarily locate the global optima. Also, we shall see that the form of the solution will give us insight into some of the relationships which hold at the optimum.

The geometric inequality

The basis for geometric programming is the *geometric inequality*, which states that the arithmetic mean of two positive numbers is always equal to or greater than their geometric mean. Thus if u_1 and u_2 represent two positive numbers, then

$$\frac{u_1 + u_2}{2} \geq (u_1 u_2)^{1/2}. \tag{5.5.1}$$

If we have four positive numbers, u_1, u_2, u_3, and u_4, the geometric inequality may be written as

$$\frac{u_1}{4} + \frac{u_2}{4} + \frac{u_3}{4} + \frac{u_4}{4} \geq u_1^{1/4} u_2^{1/4} u_3^{1/4} u_4^{1/4}. \tag{5.5.2}$$

This expression may be derived from (5.5.1) by first grouping $(u_1 + u_2)/2$ and $(u_3 + u_4)/2$ as single terms and applying the geometric inequality. The geometric inequality is then applied twice to the grouped results.

Equation (5.5.2) may be generalized to deal with the sum of n positive quantities by introducing a set of positive weights, δ_i. We then have

$$\delta_1 u_1 + \delta_2 u_2 + \cdots + \delta_n u_n \geq u_1^{\delta_1} u_2^{\delta_2} \cdots u_n^{\delta_n}, \tag{5.5.3}$$

where the weights must satisfy the "normality" condition,

$$\delta_1 + \delta_2 + \cdots + \delta_n = 1. \tag{5.5.4}$$

Duffin, Peterson, and Zener [1967] have increased the utility of the previous result by introducing a set of new variables. They let

$$t_i = \delta_i u_i. \tag{5.5.5}$$

We then have

$$t_1 + t_2 + \cdots + t_n \geq \left(\frac{t_1}{\delta_1}\right)^{\delta_1} \left(\frac{t_2}{\delta_2}\right)^{\delta_2} \cdots \left(\frac{t_n}{\delta_n}\right)^{\delta_n}. \tag{5.5.6}$$

Since the δ_i are positive numbers, the t_i now represent any positive numbers or variables, and the inequality holds when the normality condition for the δ_i are met.

We have up to this point limited our attention to a sum of individual numbers or variables. However, nothing in our discussion has really restricted us to individual variables. So long as the t_i remain positive, they can consist of a group of variables. Thus we may let

$$t_i = c_i x_1^{a_{i1}} x_2^{a_{i2}} \cdots x_k^{a_{ik}}, \tag{5.5.7}$$

where c_i and the x_i are required to be positive. The geometric inequality, as represented by (5.5.6), remains valid. The sum of the t_i terms now may be a polynomial with positive coefficients. Zener and Duffin coined the term *posynomial* to describe such an expression.

Unconstrained minimization

We are now in a position to illustrate the utility of this last result in an optimization problem by considering the situation when $t_1 = x_1$ and $t_2 = 1/x_1$. Then

$$t_1 + t_2 = x_1 + \frac{1}{x_1} \geq \left(\frac{x_1}{\delta_1}\right)^{\delta_1} \left(\frac{1}{x_1 \delta_2}\right)^{\delta_2}. \tag{5.5.8}$$

If the right-hand side of the equation is plotted against x on logarithmic coordinates (cf. Figure 5.17), we obtain a series of straight lines for those values

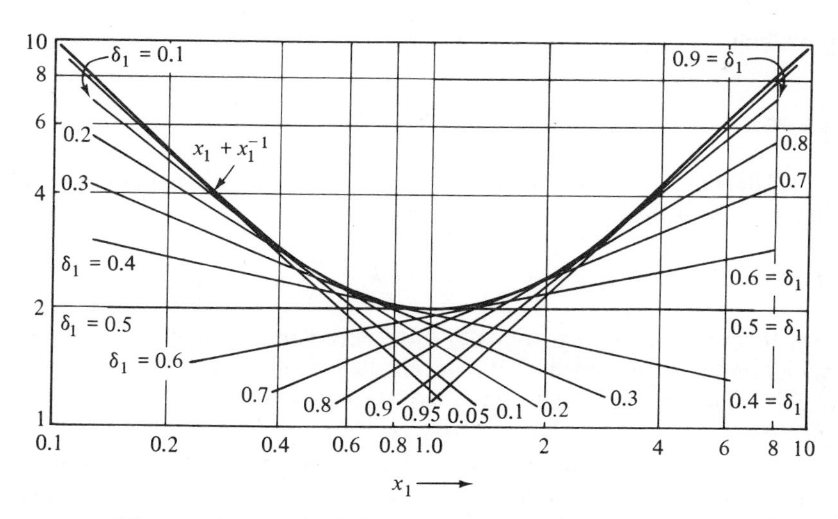

Figure 5.17. Geometric Inequality for the Function x + 1/x

of δ_1 and δ_2 allowed by the normality conditions. Each of these straight lines lies below the curve $[x_1 + 1/x_1]$, and each is tangent at one location. The line which is tangent to the minimum of the $[x_1 + 1/x_1]$ curve is that in which δ_1 and δ_2 make the right-hand portion of (5.5.6) independent of x_1. Zener and Duffin [1964] call this the "orthogonality" condition. That is, the minimum of a sum of polynomial terms may be found by determining a set of weights δ_i that will make the right-hand side of (5.5.6) independent of the x_i.

Let us now assume the t_i to be polynomials and substitute (5.5.7) for the t_i appearing on the right-hand side of (5.5.6). We obtain

$$g_o(\mathbf{X}) = t_1 + t_2 + \cdots + t_n \geq \left(\frac{c_1}{\delta_1}\right)^{\delta_1} \left(\frac{c_2}{\delta_2}\right)^{\delta_2} \cdots \left(\frac{c_n}{\delta_n}\right)^{\delta_n} x_1^{D_1} x_2^{D_2} \cdots x_k^{D_k}, \tag{5.5.9}$$

where

$$D_j = \sum_{i=1}^{n} \delta_i a_{ij}, \qquad j = 1, \ldots, k. \tag{5.5.10}$$

The left-hand side of (5.5.9) is called the *primal function* $g_o(\mathbf{X})$. If the orthogonality condition is to hold, (5.5.9) must be independent of the x_j. This is possible only if the D_j are zero. We may then write

$$t_1 + t_2 + \cdots + t_n \geq \left(\frac{c_1}{\delta_1}\right)^{\delta_1} \left(\frac{c_2}{\delta_2}\right)^{\delta_2} \cdots \left(\frac{c_n}{\delta_n}\right)^{\delta_n}, \tag{5.5.11}$$

providing

$$D_j = \sum_{i=1}^{n} \delta_i a_{ij} = 0, \qquad j = 1, \ldots, k. \tag{5.5.12}$$

If the number of terms n in the polynomial is equal to one more than the number of variables, i.e., if

$$n - 1 = k, \tag{5.5.13}$$

then there are enough equations to specify a unique set δ_i which yields the minimum of our polynomial. We do this by solving the normality equation (5.5.4) simultaneously with the k orthogonality equations generated by (5.5.12).

If the number of variables exceeds the number of terms in the polynomial, the system is overspecified and the scheme will not work. More usually, indeed almost always, the number of variables will be less than $n - 1$.

In that case the equations available are insufficient to uniquely specify the δ_i. There is now no unique value for the right-hand side of inequality (5.5.11). We know, however, that the relationship holds even when the left-hand side is at its minimum and the right-hand side at its maximum; i.e.,

$$\min[t_1 + t_2 + \cdots + t_n] \geq \max\left[\left(\frac{c_1}{\delta_1}\right)^{\delta_1}\left(\frac{c_2}{\delta_2}\right)^{\delta_2} \cdots \left(\frac{c_n}{\delta_n}\right)^{\delta_n}\right]. \quad (5.5.14)$$

When the values of δ_i are such that $t_1 + t_2 + \ldots + t_n$ is at a minimum, it can be shown that the two sides are equal (cf. Duffin, et. al. [1967]), and therefore we may write

$$\min[t_1 + t_2 + \cdots t_n] = \max\left[\left(\frac{c_1}{\delta_1}\right)^{\delta_1}\left(\frac{c_2}{\delta_2}\right)^{\delta_2} + \left(\frac{c_n}{\delta_n}\right)^{\delta_n}\right]. \quad (5.5.15)$$

The right-hand side is termed the *dual function*, $v(\boldsymbol{\delta})$, where

$$v(\boldsymbol{\delta}) = \left[\left(\frac{c_1}{\delta_1}\right)^{\delta_1}\left(\frac{c_2}{\delta_2}\right)^{\delta_2} \cdots \left(\frac{c_n}{\delta_n}\right)^{\delta_n}\right]. \quad (5.5.16)$$

Thus the minimum to the original problem is obtained by maximizing the dual over the set of δ_i which satisfy the normality and orthogonality constraints. We solve for k of the δ_i in terms of the remaining $(n - k)$ and then determine the set of $(n - k)$ values of δ_i that provides the maximum of the dual. In order to determine the optimal policy vector $(x_1, x_2, \ldots, x_n)$ from the δ_i, we must make use of the relationship that

$$t_i = v(\boldsymbol{\delta}) \cdot \delta_i. \quad (5.5.17)$$

The validity of this relationship will be demonstrated subsequently.

Constrained minimization

We have not yet indicated any way in which constraints may be incorporated. If we have a single constraint that can be put in the form

$$1 \geq g_l(\mathbf{X}) = \sum_{i=1l}^{ml} [c_{il} x_1^{a_{i1l}} x_2^{a_{i2l}} \cdots x_k^{a_{ikl}}], \quad (5.5.18)$$

where $g_l(\mathbf{X})$ is a polynomial containing terms $1l$ through ml, the constraint may be incorporated readily into our previous scheme. We shall again make use of the geometric inequality, but we shall use a series of unnormalized weights, δ'_{il}. Now, let

$$\mu_l = \delta'_{1l} + \delta'_{2l} + \cdots + \delta'_{ml}. \quad (5.5.19)$$

We then have

$$\delta_{il} = \frac{\delta'_{il}}{\mu_l}. \tag{5.5.20}$$

We now write the geometric inequality for the lth constraint and replace δ_{il} by δ'_{il}/μ_l to obtain

$$1 \geq \sum_{i=1l}^{ml} [c_{il}x_1^{a_{1il}}x_2^{a_{i2l}} \cdots x_k^{a_{ikl}}] \geq \left(\frac{c_{1l}\mu_l}{\delta'_{1l}}\right)^{\delta'_{1l}/\mu_l}\left(\frac{c_{2l}\mu_l}{\delta'_{2l}}\right)^{\delta'_{2l}/\mu_l} \cdots \left(\frac{c_{ml}\mu_l}{\delta'_{ml}}\right)^{\delta'_{ml}/\mu_l} x_1^{D_{1l}}x_2^{D_{2l}} \cdots x_k^{D_{kl}}, \tag{5.5.21}$$

where

$$D_{jl} = \sum_{i=1}^{m} \delta'_{il}a_{ijl}. \tag{5.5.22}$$

If we choose our δ'_{il} so that the exponents D_{jl} vanish, we have

$$1 \geq \left(\frac{c_{1l}\mu_l}{\delta'_{1l}}\right)^{\delta'_{1l}/\mu_l}\left(\frac{c_{2l}\mu_l}{\delta'_{2l}}\right)^{\delta'_{2l}/\mu_l} \cdots \left(\frac{c_{ml}\mu_l}{\delta'_{ml}}\right)^{\delta'_{ml}/\mu_l}. \tag{5.5.23}$$

The inequality of (5.5.23) will still hold if we take both sides to the μ_l power:

$$1 \geq \left[\left(\frac{c_{1l}}{\delta'_{1l}}\right)^{\delta'_{1l}}\left(\frac{c_{2l}}{\delta'_{2l}}\right)^{\delta'_{2l}} \cdots \left(\frac{c_{ml}}{\delta'_{ml}}\right)^{\delta'_{ml}}\right]\mu_l^{\sum_i \delta'_{il}}. \tag{5.5.24}$$

However, $\sum_i \delta'_l = \mu_l$, and therefore

$$1 \geq \left(\frac{c_{1l}}{\delta'_{1l}}\right)^{\delta'_{1l}}\left(\frac{c_{2l}}{\delta'_{2l}}\right)^{\delta'_{2l}} \cdots \left(\frac{c_{ml}}{\delta'_{ml}}\right)^{\delta'_{ml}} \cdot \mu_l^{\mu_l}. \tag{5.5.25}$$

We now form a revised dual function by multiplying the extreme sides of (5.5.11) and (5.5.25):

$$t_1 + t_2 + \cdots t_n \geq \left[\left(\frac{c_1}{\delta_1}\right)^{\delta_1}\left(\frac{c_2}{\delta_2}\right)^{\delta_2} \cdots \left(\frac{c_n}{\delta_n}\right)^{\delta_n}\left(\frac{c_{1l}}{\delta'_{1l}}\right)^{\delta'_{1l}}\left(\frac{c_{2l}}{\delta'_{2l}}\right)^{\delta'_{2l}} \cdots \left(\frac{c_{ml}}{\delta'_{ml}}\right)^{\delta'_{ml}} \cdot \mu^{\mu_l}\right]. \tag{5.5.26}$$

The minimum to our constrained minimization problem is now obtained by maximizing the revised dual. In order for the revised dual to hold, the x_k must have been eliminated from (5.5.26). This could have occurred only if

$$x_j^{D_j}x_j^{D_{jl}} = x_j^{(D_j + D_{jl})} = 1 \qquad \text{for } j = 1, 2, \ldots, k. \tag{5.5.27}$$

Hence our orthogonality constraints must be

$$D_j + D_{jl} = \sum_{i=1}^{n} \delta'_i a_{ij} + \sum_{i=1}^{m} \delta'_{il} a_{ijl} = 0 \qquad \text{for } j = 1, \ldots, k. \quad (5.5.28)$$

The procedure described can, of course, be extended to any number of constraints $g_l(\mathbf{X})$. We then have: Maximize

$$v(\boldsymbol{\delta}) = \prod_{i=1}^{n} \left(\frac{c_i}{\delta_i}\right)^{\delta_i} \left\{ \prod_{l=1}^{p} \mu_l^{\mu_l} \left[\left(\frac{c_{1l}}{\delta'_{1l}}\right)^{\delta'_{1l}} \left(\frac{c_{2l}}{\delta'_{2l}}\right)^{\delta'_{2l}} \cdots \left(\frac{c_{ml}}{\delta'_{ml}}\right)^{\delta'_{ml}} \right] \right\}, \quad (5.5.29)$$

where p indicates the number of constraints. The orthogonality constraints are now written as

$$\sum_{i=1}^{n} \delta_i a_{ij} + \sum_{l=1}^{p} \left(\sum_{i=1l}^{ml} \delta'_{il} a_{ijl} \right) = 0 \qquad \text{for } j = 1, \ldots, k, \quad (5.5.30)$$

and the normality constraints as

$$1 = \delta_1 + \delta_2 + \cdots + \delta_n \quad (5.5.31)$$

and

$$\mu_l = \delta'_{1l} + \delta'_{2l} + \cdots + \delta'_{ml}, \qquad l = 1, \ldots, p. \quad (5.5.32)$$

EXAMPLE 5.5.1

Let us illustrate the procedure by a simple problem. Suppose that we wish to find the minimum surface area of an open cylindrical tank. The tank volume is to be no less than one unit. If the tank radius is r and the height h, the problem can be written as: Minimize

$$g_o(\mathbf{X}) = \pi r^2 + 2\pi r h, \quad (5.5.33)$$

subject to

$$\pi r^2 h \geq 1. \quad (5.5.34)$$

For the objective function we have the geometric inequality

$$g_o \geq \left(\frac{\pi r^2}{\delta_1}\right)^{\delta_1} \left(\frac{2\pi r h}{\delta_2}\right)^{\delta_2}. \quad (5.5.35)$$

Our constraint is not in the required form, and therefore we rewrite it in terms of its reciprocal:

$$g_1 = \frac{1}{\pi r^2 h} \leq 1. \quad (5.5.36)$$

By use of the geometric inequality we obtain

$$1 \geq \left[\left(\frac{1}{\pi r^2 h}\right)\left(\frac{1}{\delta'_{11}}\right)\right]^{\delta'_{11}} \mu_1^{\mu_1}. \tag{5.5.37}$$

Hence our dual objective function becomes

$$v(\boldsymbol{\delta}) = \left(\frac{\pi r^2}{\delta_1}\right)^{\delta_1}\left(\frac{2\pi rh}{\delta_2}\right)^{\delta_2}\left[\left(\frac{1}{\pi r^2 h}\right)\left(\frac{1}{\delta'_{11}}\right)\right]^{\delta'_{11}} \mu_1^{\mu_1}. \tag{5.5.38}$$

The normality restraints are

$$\delta_1 + \delta_2 = 1, \tag{5.5.39}$$

$$\delta'_{11} = \mu_1, \tag{5.5.40}$$

while the orthogonality constraints are

$$2\delta_1 + \delta_2 - 2\delta'_{11} = 0, \tag{5.5.41}$$

$$\delta_2 - \delta'_{11} = 0. \tag{5.5.42}$$

Simultaneous solution of these equations yields

$$\delta_1 = \tfrac{1}{3} \qquad \delta_2 = \tfrac{2}{3}$$
$$\delta'_{11} = \tfrac{2}{3} \qquad \mu_1 = \tfrac{2}{3}.$$

In our simple example we were able to solve directly for the values of the δ. As we have indicated previously, we are usually unable to do this and we have to evaluate $\boldsymbol{\delta}$ through maximization of the dual $v(\boldsymbol{\delta})$. It would therefore be desirable if we could transform our maximization problem into one that has a unique optimum. All constraints are linear, and we would have a concave programming problem if the objective were concave. Duffin et al. [1967] have shown that a concave function can be obtained by taking the logarithm of the dual:

$$\log v(\boldsymbol{\delta}) = \sum_{i=1}^{n} \delta_i(\log c_i - \log \delta_i) + \sum_{l=1}^{p}\sum_{i=1}^{m} \delta'_{il}(\log c_i - \log \delta'_{il}) + \sum_{l=1}^{p} \mu_l \log(\mu_l). \tag{5.5.43}$$

Since the log of $v(\boldsymbol{\delta})$ is monotonic with $v(\boldsymbol{\delta})$, the maximum of log $v(\boldsymbol{\delta})$ occurs at the same position as the maximum of $v(\boldsymbol{\delta})$. Therefore, we may replace our original objective function by log $v(\boldsymbol{\delta})$. The problem of maximizing log $v(\boldsymbol{\delta})$ subject to the orthogonality and normality constriants is now a concave programming problem. The value of $g_0(\mathbf{X})$ thus obtained is the global maximum of $v(\boldsymbol{\delta})$ and hence the global minimum of our primal problem. Since the dual can be put into a form which has a unique optimum, even though the primal exhibits local optima, solution of the dual is advantageous.

Let us now examine a primal which may have local optima, and obtain the global optimum by solving the dual.

EXAMPLE 5.5.2

Consider the minimization of

$$g_0(x) = c_1 x^2 + c_2 x + c_3 x^{-3} \tag{5.5.44}$$

(cf. Zener and Duffin [1964]). The dual objective function is

$$v(\boldsymbol{\delta}) = \left(\frac{c_1}{\delta_1}\right)^{\delta_1}\left(\frac{c_2}{\delta_2}\right)^{\delta_2}\left(\frac{c_3}{\delta_3}\right)^{\delta_3} \tag{5.5.45}$$

and the normality and orthogonality conditions required are

$$\delta_1 + \delta_2 + \delta_3 = 1, \tag{5.5.46}$$

$$2\delta_1 + \delta_2 - 3\delta_3 = 0. \tag{5.5.47}$$

We cannot solve directly for all the δ_i and must therefore choose the set maximizing $v(\mathbf{x})$. We write the revised objective function as

$$\ln v(\boldsymbol{\delta}) = \delta_1 \ln\left(\frac{c_1}{\delta_1}\right) + \delta_2 \ln\left(\frac{c_2}{\delta_2}\right) + \delta_3 \ln\left(\frac{c_3}{\delta_3}\right). \tag{5.5.48}$$

Minimization of (5.5.48) subject to (5.5.46) and (5.5.47) yields the desired values of δ_i. In this simple problem we can obtain two of the δ_i in terms of the third; e.g.,

$$\delta_2 = \frac{3 - 5\delta_1}{4}, \tag{5.5.49}$$

$$\delta_3 = \frac{1 + \delta_1}{4}. \tag{5.5.50}$$

When the latter two equations are differentiated, we obtain

$$d\delta_2 = -\tfrac{5}{4}\, d\delta_1, \tag{5.5.51}$$

$$d\delta_3 = \tfrac{1}{4}\, d\delta_1. \tag{5.5.52}$$

Since $d[\ln v(\boldsymbol{\delta})]/d\delta_1 = 0$ at the optimum, we now differentiate (5.5.48) with respect to δ_1. After we replace $d\delta_2$ and $d\delta_3$ in terms of $d\delta_1$, simplify, and set the result equal to zero, we obtain

$$\ln\left\{\left(\frac{c_1}{\delta_1}\right)\left(\frac{c_2}{\delta_2}\right)^{-5/4}\left(\frac{c_3}{\delta_3}\right)^{1/4}\right\} = 0. \tag{5.5.53}$$

Since a quantity whose logarithm is zero equals 1, we get

$$\left(\frac{c_1}{\delta_1}\right)\left(\frac{c_2}{\delta_2}\right)^{-5/4}\left(\frac{c_3}{\delta_3}\right)^{1/4} = 1. \tag{5.5.54}$$

When we replace δ_2 and δ_3 in terms of δ_1, we get

$$\frac{c_1 c_3^{1/4}}{4c_2^{5/4}} = \frac{\delta_1(1+\delta_1)^{1/4}}{(3-5\delta_1)^{5/4}}. \tag{5.5.55}$$

We may solve this numerically for δ_1 if values of c_1, c_2, and c_3 are given.

Observe that the solution of Example 5.5.2 was more difficult than the solution of Example 5.5.1. One would expect that as the difference between the number of polynomial terms and number of variables increases, the difficulty of finding a solution would increase. For geometric programming problems it is usual to speak of the *degree of difficulty*, where

$$\text{degree of difficulty} = (\text{number of polynomial terms} - \text{number of variables} - 1).$$

When the degree of difficulty is zero, the orthogonality and normality conditions uniquely determine the optimum. In our last example the degree of difficulty was one. Our problem thus involved optimizing a function of one variable. In Example 5.5.1 the degree of difficulty was zero and we could solve directly for the dual variables.

Optimum values of the decision variables

The reader will have observed that whereas maximization of the dual provides the $\boldsymbol{\delta}$, $\boldsymbol{\delta}'$, and the desired value of $g_o(\mathbf{X})$, the value of $(\mathbf{X})$ is not obtained. In most problems we are really most concerned with the determination of the decision variables, x_i. In order to develop a means for obtaining them, let us return to the original geometric equality as we wrote it in (5.5.6). We had for our objective function

$$g_0(\mathbf{t}) \geq \left(\frac{t_1}{\delta_1}\right)^{\delta_1}\left(\frac{t_2}{\delta_2}\right)^{\delta_2} \cdots \left(\frac{t_n}{\delta_n}\right)^{\delta_n}, \tag{5.5.56}$$

where $\sum_{i=1}^{n} \delta_i = 1$. We can rewrite the constraint inequalities similarly, including the μ_l, since the sum of the δ' is not normalized to 1. We then have

$$g_l(\mathbf{t}) \geq \left[\left(\frac{t_{1l}}{\delta'_{1l}}\right)^{\delta'_{1l}/\mu_l}\left(\frac{t_{2l}}{\delta'_{2l}}\right)^{\delta'_{2l}/\mu_l} \cdots \left(\frac{t_{ml}}{\delta'_{ml}}\right)^{\delta'_{ml}/\mu_l}\right]\mu_l, \tag{5.5.57}$$

which can be rewritten as

$$[g_l(\mathbf{t})]^{\mu_l} \geq \left[\left(\frac{t_{1l}}{\delta'_{1l}}\right)^{\delta'_{1l}}\left(\frac{t_{2l}}{\delta'_{2l}}\right)^{\delta'_{2l}} \cdots \left(\frac{t_{ml}}{\delta'_{ml}}\right)^{\delta'_{ml}}\right]\mu_l^{\mu_l}. \tag{5.5.58}$$

We have seen that at the optimum the geometric inequality becomes an equality, and therefore the inequality relationships of (5.5.56) and (5.5.58) become equalities. Thus from (5.5.56) we have

$$[g_o(\mathbf{t})]_M = \left[\prod_{i=1}^{n}\left(\frac{t_i}{\delta_i}\right)^{\delta_i}\right]_M, \tag{5.5.59}$$

where M indicates the value at the optimum. Rearrangement of the above yields

$$1 = \frac{\left[\prod_{i=1}^{n} (t_i/\delta_i)^{\delta_i}\right]_M}{[g_o(\mathbf{t})]_M}. \tag{5.5.60}$$

This equality is satisfied if

$$\delta_i = \frac{t_i}{[g_o(\mathbf{t})]_M}. \tag{5.5.61}$$

We may show this by substitution of the above in the right-hand side of (5.5.59) to obtain

$$\prod_{i=1}^{n} ([g_o(\mathbf{t})]_M^{t_i/[g_o(t)]_M}) = [g_o(\mathbf{t})]_M^{[t_1+t_2+\cdots+t_n/[g_o(t)]]}. \tag{5.5.62}$$

But $\sum t_i = g_0(\mathbf{t})$, and therefore we have simply

$$\left[\prod_{i=1}^{n}\left(\frac{t_i}{\delta_i}\right)^{\delta_i}\right]_M = [g_o(\mathbf{t})]_M. \tag{5.5.63}$$

Similarly, for (5.5.58) to be an equality, it may be shown that

$$\delta'_{il} = \frac{\mu_k t_{il}}{[g_o(\mathbf{t})]_M}. \tag{5.5.64}$$

Thus after determining the $\boldsymbol{\delta}$ and $\boldsymbol{\delta}'$ that maximize the dual, we obtain the t_i and t_{il} from (5.5.61) and (5.5.64). We obtain the values of the decision variables by replacing the t_i and t_{il} by the original posynomial. For example,

$$\delta_1 = \frac{c_1 x_1^{a_{11}} x_2^{a_{12}} \cdots x_k^{a_{1k}}}{[g_o(\mathbf{t})]_M}. \tag{5.5.65}$$

The set of equations so obtained can be reduced to a set of linear equations by taking the logarithms of both sides of each equation. The values of the primal variables are now readily obtained by simultaneous solution of a set of any k equations that contain all of the decision variables.

EXAMPLE 5.5.3

To illustrate this procedure, let us return to the minimization of $c_1x^2 + c_2x + c_3x^{-3}$. We found previously that, at the minimum, the relationship

$$\frac{c_1c_3^{1/4}}{4c_2^{5/4}} = \frac{\delta_1(1+\delta_1)^{1/4}}{(3-5\delta_1)^{5/4}} \tag{5.5.66}$$

held. Let us consider the case where

$$(c_1, c_2, c_3) = (1, 2, 3). \tag{5.5.67}$$

Numerical solution of the above yields $\delta_1 = 0.260$, and, from this, $\delta_2 = 0.425$ and $\delta_3 = 0.315$. We now obtain

$$v(\boldsymbol{\delta}) = [g_o(\mathbf{t})]_M = 5.59, \tag{5.5.68}$$

and hence we have

$$\delta_1 = \frac{c_1x^2}{[g_o(\mathbf{t})]_M} = \frac{x^2}{5.59}, \tag{5.5.69}$$

$$\log(0.260) = 2\log x - \log 5.59, \tag{5.5.70}$$

and finally $x = 1.2$.

Application of Kuhn–Tucker conditions

The single unifying thread that has run through nearly all the areas considered has been the Kuhn–Tucker theorem. We would expect that we should now be able to apply it. Before doing so, let us first rewrite our problem in terms of transformed variables z_j, where

$$x_j = e^{z_j}. \tag{5.5.71}$$

Our objective function is then

$$g_o(\mathbf{Z}) = \sum_{i=1}^{n} c_i \exp \sum_{j=1}^{k} a_{ij}z_j, \tag{5.5.72}$$

and our constraints are

$$g_l(\mathbf{Z}) = \sum_{i=1l}^{ml} c_{il} \exp \sum_{i=1} a_{ijl}z_j \leq 1. \tag{5.5.73}$$

If we assume that the minimizing point is $(\mathbf{Z}')$ and apply the Kuhn–Tucker theorem, we conclude that a set of nonnegative Lagrange multipliers λ_l

exist such that the Lagrangian $\mathcal{L}(\mathbf{Z}', \boldsymbol{\lambda})$ has the property*

$$\nabla \mathcal{L}(\mathbf{Z}', \boldsymbol{\lambda}) = \mathbf{0}. \tag{5.5.74}$$

When we put our constraints in standard form, the Lagrangian is written as

$$\mathcal{L}(\mathbf{Z}, \boldsymbol{\lambda}) = g_o(\mathbf{Z}) + \sum_{l=1}^{p} \lambda_l [g_l(\mathbf{Z}) - 1]. \tag{5.5.75}$$

Therefore, at the minimizing point $(\mathbf{Z}', \boldsymbol{\lambda})$,

$$0 = \frac{\partial \mathcal{L}}{\partial z_q}(\mathbf{Z}', \boldsymbol{\lambda}) = \sum_{i=1}^{n} a_{iq} c_i \exp \sum_{j=1}^{k} a_{ij} z'_j + \lambda_1 \sum_{i=11}^{m1} a_{iq1} c_{i1} \exp \sum_{j=1}^{k} a_{ij1} z'_j$$
$$+ \cdots + \lambda_p \sum_{i=1p}^{mp} a_{iqp} c_{ip} \exp \sum_{j=1}^{k} a_{ijp} z'_j \qquad \text{for } q = 1, \ldots, k. \tag{5.5.76}$$

Let us now divide both sides of (5.5.76) by $g_o(\mathbf{Z}')$ and let

$$\frac{c_i \exp \sum_{j=1}^{k} a_{ij} z'_j}{[g_o(\mathbf{Z}')]} = d_i, \tag{5.5.77}$$

$$\frac{\lambda_l c_{il} \exp \sum_{j=1}^{k} a_{ijl} z'_j}{[g_o(\mathbf{Z}')]} = d_{il}. \tag{5.5.78}$$

We now have

$$\sum_{i=1}^{n} a_{iq} d_i + \sum_{i=11}^{m1} a_{iq1} d_{i1} + \cdots + \sum_{i=1p}^{mp} a_{iqp} d_{ip} = 0 \qquad \text{for } q = 1, \ldots, k. \tag{5.5.79}$$

It will be recalled that the orthogonality conditions for the dual variables were written as

$$\sum_{i=1}^{n} \delta_i a_{ij} + \sum_{l=1}^{p} \sum_{i=1l}^{ml} \delta'_{1l} a_{ijl} = 0 \qquad \text{for } j = 1, \ldots, k. \tag{5.5.80}$$

The d_i and d_{il} therefore meet the orthogonality constraint. Furthermore, since

$$\sum_{i=1}^{n} c_i \exp \sum_{j=1}^{k} a_{ij} z'_j = g_o(\mathbf{Z}'), \tag{5.5.81}$$

*Note that we have previously used the symbol z for the Lagrangian.

we have

(5.5.82) $$\sum_{i=1}^{m} d_i = 1,$$

and thus the d_i also meet the normality constraint. We may therefore equate the δ_i and d_i, and the δ_{il} and d_{il}. Hence

(5.5.83) $$\delta_i = \frac{c_i \exp \sum_{j=1}^{k} a_{ij} z'_j}{[g_o(\mathbf{Z}')]} \qquad \text{for } i = 1, \ldots, n,$$

(5.5.84) $$\delta_{il} = \frac{\lambda_l c_{il} \exp \sum_{j=1}^{k} a_{ijl} z'_j}{[g_o(\mathbf{Z}')]} \qquad \text{for } i = 1, \ldots, n.$$

From the previous definition of the weights μ_l as $\sum_i \delta'_{il}$, we have at the optimum

(5.5.85) $$\mu_l = \lambda_l \sum_i c_{il} \exp \sum_{j=1}^{k} a_{1jl} z'_j = \frac{\lambda_l g_l(\mathbf{Z}')}{g_o(\mathbf{Z}')} \qquad \text{for } l = 1, \ldots, p.$$

If the optimum lies on the intersection of all the constraints (all constraints active), then

(5.5.86) $$g_l(\mathbf{Z}') = 1 \qquad \text{for } l = 1, \ldots, p,$$

and

(5.5.87) $$\mu_l = \frac{\lambda_l}{g_o(\mathbf{Z}')} \qquad \text{for } l = 1, \ldots, p.$$

Thus when the constraints are active, our weights are simply the Lagrange multipliers divided by a constant $g_o(\mathbf{Z}')$. As we shall see in the next section, this fact can be used in developing a solution procedure.

Computational procedure

To summarize briefly, geometric programming solves the problem of minimizing

(5.5.88) $$g_o(\mathbf{X}) = \sum_{i=1}^{n} c_i x_1^{a_{i1}} x_2^{a_{i2}} \cdots x_k^{a_{ik}},$$

subject to constraints of the form

(5.5.89) $$g_l(\mathbf{X}) = \sum_{i=1l}^{ml} c_{il} x_1^{a_{i1l}} x_2^{a_{i2l}} \cdots x_k^{a_{ikl}} \leq 1,$$

where c_i, c_{il}, and x_j are all positive. The minimization problem is solved by maximizing the dual function

$$v(\boldsymbol{\delta}) = \prod_{i=1}^{n} \left(\frac{c_i}{\delta_i}\right)^{\delta_i} \left\{ \prod_{l=1}^{p} \mu_l^{\mu_l} \left[\left(\frac{c_{1l}}{\delta'_{1l}}\right)^{\delta'_{1l}} \left(\frac{c_{2l}}{\delta'_{2l}}\right)^{\delta'_{2l}} \cdots \left(\frac{c_{ml}}{\delta'_{ml}}\right)^{\delta'_{ml}} \right] \right\}, \qquad (5.5.90)$$

where the δ_i and δ'_{il} are dual variables and the μ_l are nonnegative weighting factors, which may be obtained from the values of the Lagrange multipliers and objective function at the optimum [cf. Equation (5.5.87)]. The dual variables are subject to the orthogonality and normality constraints

$$\sum_{i=1}^{n} \delta_i a_{ij} + \sum_{l=1}^{p} \sum_{i=1l}^{ml} \delta'_{il} a_{ijl} = 0 \qquad \text{for } j = 1, \ldots, k \qquad (5.5.91)$$

and

$$\sum_{i=1}^{n} \delta'_{il} - \mu_l = 0, \qquad l = 1, 2, \ldots, p, \qquad (5.5.92)$$

$$\sum_{i=1}^{n} \delta_i = 1. \qquad (5.5.93)$$

Federowicz et al. [1965] have proposed a computational procedure that makes use of the relationship between μ_l and λ_l. A set of starting values of the x_j is assumed, and a set of starting values for the dual variables may be computed from

$$\delta_i = \frac{[c_i x_i^{a_{i1}} \cdots x_k^{a_{ik}}]}{g_o(\mathbf{X})}, \qquad (5.5.94)$$

$$\delta'_{1l} = \frac{\mu_l [c_{il} x_1^{a_{i1l}} \cdots x_k^{a_{ikl}}]}{g_o(\mathbf{X})}, \qquad (5.5.95)$$

which are valid if $g_o(\mathbf{X})$ and μ_l are known. An estimate of $g_o(\mathbf{X})$ is obtained with the starting values of the primal variables. The values of μ_l are obtained using the λ_l estimated by assuming the Kuhn–Tucker theorem to hold at the starting point; i.e.,

$$\nabla g_o(\mathbf{X}) + \sum_{l=1}^{k} \lambda_l \nabla [g_l(\mathbf{X})] = \mathbf{0}. \qquad (5.5.96)$$

Note that (5.5.96) is a system of k equations in p unknowns. Since k and p will not in general be equal, a slight modification of Rosen's projection technique is used to obtain the λ_l. Let (5.5.96) be written in matrix form as

$$\nabla g_o(\mathbf{X}) + \mathbf{Q}\boldsymbol{\lambda} = \mathbf{0}, \qquad (5.5.97)$$

where

$$\boldsymbol{\lambda} = \begin{bmatrix} \lambda_1 \\ \cdot \\ \cdot \\ \cdot \\ \lambda_p \end{bmatrix}, \quad \mathbf{Q} = \begin{bmatrix} \frac{\partial g_1}{\partial x_1} & \cdots & \frac{\partial g_p}{\partial x_1} \\ \cdot & & \cdot \\ \cdot & & \cdot \\ \cdot & & \cdot \\ \frac{\partial g_1}{\partial x_k} & \cdots & \frac{\partial g_p}{\partial x_k} \end{bmatrix}, \quad \nabla[g_o(\mathbf{X})] = \begin{bmatrix} \frac{\partial g_o}{\partial x_1} \\ \cdot \\ \cdot \\ \cdot \\ \frac{\partial g_o}{\partial x_k} \end{bmatrix}.$$

The estimated values of λ are then obtained by

(5.5.98) $$\boldsymbol{\lambda}_{\text{est}} = -[\mathbf{Q}'\mathbf{Q}]^{-1}\mathbf{Q}'\nabla[g_o(\mathbf{X})].$$

If any λ_i turns out negative, the associated $\nabla(g_i)$ is set to zero in the $\mathbf{Q}$ matrix and the $\boldsymbol{\lambda}$ vector recomputed until all terms are positive. Any which are zero are set to a small positive number. The values of λ_l obtained provide μ_l, which then provide starting values for the δ_i and δ'_{il}.

The $\boldsymbol{\delta}$ obtained will not in general meet the orthogonality and normality constraints. To ensure that these conditions are met, a set of δ_i in the basis is assumed and the other δ_i are obtained in terms of the basis quantities. A linear programming problem is solved to obtain a set of δ_i and δ'_{il} that meet the constraints and yet are as close as possible to the original δ_i and δ'_{il}.

The $\boldsymbol{\delta}$ and $\boldsymbol{\delta}'$ last obtained serve as a starting point for the maximization of the logarithm of the dual function. That is, we maximize

$$v(\boldsymbol{\delta}) = \sum_{i=1}^{n} \delta_i(\log c_i - \log \delta_i) + \sum_{l=1}^{p} \sum_{i=1l}^{ml} \delta'_{il}(\log c_{il} - \log \delta'_{il}) + \sum_{l=1}^{p} \mu_l \log(\mu_l),$$

(5.5.99)

subject to the linear orthogonality and normality constraints. We may conduct the search in terms of our basis variables and solve for the other variables in terms of the basis. During this maximization with respect to the $\boldsymbol{\delta}$ and $\boldsymbol{\delta}'$, we see from the above that the μ_l need not be considered.

When it is believed that the maximization is completed, values of the polynomial terms in the primal problem are computed from (5.5.61) and (5.5.64), using the maximum value of the dual function $v(\boldsymbol{\delta})$ as an approximation of $g_o(\mathbf{X})$:

(5.5.100) $$t_i = c_i x_1^{a_{i1}} \cdots x_k^{a_{ik}} = \delta_i v(\boldsymbol{\delta}), \qquad i = 1, \ldots, n,$$

(5.5.101) $$t_{il} = c_{il} x_1^{a_{i1l}} \cdots x_k^{a_{ikl}} = \frac{\delta'_{il} v(\boldsymbol{\delta})}{\mu_l}, \qquad i = 1, \ldots, n.$$

We evaluate μ_l from $\mu_l = \sum_i \delta'_{il}$, and the primal terms can then be used to evaluate $g_o(\mathbf{t})$. If the difference $|g_o(\mathbf{t}) - v(\boldsymbol{\delta})|$ is less than some prescribed tolerance, we may conclude that the optimum has been obtained. If the optimum has not been obtained, we return to the dual program for further search.

To obtain the values of the primal variables, we note that

$$\log t_{il} = \log c_{il} + \sum_{j=1}^{ml} a_{ijl} \log(x_j) \qquad (5.5.102)$$

for each l and i. By multiple linear regression, the values of the logarithms of the variables that most closely fit can be found. The logarithms are exponentiated to obtain the primal variables.

It will be observed that to obtain the value of t_{il} we must divide by μ_l. If a constraint is inactive, the μ_l will go to zero, and such division is impossible. Whenever any of the weights approach zero, the appropriate constraint must be removed from the problem. Since such a constraint is inactive, its removal does not change the optimum

Relative advantage of geometric programming

The prime advantage of the geometric programming scheme is that it provides a global optimum for problems which can be expressed in terms of posynomials and which have constraints in the form $g_l(\mathbf{X}) \leq 1$. Since such posynomials need not be convex, the methods discussed earlier in the chapter would not necessarily locate a global optimum if directly applied. An additional advantage is that the orthogonality conditions provide relationships which can lead to useful insights into the problem This is significant when the separate terms of the polynomial refer to cost components or real physical quantities. For example in the determination of the optimum configuration of an engineering design, the objective function can often be given as: Minimize

$$g_o(\mathbf{X}) = c_1 f_1(\mathbf{X}) + c_2 f_2(\mathbf{X}) + \cdots + c_n f_n(\mathbf{X}), \qquad (5.5.103)$$

where $f_i(\mathbf{X})$ = magnitude of the ith cost component and c_i = cost per unit size of the ith component. If an electrical transformer were to be designed, the $f_i(\mathbf{X})$ might represent the size of the iron core, size of the copper wire, and magnitude of the power losses. Then c_1 would be the cost per pound of iron core, c_2 the cost per foot of wire, and c_3 the cost per kilowatt of electrical power. If the orthogonality and normality relations lead to a relationship between any two δ_i, a relationship between the cost components at the optimum is established. We can recognize this by recalling that at the optimum we would have

$$\delta_i = \frac{c_i f_i(\mathbf{X})}{[g_o(\mathbf{X})]_M}. \tag{5.5.104}$$

If, for example, we knew that $\delta_1/\delta_2 = k$, then at the optimum

$$k = \frac{[c_1 f_1(\mathbf{X})]}{[c_2 f_2(\mathbf{X})]}. \tag{5.5.105}$$

This relationship is invariant and must therefore always hold at the optimum. If the relationship between c_1 and c_2 changes over a period of time, we know how the ratio $f_1(\mathbf{X})/f_2(\mathbf{X})$ should be adjusted to maintain minimum costs.

Since geometric programming enables the user to successively approximate the optimum from the primal and the dual, upper and lower bounds to $g_o(\mathbf{X})$ are established. The optimization can thus be terminated at any preset error tolerance. This feature is also available when penalty-function methods are used.

A significant drawback to the use of geometric programming is the inability to accommodate equality constraints. Furthermore, the restriction which disallows negative terms* and inequalities of the $g_l(\mathbf{X}) \geq 1$ type can be a severe limitation. Wilde and Beightler [1967] indicate how this restriction can be overcome within a geometric programming framework. However, when this is done, a significant advantage of geometric programming—the ability to locate a global optimum—is lost.

5.6 Comparison of Nonlinear Programming Techniques

If it is necessary to solve a linear programming problem, it is clear that for a hand solution one should use the simplex method, and for a computer solution a version of the revised simplex method. No such simple guideline exists for nonlinear programming problems. No one method is clearly superior under all conditions.

The advanced state of development of linear programming makes it desirable to replace nonlinear problems with linear approximations whenever feasible. We have seen that this is readily accomplished in cases when the

---

* Duffin et al. [1967] have shown that polynomial constraint terms of the form $\Pi_j\{(t_{1j})^a/[1 - t_{2j}]^b\}$ can appear without loss of the global optimum. Functions that contain negative terms, but are not of this form, are not permissible in the original geometric programming formulation.

only nonlinearities involve maximization of a concave, monotonically increasing function (or minimization of a convex, monotonically decreasing function) of a single variable. The revised simplex method can then be used directly.

In many cases, it is found that, although the problems are almost entirely linear, the nonlinear terms do not meet the above limitations. In such situations, use of one of the other linearization techniques is desirable. Linearization procedures are rapid and allow a larger number of variables to be handled than most other techniques. Of these methods, the most powerful appears to be that of Griffith and Stewart. It places no restriction on the form of the functions in the objective or constraints and does not cause a significant increase in the number of constraints, as does the cutting-plane method. Operational programs can easily be written. When the functions appearing in the problem are highly nonlinear, linearization techniques are not always satisfactory. We have noted that in some problems convergence may not be obtained. Since linearization procedures are highly efficient when they do work, it is desirable to investigate these techniques if many similar problems are to be solved repetitively.

Experience has indicated that feasible-direction techniques are generally less powerful than the linearization or penalty-function methods. All the feasible-direction methods require a feasible starting point to be obtained. In addition, most of these methods are limited to inequality constraints or linear equality constraints. Although the gradient-projection method can be used with nonlinear equality constraints, it is very inefficient under such conditions.

Penalty-function techniques are probably the most powerful class of techniques available. These methods will usually converge to at least a local optimum, even though the functions considered meet no convexity or concavity restrictions. Although both internal and external methods work well, the external-point algorithms are simpler to program than the internal-point algorithms. Also, external-point algorithms are capable of handling equality constraints without special provisions and do not require a feasible starting point. Where highly nonlinear functions are to be treated, these methods are to be recommended. The procedure must, of course, be joined to an effective unconstrained optimization technique. Advantage may thus be taken of the continuing improvement in unconstrained minimization and maximization techniques.

It would appear that penalty-function methods are best suited to problems of moderate dimensionality. They have been used successfully in problems containing up to 100 variables.

Geometric programming has engendered a great deal of interest in many areas. The prime reason for this interest would appear to be that it is mathematically interesting, since it approaches constrained optimization from a new point of view. From a more practical viewpoint, it may be worth considering when the entire problem is expressed in terms of posynomials, since it will yield a global optimum. When the functions are in a form that must be approximated by posynomials, the difficulties of the transformation are usually such that the method is not useful. Furthermore, it is limited to inequality constraints of the form $g(\mathbf{X}) \leq 1$. The computational techniques proposed are considerably more cumbersome than those of other nonlinear programming methods.

A notable advantage of geometric programming is that it allows the transformation of a nonconvex optimization problem to a convex problem. However, it is possible to accomplish this transformation for posynomial functions without resorting to geometric programming. It will be recalled that to apply the Kuhn–Tucker theorem we introduced the transformed variables z_j, where $x_j = e^{z_j}$. Our objective function and constraints are then in the form of exponentials, as shown by (5.5.72) and (5.5.73). Duffin et al. [1967] pointed out that as long as the c_i and c_{il} are positive, these functions are convex with respect to the z_j, the transformed variables. We know that for a convex objective function subject to convex constraints an exterior-point penalty-function method will obtain the global optimum, if a feasible solution exists. Thus by utilizing this transformation of variables we can obtain the prime advantage of geometric programming without the computational difficulties. Also, the number of variables in the penalty function method (the z_j) will be considerably smaller than the number of dual variables (the δ_j and δ_{il}) that correspond to the number of terms in the dual posynomial.

Since in most realistic situations we are unable to make any statements concerning the convexity or concavity of the objective functions and constraints, we are unable to state that any solution obtained is the global optimum. This difficulty is common to each of the general techniques, the only exception being the posynomial transformation discussed above. A widely used procedure for checking for alternative optima is to start the optimization process at a number of different points. If there are only a small number of local optima, this procedure will often locate them. If the unconstrained objective function is unimodal, exterior-point penalty-function techniques will often locate the global optimum from any starting point, even though the constrained problem is neither convex nor concave. At the first penalty level, the algorithm will select a point close to the unconstrained optimum. The solutions obtained at higher penalty levels move away from this optimum only as required to satisfy the constraints. The problem of identifying the global optimum of nonlinear programming problems is one that is certain to be the subject of additional research.

We shall further consider the optimization of certain nonlinear functions in our discussion of integer programming (Chapter 6), the maximum principle (Chapters 7 and 8), and dynamic programming (Chapter 8).

REFERENCES

Beale, E. M. L., "Numerical Methods," in *Nonlinear Programming*, J. Abadie, ed., pp. 182–205. North-Holland, Amsterdam, 1967.

Carroll, C. W., "The Created Response Surface Technique for Optimizing Nonlinear Restrained Systems," *Operations Res.*, *9*, (1961), 169.

Courant, R., *Calculus of Variations and Supplementary Notes and Excercises*, revised and amended by J. Moser. New York University, 1956.

Duffin, R. J., E. L. Peterson, and C. M. Zener, *Geometric Programming—Theory and Application*, Wiley, New York, 1967.

Federowicz, A. J., D. A. Forejt, and E. Peterson, *A Computer Program to Optimize Engineering Designs by Geometric Programming*, Westinghouse Research Laboratory Report, 1965.

Fiacco, A. V., and G. P. McCormick, "The Sequential Unconstrained Minimization Technique for Nonlinear Programming: A Primal–Dual Method," *Management Sci. 10* (1964), 360.

———, "Computational Algorithm for the Sequential Unconstrained Minimization Technique for Nonlinear Programming," *Management Sci. 10* (1964). 601

———, *Nonlinear Programming—Sequential Unconstrained Minimization Techniques*, Wiley, New York, 1968.

Fletcher, R., and M. J. D. Powell, "A Rapidly Convergent Descent Method for Minimization," *Computer J. 6* (1963–1964), 163.

Griffith, R. E., and R. A. Stewart, "A Nonlinear Programming Technique for Optimization of Continuous Processing Systems," *Management Sci.*, *7* (1961), 379.

Hooke, R., and T. A. Jeeves, "Direct Search Solution of Numerical and Statistical Problems," *J. Assoc. Comp. Mach.*, *8*, No. 2 (1962), 212.

Keefer, D.L., and B. S. Gottfried, "Differential Constraint Scaling in Penalty Function Optimization," *AIIE Trans.*, *II*, No. 4 (1970), 281.

Kelley, H. J., "Method of Gradients," Chapter 6 of *Optimization Techniques: With Application to Aerospace Systems*, G. Leitman, ed., Academic Press, New York, 1962.

Kelley, J. E., "The Cutting Plane Method for Solving Convex Problems," *J. Soc. Ind. Appl. Math.*, *8*, No. 4 (1960), 2.

Klingman, W. R., and D. M. Himmelblau, "Nonlinear Programming with the Aid of a Multiple-Gradient Summation Technique," *J. Assoc. for Comp. Mach.*, *11* (1964), 400.

Lemke, C. E., "The Dual Method of Solving the Linear Programming Problem," *Naval Res. Logistics Quart.*, *1* (1954), 1.

Mugele, R. A., "A Program for Optimal Control of Nonlinear Processes," *IBM Systems J.*, 2 (Sept. 1962).

Rosen, J. B., "The Gradient Projection Method for Nonlinear Programming, Part I—Linear Constraints," *J. Soc. Ind. Appl. Math.*, *8* (1960), 181.

———, "The Gradient Projection Method for Nonlinear Programming, Part II," *J. Soc. Ind. Appl. Math.*, *9* (1961), 514.

Weisman, J., C. Wood, and L. Rivlin, "Optimal Design of Chemical Process Systems," *Chem. Eng. Progr. Symp. Ser. 61*, No. 55 (1965), 50.

Wilde, D. J., and C. S. Beightler, *Founations of Optimization*, p. 117. Prentice-Hall, Englewood Cliffs, N. J., 1967.

Wilson, R. B., "A Simplical Algorithm for Concave Programming," Ph. D. Dissertation, Harvard University, 1963.

Wolfe, P., "The Simplex Method for Quadratic Programming," *Econometrica*, *27* (1959), 382.

Zangwill, W. I., "Nonlinear Programming via Penalty Functions," *Management Sci.*, *13* (1967), 344.

Zener, C., and R. J. Duffin, "Optimization of Engineering Problems," *Westinghouse Eng.*, *24* (1964), 154.

Zoutendijk, G., *Methods of Feasible Directions*, American Elsevier, New York, 1964.

PROBLEMS

5.1. In which of the following cases can we be certain that any local optimum is also the global optimum?

(a) Minimize $y = x_1^3 + 2x_2 + 3x_3$, subject to

$$x_1 + x_2 \geq 7,$$
$$x_1 + x_2 + 2x_3 \geq 12,$$
$$x_i \geq 0, \qquad i = 1, 2, 3.$$

(b) Maximize $y = x_1 + 2x_2 + 5x_3$, subject to

$$x_1^2 + x_2^2 \leq 25,$$
$$2x_1 + x_2 + x_3 \leq 30,$$
$$x_i \geq 0, \qquad i = 1, 2, 3.$$

(c) Maximize $y = x_1 + 2x_2 + 5x_3$, subject to

$$x_1^2 + x_2^2 \geq 25,$$
$$2x_1 + x_2 + x_3 \leq 30,$$
$$x_i \geq 0, \qquad i = 1, 2, 3.$$

(d) Minimize $y = 9x_1^2 - 6x_1x_2 + x_2^2$, subject to

$$5/x_1 + e^{x_2} \leq 20,$$
$$x_i \geq 0, \qquad i = 1, 2.$$

(e) Minimize $y = 9x_1^2 - 6x_1x_2 + x_2^2$, subject to

$$x_1^3x_2 + x_1x_2^{0.5} \leq 40,$$
$$x_i \geq 0, \qquad i = 1, 2.$$

5.2. A small manufacturing corporation has a total of \$100,000 available to invest in the expansion of its three divisions. It is estimated that the profit from an investment in division I will be given by

$$p_1 = 500{,}000(1 - e^{-0.00002x_1}),$$

where x_1 is the total division I investment. Similarly, the profits from divisions II and III are given by

$$p_2 = 300{,}000(1 - e^{-0.00002x_2}),$$
$$p_3 = 400{,}000(1 - e^{-0.00002x_3}),$$

where x_2 and x_3 are the investments in divisions II and III. Determine by use of the simplex method how the corporation should distribute its investment so as to maximize its total profit.

5.3. Determine the solution to: Minimize

$$y = x_1^2 + 3x_2^2 + 1.5x_3,$$
$$2x_1 + x_2 + x_3 \geq 20,$$
$$x_1 + x_3 \geq 10,$$
$$x_i \geq 0, \qquad i = 1, 2, 3,$$

via (a) Wolfe's quadratic programming algorithm; (b) separable programming. Which of the two procedures is less time-consuming in this case?

5.4. Show that the primal quadratic programming problem: Maximize

$$y = \mathbf{C'X} + \mathbf{X'DX},$$

subject to

$$\mathbf{AX} = \mathbf{B},$$

where the components of $\mathbf{X}$ are unrestricted in sign, has a dual of the form: Minimize

$$y_D = -\mathbf{X'DX} + \boldsymbol{\lambda}'\mathbf{B},$$

subject to

$$-2\mathbf{DX} + \mathbf{A}'\boldsymbol{\lambda} \geq \mathbf{C}',$$
$$\mathbf{X} \geq \mathbf{0}.$$

(*Hint:* Use the Kuhn–Tucker conditions to establish that at the optimal solution, $\mathbf{X}_0$, $\boldsymbol{\lambda}_0$, we have $-2\mathbf{DX}_0 + \mathbf{A}'\boldsymbol{\lambda}_0 \geq \mathbf{C}'$.)

5.5. We have observed that replacement of the product x_1x_2 by y, where

$$\ln y = \ln x_1 + \ln x_2,$$

is a transformation used to produce separability. The procedure is satisfactory only if x_1 and x_2 are known to be positive. If either quantity may be zero, problems will be encountered, since $\ln 0 = -\infty$. Show how this procedure can be extended to the case in which the variables may take on zero values. (*Hint:* Let $w_i = x_i + \epsilon$, where ϵ is a known positive quantity.)

5.6. Prove that when separable programming is applied to a convex, separable program, no basis-entry restriction need be applied.

5.7. Solve the following nonlinear programming problem: Minimize

$$y = x_1^2 + 4x_1x_2 + 4x_2^2,$$

subject to

$$x_1 - x_2 \geq 3,$$
$$x_i \geq 0, \qquad i = 1, 2,$$

by
(a) Kelley's cutting-plane method. Choose the point $x_1 = 0$, $x_2 = 0$ as the starting point; (b) the method of approximation programming. Choose the point $x_1 = 2$, $x_2 = 2$ as the starting point and allow the x_i to change by no more than ± 0.5 in any one step.
Check your solutions graphically.

5.8. Resolve problem 5.3 by one or more of the following:
(a) Kelley's cutting-plane method;
(b) The method of approximation programming;
(c) Zoutendijk's method of feasible directions;
(d) Gradient projection.
For parts (a) and (b), select starting points appropriate to the procedure used. Use the feasible starting point $x_1 = 10$, $x_2 = 2$, $x_3 = 2$ for parts (c) and (d).

5.9. Use the modified method of approximation programming to: Maximize

$$y = 2x_1 + 10 \ln x_2,$$

subject to

$$x_1 + 3x_2 \leq 15,$$
$$2x_1 - x_2 \leq 10,$$
$$x_i \geq 0, \qquad i = 1, 2.$$

Approximate the objective function by a Taylor's series in which both first- and second-order terms are retained. Treat each step as a quadratic programming problem.

5.10. Determine the solution to: Maximize

$$y = 9x_1^2 - 6x_1x_2 + x_2^2 + \frac{3}{x_3^2},$$

subject to

$$g_1 = \frac{10}{x_1} + e^{x_2} - 20 \leq 0,$$

$$g_2 = -x_1 + x_3 \leq 0,$$

by converting the problem to an unconstrained minimization through the use of

(a) An exterior-point penalty-function algorithm. Begin with $k = 0.1$ and increase k by a factor of 8 for each penalty level. Select $x_1 = 0.2$, $x_2 = 1$, $x_3 = 4$ as the starting point and minimize via the method of optimal steepest descent. Terminate the problem when the $g_i \leq 0.05$.

(b) An interior-point penalty-function algorithm. Begin with $k = 4$ and reduce k by a factor of 4 for each new penalty level. Select $x_1 = 1$, $x_2 = 1$, $x_3 = 2$ as the starting point and minimize via the method of optimal steepest descent. (*Hint:* Rewrite constraints to avoid a format requiring the logarithm of a negative quantity.) Terminate the problem when y is within 5 per cent of the optimal value.

(c) Repeat (a) and (b) using the calculus to obtain the minimum at each penalty level examined.

5.11. Assume that a convex objective function $y(\mathbf{X})$ is to be minimized, subject to the single convex constraint $g(\mathbf{X}) \leq 0$, and limits on each variable $a_i \leq x_i \leq b_i$, $i = 1, \ldots, n$. Suppose that the problem is to be solved by use of an exterior penalty-function procedure. Let the modified objective function take the form

$$z(\mathbf{X}, \delta) = y(\mathbf{X}) + \delta K[g(\mathbf{X})],$$

where $\delta = 1$ when $g(\mathbf{X}) \geq 0$, and zero otherwise. Derive an expression for the *minimum* value of K which will assure that the constraint is satisfied for any allowable point $\mathbf{X}$.

5.12. Consider a minimization problem of the form: Minimize

$$y = f(\mathbf{X}),$$

subject to

$$g_i(\mathbf{X}) \leq 0, \qquad i = 1, 2, \ldots, m,$$

which is converted to an unconstrained minimization problem: Minimize

$$f(\mathbf{X}) + \sum_{i=1}^{m} \delta[g_i(\mathbf{X})]^2.$$

Show that this latter function is convex if the original objective function and constraints were convex.

5.13. Show that Equations (5.4.41) and (5.4.43), which establish bounds for the objective function when the logarithmic form of an interior penalty function is used, are valid.

5.14. (a) Use geometric programming to determine the global minimum of the function

$$g_0 = 2x_1x_2 + \frac{6x_1^2x_2^2}{x_3} + \frac{2.5x^{1.5}x_3^2}{x_2^3} + 4x_1^{0.5}$$

subject to $x_i \geq 0$, $i = 1, 2, 3$.

(b) Select any other appropriate procedure and attempt to determine the global minimum of the foregoing function. What advantage does geometric programming have in this case?

5.15. Given an objective function of the form: Minimize

$$y = \sum_{i=1}^{m} t_i - u,$$

where the t_i are posynomials, u is a single-term posynomial, and the minimum value of y is known to be negative. Show that this function can be placed in a form suitable for use with geometric programming by a suitable transformation. (*Hint:* Introduce the function t_0, where $t_0 + \sum_{i=1}^{m} t_i - u \leq 0$.)

5.16. An engineer wishes to design a set of electrical transmission lines. The capital cost of the set of lines is given by $C_1 = n(5000 + 2000d^2)$, where n is the number of lines in the set and d is the diameter of one line in inches. The equivalent capital cost of electrical power over the life of the system is given by

$$C_2 = \frac{75{,}000}{nd^2}.$$

To prevent breakage of a single line from causing a major disruption in power service, it is required that $n \geq 10$. Determine, by means of geometric programming, the values of n and d that provide the minimum cost system meeting this constraint. You may consider n a continuous variable. Check your result graphically.

5.17. By making use of the fact that

$$\log\left[\prod_i x_i^{a_i}\right] = \sum_i a_i \log x_i,$$

show that

(a) Any inequality constrained linear program can be transformed into a geometric program;

(b) The values of the x_i at the optimum of the linear program can be determined from the values of the variables at the optimum of the geometric program.

5.18. A pipeline is to transfer crude oil across a large oil refinery. The power P required to pump the oil may be computed from

$$P = 0.385 \times 10^{-12}\frac{w^3}{\rho^2 d^5},$$

where w is the oil flow in pounds/hour, ρ is the oil density in pounds/cubic foot, and d is the diameter of the line in feet. The cost incurred is given by

$$c_t = 10{,}000d^2 + 170P,$$

where the first term is the cost of the pipeline and the second is the sum of the cost of the pump and pumping power.

The oil flow is fixed at 10.0×10^6 lb/hr and the oil density at 50 lb/ft^3. In order to keep sludge from settling in the pipeline, the oil velocity v must be at least 9 ft/sec.

Determine the pipe diameter meeting the required constraints by

(a) geometric programming,

(b) an exterior-point penalty function, and

(c) approximation programming.

5.19. The oil carried by the pipeline of problem 5.18 is now to be transported at an elevated temperature. To prevent excessive heat loss to the surroundings, the line must be insulated. The hourly heat losses q are obtained from

$$q = 500/\left(\frac{L}{400} + \frac{1}{4000}\right),$$

where L is the thickness of the insulation. The cost, c_i, of the insulation and its installation is

$$c_i = 25{,}000Ld.$$

Over the life of the plant, the cost of the heat loss can be taken as $0.5q$.

(a) Assuming that the pipeline has not yet been built, set up a mathematical programming problem to determine the pipeline diameter and insulation thickness that will provide the minimum cost, yet meet the fluid velocity constraint. Can this problem be solved by geometric programming?

(b) Solve (a) by an appropriate technique.

(c) How does the solution obtained in (b) differ from what you would have obtained had you solved problem 5.18 for d and then determined the optimum L for that d?

5.20. In some inventory problems, c_i, the inventory cost per unit stored, may have the form

$$c_i = a_i - b_i x_i,$$

where x_i is the total quantity in inventory.

(a) Assume that $a_i = 1.2$, $b_i = 10^{-5}$ and restate problem 4.24 as a nonlinear programming problem.

(b) What nonlinear programming procedure do you think would be most appropriate for solution of this problem? Why?

(c) Solve this nonlinear programming problem by the procedure of your choice. Consider only the first 6 months of the year and assume the July 1 inventory is equal to that at January 1.

MATHEMATICAL PROGRAMMING EXTENSIONS

In our discussion of linear and nonlinear programming we made several tacit assumptions. The most important of these was that all our variables were continuous. In some cases this is not strictly true, since we may be dealing with integral quantities. If the number being considered is large (e.g., number of automobiles produced per year by a division of a major auto manufacturer), the error introduced by assuming the variable to be continuous is insignificant. However, there are many problems in which some or all of the variables represent a small number of discrete alternatives. We therefore need programming techniques that can provide integer solutions.

We also tacitly assumed that our computing facility was large enough to treat our programming problem as a whole. We may at times be faced with programming problems so large that it is not feasible to solve the entire problem at once. It is therefore desirable that we have a means for breaking the more common large problems into more manageable pieces.

In this chapter we shall consider how we may extend our mathematical programming algorithms so that (a) integer solutions may be obtained when needed and (b) certain types of large problems may be decomposed into smaller units.

6

6.1 The Origin of Integer Programming Problems

Before attempting to find solution procedures for mathematical programming problems in which some variables are regarded to be integers, let us examine some situations in which they may arise. One such situation is the selection of the minimum-length path from a series of possible alternatives. Consider a segment of a road map, as shown in Figure 6.1a. Let the lettered nodes correspond to fixed points on the map where two or more paths intersect, and let the numerical values adjacent to the various paths represent the average travel times between nodes.

Our objective will be to travel from point A to point G in a minimum length of time. Notice that there are several alternative routes that we can choose—some being faster than others. Furthermore, the route that traverses the least distance will not necessarily be the fastest. We seek the particular route that will minimize the *total* travel time between nodes A and G. Problems of this type are appropriately called *minimum-path* or *minimum-trajectory* problems.

For simplicity let us redraw the road network as shown in Figure 6.1b. We have not lost any necessary information by doing this, since we are main-

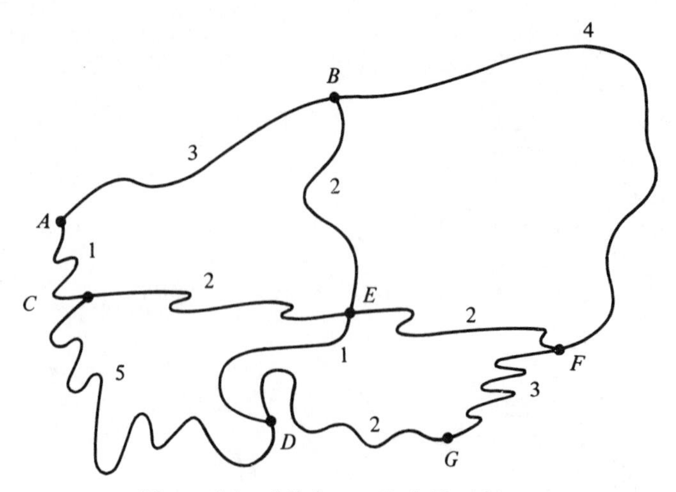

Figure 6.1a. Minimum Path Problem

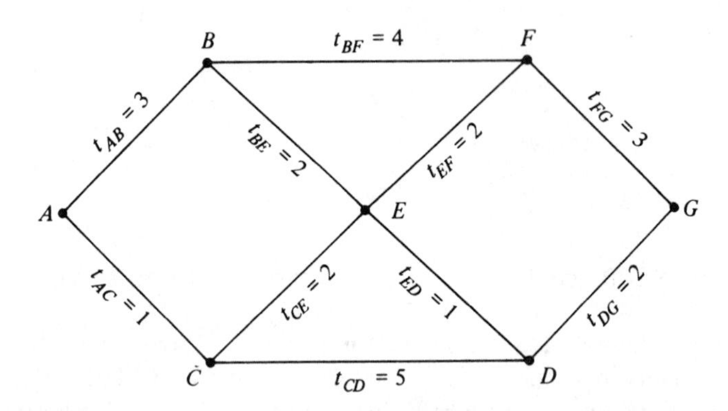

Figure 6.1b. Integer Programming Formulation of a Minimum Path Problem

taining the correct connections between the nodes and we still know the average driving time along each path. Thus the true shape of the paths is immaterial.

To proceed with this problem, let us introduce a set of variables, δ_{IJ}, which will indicate whether we have chosen to follow the path between the adjacent nodes I and J. If we choose to follow the path between I and J, δ_{IJ} is one; otherwise, δ_{IJ} is zero. If t_{IJ} represents the travel time from node I to node J, the quantities $\delta_{IJ}t_{IJ}$ then represent the time actually spent going from I to J. Our problem then becomes: Minimize

$$\begin{aligned}(6.1.1) \qquad & \delta_{AB}t_{AB} + \delta_{AC}t_{AC} + \delta_{BE}t_{BE} + \delta_{CE}t_{CE} + \delta_{BF}t_{BF} + \delta_{EF}t_{EF} \\ & + \delta_{CD}t_{CD} + \delta_{ED}t_{ED} + \delta_{FG}t_{FG} + \delta_{DG}t_{DG},\end{aligned}$$

subject to

$$\delta_{IJ} = 0 \quad \text{or} \quad \delta_{IJ} = 1. \tag{6.1.2}$$

(Note that we use capital letters for subscripts, to correspond to the nodes in the network.)

We cannot allow all the δ_{IJ} to be zero. To prevent this, we observe that if we enter a node, we must leave that node by one of the exit paths. Thus we can write for node A,

$$1 = \delta_{AB} + \delta_{AC}, \tag{6.1.3}$$

and for node B,

$$\delta_{AB} = \delta_{BE} + \delta_{BF}. \tag{6.1.4}$$

Thus if δ_{AB} is nonzero, either δ_{BF} of δ_{BE} will be nonzero. We can write similar constraint equations for all other nodes.

A program of this type is an *all-integer* programming problem, since all the variables must be integers. It may also be characterized as a *zero–one* integer programming problem, since we have restricted each of our variables to be either zero or one.

Many problems of interest are *mixed-integer programming problems* in which only some of the variables are required to be integers. An interesting example of such a situation is the fixed-charge problem. Suppose that a manufacturing plant consisting of a number of different units is being constructed. We may find associated with each unit certain minimum fixed costs (e.g., minimum land acquisition and preparation costs), and thereafter costs vary linearly with capacity. Thus

$$c_i = a_i + b_i x_i, \tag{6.1.5}$$

where c_i is the cost of unit i, x_i is the capacity of unit i, and a_i and b_i are constants. However, we do not incur any cost if we do not construct that particular unit ($x_i = 0$). Therefore, we write our costs as

$$c_i = a_i\delta_i + b_i x_i, \tag{6.1.6}$$

where

$$\begin{aligned} \delta_i &= 1 \qquad \text{if } x_i > 0, \\ \delta_i &= 0 \qquad \text{if } x_i = 0. \end{aligned}$$

To put this in the form of a mixed-integer programming problem, we first determine an upper bound u_i for each x_i. We then require

$$x_i - \delta_i u_i \leq 0, \tag{6.1.7}$$

where

$$\delta_i = 0 \qquad \text{or} \qquad 1. \tag{6.1.8}$$

If x_i is positive, this constraint can only be met if δ_i is 1. If $x_i = 0$, then δ_i

may assume the value zero, and minimization of

$$\sum_i c_i$$

will force it to do so.

In both of the previous examples we have used integer variables to represent yes–no decisions (e.g., to take a particular path, construct a particular unit). The zero is used to represent the no decision, and the one represents a yes decision. The binary variables δ_i thus allow us to include logical decisions within our program. A zero value for δ_i indicates that we are not to pursue the particular course of action, whereas a value of one indicates that the line of action is to be taken.

This concept is of considerable importance in the optimization of engineering systems. In many such problems, the system configuration is assumed to be fixed. The optimization then determines the optimum number and sizes of the various components. If a different configuration is to be considered, an entirely new problem is formulated. However, this procedure need not be followed. We can consider a number of alternative configurations within a given problem by essentially the same formulation as that used for the fixed-charge problem. To ensure that only one of the possible alternatives will be followed, we add the additional constraint that

$$\sum_{i=1}^{n} \delta_i = 1. \tag{6.1.9}$$

The problem can thus take on some of the characteristics of a true system design, since it selects the optimum system configuration as well as size.

Through the use of binary variables, Equation (6.1.9) expresses the logical relation that only one of a given set of possibilities may be chosen (one and only one binary variable equals 1). There are many other possibilities. Some of the more important logical relationships which can be expressed in terms of binary variables are

1. At most one variable equals 1. (At most one course of action will be taken.)

$$\sum_{i=1}^{n} \delta_i \leq 1. \tag{6.1.10}$$

2. At least one variable equals 1. (At least one course of action will be taken.)

$$\sum_{i=1}^{n} \delta_i \geq 1. \tag{6.1.11}$$

3. One and only one variable equals 0. (One course of action will be omitted.)

$$\sum_{i=1}^{n} \delta_i = n - 1. \tag{6.1.12}$$

4. At most one variable equals 0. (At most one course of action will be omitted.)

$$\sum_{i=1}^{n} \delta_i \geq n - 1. \tag{6.1.13}$$

5. At least one variable equals 0. (At least one course of action will be omitted.)

$$\sum_{i=1}^{n} \delta_i \leq n - 1. \tag{6.1.14}$$

The above expressions are easily generalized to allow relations of the form

6. Exactly k variables equal 1. (Exactly k actions will be taken.)

$$\sum_{i=1}^{n} \delta_i = k. \tag{6.1.15}$$

7. At most k variables equal 1. (At most k actions will be taken.)

$$\sum_{i=1}^{n} \delta_i \leq k. \tag{6.1.16}$$

8. At most k variables equal 0. (At most k actions will be omitted.)

$$\sum_{i=1}^{n} \delta_i \geq n - k. \tag{6.1.17}$$

Other types of logical relationships can also be expressed by use of zero–one variables. For example, problems where statements contain either–or requirements give rise to what are called *disjunctive constraints*. These are a set of constraints at least one of which must be satisfied.

The minimum-batch-size problem is an example of such a situation. In many production problems we wish to avoid uneconomic small batches. We may choose not to manufacture a particular item, but if we do decide to produce it, our batch size must be at least k_i. Thus if x is the batch size, we require

$$x_i = 0 \quad \text{or} \quad x_i \geq k_i. \tag{6.1.18}$$

Such a situation can be represented by the constraints

$$x_i \leq h_i\delta_i, \tag{6.1.19}$$
$$x_i \geq k_i\delta_i, \tag{6.1.20}$$

where δ_i is a zero–one binary variable and h_i is an upper limit on x_i. Constraints of this nature give rise to disconnected feasible regions. Any method that requires a convex feasible region could not be used for solution of such a problem.

Logical expressions are not the only source of integer variables. For example, integer programming problems arise in the assignment of scarce resources. Consider, for example, a problem involving the shipment of merchandise within a network of several cities. Suppose that at cities 1, 2, and 3, we have 5, 10, and 20 truckloads of merchandise waiting to be shipped. At cities 4, 5, and 6, we have 7, 4, and 8 trucks available. If we know the cost of dispatching a truck between any two cities is c_{ij}, then we wish to minimize the cost of dispatching the empty trucks to pick up the waiting merchandise. That is, our objective function is: Minimize

$$\sum_{i=4}^{6} \sum_{j=1}^{3} c_{ij}x_{ij}, \tag{6.1.21}$$

where x_{ij} represents the number of trucks dispatched from location i to location j. At city 4, where trucks are to originate, we have the constraint

$$\sum_{j=1}^{3} x_{4j} \leq 7. \tag{6.1.22}$$

Similar constraints are written for the other originating cities. For receiving city 1, we have the constraint

$$\sum_{i=4}^{6} x_{i1} = 5. \tag{6.1.23}$$

Equality constraints of the same form are written for cities 2 and 3. Since we cannot dispatch a fraction of a truck, the x_{ij} must all be integers. This is therefore another example of an *all-integer programming problem.* Here the variables are, however, not restricted to zero–one values.

All the problems considered would be the usual linear programming problem if it were not for the integer restrictions. Obviously, integer programming problems can arise in nonlinear programming as well. Consider the design of a segment of a chemical processing plant. Let us assume that we are to react two given streams, A and B, to form product C (cf. Figure 6.2). The rate of reaction is a nonlinear function of pressure P, temperature T, the concentration of the reacting streams, and the time they are in contact. We wish to maximize the yield of C, which requires that we maximize the reaction rate. The reaction is to take place in vessels whose size is fixed at a given volume V. The total volume of the system, upon which the concentrations and the contact time depend, varies only with the number n of reactors.

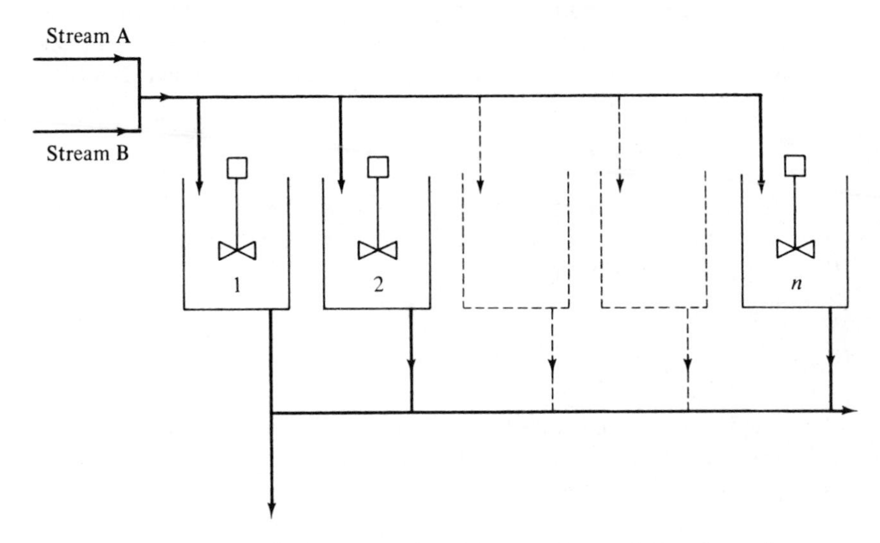

Figure 6.2. Chemical Process Involving an Integral Number of Reactor Vessels

Hence we must maximize a function

$$f(P, T, n), \tag{6.1.24}$$

where n is an integer.

Mixed-integer nonlinear programming problems of this nature arise frequently in process optimization. The physical relationships often are highly nonlinear. Furthermore, it is quite usual for various equipment units (e.g., pumps, heat exchangers, distillation towers, turbines) to have maximum size limitations or to be available in only discrete sizes. Whenever this is the case, reference to the total capacity will necessitate the use of additional integer variables. Such process design problems may also include a number of binary variables introduced through logical constraints.

6.2 Integer Programming Problems Solvable by Normal Linear Programming Techniques

In any integer programming problem, there is always the possibility in the solution obtained by our usual techniques that variables required to be integers will be sufficiently close to integers to be acceptable as the answer without significant rounding. Although this is unlikely in the general case, there are several important problem categories so structured that an integer solution will be obtained. We shall examine two such problem categories.

Path-length problems

Let us return to the problem of determining the path yielding the minimum travel time between two nodes. We may restate the objective function [Equation (6.1.1)] more generally:
Minimize

$$\sum_I \sum_J \delta_{IJ} t_{IJ}, \tag{6.2.1}$$

where, as before, δ_{IJ} indicates travel from I to J when it is nonzero and t_{IJ} is the travel time between I and J. We can also generalize our constraints at the beginning node, node A, we have

$$1 = \sum_J \delta_{AJ}. \tag{6.2.2}$$

At the end node, node N, we have

$$1 = \sum_I \delta_{IN}. \tag{6.2.3}$$

At an interior node, node K,

$$\sum_I \delta_{IK} - \sum_J \delta_{KJ} = 0. \tag{6.2.4}$$

Let us remove the integer restriction on δ_{IJ} and simply require that

$$0 \leq \delta_{IJ} \leq 1, \tag{6.2.5}$$

for all I and J.

We can easily reason that if any unique minimum path exists, our normal linear programming procedures will locate this optimum and the δ_{IJ} determined will be zero or one. If there is flow of a partial unit over a complete path other than the optimum, the objective function can be improved by removing the fractional travel in the nonoptimal path. Similarly, if over any segment of the path we were to have a fractional flow in a nonoptimal path, we improve the objective function by removing the flow and adding it to the optimal path. Then the global optimum must occur when the δ_{IJ} in the optimal path are one and all other δ_{IJ} are zero. No integer restrictions are therefore required, and the problem can be solved by the simplex procedure. Because of the special nature of this problem, shortcut methods can be devised, and a number are available (see Dreyfus [1969]).

Problems of identifying the longest or most time-consuming segment in a minimum-time problem can be solved in the manner indicated above. Identification of such "bottlenecks" is of major practical importance. In the scheduling of a large industrial project, for example, one is confronted with the scheduling of a number of interrelated activities. Some of these may be carried on simultaneously; others must be sequential. We can arrrange these activities into a network (like the path of Figure 6.1b), which shows the required interrelationship of the sequential activities. We replace the travel times by the times required to complete the various activities. The shortest time in which we can complete the project is then determined by identifying the longest path through the network. The longest path is called the *critical path.* Critical-path analyses are now an important managerial technique.

Transportation and allocation problems

In Section 6.1 we considered the dispatching of available trucks in a manner that would minimize costs to several cities where merchandise was stored. This is an example of a quite general class of problem called the *transportation problem.* In its original form, it is the determination of how best to ship some homogeneous commodity from a group of m cities to another group of n cities where the same commodity is needed. Although the problem is usually stated in terms of commodity shipment, the same formulation applies to many different allocation problems.

We shall refer to origins of the commodities as sources and the destinations as sinks. As before, the variables x_{ij} shall indicate the number of units moved from source i to sink j. For each such movement we shall assign a cost, c_{ij}. Initially, we shall assume that the demands, represented by b_j, at our sinks are just equal to the supplies, represented by d_i, at our sources. We can now state our problem as that of minimizing the shipping cost

$$\sum_{i=1}^{m} \sum_{j=1}^{n} c_{ij}x_{ij}, \tag{6.2.6}$$

subject to satisfying all demands by requiring that

$$\sum_{i=1}^{m} x_{ij} = b_j \qquad \text{for } j = 1, \ldots, n, \tag{6.2.7}$$

and using all available sources by requiring that

$$\sum_{j=1}^{n} x_{ij} = d_i \qquad \text{for } i = 1, \ldots, m. \tag{6.2.8}$$

We also require

$$x_{ij} \geq 0 \tag{6.2.9}$$

for all i and j.

Let us consider the simple situation where $m = n = 3$. Our constraint equations are then

$$
\begin{aligned}
&(6.2.10) \quad x_{11} + x_{12} + x_{13} &&= d_1 \\
&(6.2.11) \quad x_{21} + x_{22} + x_{23} &&= d_2 \\
&(6.2.12) \quad x_{31} + x_{32} + x_{33} &&= d_3 \\
&(6.2.13) \quad x_{11} + x_{21} + x_{31} &&= b_1 \\
&(6.2.14) \quad x_{12} + x_{22} + x_{32} &&= b_2 \\
&(6.2.15) \quad x_{13} + x_{23} + x_{33} &&= b_3.
\end{aligned}
$$

We thus apparently have $m + n$ constraint equations. However, not all the equations are independent. The sum of the first three equations equals the sum of the last three:

$$\left(\sum_j b_j = \sum_i d_i\right).$$

Therefore, if we subtract Equations (6.2.10) and (6.2.11) from the sum of (6.2.13), (6.2.14), and (6.2.15), we get Equation (6.2.12). Hence we really only have $(m + n - 1)$ independent constraints. From our previous studies of linear programming we know that we can therefore have at most $(m + n - 1)$ variables in our solution.

The information from Equations (6.2.10)–(6.2.15) can be most succinctly conveyed by forming a tableau as shown in Table 6.1. The column sums represent Equations (6.2.13)–(6.2.15); the row sums represent Equations (6.2.10)–(6.2.12). We may now show that in at least one row or column of Table 6.1 there will be only one nonzero element. Assume that each row contained at least two nonzero elements, and each column contained at least

Table 6.1. MATRIX REPRESENTATION OF TRANSPORTATION PROBLEM

$i \backslash j$	1	2	3	
1	x_{11}	x_{12}	x_{13}	d_1
2	x_{21}	x_{22}	x_{23}	d_2
3	x_{31}	x_{32}	x_{33}	d_3
	b_1	b_2	b_3	

two nonzero elements. Since there are m rows, we would have at least $2m$ variables in the basis. We then can say that the number of variables in the basis $\geq 2m$. Similarly, since there are n columns, we would have the number of variables in the basis $\geq 2n$. Now suppose that $n < m$. If so, $2m > n + m$, and we would have more than $n + m$ variables in the basis. We are allowed only $(m + n - 1)$ variables in the basis as we have only $(m + n - 1)$ independent constraints. Now suppose that $m \leq n$; then $2n \geq m + n$ and we still would have at least $m + n$ variables in the basis. Thus for any relative magnitude of m and n we have an impossible situation if all rows and columns have at least two nonzero elements. Also, each row and column must have at least one nonzero element or we could not meet our constraint equations. We thus conclude that at least one row or column will contain just one nonzero element.

By making use of the fact that a single row or column will have only one nonzero element, we can determine a basic feasible solution. To illustrate this, let us assume the following values for the source and sink requirements:

$$\begin{aligned} d_1 &= 10, & b_1 &= 5, \\ d_2 &= 15, & b_2 &= 26, \\ d_3 &= 31, & b_3 &= 25. \end{aligned} \tag{6.2.16}$$

Let us arbitrarily assume that $x_{11} = 5$. Referring to Table 6.2, we see that x_{21} and x_{31} are zero. To meet requirement d_1, we can let x_{12} equal 5 without exceeding b_2. Now we assign 15 to x_{22}, which is the maximum number of units that will not violate a requirement. The remainder of the table is then readily filled in.

Table 6.2. INITIAL BASIC FEASIBLE SOLUTION FOR SAMPLE TRANSPORTATION PROBLEM

i \ j	1	2	3	
1	5	5	0	$d_1 = 10$
2	0	15	0	$d_2 = 15$
3	0	6	25	$d_3 = 31$
	$b_1 = 5$	$b_2 = 26$	$b_3 = 25$	

The procedure we have just followed can always be used to determine the values of all the nonzero decision variables, once a value is assigned to one of them. If the values of the a_i and b_j are integers, then the variable in the row or column with only one nonzero value must have an integer value. All other variable values can be obtained from that value by addition and subtraction of integers and therefore must be integers themselves. We therefore conclude that *if the d_i and b_i are nonnegative integers, then every basic feasible solution has integral values.*

We can solve the transporatation problem by use of the simplex method, and we know from the above argument that each solution obtained will have integral values. The simplex procedure is, however, rarely used since the structure of the transportation-problem matrix is such that a much simpler procedure is available. The simplified procedure, which is due to Dantzig [1951], is based upon the use of the *simplex multipliers.* It may be shown that a set of constants (π_k, the simplex multipliers) can be found such that when the l constraints of a linear programming problem are multiplied by π_1 through π_l, respectively, we obtain for each variable x_{ik} in a basic feasible solution,

$$-c_{ik} + \sum_{i=1}^{l} a_{ik}\pi_i = 0 \qquad \text{for } k = 1, 2, \ldots, l. \tag{6.2.17}$$

Here the a_{ik} are the elements of the constraint matrix ($\mathbf{AX} = \mathbf{B}$) and variables 1 through l are in the basis. In the transportation matrix, the elements of the $\mathbf{A}$ matrix are zero or one so that equation (6.2.17) reduces to

$$-c_{ij} + \pi_i + \pi_{i+j} = 0 \tag{6.2.18}$$

for each x_{ij} in the solution. It is more convenient to represent the π_i obtained from the rows of our matrix (Table 6.1) by u_i and those from the columns by v_j. If we do so, Equation (6.2.18) becomes

$$c_{ij} = u_i + v_j \tag{6.2.19}$$

for basic x_{ij}.

Dantzig has shown that we have a minimum feasible solution if for every x_{ij} not in the solution

$$u_i + v_j - c_{ij} \leq 0. \tag{6.2.20}$$

If we have a trial set of u_i and v_j, we can obtain an improved solution by selecting as the new variable entering the basis that variable having the maximum $(u_i + v_j - c_{ij})$.

Let us return to our earlier example (tableau of Table 6.2) to see how we may obtain a trial set of u_i and v_j. We had obtained $(m + n - 1)$ variables in the solution. Now let us assume that we are given the cost matrix of Table

6.3 and, with that information, write Equation (6.2.19) for each of the basic variables:

Table 6.3. COST MATRIX FOR SAMPLE TRANSPORTATION PROBLEM

i \ j	1	2	3
1	3	2	3
2	3	2	4
3	1	2	1

$$
\begin{aligned}
u_1 + v_1 &= c_{11} = 3, \\
u_1 + v_2 &= c_{12} = 2, \\
u_2 + v_2 &= c_{22} = 2, \\
u_3 + v_2 &= c_{32} = 2, \\
u_3 + v_3 &= c_{33} = 1.
\end{aligned}
\tag{6.2.21}
$$

We have written five equations, but we have six unknowns. We therefore arbitrarily let one of the variables equal its corresponding c_{ij} (i.e., let $u_3 = 1$). We then obtain $v_3 = 0$, $v_2 = 1$, $u_2 = 1$, $u_1 = 1$, and $v_1 = 2$. We now calculate $(u_i + v_j - c_{ij})$ for all the variables not in the basis and enter these values (indirect benefits) in the matrix of Table 6.4. The *'s indicate variables in the basis.

Table 6.4. INDIRECT-BENEFIT MATRIX FOR FIRST SOLUTION

i \ j	1	2	3
1	*	*	−2
2	0	*	−3
3	2	*	*

We observe that we have a single positive value in the (3, 1) location of the indirect-benefit matrix. We therefore wish to introduce x_{31} into the basis. We do so at some unknown level. We then adjust all the values of the tableau of Table 6.2 so that we maintain the constraint requirements. When we do so we obtain Table 6.5. In this table θ represents the unknown level of variable x_{31}. We wish, if possible, to eliminate only one variable and produce no negative x_{ij}. It is therefore obvious that the maximum value we can assign to θ is 5. We now redetermine u_i and v_j for the new basis; $u_1 = 1, u_2 = 1, u_3 = 1$, $v_1 = 0$, $v_2 = 1$, and $v_3 = 0$. With these values, the indirect-benefit matrix is re-evaluated (cf. Table 6.6). It is seen that in all cases the indirect benefits ≤ 0, and therefore we have obtained the minimum-cost solution. Normally, we would not be so fortunate and one or more values of the indirect-benefit matrix would be positive. We would then have to continue the process, bringing the variable with the most positive indirect benefit into the basis at each step, until we have satisfied our optimality condition. Note that each time we establish a new basis we must redetermine the u_i and v_j.

Table 6.5. PERTURBED TRANSPORTATION-PROBLEM MATRIX

i \ j	1	2	3	
1	$5 - \theta$	$5 + \theta$		$d_1 = 10$
2		15		$d_2 = 15$
3	θ	$6 - \theta$	25	$d_3 = 31$
	$b_1 = 5$	$b_2 = 26$	$b_3 = 25$	

Our problem could have been made more difficult if the demands and requirements had been such that x_{32} in Table 6.2 had been 5. When we introduced x_{31} into the basis at a level of 5, we would have eliminated x_{32} as well as x_{11}. We would then have a degenerate solution. To maintain a solution with exactly $(m + n - 1)$ variables in the basis, we would have retained x_{32} (c_{32} lower than c_{11}) in the basis, but at a value of zero. Computations would proceed as before. That is, we would evaluate our u_i, v_j, and indirect cost

Table 6.6. INDIRECT-BENEFIT MATRIX FOR MINIMUM FEASIBLE SOLUTION

i \ j	1	2	3
1	−2	*	−2
2	−2	*	−3
3	*	*	*

benefits as if x_{32} were in the basis. Subsequently, x_{32} could return to the basis at some positive level.

We have not yet dealt with the problem in which the availability at the sources does not match the demands at the sinks. An example of such a problem is that cited in Section 6.1 in which we had more trucks available for dispatch than were needed. We may easily handle a problem of this nature by creating a fictitious sink, $(n + 1)$. The cost of shipping to this fictitious sink from any location would be zero. The demand, b_{n+1}, at this sink would be set at the difference between the total number of units available and the total number required (for the problem of Section 6.1, $b_{n+1} = 2$). Similarly, if we had a problem in which the number of units required were greater than those available, we would set up a fictitious source with zero shipping cost. The units that appeared to be shipped from this source would represent unfilled demand.

6.3 Cutting-plane Methods for Solution of Integer Linear Programming Problems

All-integer problems

We have seen that the solutions of certain types of problems, when properly formulated, are automatically integers. We know, of course, that the solutions to the usual linear programming problem will not be in terms of integers. For example, let us consider the very simple problem of: Maximize

$$x_1 + x_2, \tag{6.3.1}$$

subject to

$$\begin{aligned} 3x_1 + 2x_2 &\leq 10, \\ x_1 + 4x_2 &\leq 11, \end{aligned} \tag{6.3.2}$$

and

$$x_1, x_2 \geq 0. \tag{6.3.3}$$

The graphical representation of this problem, shown in Figure 6.3, indicates that the constrained maximum occurs at $\mathbf{X} = (1.8, 2.3)$. Now let us assume that x_1 and x_2 represent integral quantities. The feasible integer solutions to this problem are represented by the circles in Figure 6.3. The constrained maximum is obviously at (2, 2), which is within the feasible area (shaded region), and not at the intersection of the original constraints. If we are to make use of our previous procedures for solving linear programming problems, we shall have to introduce one or more additional constraints to express the integer requirement.

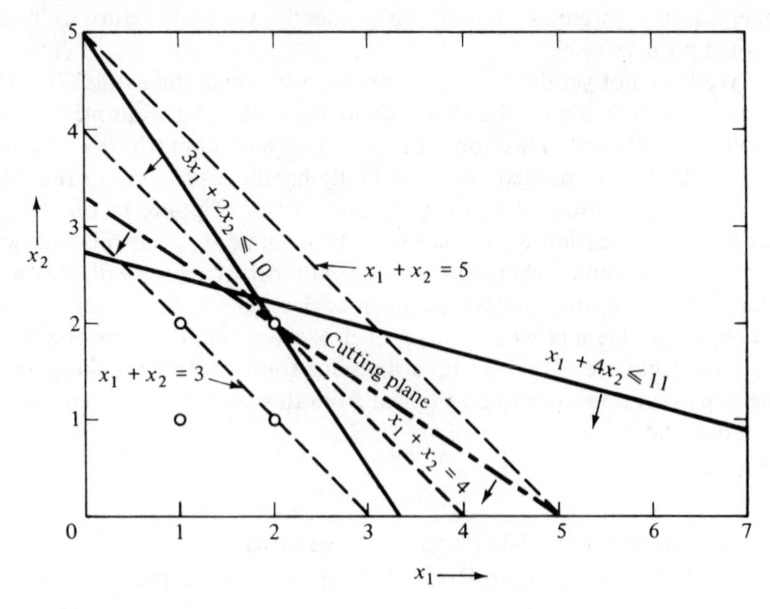

Figure 6.3. All-Integer Programming Problem Solved with Use of Cutting Plane

Before proceeding, let us express our constraints as equalities by introducing the slack variables x_3 and x_4:

(6.3.4) $$3x_1 + 2x_2 + x_3 = 10,$$

(6.3.5) $$x_1 + 4x_2 + x_4 = 11.$$

Since all the coefficients and the right-hand sides are integers, it is obvious that if x_1 and x_2 are integers, then x_3 and x_4 will also be integers. We thus have an *all*-integer programming problem. If one or more of the coefficients, or one or more of the right-hand sides, had not been integers, then the slack variables would not necessarily be integers, even though x_1 and x_2 were. We would then have a *mixed*-integer programming problem.

Let us now consider how we may add an additional constraint to our problem, expressing our integer requirement. We have seen that our maximum-integer solution does not lie on the intersection of the constraint lines. Hence we know one of the slacks must be nonzero. Also, we know from the foregoing discussion that these slack variables must be integers. Since 1 is the smallest nonzero integer, we know x_3 or x_4 must be equal to or greater than 1. We can write this as an additional constraint:

(6.3.6) $$x_3 + x_4 \geq 1.$$

By making use of Equations (6.3.4) and (6.3.5), we can solve for x_3 and x_4 in terms of x_1 and x_2, so that our additional constraint can be written as

$$3x_2 + 2x_1 \leq 10. \tag{6.3.7}$$

We now introduce this additional constraint (cf. Figure 6.3). Since this additional constraint "cuts" off a portion of the previously feasible region, it is called a cutting plane. We observe that our new maximum feasible solution, which lies on the intersection of the cutting plane and one of the original constraints, has all-integer values ($x_1 = 2$, $x_2 = 2$).

In the simple illustrative problem it was easy to obtain a single cutting plane that provided an integral solution. In more complicated problems, a number of cutting planes may be required, and their determination may not be obvious. A systematic way of determining these cuts is therefore needed. Such a procedure has been devised by Gomory [1958], who also showed that an optimal solution would be obtained in a finite number of steps.

We begin the development of Gomory's algorithm by assuming that we have solved the original linear programming problem and obtained a nonintegral solution vector $\mathbf{P}_0$. From Section 4.4 we know that we can express $\mathbf{P}_0$ in terms of the other vectors $\mathbf{P}_i$; i.e.,

$$\mathbf{P}_0 = x_1\mathbf{P}_1 + x_2\mathbf{P}_2 + \cdots + x_n\mathbf{P}_n. \tag{6.3.8}$$

No more than m of the x_i will have nonzero values, where m is the number of constraints (rows) in the problem. Let us consider only the ith component of the basic optimal solution. We can obtain this component from the ith row of Equation (6.3.8):

$$p_{i0} = x_1 p_{i1} + x_2 p_{i2} + \cdots + x_n p_{in}. \tag{6.3.9}$$

Also, we can make use of the identity matrix for the first m columns (assuming the first m variables are in the basis) to write

$$p_{i0} = x_i + \sum_{j=m+1}^{n} x_j p_{ij} \tag{6.3.10}$$

or

$$x_i = p_{i0} - \sum_{j=m+1}^{n} x_j p_{ij}. \tag{6.3.11}$$

Any positive real number can be written in terms of its integer part u, and the remaining fractional part r. If we do so, we have

$$p_{i0} = u_{i0} + r_{i0}, \tag{6.3.12}$$

$$p_{ij} = u_{ij} + r_{ij}. \tag{6.3.13}$$

In the above equations u_{i0} and u_{ij} are the largest integers that do not exceed p_{i0} and p_{ij}, and r_{i0} and r_{ij} are fractional remainders such that $0 < r_{i0} < 1$, $0 < r_{ij} < 1$. By substituting these expressions into Equation (6.3.11), we obtain

$$x_i = \left[u_{i0} - \sum_{j=m+1}^{n} x_j u_{ij}\right] + \left[r_{i0} - \sum_{j=m+1}^{n} x_j r_{ij}\right]. \tag{6.3.14}$$

Now suppose that we construct a basic feasible solution in which all the basic variables are integers. To do so we shall have to introduce additional slack or surplus variables, since some constraints will no longer be met exactly. We can use Equation (6.3.14) to devise a restriction on these variables much in the same manner as we did earlier for our illustrative problem. If the solution is integral, then x_i and $[u_{i0} - \sum_{j=m+1}^{n} x_j u_{ij}]$ will be integers. This means that $[r_{i0} - \sum_{j=m+1}^{n} x_j r_{ij}]$ must also be an integer. Furthermore, since both the x_j and r_{ij} are positive, $[\sum_{m+1}^{n} x_j r_{ij}]$ is positive. We also know that $0 < r_{i0} < 1$, and therefore if $[r_{i0} - \sum_{j=m+1}^{n} x_j r_{ij}]$ is to be an integer, it must be a nonpositive integer. We therefore introduce the constraint

$$r_{i0} - \sum_{j=m+1}^{n} x_j r_{ij} \leq 0. \tag{6.3.15}$$

This is the Gomory constraint or "cut."

Table 6.7. SIMPLEX SOLUTION OF ORIGINAL LINEAR PROGRAMMING PROBLEM OF FIGURE 6.3

	c		1	1	0	0
Basis		$\mathbf{P}_0$	$\mathbf{P}_1$	$\mathbf{P}_2$	$\mathbf{P}_3$	$\mathbf{P}_4$
$\mathbf{P}_2$	1	2.3	0	1	−0.1	0.3
$\mathbf{P}_1$	1	1.8	1	0	0.4	−0.2
$(y_j - c_j)$		4.1	0	0	0.3	0.1

Let us now see how we may apply the Gomory procedure to the problem illustrated in Figure 6.3. We first solve the linear programming problem without reference to the integer restrictions and obtain the tableau shown in Table 6.7. For the moment, let us arbitrarily choose the $\mathbf{P}_2$ row to generate a cut. By following Equation (6.3.15), we obtain from the first row

$$0.3 - 0.9x_3 - 0.3x_4 \leq 0. \tag{6.3.16}$$

Note that we have defined all r_{ij} to be positive. Therefore r_{13} cannot take on the value -0.1. We must choose the next lowest integer for u_{13} so that $r_{13} = 0.9$.

The significance of this last constraint may be understood by using Equations (6.3.4) and (6.3.5) to express x_3 and x_4 in terms of x_1 and x_2. When we do so, we find that our cut is equivalent to imposing the constraint $x_1 + x_2 \leq 4$. From Figure 6.3 we see that this additional constraint can provide us with an integer solution.

To utilize the additional constraint in our numerical procedure, we express it as an equality by adding the slack variable x_5:

$$0.3 - 0.9x_3 - 0.3x_4 + x_5 = 0. \tag{6.3.17}$$

This constraint is now annexed to our previous tableau by adding an additional row and column, as shown in Table 6.8. The new basis vectors are $\mathbf{P}_2$, $\mathbf{P}_1$, and $\mathbf{P}_5$.

Table 6.8. BASIC SOLUTION WITH ADDED GOMORY CONSTRAINT

	c		1	1	0	0	0
Basis		$\mathbf{P}_0$	$\mathbf{P}_1$	$\mathbf{P}_2$	$\mathbf{P}_3$	$\mathbf{P}_4$	$\mathbf{P}_5$
$\mathbf{P}_2$	1	2.3	0	1	-0.1	0.3	0
$\mathbf{P}_1$	1	1.8	1	0	0.4	-0.2	0
$\mathbf{P}_5$	0	-0.3	0	0	-0.9	-0.3	1
$(y_j - c_j)$		4.1	0	0	0.3	0.1	0

It will be observed that in Table 6.8 all the $(y_j - c_j)$ values are positive, thus giving the impression of a maximum. However, since one of the $\mathbf{P}_0$ values is negative, we know the solution is infeasible. We thus have an infeasible basic solution. We know that we want to eliminate $\mathbf{P}_5$ from the basis so that our new constraint may be active. However, since all the $(y_j - c_j)$ values are positive, we have no way of choosing a new vector to enter the basis.

This difficulty is identical to that encountered in Section 5.2, when we considered Kelley's cutting-plane procedure. At that point we observed that we could surmount our difficulty by considering the dual. Although the primal problem is infeasible, the dual problem is feasible but not optimal. Since the dual optimality criterion is not met, we can determine the new vectors to enter the basis when we consider the dual.

By consideration of the properties of the dual, Lemke [1954] has developed a very clever algorithm that can be applied directly to the *primal* in situations such as the one under discussion. His *dual-simplex* algorithm starts with a primal infeasible but dual feasible solution. The number of negative variables is then decreased while maintaining the optimality criterion. The solution is obtained when there are no longer any negative variables.

The dual-simplex rules for establishing a new basis are simple. The vector which leaves the basis is that for which x_{l0} is the most negative. The vector entering the basis is that vector for which $(y_j - c_j)/p_{lj}$ is a maximum, providing $p_{lj} < 0$. If we apply these rules to Table 6.8. we see that x_{50} is negative, and therefore $\mathbf{P}_5$ is to be eliminated from the basis. Table 6.9 shows the tableau obtained after the basis transformation is made. All negative values have been eliminated from the $\mathbf{P}_0$ column, and hence no additional iterations are required. Also we see that we have obtained the desired integer solution to the problem ($x_1 = 2$, $x_2 = 2$).

Table 6.9. OPTIMAL SOLUTION OF ALL-INTEGER PROBLEM

	c		1	1	0	0	0
Basis		$\mathbf{P}_0$	$\mathbf{P}_1$	$\mathbf{P}_2$	$\mathbf{P}_3$	$\mathbf{P}_4$	$\mathbf{P}_5$
$\mathbf{P}_2$	1	2.0	0	1	−1.0	0	1.0
$\mathbf{P}_1$	1	2.0	1	0	1.0	0	−0.666
$\mathbf{P}_4$	0	1.0	0	0	3.0	1	−3.33
$(y_j - c_j)$		4.0	0	0	0.0	0	0.333

In most cases only one or two iterations will be required to obtain a primal feasible solution after annexing a Gomory cut to the problem. However, for more complicated problems a single cut is rarely sufficient to provide an all-integer solution. In that event, we may again choose another row having a nonintegral value to generate a new cut. The additional constraint is annexed to our problem and the problem (now with two Gomory constraints) is resolved using the dual-simplex algorithm. The process is repeated, annexing a new constraint each time, as many times as required to obtain an integral solution. The number of constraints in the problem is kept within bounds by noting that if any of the slack variables for the Gomory constraints appear in the solution to the kth problem, then problem $k + 1$ can be formulated by eliminating that constraint and variable (striking the appropriate row and column from the matrix). Since such a constraint is nonbinding, we obtain the same solution with or without the constraint.

We have noted that the foregoing cutting-plane procedure can be shown to converge after determining a finite number of cuts. To guarantee this convergence, it is necessary to modify the problem slightly so that the c_j are integers. Thus y will be an integer when the x_j are integers. We are also required to annex a number of redundant rows to our simplex tableau. When all the rows are arranged in a particular order (see Hadley [1964]), the first row with a noninteger value in the $\mathbf{P}_0$ column is chosen for generating the next cut. In practice, this procedure is inconvenient and rarely used. It is generally assumed that convergence will be speeded by choosing that cut which bites as deeply as possible. This is usually taken to mean the selection of the row that will give the largest r_{i0}.

Mixed-integer problems

Mixed-integer problems are more likely to be encountered than the all-integer problems we have just examined. Problems in which only a few of the variables (e.g., those representing the number or sizes of various pieces of equipment) must be integers are quite common in the process industries. Also, we have seen that even when all the variables are required to be integers, we obtain a mixed-integer problem unless all coefficients and right-hand sides are integers.

Gomory [1960] has extended his cutting-plane method to the mixed-integer problem. The basic algorithm is the same as for the all-integer case. The linear programming problem, without any integer constraints, is again solved. We then add an additional constraint (cut) and solve the augmented problem by the dual-simplex algorithm. As in the all-integer problem, we continue adding constraints until all the integer requirements are satisfied. The algorithm differs from the all-integer algorithm only in the method of determining the cutting planes.

Let us first consider the case in which a single variable, x_i, is constrained to be an integer. From Equation (6.3.11) we know that we can write x_i as

$$x_i = p_{i0} - \sum_{j=m+1}^{n} x_j p_{ij} \tag{6.3.18}$$

if x_i is in the basis. (If x_i is not in the basis, it has a zero value and hence satisfies the integer requirement.) We shall again express p_{i0} in terms of an integer part u_{i0} and a fractional remainder r_{i0} to obtain

$$r_{i0} - \sum_{j=m+1}^{n} x_j p_{ij} = x_i - u_{i0}. \tag{6.3.19}$$

For any feasible solution to the mixed-integer problem, the right-hand side of Equation (6.3.19) must be an integer; we then have

(6.3.20) $$r_{i0} - \sum_{j=m+1}^{n} x_j p_{ij} = \text{integer}.$$

It is useful to divide the p_{ij} into two subsets. Let the set J^+ include all $p_{ij} \geq 0$ and the subset J^- all $p_{ij} < 0$. We may then rewrite our last result as

(6.3.21) $$\sum_{J^+} x_j p_{ij} + \sum_{J^-} x_j p_{ij} = k + r_{i0},$$

where k represents a positive or negative integer.

Now consider the case where $\sum_J x_j p_{ij}$ is zero or positive. Then the inequality

(6.3.22) $$\sum_{J^+} x_j p_{ij} + \sum_{J^-} x_j p_{ij} \geq r_{i0}$$

obviously holds, since in this case k must be a nonnegative integer. Furthermore, since the second term on the left-hand side is negative, we can write

(6.3.23) $$\sum_{J^+} x_j p_{ij} \geq r_{i0}.$$

If we assume now that $\sum_J x_j p_{ij}$ is negative, k must be a negative integer. Hence k must be equal to or less than -1, and we can write, by reasoning similar to the above,

(6.3.24) $$\sum_{J^-} x_j p_{ij} \leq r_{i0} - 1.$$

If we multiply Equation (6.3.24) by $r_{i0}/(r_{i0} - 1)$, we obtain

(6.3.25) $$\sum_{J^-} \left(\frac{r_{i0}}{r_{i0} - 1}\right) x_j p_{ij} \geq r_{i0}.$$

Since Equation (6.3.23) or (6.3.25) must be satisfied, and the left-hand sides of the inequalities are positive or zero, in every case we must satisfy

(6.3.26) $$\sum_{J^+} x_j p_{ij} + \sum_{J^-} \left(\frac{r_{i0}}{r_{i0} - 1}\right) x_j p_{ij} \geq r_{i0}.$$

This inequality furnishes our additional constraint when only a single variable must be an integer. We sequentially apply these constraints until we obtain an integer value for x_i.

To show the application of this constraint, let us again consider the problem illustrated in Figure 6.3. Now, however, let us assume that only x_2 is constrained to be an integer. From the results of the solution of the linear programming problem without the integer constraint (Table 6.7), we evaluate the Gomory constraint, Equation (6.3.26), as

$$[0.3x_4] + \left[\left(\frac{0.3}{0.3 - 1.0}\right)(-0.1)x_3\right] \geq 0.3. \qquad (6.3.27)$$

We may express this constraint in terms of x_1 and x_2 by use of Equations (6.3.4) and (6.3.5) to obtain

$$0.429x_1 + 1.286x_2 \leq 3.43. \qquad (6.3.28)$$

The effect of this constraint is shown in Figure 6.4, where it is seen that an integer solution is obtained.

Although a single Gomory cut provided an integer solution in our simple problem, this will not necessarily be the case in more complex problems. We therefore wish to obtain more powerful cuts. It is possible to do so when more than one variable is constrained to be an integer. This additional information may be used to deduce a stronger inequality than Equation (6.3.26). Before indicating this result, it is convenient to rewrite Equation (6.3.26) as

$$\sum_j (d_j)x_j \geq r_{i0}, \qquad (6.3.29)$$

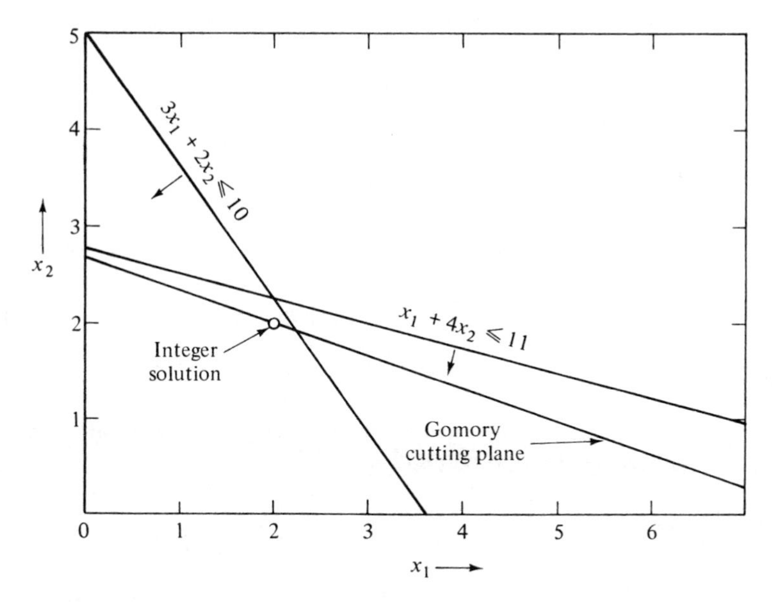

Figure 6.4. Application of Gomory's Mixed-Integer Cutting-Plane Algorithm

where

$$d_j = p_{ij} \qquad \text{if } p_{ij} \geq 0 \text{ and } x_j \text{ unconstrained (to be integer),}$$

$$d_j = \left(\frac{r_{i0}}{r_{i0} - 1}\right) p_{ij} \qquad \text{if } p_{ij} < 0 \text{ and } x_j \text{ unconstrained.}$$

We now formulate our problem as the determination of d_j for those variables constrained to be integers.

Let R represent the subset of variables constrained to be integers, and R^* the subset of variables not so constrained. We then can rewrite Equation (6.3.19) as

$$r_{i0} - \sum_R x_j p_{ij} - \sum_{R^*} x_j p_{ij} = x_i - u_{i0}. \tag{6.3.30}$$

Now for the integer-constrained variables, let us write

$$p_{ij} = u_{ij} + r_{ij}, \tag{6.3.31}$$

where u_{ij} is the largest integer smaller than p_{ij}, and r_{ij} is the remainder. This gives

$$r_{i0} - \sum_R x_j r_{ij} - \sum_{R^*} x_j p_{ij} = x_i - u_{i0} + \sum_R x_j u_{ij} = \text{integer.} \tag{6.3.32}$$

If we now introduce the two subsets J^{+*} and J^{-*} corresponding respectively to the positive and negative p_{ij} in R^*, and follow a line of reasoning similar to that used previously, we obtain

$$\sum_R r_{ij} x_j + \sum_{J^{+*}} p_{ij} x_j + \sum_{J^{-*}} \left(\frac{r_{i0}}{r_{i0} - 1}\right) p_{ij} x_j \geq r_{i0} \tag{6.3.33}$$

as our additional constraint. Note that if all variables were constrained, the last two terms on the left would be zero. The constraint would then be identical to the one previously derived for the all-integer case.

We could have begun our derivation by writing

$$p_{ij} = u'_{ij} - (1 - r_{ij}), \tag{6.3.34}$$

where u'_{ij} is the smallest integer greater than p_{ij}. If we had done so, we would have obtained as our additional constraint

$$\sum_R \left(\frac{r_{i0}}{1 - r_{i0}}\right)(1 - r_{ij}) x_j + \sum_{J^{+*}} p_{ij} x_j + \sum_{J^{-*}} \left(\frac{r_{i0}}{r_{i0} - 1}\right) p_{ij} x_j \geq r_{i0}. \tag{6.3.35}$$

Since we can apply either Equation (6.3.33) or (6.3.35), we would like to use whichever expression will give the deepest cut. We may obtain the best cut by choosing the coefficient of x_j, for those x_j in the subset R, as

$$\left(\frac{r_{i0}}{1 - r_{i0}}\right)(1 - r_{ij})x_j \qquad \text{if} \quad r_{ij} > r_{i0},$$

and

$$r_{ij}x_j \qquad \text{if} \quad r_{ij} \leq r_{i0}.$$

This choice will produce a hyperplane that will cut as deeply as possible.

From the similarity of Equations (6.3.33), (6.3.35), and (6.3.26), we can readily see that we may write them all in the form of Equation (6.3.29). We thus summarize our results by representing our additional constraint as

$$\sum_{j=m+1}^{n} (d_j)x_j \geq r_{i0} \tag{6.3.36}$$

where

$$\begin{aligned}
d_j &= p_{ij} && \text{if } p_{ij} \geq 0 \text{ and } x_j \text{ unconstrained (to be integer),}\\
d_j &= \left(\frac{r_{i0}}{r_{i0} - 1}\right)p_{ij} && \text{if } p_{ij} < 0 \text{ and } x_j \text{ unconstrained (to be integer),}\\
d_j &= \left(\frac{r_{i0}}{1 - r_{i0}}\right)(1 - r_{ij}) && \text{if } r_{ij} > r_{i0} \text{ and } x_j \text{ constrained (to be integer),}\\
d_j &= r_{ij} && \text{if } r_{ij} \leq r_{i0} \text{ and } x_j \text{ constrained (to be integer).}
\end{aligned}$$

We now multiply both sides of (6.3.36) by -1 to obtain

$$\sum_{j=m+1}^{n} (-d_j)x_j \leq -r_{i0}. \tag{6.3.37}$$

This last constraint is converted to an equality by the addition of a slack variable and is then added to the original constraint set. Since we now have a negative right-hand side, we can proceed in exactly the same fashion as in the all-integer case.

Gomory [1960] has shown that his cutting-plane method will produce an optimal solution, if one exists, with the introduction of a finite number of cutting planes. The finiteness proof requires that we impose conditions essentially the same as those required to show finiteness in the all-integer case.

Performance of cutting-plane methods

Performance of the cutting-plane technique has been encouraging for small problems, but disappointing in larger problems. Although Gomory has shown that both his all-integer and mixed-integer algorithms converge in a finite number of cycles if a feasible solution exists, some large problems have been run for well over a thousand cycles without obtaining convergence. Another difficulty is the large number of additional constraints that may be required to establish the integer restrictions. This can lead to problems too large for present computational facilities. These deficiencies have led to an interest in modified cutting plane procedures as well as other techniques. A summary of more recent work with cutting plane algorithms is given by Geoffrion and Marsten [1972].

6.4 Branch-and-bound Methods

Integer linear programming problems

Consider a linear programming problem in which only one of the variables, x_j, is required to be an integer. If x_j were constrained to lie between zero and k, we could obtain the optimum solution to our mixed-integer problem by solving $k + 1$ linear programming problems in which x_j was successively set at 0, 1, 2, etc. For a problem of this nature, this direct enumeration method might be the simplest means of solution. However, if we were to have l variables all constrained to be integers between zero and k, the direct enumeration of all-integer combinations would require the solution of $(k + 1)^l$ problems. Obviously, this rapidly becomes impractical as the number of integer variables increases.

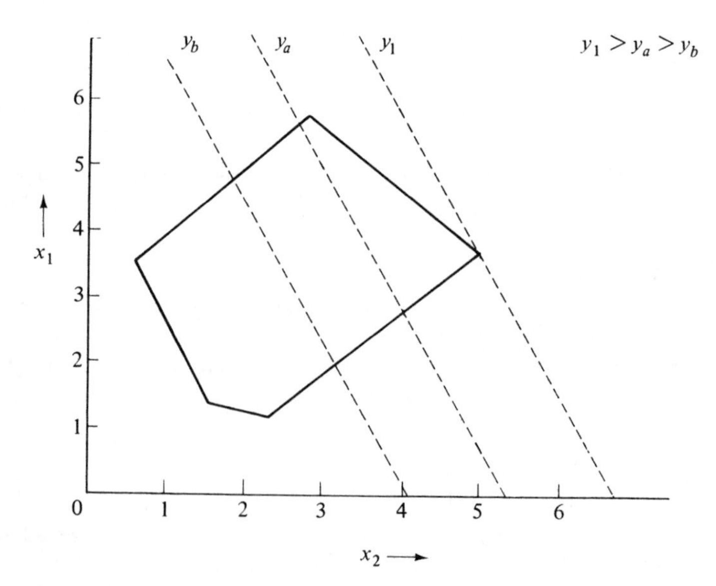

Figure 6.5. Geometric Representation of Simple Linear Program Prior to Imposition of Constraints

Branch-and-bound methods are essentially direct enumeration techniques that include a means for excluding from consideration a large number of the possible integer combinations. To show how this may be accomplished, let us examine the simple two-variable problem shown in Figure 6.5. Here we seek to: Maximize

$$y = \mathbf{C}'\mathbf{X} \tag{6.4.1}$$

subject to

$$\mathbf{AX} \leq \mathbf{B} \tag{6.4.2}$$

where x_1 is an integer but x_2 is continuous. The feasible region for the linear programming problem without the integer restriction is the shaded area within the polygon. The optimal solution of the noninteger problem occurs at (3.7, 5.1), where the objective function has a value y_1. Now let us create two new problems. In the first of these we shall restrict x_1 so that $x_1 \leq 3$ and in the second require that $x_1 \geq 4$. We have not excluded any feasible region, since x_1 cannot lie between 3 and 4 or the solution would be noninteger. If we solve these two problems, we obtain y_2 and y_3 for the objective functions (cf. Figure 6.6). In both cases the solutions are at integer values, and hence they meet our integer constraint. Since no feasible solution to our original integer problem can have a solution greater than $\max[y_2, y_3]$, the greater of these two values, y_2, is clearly the optimal integer solution. Note that the optimum occurs at $x_1 = 4$ and $4 < x_2 < 5$.

Now let us suppose that we wish to restrict both x_1 and x_2 to be integers. Since x_2 was not an integer in our previous solutions, we must apply addi-

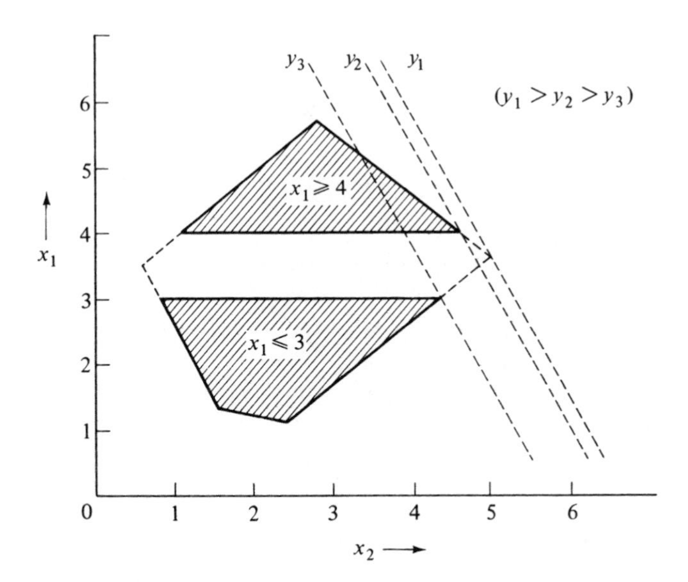

Figure 6.6. Simple Mixed-Integer Program with Bounding Constraints

tional conditions. From each of the previous problems we can create two additional problems in which x_2 is restricted so that $x_2 \leq 4$ or $x_2 \geq 5$. These problems may be represented on the "branching" decision tree shown in Figure 6.7. As we add restrictions, we can only decrease the objective function. Thus y_2 serves as an upper "bound" for all branches stemming from node 2 and y_3 serves as an upper "bound" for all branches stemming from node 3. The meaning of branch-and-bound is now obvious.

Upon examination of the additional problems generated, it is seen from Figure 6.8 that the problems in which $x_2 \geq 5$ are infeasible. We may there-

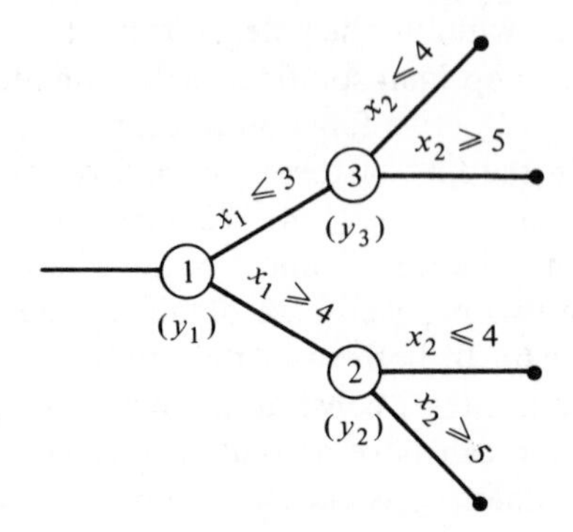

Figure 6.7. First Branching Decision Tree for Integer Programming Program

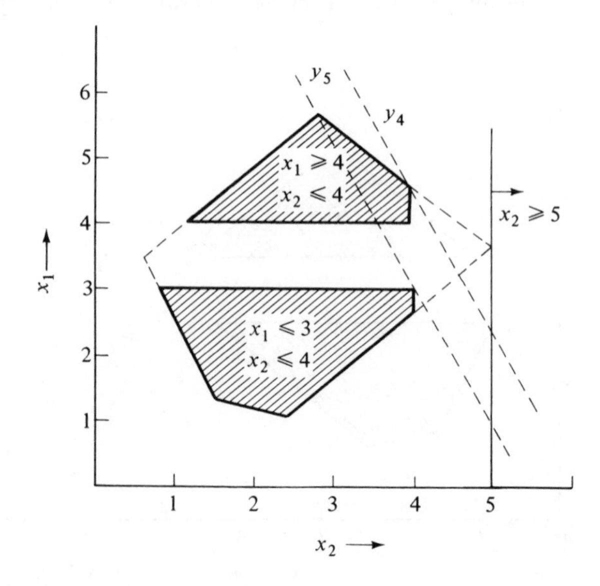

Figure 6.8. All-Integer Program with Two Sets of Bounding Constraints

fore terminate these branches, since we know that addition of any more restrictions certainly cannot restore feasibility. The other two problems generated yield feasible answers. From Fig. 6.8 we see that y_4 is the greater objective function, yielding an optimal value of x_1 between 4 and 5. We therefore generate two new problems from node 4 by setting $x_1 \leq 4$ and $x_1 \geq 5$ (cf. tree of Figure 6.9). We observe that since we have previously set $x_1 \geq 4$ in this branch, the first of these restrictions is equivalent to setting $x_1 = 4$. Solution of these problems yields y_6 and y_7, respectively, for the objective functions, as shown in Figure 6.10. Since y_7 is greater than y_6, and

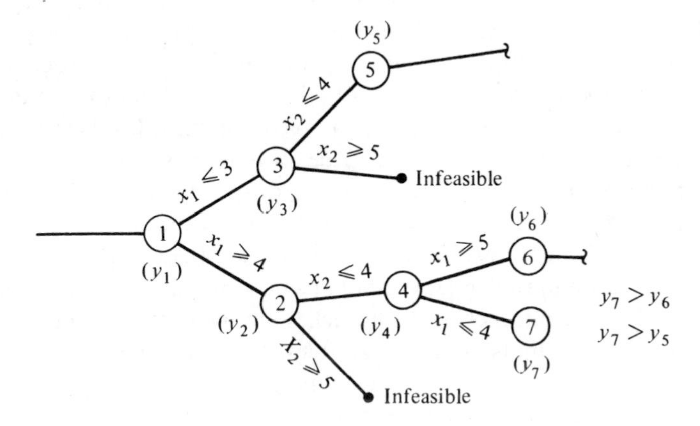

Figure 6.9. Final Branching Decision Tree

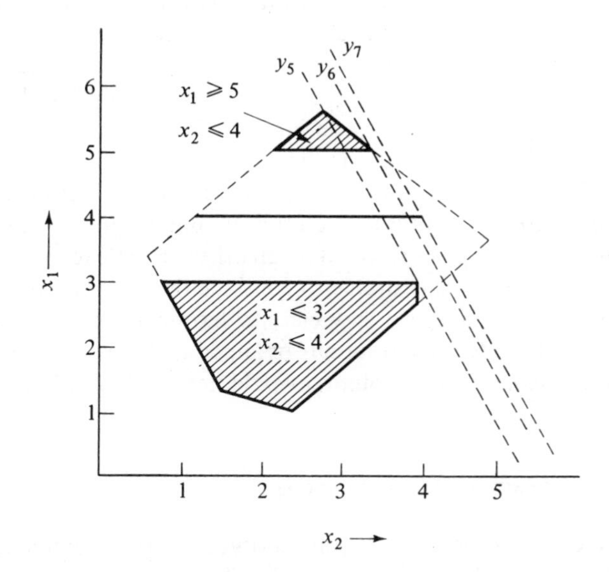

Figure 6.10. Final Solution of All-Integer Programming Problem by Branch-and-Bound

since we know that all branches stemming from node 6 will have lower objective functions than 6, we may terminate the branch. Also, we find that y_7 is greater than y_5 and hence we may also terminate the branch from node 5. Our only active branch is now at node 7. We now observe that the node 7 solution is an integer solution ($x_1 = 4, x_2 = 4$). Since all-integer solutions stemming from this node must have objective functions no greater than y_7, we recognize that we have located the optimal integer solution.

The branch-and-bound technique was first applied to integer linear programming problems by Land and Doig [1960]. The method used in solving the last illustrative problem is a modification of the Land and Doig method, developed by Dakin [1965]. The original linear programming problem is first solved by the simplex method without regard to the integer restraints. If the variables that are required to be integers are found to be integers in this solution, the optimum solution has been obtained. If this is not the case, then one of the variables required to be an integer is sequentially set at the integer above and below the value obtained from the original problem. The solutions to these two new problems, obtained by the simplex method, then serve as bounds to the problems branching from these nodes (upper bounds in a maximization problem, lower bounds in a minimization problem). Additional problems from each of the branches are then created by placing upper and lower integer limits on other variables. The branching process continues until all except one branch can be terminated by (a) showing the branches to be infeasible, or (b) showing the branch to have a less desirable objective function than another feasible branch.

Branch-and-bound algorithms are most useful for mixed-integer programming problems. Computational results have compared very favorably with cutting-plane methods. Dakin has reported that his modification of the Land and Doig algorithm was able to solve a 93-variable problem, 80 of the variables being constrained to be integer. The Gomory procedure was unable to solve this same problem in 2000 cycles.

In a large integer programming problem, a number of feasible solutions may be found before the optimum solution is found. Since at any stage in the computation we can place a bound on all other possible feasible solutions, we can decide whether it is desirable to continue the calculation. If the possible improvement is small, we may elect to terminate the calculation and choose the best of the feasible solutions obtained thus far. If a cutting-plane technique were used, a feasible solution would not be obtained until the final cycle.

Application to separable programming

An interesting application of mixed-integer programming occurs in connection with separable programming. In Chapter 5 we replaced separable nonlinear functions of the form

$$\sum_{j=1}^{n} f_j(x_j) \tag{6.4.3}$$

by linear approximations

$$\sum_{j=1}^{n} \sum_{l=1}^{sj} \alpha_{lj} k_{lj}, \tag{6.4.4}$$

where the α_{lj} were new variables and the k_{lj} were constants. We solved the resultant linear problem with a modified simplex method in which we allowed no more than two α_{lj} having the same index j in the solution, and we required that they be adjacent. We could not guarantee that the solution obtained was anything other than a local optimum of the approximating problem. If we formulate the approximating problem as a mixed-integer problem, however, the solution to the mixed-integer problem can be shown to be the global optimum.

In the mixed-integer formulation we abandon our restrictions on the variables entering the basis. We impose equivalent restrictions by defining a new set of variables δ_{ij}, which are defined, for each j, by

$$\alpha_{1j} \leq \delta_{1j}, \tag{6.4.5}$$

$$\alpha_{2j} \leq \delta_{1j} + \delta_{2j}, \tag{6.4.6}$$

$$\alpha_{3j} \leq \delta_{2j} + \delta_{3j}, \tag{6.4.7}$$

$$\cdots\cdots\cdots\cdots\cdots\cdots$$

$$\alpha_{sj-1} \leq \delta_{sj-2} + \delta_{sj-1}, \tag{6.4.8}$$

$$\alpha_{sj} \leq \delta_{sj-1}, \tag{6.4.9}$$

with the restriction that

$$\sum_{l=1}^{sj-1} \delta_{lj} = 1 \tag{6.4.10}$$

and that the δ_{lj} be zero–one integers. Since the sum of the δ_{lj} equals 1, only one δ_{lj} will have a nonzero value for any given j. We see that any given δ_{lj} appears in only two equations, and that these are adjacent. This means that only two adjacent α_{lj} can appear in the basis if all restrictions are met. We have thus restated our basis-entry restrictions in terms of integer variables. If we now solve this mixed-integer problem by a branch-and-bound algorithm, we obtain the global optimum to our approximating problem. If our linear approximations are valid, the global optimum of the approximating problem may be expected to be close to the global optimum of our original problem, even though the original problem was *not* a convex or concave programming problem.

Integer nonlinear programming problems

In describing the basic branch-and-bound technique, we did not link it to a particular method for solving our optimization problem. Suppose that instead of having to solve a linear programming problem, prior to imposition of the integer constraints, we were required to solve a convex nonlinear problem. We could do this by one of the methods described in Chapter 5. Since the problem is known to be convex, the solution obtained will be the global optimum. Now we create two new problems by successively imposing on the variables required to be integer the constraints

$$x_j \leq k \tag{6.4.11}$$

and

$$x_j \geq k + 1, \tag{6.4.12}$$

where k is the integer just below the value of x_j found originally. If the latter constraint is restated as

$$\frac{1}{x_j} \leq \frac{1}{k + 1}, \tag{6.4.13}$$

then both of the new problems will be convex. The solutions obtained for these problems will again be globally optimal; therefore, they will represent lower bounds to the solutions of all subsequent problems branching from the respective nodes. We may therefore proceed with branch-and-bounding in the same manner as we did in our linear problem, providing we express our constraints in the form of Equations (6.4.11) or (6.4.13). The final solution obtained will be the global optimum of the original integer problem.

If the original nonlinear programming problem were not convex, we could still apply our branch-and-bound procedure. However, we could no longer be certain that the solution obtained at any step was the global optimum for the set of constraints being considered. Hence we could not be certain that the objective function found represented the real bound on the problems stemming from that node. We might resolve each such problem several times, beginning each such resolution at a different starting point. This would increase the confidence in the minimum (or maximum) solution obtained. However, the possibility of establishing a false bound for a given branch cannot be completely eliminated. Our difficulty in determining the global optimum to a nonconvex problem is thus heightened when integer restraints are imposed.

The branch-and-bound technique is again most useful in nonlinear problems in which only a few of the variables are integers. When many of the variables are integers, the number of nonlinear problems that must be solved becomes too great for practical use. A good method for dealing with such situations remains to be determined.

Zero–one integer linear programming

Zero–one integer linear programs can be solved by the branch-and-bound technique described earlier. At each stage we would have to obtain our bounds by means of the simplex method. Balas [1965] has made use of the properties of zero–one problems to derive a branch-and-bound method that does not require the simplex method. Balas describes his algorithm as "additive," since it uses only additions and subtractions. The Balas method is worth considering if the problem is one involving only zero–one variables.

6.5 Decomposition of Large Programming Problems

In reading the preceding chapters, one might gain the impression that optimization theory is applied primarily to problems with a few variables and a small number of constraints. The illustrative examples are almost all of this nature. However, such an impression of the real application of optimization theory would be far from the truth. The illustrative examples have, of course, been chosen to illustrate the methods presented with a minimum of confusing detail. In real optimization problems, one is almost always faced with a significant number of variables and constraints. If the problem is very large ("large" being a very subjective term here), satisfactory solutions usually cannot be obtained using the techniques that were presented earlier.

Many realistic optimization problems fall into this category. We have previously noted that some optimization problems become so large and complex that our usual techniques find difficulty in dealing with the problem as a whole. When this occurs, one naturally considers how the complex optimization problem can be reduced to a series of less complex subproblems that can be optimized separately. The success of such an effort depends on the mathematical structure of the system being optimized.

Decomposable linear programs

We shall begin our consideration of decomposition techniques with linear programming problems. In many large linear programming problems the matrix of coefficients (which we have denoted as the vector **A**) contains a large number of zeros. In Section 6.2 we examined the transportation problem, which had a particularly sparse matrix. We were able to take advantage of this to develop a rapid solution procedure for this problem. There are also other characteristics of coefficient matrices that are of major significance. The most important of these is the *block angular* matrix configuration shown in Figure 6.11. In this figure the shaded areas contain the only nonzero elements in the constraint matrix.

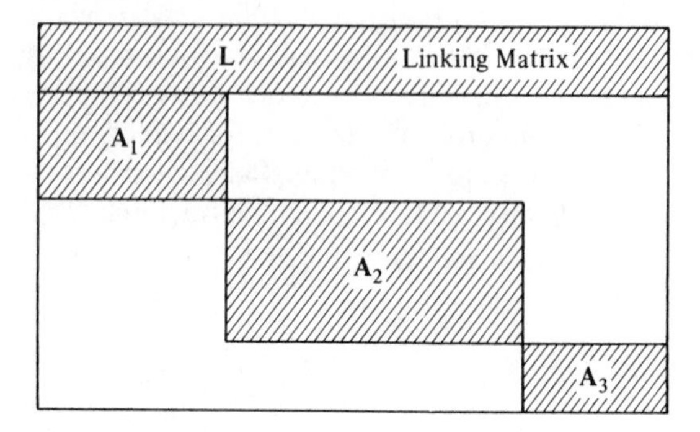

Figure 6.11. Block-Angular Matrix

The significance of the block-angular matrix may be seen by considering the operation of a multidivisional corporation. The operation of any division generally can be represented mathematically by a number of variables and constraints which apply only within that division. We would expect that only a few of the variables affecting the division's operation will affect the corporation as a whole. We may gather the constraints involving variables that are internal to division i and list these sequentially. If we do so, we find that we can describe these constraints through the matrix $\mathbf{A}_i$ (cf. Figure 6.11). Those constraints involving variables appearing in constraints affecting other divisions we locate in matrix **L**. Since **L** links the operations of several divisions, we call it the *linking matrix*. Many of the very large linear programming problems that have been studied have been generated by the optimization of the behavior of such a multidivisional corporation.

Whenever our problem exhibits the block-angular structure shown in Figure 6.11, we can decompose the problem into a series of smaller subproblems. Each subproblem corresponds to a problem generated by consideration of a single $\mathbf{A}_i$ matrix. By repetitive solution of these subproblems it is possible to obtain an optimal solution to our original problem. The technique by which this is accomplished is due to Dantzig and Wolfe [1960] and is called the *method of decomposition*.

Mathematical basis

Let us return to the matrix of Figure 6.11. For convenience, the linking matrix, which is located in the first m_0 rows, will be partitioned into p submatrices. We shall give each submatrix $\mathbf{L}_i$ the same index as the $\mathbf{A}_i$ matrix in the corresponding position. This nomenclature is shown in Figure 6.12.

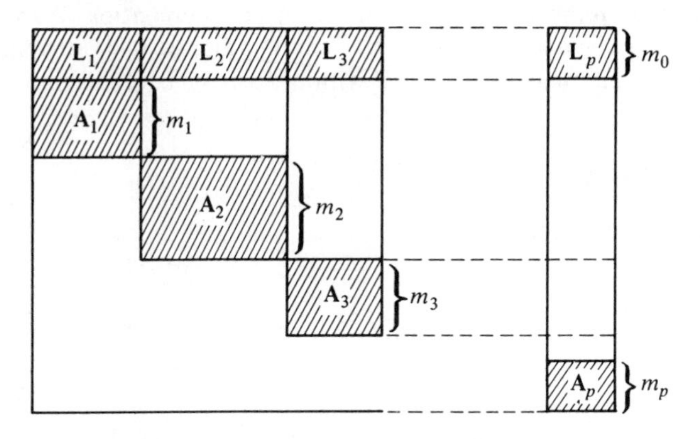

Figure 6.12. Block-Angular Matrix Arranged for Solution by the Method of Decomposition

Let us assume that our original problem can be stated as

$$\text{Problem A}\quad\begin{cases}\text{Maximize} & y = \mathbf{C}'\mathbf{X}, \qquad (6.5.1)\\ \text{subject to} & \mathbf{AX} \leq \mathbf{B} \qquad (6.5.2)\\ & \mathbf{X} \geq \mathbf{0}. \qquad (6.5.3)\end{cases}$$

If we partition $\mathbf{X}$ in the same manner that we partition $\mathbf{A}$, and let $\mathbf{X}_i$ represent those variables associated with $\mathbf{A}_i$, then we can rewrite our problem as: Maximize

$$y = \sum_{i=1}^{p} \mathbf{C}_i'\mathbf{X}_i, \qquad (6.5.4)$$

subject to

$$\sum_{i=1}^{p} \mathbf{L}_i\mathbf{X}_i \leq \mathbf{B}_0, \qquad (6.5.5)$$

$$\left.\begin{aligned}\mathbf{A}_i\mathbf{X}_i &\leq \mathbf{B}_i \qquad (6.5.6)\\ \mathbf{X}_i &\geq \mathbf{0} \qquad (6.5.7)\end{aligned}\right\} \text{ for } i = 1, 2, \ldots, p.$$

Now let us consider a particular set of constraints of Equation (6.5.6). The set of constraints $\mathbf{A}_i\mathbf{X}_i \leq \mathbf{B}_i$ defines a portion of the solution space; this portion we designate as S_i. It is reasonable to assume that in a real problem the constraints of each set (physically, the constraints which apply to each particular division) will lead to a bounded solution. If so, S_i is a convex polyhedron.

A convex polyhedron has the property that any point within the polyhedron can be represented by a convex linear combination of the external points of the polyhedron. If we let $\mathbf{X}_{ij}$ represent one of the l_i extremal points of polyhedron S_i, then any point $\mathbf{X}_i$ within the polyhedron is represented by

$$\mathbf{X}_i = \sum_{j=1}^{l_i} \mu_{ij}\mathbf{X}_{ij}, \tag{6.5.8}$$

where the μ_{ij} are scalars such that

$$\sum_{j=1}^{l_i} \mu_{ij} = 1 \tag{6.5.9}$$

and

$$\mu_{ij} \geq 0 \qquad \text{for } j = 1, 2, \ldots, l_i. \tag{6.5.10}$$

A similar set of equations holds for each S_i (i.e., for each value of i). We can now rewrite our maximization problem [Equations (6.5.4)–(6.5.7)] as

$$\text{Problem B} \begin{cases} \text{Maximize} & y = \sum_{i=1}^{p} \sum_{j=1}^{l_i} \mu_{ij}[\mathbf{C}_i'\mathbf{X}_{ij}], & (6.5.11) \\ \text{subject to} & \sum_{i=1}^{p} \sum_{j=1}^{l_i} \mu_{ij}[\mathbf{L}_i\mathbf{X}_{ij}] \leq \mathbf{B}_0, & (6.5.12) \\ & \sum_{=1}^{l_i} \mu_{ij} = 1 \qquad \text{for } i = 1, 2, \ldots, p, & (6.5.13) \\ & \mu_{ij} \geq 0 \qquad \text{for all } i \text{ and } j. & (6.5.14) \end{cases}$$

We must recognize that the $\mathbf{X}_{ij}$ do *not* represent a set of variables, but rather each $\mathbf{X}_{ij}$ represents a set of specific values that describes one of the vertices of S_i. With that in mind, we see that Equations (6.5.11)–(6.5.14) represent a new linear problem in the variables μ_{ij}. When we solve this problem and obtain the maximizing μ_{ij}, we can then determine the desired $\mathbf{X}_i$ from Equation (6.5.8).

The method of decomposition is based on the transformation just described. By determination of a set of feasible solutions to Equation sets (6.5.6) and (6.5.7), we generate a series of $\mathbf{X}_{ij}$. With these $\mathbf{X}_{ij}$ we determine the μ_{ij} that yield the optimum solution to the overall problem. By this procedure we have drastically reduced the number of equations that must be manipulated simultaneously. Our original problem (cf. Figure 6.12) contained

$\sum_{i=1}^{p} m_i$ equations; our revised problem contains only $(m_0 + p)$ equations. We may have simultaneously increased the number of variables from $\sum_{i=1}^{p} n_i$ to $\sum_{i=1}^{p} l_i$. However, the Dantzig–Wolfe algorithm generally does not require the evaluation of all the extremal points of the S_i. The algorithm provides a systematic scheme for determination of the $\mathbf{X}_{ij}$ so that usually only a fraction of the possible $\mathbf{X}_{ij}$ must be found. The actual number of variables required for solution of a given problem is usually considerably smaller than $\sum_{i=1}^{p} l_i$.

Generation of proposals

Let us return to our example of a multidivisional corporation. The constraints that act on division i alone we designate by means of

$$\mathbf{A}_i\mathbf{X}_i \leq \mathbf{B}_i. \tag{6.5.15}$$

The $\mathbf{X}_i$ represent the amount of the various outputs produced and the $\mathbf{B}_i$ represent limits on the various requirements imposed. If the division were operating independently, it would simply maximize its profits. If $\mathbf{C}_i$ is the vector representing the profits from each of the outputs of division i, the division would then have as its objective: Maximize

$$\mathbf{C}_i'\mathbf{X}_i, \tag{6.5.16}$$

subject to Equation (6.5.15). Solution of this linear programming problem would provide one of the external points of S_i(one of the $\mathbf{X}_{ij}$). We shall call such a solution a *proposal.* Every division can generate such a proposal based on its divisional constraints and costs. We would then have a proposal set for evaluation.

Since none of the divisions has considered the effect of its activities on the other divisions, a proposal set obtained as above will certainly not be optimal for the corporation. In fact, it is quite likely that such a proposal would not even be feasible on a corporate basis since the constraints of the linking matrix are probably unsatisfied. In order to force each division to conform to corporate constraints and costs, the division is furnished with a revised objective function. That is, we require that each division: Maximize

$$(\mathbf{C}_i - \mathbf{R}_i)'\mathbf{X}_i \tag{6.5.17}$$

where $\mathbf{R}_i$ represents the penalty assigned because of the burdens the original division plan imposed on the corporation. The divisions now redetermine their optimal plans, using their revised objective function. This new proposal set is evaluated, and modified penalties are assigned to the divisions. The entire process is repeated as many times as needed to arrive at a converged set of decisions that is optimal for the entire corporation.

There are only a finite number of options—basic feasible solutions to the divisional constraints—that can be tried. Furthermore, as we shall see, the penalties are designed to reduce corporate costs at each step. Therefore, Dantzig and Wolfe [1960] are able to show that their algorithm will converge in a finite number of cycles.

Our overall strategy is thus to obtain a proposal set from the divisions, determine how the division proposals affect corporate profitability, assign divisional penalties, $\mathbf{R}_i$, accordingly, and then generate new proposals. In the following sections we shall develop the details of how this may be done.

The executive program

Let us now examine how we may obtain the penalties $\mathbf{R}_i$ needed to move the divisional solutions in the directions desired. To do so, we must look at the corporate costs and the corporate constraints defined by the linking matrix **L**. These costs and constraints are given in terms of our revised problem (Problem B) expressed by Equations (6.5.11)–(6.5.14). We call this problem the *executive program.*

Each proposal generated by a division provides a particular $\mathbf{X}_{ij}$. The corresponding μ_{ij} represents the weight to be assigned to that proposal or extremal point. In the early stages of the solution we have only a few proposals to consider. We shall generally not have determined enough of the extremal points ($\mathbf{X}_{ij}$) to allow a truly optimal solution to be determined. Recall that we stated that any point $\mathbf{X}_i$ could always be described in terms of all the extremals of S_i. We did not say any such point could necessarily be described by some arbitrary portion of the extremal points. The function of the executive program is, then, to establish penalty sets, which, when communicated to the divisions, will cause the generation of the additional extremal points needed to establish the optimum.

For this we need to consider briefly the dual of the executive program. In Chapter 2, we observed that the negatives of the optimal values of the Lagrange multipliers could be interpreted in terms of shadow prices. The shadow price is equivalent to the marginal profit value of the kth requirement. That is, $-\lambda_k$ tells us how much the total profit would be increased if the limit on requirement k, b_k, could be increased by one unit. Since $w_k = -\lambda_k$, we have

$$w_k = \frac{\partial y}{\partial b_k}. \qquad (6.5.18)$$

The dual slack variables may be interpreted as the loss incurred by production of one unit that is not in the optimal product line.

The marginal profit interpretation of the ordinary dual variables can be extended to nonoptimal conditions. We then can say that w_k represents the increase in profit which could be earned by increasing the amount of requirement k when resource k is used in the manner specified by the current basic feasible solution. We shall use this interpretation to establish the penalties to be applied to the divisional proposals. To distinguish between the dual variables, we shall designate the dual variables originating from the corporate constraints as π_k. Those dual variables originating from our constraints on the sum of the weights will be designated as $\hat{\pi}_i$. If Problem B is rewritten in terms of the dual variables, we have: Minimize

$$\mathbf{B}'\boldsymbol{\pi} + \sum_{i=1}^{p} \hat{\pi}_i \tag{6.5.19}$$

subject to

$$\boldsymbol{\pi}'[\mathbf{L}_i\mathbf{X}_{ij}] + \hat{\pi}_i \geq [\mathbf{C}_i'\mathbf{X}_{ij}] \qquad \text{for } i = 1, \ldots, p, \quad \text{for } j = 1, \ldots, l, \tag{6.5.20}$$

where $\boldsymbol{\pi}$ is a vector whose elements are the π_k. The π_k are called *simplex multipliers*. We referred to the simplex multipliers in our development of the transportation algorithm (cf. Section 6.2).

Let us consider the corporate constraint on requirement k. We may write this constraint as

$$a_{k1}x_1 + a_{k2}x_2 + \cdots + a_{kq}x_q + \cdots + a_{kn}x_n \leq b_k. \tag{6.5.21}$$

We have considered x_q as the quantity of product q that is produced. Since the kth constraint sums up all requirements of type k, we must consider a_{kq} as the amount of requirement k needed to produce a unit of product q. We have just determined that π_k is the cost or value to the corporation of a unit of requirement k. The cost of all requirements of type k is then

$$\sum_q \pi_k a_{kq} x_q. \tag{6.5.22}$$

The total value of all corporate requirements is obtained by summing over all corporate constraints. That is, the total value of all the corporate requirements is given by

$$\sum_k \sum_q \pi_k a_{kq} x_q. \tag{6.5.23}$$

Each division accounts for a portion of these costs. The portion contributed by division i is simply that portion of (6.5.23) due to division i variables. We may write this as

$$\sum_k \sum_{q_i} \pi_k a_{kq} x_q, \tag{6.5.24}$$

where $\sum_{q_i}$ indicates that the summation is restricted to variables associated with matrix $\mathbf{A}_i$.

The contribution that division i makes to corporation profits is $\mathbf{C}_i'\mathbf{X}_i$ less the value of the corporate resources which it uses. We have just shown that the total value of these resources is given by (6.5.24). Therefore, we may consider that the net profit earned by division i is $\mathbf{C}_i'\mathbf{X}_i$ less the value of its resources. We therefore would be justified in saying that the proper objective function for division i is

$$y_i = \mathbf{C}_i'\mathbf{X}_i - \sum_k \sum_{q_i} \pi_k a_{kq} x_q \tag{6.5.25}$$

or, more compactly, as

$$y_i = (\mathbf{C}_i - \mathbf{R}_i)'\mathbf{X}_i, \tag{6.5.26}$$

where $\mathbf{R}_i = \boldsymbol{\pi}'\mathbf{A}_i$. We may consider $\mathbf{R}_i$ as the penalty to be subtracted from divisional profit to account for the corporate expenditures associated with division i.

Once a feasible solution has been found, both the μ_{ij} and the π_k will be known, and a set of penalties can then be determined for each division from Equation (6.5.23). This provides a new objective function for each division. The divisions then determine a new set of proposals by solving Equations (6.5.5)–(6.5.7) with these revised objective functions. The executive program determines if any of the new divisional proposals can be used to improve the corporation's profits. If so, the executive program finds a new set of π_k, resulting in still another set of divisional objective functions, and the cycle is continued. If none of the divisional proposals can be used to improve corporation profits, an optimal solution has been located. The μ_{ij} that the executive program determined at the previous solution are then optimal.

Test for optimality

In discussing the role of the simplex multipliers π_k in determining the penalties to be applied to divisional proposals, we ignored the multipliers associated with the constraints requiring $\sum_j \mu_{ij}$ to be 1. We shall use these multipliers, which we have designated by $\hat{\pi}_i$, to determine whether an optimum solution has been reached.

Let us first indicate how each of these constraints can be interpreted economically. The condition $\sum_j \mu_{ij} = 1$ requires that we choose weights such that the master program uses a true average of the ith division's proposed solutions. Just as with any other constraint, the simplex multiplier associated with this constraint must be interpreted as the marginal profit obtained by increasing the constraint limit by one unit. Our limit of unity represents all the company resources made available to division i. Increasing this limit by one unit ($\sum_j \mu_{ij} = 2$) would mean a 100 per cent increase in the company resources to division i. Then the $\hat{\pi}_i$ may be taken as the relative marginal profitability of the transfer of additional resources to division i.

The contribution that the ith division could make to corporate profitability by its *new* proposal is $(\mathbf{C}_i - \mathbf{R}_i)'\mathbf{X}_i$. That is, the relative marginal profitability of the transfer of resources to division i in accordance with its new proposal is $(\mathbf{C}_i - \mathbf{R}_i)'\mathbf{X}_i$. But, from the previous solution of the executive program, we know that $\hat{\pi}_i$ is the relative marginal profitability for the transfer of resources to division i in accordance with the weighted average of the proposals in the basis. The new proposal can therefore improve profitability as long as any

$$(\mathbf{C}_i - \mathbf{R}_i)' \mathbf{X}_i > \hat{\pi}_i. \tag{6.5.27}$$

That is, if the new proposal is incorporated into the basis, a more profitable solution is obtained. Hence, if inequality (6.5.27) holds, the solution is nonoptimal. However, when all

$$(\mathbf{C}_i - \mathbf{R}_i)' \mathbf{X}_i \leq \hat{\pi}_i, \tag{6.5.28}$$

no improvement in the objective function can be obtained by incorporating the new proposal into the basis. Our last allocation of resources was satisfactory and hence our previous solution was optimal.

The complete cycle

We have now assembled all the pieces needed for a complete cycle of the decomposition procedure. Let us summarize the procedure to be followed.

1. A feasible solution to the divisional and corporate constraints is determined.
2. The dual of the executive program [Equations (6.5.19) and (6.5.20)] is solved for a set of simplex multipliers π_k and a set of $\hat{\pi}_i$.
3. The simplex multipliers are used to establish revised objective functions for each of the divisions [Equation (6.5.25)], and new proposals are obtained by maximizing these subject to Equations (6.5.6) and (6.5.7).
4. The marginal profitability of the proposals is compared [Equation (6.5.27)] with the $\hat{\pi}_i$ from the previous solution of the executive program. If the optimality conditions [Equation (6.5.28)] are met, the solution has been found; if not, we proceed to step 5.
5. The current and previous proposals are considered through the dual of the executive program. Equation (6.5.19) is minimized subject to Equation set (6.5.20) to obtain a revised set of simplex multipliers, π_k.
6. We return to step 3 and continue the cycle until the optimality condition [Equation (6.5.28)] is met.
7. When the condition for optimality is met, we solve the executive program primal [Equations (6.5.11)–(6.5.14)] for the μ_{ij}. The final values of the $\mathbf{X}_i$ are then obtained from (6.5.8).

It should be observed that each time a new set of divisional proposals is supplied, the number of weights μ_{ij} in the executive program must be increased correspondingly. However, we need not solve for the values of μ_{ij} until the optimum has been reached, as we are not concerned with the intermediate values of μ_{ij}. We only require the values of the π_k and $\hat{\pi}_i$. The increase in the number of μ_{ij} in the executive program will be seen as an increase in the number of constraints in the dual. When the optimal solution has been found, the μ_{ij} are used to instruct the various divisions as to the production policy they should follow.

EXAMPLE 6.5.1

As a simple illustration of the decomposition procedure, let us consider an example similar to one originated by Baumol and Fabian [1964]. Suppose that a corporation which is composed of two manufacturing divisions wishes to maximize its profits. Division 1 can produce two commodities whose outputs we shall represent by x_1 and x_2. Division 2 can produce three commodities whose outputs will be represented by x_3, x_4, and x_5. Both divisions make use of one corporate resource (viz., capital) whose supply is limited to b_1. In addition, the output from division 1 is limited by the capacities, b_2 and b_3, of two of its own resources. We shall also assume that division 2 is limited by the capacity b_4 of one of its resources.

The state of affairs we have just described can be represented by the following problem: Maximize

$$y = c_1x_1 + c_2x_2 + c_3x_3 + c_4x_4 + c_5x_5, \tag{6.5.29}$$

subject to the corporate constraint

$$a_{11}x_1 + a_{12}x_2 + a_{13}x_3 + a_{14}x_4 + a_{15}x_5 \leq b_1, \tag{6.5.30}$$

division 1 constraints

$$a_{21}x_1 + a_{22}x_2 \leq b_2, \tag{6.5.31}$$

$$a_{31}x_1 + a_{32}x_2 \leq b_3, \tag{6.5.32}$$

division 2 constraint

$$a_{43}x_3 + a_{44}x_4 + a_{45}x_5 \leq b_4, \tag{6.5.33}$$

and

$$x_i \geq 0 \qquad \text{for } i = 1, 2, \ldots, 5. \tag{6.5.34}$$

We see that the constraint matrix has the block-angular form required for application of the decomposition algorithm. We shall assume that the specific objective function to be optimized is given by: Maximize

$$y = x_1 + 2x_2 + 3x_3 + x_4 + 2x_5, \tag{6.5.35}$$

and the corporate constraint is

$$4x_1 + x_2 + 2x_3 - 5x_4 + x_5 \leq 15. \tag{6.5.36}$$

Let us also assume that we have an initial feasible solution given by

$$\mathbf{X}_{11} = \begin{bmatrix} 7 \\ 4 \end{bmatrix}, \qquad \mathbf{X}_{21} = \begin{bmatrix} 0 \\ 12 \\ 0 \end{bmatrix}.$$

Since the executive program is written in terms of the weights μ_{ij}, we first express each of the original variables in terms of the μ_{i1} and the divisional solutions. That is, we let

$$\begin{aligned} x_1 &= 7(\mu_{11}), \qquad x_2 = 4(\mu_{11}), \\ x_3 &= 0(\mu_{21}), \qquad x_4 = 12(\mu_{21}), \qquad x_5 = 0(\mu_{21}). \end{aligned} \tag{6.5.37}$$

The executive program is then: Maximize

$$\begin{aligned} y &= 7\mu_{11} + 2(4\mu_{11}) + 3(0) + (12\mu_{21}) + 2(0) \\ &= 15\mu_{11} + 12\mu_{21}, \end{aligned} \tag{6.5.38}$$

subject to

$$\begin{aligned} 4(7\mu_{11}) + (4\mu_{11}) + 2(0) - 5(12\mu_{21}) + (0) \\ = 32\mu_{11} - 60\mu_{21} \leq 15, \end{aligned} \tag{6.5.39}$$

$$\mu_{11} = 1, \tag{6.5.40}$$

$$\mu_{21} = 1. \tag{6.5.41}$$

Although the values of μ_{11} and μ_{21} are known, we need the values of $\hat{\pi}_1$ and $\hat{\pi}_2$ in order to evaluate the next proposals. We shall therefore solve the dual,* which we write as: Minimize

$$15\pi_1 + \hat{\pi}_1 + \hat{\pi}_2, \tag{6.5.42}$$

subject to

$$32\pi_1 + \hat{\pi}_1 + 0 \geq 15, \tag{6.5.43}$$

$$-60\pi_1 - 0 + \hat{\pi}_2 \geq 12. \tag{6.5.44}$$

Since $\hat{\pi}_1$ and $\hat{\pi}_2$ originate from equality constraints, their signs are unrestricted. We shall therefore need to replace each of the $\hat{\pi}_i$ in terms of the difference of two quantities. We then have: Minimize

$$15\pi_1 + (w_2 - w_3) + (w_4 - w_5), \tag{6.5.45}$$

* When slack vectors are present in each constraint of the primal program, the primal program solution contains the dual variables in the $(y_j - c_j)$ positions under the slack vector columns. However, when some of the constraints are equalities, as in this case, all the dual variables do not appear directly in the simplex solution of the primal.

subject to

$$\text{(6.5.46)} \qquad 32\pi_1 + (w_2 - w_3) \qquad\qquad - w_6 \qquad = 15,$$

$$\text{(6.5.47)} \qquad -60\pi_1 \qquad + (w_4 - w_5) \qquad - w_7 = 12,$$

where w_6 and w_7 are surplus variables. Our initial simplex tableau is given by Table 6.10. We observe that our optimality criterion is met in the initial tableau. We then have from the values in the $\mathbf{P}_0$ column

$$\text{(6.5.48)} \qquad \hat{\pi}_1 = w_2 - w_3 = 15 - 0 = 15,$$

$$\text{(6.4.49)} \qquad \hat{\pi}_2 = w_4 - w_5 = 12 - 0 = 12,$$

$$\text{(6.5.50)} \qquad \pi_1 = 0.$$

The zero value for π_1 is to be expected since the initial solution is feasible. No penalty can therefore be imposed for violating corporate feasibility. However, we now return to the divisions and resolve the divisional problems without the necessity of meeting the corporate constraint. We may expect that these new solutions will provide an improved objective function.

Table 6.10. SIMPLEX TABLEAU FOR EXECUTIVE PROGRAM FOLLOWING INITIAL FEASIBLE SOLUTION

	c		15	1	−1	1	−1	0	0
Basis		$\mathbf{P}_0$	$\mathbf{P}_1$	$\mathbf{P}_2$	$\mathbf{P}_3$	$\mathbf{P}_4$	$\mathbf{P}_5$	$\mathbf{P}_6$	$\mathbf{P}_7$
$\mathbf{P}_2$	1	15	32	1	−1	0	0	−1	0
$\mathbf{P}_4$	1	12	−60	0	0	1	−1	0	−1
$(y_j - c_j)$		27	−43	0	0	0	0	−1	0

Let us assume that the new divisional proposals are

$$\mathbf{X}_{12} = \begin{bmatrix} 6 \\ 5 \end{bmatrix}, \qquad \mathbf{X}_{22} = \begin{bmatrix} 0 \\ 0 \\ 9 \end{bmatrix}.$$

We first consider our optimality criteria and compare the $(\mathbf{C}_i - \mathbf{R}_i)'\mathbf{X}_i$ with the $\hat{\pi}_i$ obtained previously. We have

$$\text{(6.5.51)} \qquad (\mathbf{C}_1 - \mathbf{R}_1)'\mathbf{X}_{12} = x_1 + 2x_2 - \pi_1[4x_1 + x_2] = 16,$$

$$\text{(6.5.52)} \qquad (\mathbf{C}_2 - \mathbf{R}_2)'\mathbf{X}_{22} = 3x_3 + x_4 + 2x_5 - \pi_1[2x_3 - 5x_4 + x_5] = 18.$$

We see that $(\mathbf{C}_1 - \mathbf{R}_1)'\mathbf{X}_{12} > \hat{\pi}_1$ and $(\mathbf{C}_2 - \mathbf{R}_2)'\mathbf{X}_{22} > \hat{\pi}_2$, and therefore we are not yet at the optimum. We must resolve the executive program.

Again we express the original variables in terms of the μ_{ij} to obtain

$$x_1 = 7\mu_{11} + 6\mu_{12}, \tag{6.5.53}$$

$$x_2 = 4\mu_{11} + 5\mu_{12}, \tag{6.5.54}$$

$$x_3 = 0\mu_{21} + 0\mu_{22}, \tag{6.5.55}$$

$$x_4 = 12\mu_{21} + 0\mu_{22}, \tag{6.5.56}$$

$$x_5 = 0\mu_{21} + 9\mu_{22}. \tag{6.5.57}$$

Our executive program is now: Maximize

$$\begin{aligned} y &= (7\mu_{11} + 6\mu_{12}) + 2(4\mu_{11} + 5\mu_{12}) + 3(0) + (12\mu_{21}) + 2(9\mu_{22}) \\ &= 15\mu_{11} + 16\mu_{12} + 12\mu_{21} + 18\mu_{22} \end{aligned} \tag{6.5.58}$$

subject to

$$\begin{aligned} &4(7\mu_{11} + 6\mu_{12}) + (4\mu_{11} + 5\mu_{12}) + 2(0) - 5(12\mu_{21}) + (9\mu_{22}) \\ &= 32\mu_{11} + 29\mu_{12} - 60\mu_{21} + 9\mu_{22} \leq 15 \end{aligned} \tag{6.5.59}$$

$$\mu_{11} + \mu_{12} = 1 \tag{6.5.60}$$

$$\mu_{21} + \mu_{22} = 1 \tag{6.5.61}$$

$$\mu_{ij} \geq 0 \quad j = 1, 2 \quad i = 1, 2. \tag{6.5.62}$$

Our dual program becomes

$$32\pi_1 + \hat{\pi}_1 \geq 15, \tag{6.5.63}$$

$$29\pi_1 + \hat{\pi}_1 \geq 16, \tag{6.5.64}$$

$$-60\pi_1 + \hat{\pi}_2 \geq 12, \tag{6.5.65}$$

$$9\pi_1 + \hat{\pi}_2 \geq 18. \tag{6.5.66}$$

Solution of the dual gives us $\pi_1 = \frac{2}{23}$, and we may now determine y_1, the revised objective function for division 1. From (6.5.26) we have

$$\begin{aligned} y_1 = (\mathbf{C}_1 - \mathbf{R}_1)'\mathbf{X}_1 &= c_1x_1 + c_2x_2 - \pi_1a_{11}x_1 - \pi_1a_{12}x_2 \\ &= x_1 + 2x_2 - \tfrac{2}{23}[4x_1 + x_2]. \end{aligned} \tag{6.5.67}$$

Similarly, for division 2 the revised objective function is

$$\begin{aligned} y_2 = (\mathbf{C}_2 - \mathbf{R}_2)'\mathbf{X}_2 &= c_3x_3 + c_4x_4 + c_5x_5 - \pi_1a_{13}x_3 - \pi_1a_{14}x_4 - \pi_1a_{15}x_5 \\ &= 3x_3 + x_4 + 2x_5 - \tfrac{2}{23}[2x_3 - 5x_4 + x_5]. \end{aligned} \tag{6.5.68}$$

We would now return to the divisional programs, using Equations (6.5.67) and (6.5.68) as the respective objective functions, to obtain new proposals. The cycle would continue until the new proposals can provide no further improvement in the corporation objective function. Note that each time we return to the executive program an additional μ_{1j} and μ_{2j} must be added to the executive-program variables.

Although we have used the simplex method for solution of our problem, a solution of a large-size problem on a digital computer would most likely be accomplished by means of the revised simplex procedure. The reader will recall (cf. Section 4.7) that when the revised simplex method is used, we do not compute the values of the variables at each iteration since the calculations are carried out using $\mathbf{B}^{-1}$, the inverse of the basis matrix ($\mathbf{B}$ is the matrix obtained from those columns in the $\mathbf{A}$ matrix that correspond to vectors in the solution). Thus $\mathbf{B}^{-1}$ is known at each step of the procedure. It may be shown that the dual variable vector $\mathbf{W}$ is obtained from

$$\mathbf{W} = \mathbf{C}'_B \mathbf{B}^{-1} \tag{6.5.69}$$

where $\mathbf{C}_B$ is the cost vector containing only those cost coefficients corresponding to the variables in the basis $\mathbf{B}$. If we wished, we could then solve the primal of the executive program directly with the revised simplex method. Since the π_k ahd $\hat{\pi}_i$ are the dual variables, we could then obtain them from Equation (6.5.69).

The initial solution

In the example just considered it was assumed that initial solutions for the subproblems and the executive problem were available. Real life is seldom as obliging as a textbook example. We observed earlier that this initial solution must be feasible, i.e., must meet the executive-program constraints. The necessity for this requirement may be made apparent by referring to the initial executive program of Example 6.5.1. We see that there is only one set of weights, and therefore $\mu_{i1} = 1$ for all i [Equations (6.5.40) and (6.5.41)]. The executive-program solution is thus identical to the first proposal furnished to it. Since the executive-program solution must meet the executive-program constraints, the initial proposal must also meet these constraints.

In some cases an obviously feasible solution can be provided by means of the slack variables. It is quite possible that in many problems slack variables will appear in only some constraints. In that case, artificial variables must be added and Phase I of the simplex or revised simplex method applied. The Phase I procedure is first applied to each subproblem to determine a feasible solution for each subproblem. The Phase I procedure can then be applied to the entire problem in conjunction with the decomposition algorithm to obtain an initial feasible solution.

We may find that one of the subproblems has no feasible solution. If that is the case, the problem as a whole has no feasible solution. Another possibility is that one or more of the subproblems is unbounded. Earlier we based the decomposition method on the assumption that all subproblems would be bounded. Orchard-Hays [1968] suggests that in the event of an unbounded ith subproblem an artificial bound be applied. This can be done by the addition of a row

$$\sum_{q_i} x_q \leq G \tag{6.5.70}$$

where G is an arbitrarily large positive number.

6.6 Other Applications of the Decomposition Principle

As we have just seen, a large linear programming problem having a coefficient matrix that can be put in block-angular form can be broken into more manageable subproblems by means of the decomposition principle. We found the decomposition principle particularly useful in the economic analysis of a multidivision enterprise. Since the subproblems corresponded to the optimization of the operation of the several divisions, we could find an economic interpretation for the relationship between the executive program and the subproblem.

Many other problems of a similar nature can also be readily handled by this technique. For example, consider a large multidivision petroleum company. Each division must ship crude and refined products between a number of locations. Each division can determine its best shipment procedure by solving a transportation problem. However, if some overall corporation constraint on the availability of the transport (e.g., tankers) is imposed, the divisional plans may not be feasible. We can readily apply the decomposition principle here by imposing the transport restriction through the executive program. Each division will now solve a series of transportation problems with differing cost matrices.

Other matrix forms

The transportation problem has a coefficient matrix that would fit within our description of a block-angular matrix. However, the decomposition principle can be applied in some cases in which this restriction is not met. We can apply the decomposition principle to *any* matrix, provided we are content to split the matrix into two parts. Consider, for example, the coefficient matrix shown schematically in Figure 6.13. Even though the entire matrix contains nonzero elements, we can still partition it horizontally into a linking matrix **L** and a single subproblem **A**. Since the linking matrix may

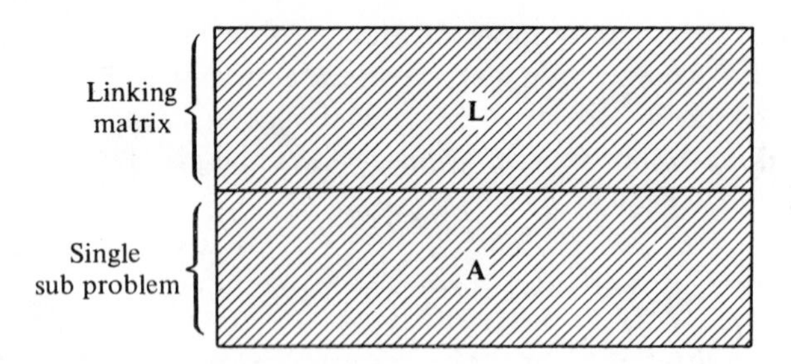

Figure 6.13. Horizontal Partitioning of Matrix for Use with Decomposition Algorithm

contain all the variables of the subproblem, we violate no rule by this partition. The position at which we partition the matrix is entirely arbitrary, and thus **L** need not include more than one constraint.

If the **A** matrix obtained by the horizontal splitting is too large, we may again partition it into two pieces. The partitioning may be continued until the subproblems are of manageable size. It should be observed, however, that unless the submatrices formed by partitioning have special structures which make the solution easy, the decomposed problem will require more calculations than the original problem. Horizontal partitioning is therefore normally used only when it is impossible to handle the original single problem.

As another example of a matrix type to which the decomposition principle can be advantageously applied, let us consider the "bi-angular" matrix of Figure 6.14. We first partition the problem into a linking matrix **L** and a single subproblem **A**. We can then rewrite the subproblem in terms of the dual. The dual, however, will have a block-angular form that can be subsequently partitioned into a set of p subproblems.

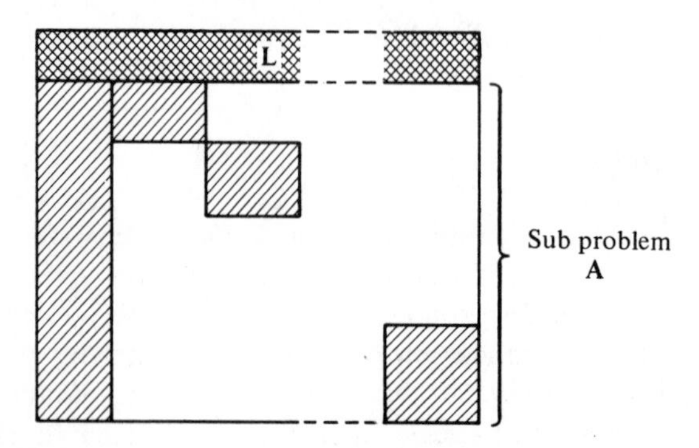

Figure 6.14. Biangular Matrix (Shaded Areas Represent Non-Zero Elements)

We may find some matrices that are very nearly in block-angular form except for a few variables which appear in two subproblems. Suppose that x_j appears in both $\mathbf{A}_1$ and $\mathbf{A}_2$. We could proceed by leaving x_j in $\mathbf{A}_1$, but wherever it appears in $\mathbf{A}_2$ we replace it by the new variable x_{n+1}. We then add to the executive program the constraint

$$x_j - x_{n+1} = 0. \tag{6.6.1}$$

We shall have thus eliminated x_j from all but one subproblem and transformed the matrix into block-angular form.

Nonlinear problems

The decomposition principle may actually be applied to a still broader class of problems than we have heretofore indicated. Suppose that we have the problem: Maximize

$$y = (s_1x_1 + s_2x_2 + \cdots + s_qx_q) + (s_{q+1}x_{q+1} + s_{q+2}x_{q+2} + \cdots + s_nx_n), \tag{6.6.2}$$

subject to the linear overall constraints

$$a_{j1}x_1 + a_{j2}x_2 + \cdots + a_{jq}x_q + a_{j(q+1)}x_{q+1} + \cdots + a_{jn}x_n \leq b_j, \quad j = 1, 2, \ldots, m, \tag{6.6.3}$$

and to the convex (nonlinear) constraints

$$f_j(x_1, x_2, \ldots, x_q) \leq b_j \qquad j = m + 1, \ldots, m + r, \tag{6.6.4}$$

$$f_j(x_{q+1}, x_{q+2}, \ldots, x_n) \leq b_j \qquad j = m + r + 1, \ldots, m + r + t. \tag{6.6.5}$$

This problem is very much like the ones previously considered. However, we have now introduced nonlinearities into the constraints of the subproblems. The overall constraints and the objective function are still linear.

The introduction of nonlinearities into our subproblem constraints really makes no difference to the decomposition procedure. All the special decomposition calculations only involve the objective function and overall constraints, and these are still linear. If the divisional constraints yield bounded convex feasible regions, then they can provide optimal solutions that can be averaged by the executive program. Our decomposition method works in the same manner as in the linear case, except that we must use an appropriate nonlinear programming technique to solve the subproblems.

We observed that we can extend the range of applicability of the decomposition procedure by using the technique of defining new variables when the constraint matrix is not quite in block-angular form [cf. Equation (6.6.1)]. We may also use this technique when some of the terms of the objective function are nonlinear functions, [$f(\mathbf{X})$]. For a maximization problem, we simply replace the nonlinear term in the objective function by x_{n+1} and add the nonlinear constraint

$$x_{n+1} - f(\mathbf{X}) \leq 0 \tag{6.6.6}$$

to the ith subproblem. We should note that in order for this technique to yield a convex constraint set the nonlinear term in the objective function must be concave (for a maximization problem).

Application to separable programming

The decomposition principle is also significant if we attempt to solve a nonlinear problem by separable programming. Although we need not apply the decomposition algorithm directly, we can make use of the basic principle as described below.

In Chapter 5 we applied separable programming when our objective function and constraints were nonlinear separable functions of the independent variables. That is, when we could state our problem as: Maximize (Minimize)

$$y = \sum_{j=1}^{n} f_j(x_j), \tag{6.6.7}$$

subject to

$$\sum_{j=1}^{n} g_{ij}(x_j)\begin{Bmatrix} > \\ = \\ < \end{Bmatrix} b_i, \qquad i = 1, 2, \ldots, m. \tag{6.6.8}$$

We linearized the problem by representing each nonlinear function as a series of linear functions between grid points. Our linearized problem was written in terms of the new linear variables α_{lj}, $0 \leq \alpha_{lj} \leq 1$, which represented the weights applied to the various grid points. We obtained: Maximize (Minimize)

$$y = \sum_{j=1}^{n} \sum_{l=1}^{s_j} \alpha_{lj} k_{lj}, \tag{6.6.9}$$

subject to

$$\sum_{j=1}^{n} \sum_{l=1}^{s_j} \alpha_{lj} k_{lj}\begin{Bmatrix} > \\ = \\ < \end{Bmatrix} b_j, \qquad i = 1, 2, \ldots, m, \tag{6.6.10}$$

$$\sum_{l=1}^{s_j} \alpha_{lj} = 1, \qquad j = 1, 2, \ldots, n, \tag{6.6.11}$$

and the usual nonnegativity requirement. We also required that for any j no more than two α_{lj} could be positive and that these two be adjacent. However, when the problem is convex, we saw that these latter requirements need not be imposed explicitly.

Often when we provide a fine enough grid for accurate representation of the nonlinear functions, we find that we have introduced a very large number of α_{lj}. Wolfe [1962] has described a decomposition-type procedure that leads to smaller-size linear problems when the original problem is separable and *convex*. We first construct a coarse grid for each of the variables. By selecting only a small number of grid points, we keep our linear problem at a reasonable level. We obtain the optimum primal and dual solution to this problem by the usual methods of separable programming. Let us designate the dual solution as $\hat{\pi}, \pi_1, \pi_2, \ldots, \pi_m$, where $\hat{\pi}$ corresponds to the constraint of Equation (6.6.11) and π_1 through π_m correspond to the constraints of Equation (6.6.10). We shall use this information to determine what additional grid point should be added to improve the accuracy of our approximating solution.

Let us think of our linearized problem of Equations (6.6.9)–(6.6.11) as the executive program in the decomposition algorithm. The grid points selected for initial evaluation of the separable functions are analogous to a series of proposals from a subproblem. We must establish a subproblem that will generate an improved proposal (additional grid point).

We may consider our original subproblem to be described by Equations (6.6.7) and (6.6.8). We then obtain our subproblem objective function from Equation (6.5.26). For a maximization problem, the objective function is: Maximize

$$y = \mathbf{C}'\mathbf{X} - \mathbf{R}'\mathbf{X}. \tag{6.6.12}$$

We rewrite this for our present case by observing that the $\mathbf{C}'\mathbf{X}$ corresponds to $\sum_{j=1}^{n} f_j(x_j)$, whereas $\mathbf{R}'\mathbf{X}$ corresponds to $\sum_{i=1}^{m} \sum_{j=1}^{n} \pi_i g_{ij}(x_j)$. Therefore, for a maximization problem, we have: Maximize

$$y = \sum_{j=1}^{n} f_j(x_j) - \sum_{i=1}^{m} \sum_{j=1}^{n} \pi_i g_{ij}(x_j). \tag{6.6.13}$$

Our revised subproblem need have no constraints, since our constraints are contained within the executive program. We therefore determine the value $\mathbf{X}$ that maximizes (6.6.13) by one of the unconstrained optimization procedures of Chapter 3. The value of $\mathbf{X}$ so obtained is the new proposal to the executive program. We then add to the executive program problem [Equations (6.6.9)–(6.6.11)] a new set of $\alpha_{lj} k_{lj}$ corresponding to this new grid point. We resolve the augmented linear problem and continue the cycle.

We have really transferred the work of solving the original problem from our linearized executive problem to our subproblem. Whether this is useful depends in large measure on the nature of the problem. We have produced an unconstrained minimization problem, which is easier to solve than our original problem. However, we could have accomplished the same thing by use of a penalty-function technique. If the problem format is such that the unconstrained minimization obtained by decomposition is particularly easy to solve, then the decomposition procedure may be worth considering.

EXAMPLE 6.6.1

As an illustration of the application of the decomposition method to non-block-angular matrices with nonlinear terms, consider a problem similar to Example 6.5.1. Let our problem be: Maximize

$$y = x_1 + x_2 - 3x_3^2 + x_4 + 2x_5, \tag{6.6.14}$$

subject to

$$4x_1 + x_2 + 2x_3 - 5x_4 + x_5 \leq 15, \tag{6.6.15}$$

$$\begin{cases} 3x_1 + x_2^2 \leq 30, & (6.6.16) \\ x_1^2 + 2x_2 \leq 60, & (6.6.17) \end{cases}$$

$$x_2 + x_3^4 + 2x_4 + x_5^2 \leq 55. \tag{6.6.18}$$

Our problem differs from Example 6.5.1 in that (a) the constraint matrix is not in block-angular form, (b) there is a nonlinear term in the objective function, and (c) there are nonlinear terms in the divisional constraints. The last difference, item (c), causes no difficulty since the divisional constraints are convex. However, items (a) and (b) must be remedied before proceeding.

We define a new variable x_6 such that

$$x_6 - x_2 = 0, \tag{6.6.19}$$

and rewrite Equation (6.6.18) as

$$x_6 + x_3^4 + 2x_4 + x_5^2 \leq 55, \tag{6.6.20}$$

and use Equation (6.6.19) as an additional corporate constraint. The constraint matrix is now in block-angular form.

We may remove the nonlinear term in the objective function by rewriting (6.6.14) as: Maximize

$$y = x_1 + x_2 - x_7 + x_4 + 2x_5 \tag{6.6.21}$$

and adding the constraint

$$-x_7 + 3x_3^2 \leq 0 \tag{6.6.22}$$

to the division-2 constraint set. If the objective function had been $x_1 + x_2 + 3x_3^2 + x_4 + 2x_5$, this procedure would not have been satisfactory, since the additional constraint obtained instead of (6.6.22) would not have been convex.

With the indicated revisions, our problem can now be solved by the method of decomposition in conjunction with one of the nonlinear programming procedures of Chapter 5.

6.7 Conclusion

Our ability to describe complex systems in terms of mixed- or all-integer programming problems has perhaps outstripped our ability to solve such problems conveniently. Attempts to develop improved algorithms furnish the subject matter of much recent literature. For example, Echols and Cooper [1968] have used direct search to solve all-integer linear-programming problems. The search was constrained to take only integer steps, and a heuristic procedure was provided to deal with other constraints. There is no guarantee that the solution obtained will be a global optimum. Algorithms of other types have been presented by Benders [1962], Driebeek [1966], and Raghavachari [1969], among others. Geoffrion and Marsten [1972] summarize much of this work.

In Chapter 8 we shall examine dynamic programming, and we shall see that dynamic programming is applicable to integer programming problems. However, we shall also find that dynamic programming is applicable only to problems which can be formulated in a particular manner. When it is applicable, its use should be considered.

Decomposition procedures are of somewhat less current interest than integer programming. The need for a decomposition procedure first arose in linear programming. The phenomenal success of the linear programming algorithms led to their application to systems of increasing size, and the problem sizes soon outstripped available computer capacities. This difficulty was admirably circumvented by the decomposition principle. The situation is now changed. Computer capacities have increased greatly so that operating linear programming systems can handle problems with more than 10,000 variables and thousands of constraints. If a problem can be run as a single unit, the computation time is nearly always less than if it is decomposed. Therefore, it is only in exceptional cases that the decomposition principle need be applied to purely linear problems if a sufficiently large computer is available to accommodate the entire problem.

Once we enter the realm of nonlinear programming, we are not so fortunate. Presently available nonlinear programming algorithms appear to work best with problems of moderate dimensionality. Segmenting large nonlinear problems may, therefore, often be desirable. When the problem structure is such that the method of decomposition is not applicable, such techniques as partition programming (Rosen, cf. [1963]) and multi-level programming (Lasdon and Schoeffler [1964]) should be considered. For a complete coverage of large scale systems techniques, the reader is referred to Lasdon [1970].

REFERENCES

Balas, E., "An Additive Algorithm for Solving Linear Programs with 0–1 Variables," *Operations Research, 13* (1965), 517.

Benders, J. F., "Partitioning Procedures for Solving Mixed-Variables Programming Problems," *Numerishe Mathematik, 4* (1962), 238.

Baumol, W. J. and T. Fabian, "Decomposition, Pricing for Decentralization and External Economies," *Management Science, 11* (1964), 1.

Dakin, R. J., "A Tree-Search Algorithm for Mixed-Integer Programming Problems," *Computer Journal, 8* (1965), 250.

Dantzig, G. B., "Application of the Simplex Method to a Transportation Problem," Chapter 23 of "Activity Analyses of Production and Allocation," T. Koopmans, ed., John Wiley & Sons, Inc., N. Y. (1951).

Dantzig, G. B. and P. Wolfe, "Decomposition Principle for Linear Programs," *Operations Research, 8* (1960), 101.

Dreyfus, S. E., "An Appraisal of Some Shortest-Path Algorithms," *Operations Research, 17,* (1969) 395.

Driebeek, N. J., "An Algorithm for Solution of Mixed Integer Programming Problems," *Management Science, 12* (1966), 576.

Echols, R. E. and L. Cooper, "Solution of Integer Linear Programming Problems by Direct Search", *J. Assoc. for Computing Machinery, 15,* No. 1 (1968), 75.

Geoffrion, A. M. and R. E. Marsten, "Integer Programming Algorithms: A Framework and State of the Art Survey," *Management Science, 18* (1972), 7.

Gomory, R. E., "Outline of an Algorithm for Integer Solutions to Linear Programs)," *Bulletin of the American Mathematical Society, 64* (1958), 275.

Gomory, R. E., "An Algorithm for the Mixed-Integer Problem," Report P-1885, The Rand Corp. (1960).

Hadley, G., "Nonlinear and Dynamic Programming," Addison-Wesley Publishing Co., Reading, Mass. Chapter 8, p. 276 (1964).

Land, A. H. and A. Doig, "An Automatic Method of Solving Discrete Programming Problems," *Econometrica, 28* (1960), 497.

Lasdon, L. S., "Optimization Theory for Large Systems," Macmillan, New York, 1970.

Lasdon, L. S. and J. D. Schoeffler, "Decentralized Plant Control," *Proceedings of the Instrument Society of America*, Vol. 19, Part 3: Advances in Instrumentation, Paper 3.3–3.64 (1964).

Lemke, C. E., "The Dual Method of Solving the Linear Programming Problem," *Naval Research Logistics Quarterly 1* (1954), 1.

Orchard-Hays, W., *Advanced Linear Programming Computing Techniques*, p. 273. New York: McGraw-Hill Book Co., 1968.

Raghavachari, M., "On Connections Between Zero-One Integer Programming and Concave Programming Under Linear Constraints," *Operations Research, 17* (1969), 680.

Rosen, J. B., "Convex Partition Programming," in *Recent Advances in Mathematical Programming*, p. 159, R. L. Graves and P. Wolfe, eds., McGraw-Hill Book Co., New York (1963).

Wolfe, P., "Recent Developments in Nonlinear Programming", in *Advances in Computers*, Vol. 3, p. 155, F. L. Alt and M. Rubinoff, eds. New York: Academic Press (1962).

PROBLEMS

6.1. Construct a linear programming problem that has an optimum feasible solution, providing all variables are continuous, but has no feasible solution when all variables must be integers.

6.2. Determine the necessary conditions for $y(x)$ to have a local minimum at x_0 when x is defined only over the set of positive integers.

6.3. A mining corporation has six abandoned mines and is interested in reopening them. The capital costs c_i of reopening each mine, as well as the annual profits p_i from each mine, have been estimated by the corporation's engineers. The corporation has only T dollars available for capital investment, and this sum is inadequate to reopen all the mines. Set up an all-integer programming problem that would allow the corporation to decide which mines should be reopened.

6.4. A manufacturer estimates that the increase in annual profit, Δp, gained by increasing his annual capacity, is given by

$$\Delta p = k(s)^{0.7},$$

where k is a constant and s is the increase in capacity. One way he can increase his capacity is by adding new machinery. The production process requires operations on machines of type A and type B. Each machine type is available in only one size. Type A machines cost c_1 dollars each, and each is capable of producing r_1 units of product annually. Type B machines cost c_2 dollars per machine and each machine is capable of processing r_2 units annually. Corporation policies require that the annual profit on any investment be no less than 10 per cent. Develop a mixed-integer programming model that would allow the manufacturer to determine how many new machines of each type he should purchase.

6.5. A salesman wants to visit each one of four cities. He wishes to begin and end his tour at his home base, city 1. The distance between cities is known. He is to visit all four cities on his tour and visit no city more than once (except city 1 to which he returns). Derive an integer programming model that would allow him to select the shortest tour. (This problem is a version of the well-known "traveling salesman" problem. Make certain that your formulation rules out disconnected cycles.)

6.6. The construction of a nuclear power plant requires a long sequence of coordinated activities. The major activities required, the events required before they begin, and the lengths of time for each activity are shown in the accompanying table.

Activity	*Activities Which Must Be Completed Before Beginning*	*Length of Time Required (months)*
(a) Selection of plant parameters	none	3
(b) Nuclear and thermal design of core	a	9
(c) Mechanical design of core	b	8
(d) Procurement of enriched fuel	a	13
(e) Fuel fabrication	d, b	14
(f) Primary system design	a	10
(g) Reactor-vessel and steam-generator procurement	a, b, f	14
(h) Turbine procurement	a, b, f	20
(i) Foundation construction	f	6
(j) Primary-system construction	f, i	7
(k) Installation of reactor vessel and steam generator	g, j	3
(l) Construction of vapor container	k	3
(m) Construction of secondary system	k	4
(n) Installation of turbine	h	2
(o) Plant testing	l, m, k, n	1
(p) Fuel loading	o, e	1
(q) Power generation	k	1

By use of the simplex algorithm, determine the minimum time in which a plant can be constructed.

6.7. Solve problem 4.20 by using the transportation algorithm.

6.8. Show that Equation (6.2.17) may be derived from the Lagrangian.

6.9. (a) Use the method of Griffith and Stewart to solve the following transportation problem: Minimize

$$y = 2x_{11} + 3x_{11}^2 + 1.5x_{12} + 4.6x_{12}^2 + 3.2x_{13} + 1.8x_{13}^2$$
$$+ 2.2x_{21}^2 + 3.1x_{22} + 0.9x_{22}^2 + 4x_{23} + x_{23}^2,$$

subject to

$$x_{11} + x_{12} + x_{13} = 2000,$$
$$x_{21} + x_{22} + x_{23} = 2500,$$
$$x_{11} + x_{21} = 1000,$$
$$x_{12} + x_{22} = 1800,$$
$$x_{13} + x_{23} = 1700.$$

(b) Show that solution by the method of Griffith and Stewart of any transportation problem having the form: Minimize

$$y = \sum_i \sum_j c_{ij}x_{ij} + c_{ij}x_{ij}^2,$$

subject to

$$\sum_{j=1}^{l} x_{ij} = b_i, \qquad i = 1, \ldots, m,$$

$$\sum_{i=1}^{m} x_{ij} = b_j, \qquad j = 1, \ldots, l,$$

where the b_i and b_j are integers, will lead to integer values for the x_{ij}.

6.10. (a) Solve the following problem by use of Gomory's cutting-plane procedure: Maximize

$$y = x_1 + 2x_2,$$

subject to

$$2x_1 - x_2 \leq 12,$$
$$-x_1 + x_2 \leq 2.8,$$
$$1.2x_1 + 3x_2 \leq 24,$$
$$x_1 = \text{nonnegative integer},$$
$$x_2 = \text{nonnegative integer}.$$

(b) Resolve (a), restricting only x_1 to be an integer. Again use Gomory's cutting-plane procedure.

6.11. Assume that the mining company of problem 6.3 can invest only $850,000 and is presented with the following data:

Mine No.	*Cost to Reopen* ($)	*Annual Profit* ($)
1	450,000	60,000
2	250,000	35,000
3	400,000	45,000
4	200,000	25,000
5	120,000	20,000
6	225,000	30,000

Determine the allocation of funds to be made by means of (a) the branch-and-bound procedure; (b) Gomory's cutting-plane procedure. Which of these algorithms is most efficient for this problem?

6.12. The manufacturer in problem 6.4 estimates that $k = 200$, $c_1 = 4000$, $c_2 = 2000$, $r_1 = 20$, and $r_2 = 13$. Determine the number of machines of each type to be purchased by means of a branch-and-bound algorithm.

6.13. Lawler and Wood [*Operations Res.*, *14*, 699 (1966)] describe the basic operation of the branch-and-bound procedure as that of replacing an original problem [problem (0)] by a new problem [problem (1)] which is easier to solve and whose optimal solution always bounds the original problem. They state that "If **X** is the optimum solution to problem (1) it is also the optimum solution to problem (0) if: (a) it is a feasible solution to problem (0), and (b) the objective function of problem (1) equals the objective function of problem (0)." Show that the minimization of a convex function subject to convex constraints by an exterior-point penalty function may be considered a branch-and-bound procedure and that the above theorem holds at the optimum.

6.14. (a) Formulate a separable programming problem that will locate the *absolute* minimum of

$$y = 2x_1^3 + 3x_2 + 4x_3 - 5x_1x_2x_3,$$

subject to

$$0 \leq x_i \leq 3, \quad i = 1, 2, 3.$$

(b) Solve (a) assuming that it is sufficient to determine the values of each x_i at the minimum within ± 0.25.

(c) Show how the absolute minimum of the objective function can be obtained by means of geometric programming.

6.15. (a) Indicate how the Land-and-Doig algorithm can be modified to solve the traveling-salesman problem (problem 6.5) by including the prohibition against disconnected cycles within the branch-and-bound procedure. (Do not add inequality constraints to exclude disconnected cycles.)

(b) Determine the tour requiring the minimum distance traversed if the travel distances are as shown.

	Distance Between Cities (d_{ij})			
From City \ *To City*	*1*	*2*	*3*	*4*
1	–	20	35	10
2	20	–	15	8
3	35	15	–	13
4	10	8	13	–

The tour is to begin and end at city 1.

6.16. Solve the investment-portfolio example described in problem 4.23 when we add the following constraints:

(a) Each stock in the portfolio must account for at least 5 per cent of the portfolio.

(b) Each stock in the portfolio must account for no more than 10 per cent of the portfolio.

Use an appropriate integer-programming algorithm.

6.17. Show that any integer-programming problem can be expressed in terms of zero–one variables.

6.18. After an appropriate transformation, decompose the following problem into a linking matrix and two subproblems: Minimize

$$10x_1 + 15x_2 - 9x_3 + x_4 - 20x_5 + 3x_6,$$

subject to

$$\begin{aligned} x_1 + 2x_2 + 1.5x_3 + 2x_4 &\leq 10 \\ 2x_1 + 3.5x_2 + 1.5x_3 + 3x_4 &\leq 18 \\ -x_1 - 2.5x_2 + 2.5x_3 &\leq 7 \\ 1.6x_1 + x_2 + x_5 + 2x_6 &\leq 15 \\ 3x_1 + 2x_2 + 1.5x_5 + x_6 &\leq 25 \\ x_i \geq 0, \quad i = 1, 2, \ldots, 6. \end{aligned}$$

6.19. An oil refiner is required to ship gasoline from refineries A and B to cities 1, 2, 3, and 4. At refinery A he has available 18 tank-carloads of gasoline and at refinery B, 16 tank-car loads. He must supply 4 tank-car loads to city 1, 14 to city 2, 6 to city 3, and 10 to city 4. The shipping costs per tank car are given by

$$\mathbf{C}_A = \begin{bmatrix} 1.5 \\ 3 \\ 3 \\ 2.5 \end{bmatrix}, \qquad \mathbf{C}_B = \begin{bmatrix} 4 \\ 0.5 \\ 1.5 \\ 3 \end{bmatrix}.$$

Most of the cities are linked to the refineries by rail. However, shipments between refinery A and city 3 must be over the road by tank truck, as must shipments between refinery B and city 2. The oil company has only 17 tank trucks available for use in gasoline shipment.

Use the method of decomposition to determine the minimum-cost shipping procedure subject to the overall limitation on tank trucks and the availability and needs for gasoline. Assume that the capacity of a tank truck equals that of a tank car.

6.20. (a) Formulate problem 4.24 in a manner suitable for solution by the method of decomposition.

(b) Consider only the first 6 months of the year and solve this problem by the method of decomposition. You may assume that the inventory at the end of June is the same as at the end of January.

6.21. Solve problem 6.18 by the method of decomposition.

6.22. The designers of a missile guidance system have as their objective the design of a minimum-weight system subject to the requirement that the overall cost of the system shall not exceed $\$5 \times 10^6$. To speed their work, the designers have broken the system into two subsystems. The engineers working on subsystem 1 find the cost of this subsystem to be given by $10^7/w_1^{0.5}$, where w_1 = weight (in hundreds of pounds) of subsystem 1. The designers of subsystem 2 find that the cost of their subsystem is expressed by $(1.5 \times 10^7)/w_2 - 3 \times 10^6 w_2^{0.2}$, where w_2 = weight (in hundreds of pounds) of subsystem 2.

(a) Determine the optimum values of w_1 and w_2, considering the entire problem as a single unit. Use separable programming or any other applicable nonlinear programming technique.

(b) Use the method of decomposition to solve this problem.

6.23. A manufacturer is faced with a seasonable demand for his product as shown in the table.

Season	*Demand*
Winter	6,000
Spring	15,000
Summer	20,000
Autumn	7,000

The manufacturer can produce only 10,000 units per season on a normal production basis. Additional production during a season costs an additional \$4 per unit. The cost of carrying inventory for a season is \$1.50 per unit. The manufacturer estimates his seasonal inventory costs, using the average of the inventories on hand at the beginning and the end of the season.

The manufacturer believes that the demand for his product will be stable for a number of years. He therefore wishes to establish a production policy such that his *equilibrium* costs (costs when system inventory varies during the year but is the same from year to year) are a minimum. Determine the optimum policy by means of the method of decomposition.

OPTIMIZATION OF FUNCTIONALS

In Chapter 2 we discussed the application of classical calculus to the optimization of functions that are continuous and differentiable. We saw that the establishment of necessary and sufficient conditions for an extremum becomes quite involved, particularly when the dimensionality of a problem becomes high or constraints are present. Moreover, we saw that the application of the necessary conditions can introduce severe computational difficulties, even with simple unconstrained functions. Nonetheless, our examination of the classical calculus was illuminating from the standpoint of understanding the nature of the beast. In addition, the background that we obtained was useful in the subsequent development of more efficient computational tools.

In this chapter we shall again turn to the classical approach for essentially the same reasons as in Chapter 2. Now, however, we shall examine a considerably more complex problem—that of extremizing a *function of an unknown function*. If done properly, we shall see that the unknown function can be determined along the way. For example, suppose that we seek some function $u(t)$ which will cause the integral of a known function $\phi\{t, u(t)\}$ to be extremized. This integral, which is a function of the independent variable t and the unknown *dependent* variable $u(t)$, is known as a *functional*. The

7

problem of optimizing a functional is often referred to as the *variational problem*, and the branch of calculus which treats this problem is called the *calculus of variations.*

The variational problem with differential constraints is a particularly important topic in the areas of automatic-control theory and optimal engineering design. Therefore, it is not surprising that this problem has received a great deal of attention. Recently L. S. Pontryagin and other Soviet mathematicians have developed a comprehensive theory, called the *maximum principle*, for solving optimization problems of this type. We shall discuss the maximum principle in this chapter, presenting it first as an outgrowth of the calculus of variations, and then extending it to a broader class of problems, which the calculus of variations is not equipped to handle.

7.1 Necessary Conditions: One Unknown Function and One Independent Variable

Let us begin our discussion of functional optimization by considering the extremization of an integral involving one unknown function. That is, let

$$y = \text{Max(Min)} \int_{t_0}^{t_f} \phi(u, u', t)\, dt, \tag{7.1.1}$$

where u is an unknown function of t, which we shall assume to be continuous and to have a continuous second derivative with respect to t. We shall refer to the first derivative as simply u'. Let us assert that the boundary conditions are known; i.e., $u(t_0) = u_0$, $u(t_f) = u_f$. We seek the *particular* function $u(t)$ that satisfies the boundary conditions and extremizes the above integral. Such extremizing functions are frequently called *extremals*.

Suppose that we represent *any* continuous and differentiable function $\bar{u}(t)$ as

$$\bar{u}(t) = u(t) + \epsilon\eta(t), \qquad t_0 \le t \le t_f, \tag{7.1.2}$$

where $u(t)$ is the *particular* function that extremizes the integral and satisfies the boundary conditions; i.e., $u(t_0) = u_0$ and $u(t_f) = u_f$; and $\eta(t)$ is an *arbitrary* function with continuous second derivatives and vanishing boundary conditions; i.e., $\eta(t_0) = \eta(t_f) = 0$. The parameter ϵ will represent some small arbitrary quantity. Thus Equation (7.1.2) represents some arbitrary function whose path is close to the optimal path $u(t)$ within the interval $t_0 < t < t_f$. This situation is shown graphically in Figure 7.1.

We can now represent our problem as

$$y = \text{I}(0) = \text{Max(Min)} \{\text{I}(\epsilon)\}, \tag{7.1.3}$$

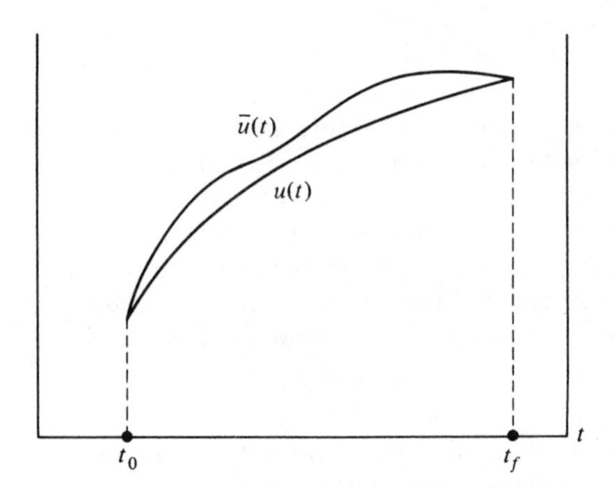

Figure 7.1. Arbitrary Function Satisfying Boundary Conditions and Having Path Close to Optimal Path

where

$$I(\epsilon) = \int_{t_0}^{t_f} \phi(u + \epsilon\eta, u' + \epsilon\eta', t)\, dt \tag{7.1.4}$$

and

$$I(0) = \int_{t_0}^{t_f} \phi(u, u', t)\, dt \tag{7.1.5}$$

and η' is the first derivative of $\eta(t)$. Notice that the left-hand side of Equation (7.1.4) depends only upon ϵ, because introduction of the limits of integration has eliminated any dependence upon t.

Since we wish to extremize $I(\epsilon)$, we have the necessary condition

$$\lim_{\epsilon \to 0} \frac{I(\epsilon) - I(0)}{\epsilon} = 0. \tag{7.1.6}$$

Equation (7.1.6) is sometimes written as

$$\epsilon \left\{ \lim_{\epsilon \to 0} \frac{I(\epsilon) - I(0)}{\epsilon} \right\} = \delta I = 0. \tag{7.1.7}$$

The term δI is called the *first variation* of the integral $I(\epsilon)$. Hence a necessary condition for the extremization of $I(\epsilon)$ is that the first variation vanish.

From Equations (7.1.3) and (7.1.4), we can express (7.1.6) as

$$\lim_{\epsilon \to 0} \int_{t_0}^{t_f} \left[\frac{\phi(u + \epsilon\eta, u' + \epsilon\eta', t) - \phi(u, u', t)}{\epsilon} \right] dt = 0. \tag{7.1.8}$$

Expanding the numerator in a Taylor series about $\epsilon = 0$ (providing ϕ has continuous first partial derivatives), we have

$$\phi(u + \epsilon\eta, u' + \epsilon\eta', t) = \phi(u, u', t) + \left[\frac{\partial \phi}{\partial(u + \epsilon\eta)} \frac{\partial(u + \epsilon\eta)}{\partial \epsilon} \right]_{\epsilon=0} \epsilon + \left[\frac{\partial \phi}{\partial(u' + \epsilon\eta')} \frac{\partial(u' + \epsilon\eta')}{\partial \epsilon} \right]_{\epsilon=0} \epsilon + 0(\epsilon^2) \tag{7.1.9}$$

or

$$\frac{\phi(u + \epsilon\eta, u' + \epsilon\eta', t) - \phi(u, u', t)}{\epsilon} = \frac{\partial \phi}{\partial(u + \epsilon\eta)} \eta \bigg|_{\epsilon=0} + \frac{\partial \phi}{\partial(u' + \epsilon\eta')} \eta' \bigg|_{\epsilon=0} + 0(\epsilon) = \frac{\partial \phi}{\partial u} \eta + \frac{\partial \phi}{\partial u'} \eta' + 0(\epsilon). \tag{7.1.10}$$

Substituting this result into Equation (7.1.8), multiplying by ϵ, and ignoring the higher order terms, we have

$$\delta I = \epsilon \int_{t_0}^{t_f} \left[\frac{\partial \phi}{\partial u} \eta + \frac{\partial \phi}{\partial u'} \eta' \right] dt = 0. \tag{7.1.11}$$

Let us now integrate the second term in the integrand by parts:

$$\int_{t_0}^{t_f} \frac{\partial \phi}{\partial u'} \eta' \, dt = \frac{\partial \phi}{\partial u'} \eta \Big|_{t_0}^{t_f} - \int_{t_0}^{t_f} \eta \frac{d}{dt} \frac{\partial \phi}{\partial u'} \, dt, \tag{7.1.12}$$

providing $(d/dt)(\partial \phi / \partial u')$ exists and is integrable. Recall, however, that $\eta(t_0) = \eta(t_f) = 0$, so that the first term on the right-hand side vanishes.* Hence

$$\int_{t_0}^{t_f} \frac{\partial \phi}{\partial u'} \eta' \, dt = - \int_{t_0}^{t_f} \eta \frac{d}{dt} \frac{\partial \phi}{\partial u'} \, dt. \tag{7.1.13}$$

Therefore, Equation (7.1.11) becomes

$$\delta I = \epsilon \int_{t_0}^{t_f} \left[\frac{\partial \phi}{\partial u} - \frac{d}{dt} \frac{\partial \phi}{\partial u'} \right] \eta \, dt = 0. \tag{7.1.14}$$

Now observe that Equation (7.1.14) is valid for *any arbitrary* function $\eta(t)$ which is subject to the restrictions stated earlier. Hence it must be true that

$$\frac{d}{dt} \frac{\partial \phi}{\partial u'} - \frac{\partial \phi}{\partial u} = 0 \tag{7.1.15}$$

at all points within the interval $t_0 < t < t_f$. Equation (7.1.15) is a second-order, ordinary differential equation which $u(t)$ must satisfy in order that our original integral be extremized.

Equation (7.1.15) is known as the *Euler–Lagrange* equation of variational calculus. Note that this condition is necessary, but not sufficient, to assure the existence of an extremum. It can be shown (cf. Elsgolc [1961]) that a sufficient condition for a minimum (maximum) is that $\partial^2 \phi / \partial (u')^2 \geq 0$ (≤ 0).

Equation (7.1.15) can be expressed in a more convenient form when ϕ does not depend explicitly on t. Under these conditions we note that

$$\frac{d}{dt}\left(u' \frac{\partial \phi}{\partial u'} - \phi \right) = u'' \frac{\partial \phi}{\partial u'} + u' \frac{d}{dt} \frac{\partial \phi}{\partial u'} - u' \frac{\partial \phi}{\partial u} - u'' \frac{\partial \phi}{\partial u'}. \tag{7.1.16}$$

---

* If this condition is not stated at one of the end points, then it is necessary that $\partial \phi / \partial u' = 0$ at that end point.

Simplifying,

$$\frac{d}{dt}\left(u'\frac{\partial\phi}{\partial u'} - \phi\right) = u'\left(\frac{d}{dt}\frac{\partial\phi}{\partial u'} - \frac{\partial\phi}{\partial u}\right) = 0, \tag{7.1.17}$$

in view of Equation (7.1.15). Since the above expression is an exact differential, we can integrate once, yielding

$$u'\frac{\partial\phi}{\partial u'} - \phi = c_1, \tag{7.1.18}$$

where c_1 is a constant of integration.

Equation (7.1.18) is an alternative form of the Euler–Lagrange equation. When it is applicable, its use is more advantageous than Equation (7.1.15), since it is a first-order differential equation.

It should be pointed out that both forms of the Euler–Lagrange equation appear deceptively simple. In practice the solution of either of these differential equations and the evaluation of the constants of integration can only be carried out for a small number of simple problems. We shall see the kinds of difficulties that can arise in the following example.

EXAMPLE 7.1.1 THE BRACHISTOCHRONE

As an example in the use of the Euler–Lagrange equation, let us consider a classical problem in the calculus of variations. Determine the shape of a rigid wire with fixed ends, $u(t_0) = u_0$, $u(t_f) = u_f$, such that a frictionless bead which slides over the wire will fall from u_0 to u_f under the force of gravity in the shortest time, τ.

In this problem let u be measured vertically down, and let $t_0 = 0$, $u(t_0) = 0$, $t_f = 1$, $u(t_f) = 1$. We wish to minimize

$$\mathrm{I} = \int_{\tau_0}^{\tau_f} d\tau = \int_{s_0}^{s_f} \frac{ds}{v}, \tag{7.1.19}$$

where s represents the length of the path and v represents velocity. We can write

$$ds = \sqrt{1 + (u')^2}\, dt \tag{7.1.20}$$

and, from elementary physics,

$$v = \sqrt{2gu}. \tag{7.1.21}$$

Hence our problem is to determine

$$y = \min \frac{1}{\sqrt{2g}} \int_0^1 \sqrt{\frac{1 + (u')^2}{u}}\, dt \tag{7.1.22}$$

with $u(0) = 0$ and $u(1) = 1$.

For simplicity let us choose $g = \frac{1}{2}$, so that the function ϕ becomes simply

$$\phi = \sqrt{\frac{1 + (u')^2}{u}}. \tag{7.1.23}$$

Since t does not appear explicitly in this expression, we can utilize the first-order from of the Euler–Lagrange necessary condition, Equation (7.1.18), which yields

$$(u')^2\Big\{u[1 + (u')^2]\Big\}^{-1/2} - \left[\frac{1 + (u')^2}{u}\right]^{1/2} = c_1. \tag{7.1.24}$$

Rearranging,

$$-\Big\{u[1 + (u')^2]\Big\}^{-1/2} = c_1, \tag{7.1.25}$$

or

$$\frac{du}{dt} = \sqrt{\frac{1}{uc_1^2} - 1}. \tag{7.1.26}$$

Separating variables,

$$dt = \frac{\sqrt{u}\,du}{\sqrt{(1/c_1^2) - u}}. \tag{7.1.27}$$

Equation (7.1.27) can be integrated if we introduce the substitution

$$u = \frac{1}{c_1^2}\sin^2\theta = \frac{1}{2c_1^2}(1 - \cos 2\theta), \tag{7.1.28}$$

which yields

$$t = \frac{1}{c_1^2}\left(\theta - \frac{1}{2}\sin 2\theta\right) + c_2.$$

Since $t = 0$ when $\theta = 0$, we see that $c_2 = 0$, providing $c_1^2 > 0$. Moreover, we can invoke the condition that $u = 1$ when $t = 1$, resulting in the expression

$$2\theta_f - 1 = \sin 2\theta_f - \cos 2\theta_f. \tag{7.1.29}$$

Note that the above equation must be solved numerically for θ_f. When we do so (by means of an iterative numerical procedure), we obtain $\theta_f = 1.206$. Substitution of this result into Equation (7.1.28) yields $c_1^2 = \sin^2(1.206) = 0.873$. Hence the solution that we seek is

$$u = 0.573(1 - \cos\tau), \tag{7.1.30}$$

$$t = 0.573(\tau - \sin\tau), \tag{7.1.31}$$

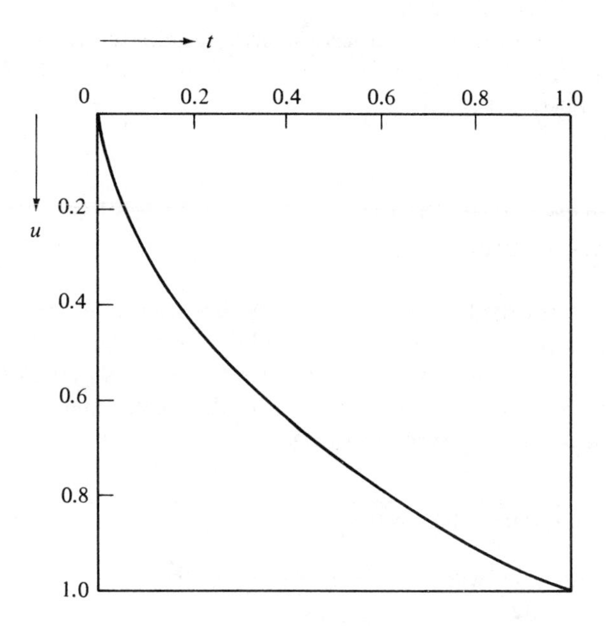

Figure 7.2. The Brachistochrone

and $\tau_f = 2.412$, where we have replaced 2θ by τ. The shape of the curve $u(t)$ is shown in Figure 6.2.

Notice that this solution was not easy to obtain, since we found it necessary to utilize a rather tricky substitution to carry out the integration, and we then had to resort to a numerical procedure to evaluate the constants of integration. All this trouble to solve a relatively simple problem. Difficulties such as these are, unfortunately, characteristic of the calculus of variations. In fact, the Euler-Lagrange equations for many practical problems cannot be integrated in closed form at all. Even if the Euler-Lagrange equations can be integrated, the evaluation of the constants of integration is, as a practical matter, a most difficult task. Thus we see some realistic limitations to the use of the Euler–Lagrange equation in the optimization of simple functionals.

There is one case of some interest in which the Euleur–Lagrange equations do simplify appreciably. If ϕ does not depend on u', but simply on u and t, then Equation (7.1.15) can be written as

$$\frac{\partial \phi}{\partial u} = 0. \tag{7.1.32}$$

Solution of Equation (7.1.32) provides the desired function. Under these conditions we may consider our problem as a succession of ordinary optimization problems that must be solved for every value of t between t_0 and t_f.

It is fortunate when this situation arises, since Equation (7.1.32) will simply be an algebraic equation that can be solved more readily than the differential equation obtained in the general case.

7.2 Necessary Conditions: Problems of Higher Dimensionality

The ideas presented in Section 7.1 can be extended to problems of higher dimensionality with little difficulty. We shall not repeat the details of the derivations here, as they may be found in several textbooks that treat the calculus of variations. The appropriate forms of the Euler–Lagrange equations are presented below, however, for future reference.

Several independent variables

Suppose that we wish to optimize a functional which contains several independent variables $t_1, t_2, \ldots, t_n$; i.e., let

$$y = \text{Max(Min)} \int_{t_{1_0}}^{t_{1_f}} \int_{t_{2_0}}^{t_{2_f}} \cdots \int_{t_{n_0}}^{t_{n_f}} \phi(u, u', t_1, t_2, \ldots, t_n)\, dt_1, dt_2, \ldots, dt_n, \tag{7.2.1}$$

where the boundary conditions are known. If $u(t_1, t_2, \ldots, t_n)$ is a continuously varying function of the independent variables $t_1, t_2, \ldots, t_n$, and if the second partial derivatives of u exist and are continuous, then, by following a procedure similar to that of Section 7.1, we can show that a necessary condition that the integral be extremized is that

$$\frac{\partial}{\partial t_1}\frac{\partial \phi}{\partial u_{t_1}} + \frac{\partial}{\partial t_2}\frac{\partial \phi}{\partial u_{t_2}} + \cdots + \frac{\partial}{\partial t_n}\frac{\partial \phi}{\partial u_{t_n}} - \frac{\partial \phi}{\partial u} = 0, \tag{7.2.2}$$

where u_{t_i} refers to the partial derivative of u with respect to t_i. Equation (7.2.2) is the Euler–Lagrange equation for a problem containing one unknown function and several independent variables. Notice that Equation (7.2.2) is a *partial* differential equation, thus compounding the difficulties involved in integration.

Several unknown functions

Now suppose that we have a functional which depends upon several single-variable unknown functions $u_1(t), u_2(t), \ldots, u_n(t)$. We want to obtain the optimum

$$y = \text{Max(Min)} \int_{t_0}^{t_f} \phi(u_1, u_2, \ldots, u_n, u_1', u_2', \ldots, u_n', t)\, dt. \tag{7.2.3}$$

Let each of the u_i and the u_i' satisfy the continuity and differentiability conditions described in Section 7.1. Furthermore, we shall again assume that the boundary conditions are known. Under these conditions, the analysis of Section 7.1 can be generalized to yield the following set of necessary conditions for an optimum:

$$\begin{aligned} \frac{d}{dt}\frac{\partial \phi}{\partial u_1'} - \frac{\partial \phi}{\partial u_1} &= 0, \\ \frac{d}{dt}\frac{\partial \phi}{\partial u_2'} - \frac{\partial \phi}{\partial u_2} &= 0, \\ &\vdots \\ \frac{d}{dt}\frac{\partial \phi}{\partial u_n'} - \frac{\partial \phi}{\partial u_n} &= 0. \end{aligned} \tag{7.2.4}$$

Notice that we now have a *set* of Euler–Lagrange equations, the number of equations being equal to the number of unknown functions.

EXAMPLE 7.2.1

Let us apply the appropriate form of the Euler–Lagrange equation to extremization of the integral

$$\mathrm{I} = \int_{t_{1_0}}^{t_{1_f}} \int_{t_{2_0}}^{t_{2_f}} \int_{t_{3_0}}^{t_{3_f}} \phi(u, t_1, t_2, t_3)\, dt_1, dt_2, dt_3, \tag{7.2.5}$$

where $\phi(u, t_1, t_2, t_3)$ is expressed as

$$\phi(u, t_1, t_2, t_3) = \left(\frac{\partial u}{\partial t_1}\right)^2 + \left(\frac{\partial u}{\partial t_2}\right)^2 + \left(\frac{\partial u}{\partial t_3}\right)^2 - 2uF(t_1, t_2\, t_3). \tag{7.2.6}$$

We shall assume that $u(t_{10}, t_{20}, t_{30}) = u_0$ and $u(t_{1f}, t_{2f}, t_{3f}) = u_f$ are known, and that $F(t_1, t_2, t_3)$ is a known, piecewise continuous function of (t_1, t_2, t_3).

Taking the partial derivative of ϕ with respect to u_{t_i}, we have

$$\frac{\partial \phi}{\partial u_{t_i}} = 2\frac{\partial u}{\partial t_i}$$

and

$$\frac{\partial}{\partial t_i}\frac{\partial \phi}{\partial u_{t_i}} = 2\frac{\partial^2 u}{\partial t_i^2}. \tag{7.2.7}$$

Substitution into Equation (7.2.2) and division by a factor of 2 thus yields

$$\frac{\partial^2 u}{\partial t_1^2} + \frac{\partial^2 u}{\partial t_2^2} + \frac{\partial^2 u}{\partial t_3^2} + F(t_1, t_2, t_3) = 0. \tag{7.2.8}$$

Many scientists and engineers will recognize this expression as describing the steady-state temperature distribution in a solid with a distributed heat source.

We shall not consider the integration of this partial differential equation, as this is a subject unto itself. It is noteworthy, however, that in many physical situations partial differential equations are considerd as an alternative means of representing a problem which involves optimization of a functional.

EXAMPLE 7.2.2

Find the shortest distance between the points (x_0, y_0, z_0) and (x_f, y_f, z_f). In other words: Minimize

$$S = \int_{(x_0, y_0, z_0)}^{(x_f, y_f, z_f)} \{(dx)^2 + (dy)^2 + (dz)^2\}^{1/2}. \tag{7.2.9}$$

This can be written parametrically as: Minimize

$$S = \int_{t_0}^{t_f} \{1 + (y')^2 + (z')^2\}^{1/2} \, dt. \tag{7.2.10}$$

Computation of the partial derivatives required in Equations (7.2.4) yields

$$\frac{\partial \phi}{\partial y'} = \{1 + (y')^2 + (z')^2\}^{-1/2} y', \qquad \frac{\partial \phi}{\partial y} = 0, \tag{7.2.11}$$

$$\frac{\partial \phi}{\partial z'} = \{1 + (y')^2 + (z')^2\}^{-1/2} z', \qquad \frac{\partial \phi}{\partial z} = 0. \tag{7.2.12}$$

Hence the Euler–Lagrange equations become

$$\frac{d}{dt}\{y'[1 + (y')^2 + (z')^2]^{+1/2}\} = 0, \tag{7.2.13}$$

$$\frac{d}{dt}\{z'[1 + (y')^2 + (z')^2]^{+1/2}\} = 0. \tag{7.2.14}$$

Now the general equation of a plane, $ax + by + cz - d = 0$, will satisfy the above two differential equations. Hence a line can always be found that will satisfy the above expressions and also the boundary conditions. We conclude, therefore, that the shortest distance between two points is a straight line.

7.3 Integral Constraints

As we have seen in earlier chapters, many optimization problems involve auxiliary conditions that must be satisfied by the optimal solution. These constraints restrict the space from which we choose an optimal solution. Hence the constrained optimum may yield a less desirable value than the same problem without the constraints. Furthermore, the presence of one or more constraints introduces additional difficulties in obtaining the optimal solution.

Many problems that require optimization of a functional also involve one or more constraints. As the algebraic functions of the earlier chapters were accompanied by algebraic constraints, so are functional optimization problems frequently accompanied by functional constraints. A particularly common problem of this type involves the optimization of an integral

$$\mathrm{I} = \int_{t_0}^{t_f} \phi(u, u', t)\, dt, \tag{7.3.1}$$

subject to an integral constraint of the form

$$G = \int_{t_0}^{t_f} \psi(u, u', t)\, dt = 0. \tag{7.3.2}$$

Problems of this type are frequently referred to as *isoperimetrics*. As before, $u(t)$ is an unknown optimizing function with known end points, which is assumed to have continuous first and second derivatives with respect to t.

This problem can be handled in a manner quite analogous to constrained algebraic functions through the introduction of a Lagrange multiplier. That is, we form the augmented integral

$$L = \mathrm{I} + \lambda G \tag{7.3.3}$$

$$= \int_{t_0}^{t_f} \phi(u, u'\ t)\, dt + \lambda \int_{t_0}^{t_f} \psi(u, u', t)\, dt \tag{7.3.4}$$

$$= \int_{t_0}^{t_f} \{\phi(u, u', t) + \lambda\psi(u, u', t)\}\, dt, \tag{7.3.5}$$

where λ, the Lagrange multiplier, is an undetermined constant.

Application of the Euler–Lagrange equation to the integrand of the last expression results in

$$\frac{d}{dt}\left(\frac{\partial \phi}{\partial u'} + \lambda \frac{\partial \psi}{\partial u'}\right) - \left(\frac{\partial \phi}{\partial u} + \lambda \frac{\partial \psi}{\partial u}\right) = 0. \tag{7.3.6}$$

This expression, together with Equation (7.3.2), enables solution for both the Lagrange multiplier λ and the extremal $u(t)$.

As in the case of unconstrained functions, the Euler–Lagrange equation can be simplified when t does not appear explicitly in either the ϕ or the ψ function. Under these conditions the Euler–Lagrange equation becomes

$$u' \frac{\partial}{\partial u'}(\phi + \lambda\psi) - (\phi + \lambda\psi) = c_1, \tag{7.3.7}$$

where c_1 is an undetermined constant of integration. We shall not present a derivation of this equation, the logic being essentially identical to the unconstrained case.

EXAMPLE 7.3.1

Determine the shape of a curve $u(t) > 0$, $0 \leq t \leq 1$, whose length S is fixed and whose end points are known; i.e., $u(0) = 0$ and $u(1) = 1$, such that the area beneath this curve is minimized. In other words: Minimize

$$\text{I} = \int_0^1 u\,dt, \tag{7.3.8}$$

subject to

$$\int_0^1 [1 + (u')^2]^{1/2}\,dt = S, \tag{7.3.9}$$

with $u(0) = 0$ and $u(1) = 1$.

The Lagrangian can be written

$$L = \int_0^1 u\,dt + \lambda\left\{\int_0^1 [1 + (u')^2]^{1/2}\,dt - S\right\}, \tag{7.3.10}$$

$$= \int_0^1 \{u + \lambda[1 + (u')^2]^{1/2}\}\,dt - \lambda S. \tag{7.3.11}$$

For simplicity, let $z = L + \lambda S$; i.e.,

$$z = \int_0^1 \{u + \lambda[1 + (u')^2]^{1/2}\}\,dt. \tag{7.3.12}$$

Since the length of the curve, S, is a known constant, optimization of Equation (7.3.12) will yield the same extremizing function as optimization of Equation (7.3.11). The actual minimum area can then be determined either from the expression

$$A_{\text{min}} = z_{\text{min}} - \lambda S \tag{7.3.13}$$

or by direct substitution of the extremal $u(t)$ into Equation (7.3.8).

Proceeding with the mechanics of the solution, we see that t does not appear explicitly in the integrand of Equation (7.3.12). Hence the appropriate form of the Euler–Lagrange equation for this problem is Equation (7.3.7). Application of Equation (7.3.7) to this problem yields

$$u + \lambda[1 + (u')^2]^{1/2} - \lambda(u')^2[1 + (u')^2]^{-1/2} = c_1, \tag{7.3.14}$$

which can be rearranged to give

$$\lambda[1 + (u')^2]^{-1/2} = c_1 - u. \tag{7.3.15}$$

By separating variables, we obtain

$$\int dt = \int \left[\left(\frac{\lambda}{c_1 - u} \right)^2 - 1 \right]^{-1/2} du + c_2, \tag{7.3.16}$$

or

$$\int dt = -\int (u - c_1)[\lambda^2 - (u - c_1)^2]^{-1/2} \, du + c_2. \tag{7.3.17}$$

Integration and subsequent rearrangement yield

$$u = c_1 \pm \sqrt{\lambda^2 - (t - c_2)^2}. \tag{7.3.18}$$

Substitution of Equation (7.3.18) into the two boundary conditions, i.e., $u(0) = 0$ and $u(1) = 1$, results in

$$\begin{aligned} c_1^2 &= \lambda^2 - c_2^2, \\ (c_1 - 1)^2 &= \lambda^2 - (c_2 - 1)^2, \end{aligned} \tag{7.3.19}$$

which yields

$$\begin{aligned} c_1 &= \tfrac{1}{2}\{1 \pm \sqrt{2\lambda^2 - 1}\}, \\ c_2 &= -\tfrac{1}{2}\{1 \mp \sqrt{2\lambda^2 - 1}\}. \end{aligned} \tag{7.3.20}$$

From Equations (7.3.18) and (7.3.20) we see that the problem is solved in terms of the unknown constant λ.

It remains to determine a specific value for λ. This can be done if we make use of the original constraining equation (7.3.9). Let us differentiate Equation (7.3.18) with respect to t and substitute the result into Equation (7.3.9). Integration and rearrangement yield

$$\sin^{-1}\left(-\frac{c_2}{\lambda}\right) - \sin^{-1}\left(\frac{1 - c_2}{\lambda}\right) = \pm\frac{S}{\lambda}. \tag{7.3.21}$$

For a known value of S, Equations (7.3.20) and (7.3.21) then constitute three independent expressions in the three unknowns: c_1, c_2, and λ.

Some reflection reveals that this problem is only meaningful for $\sqrt{2} < S < 2$. Let us choose a value of $S = \frac{3}{2}$. Solving Equations (7.3.20) and (7.3.21), we obtain values of $\lambda = 1.2104$, $c_1 = 1.1947$, $c_2 = -0.1928$. (These are not the only values that satisfy the three nonlinear algebraic equations; however, these values are the only ones consistent with the given minimization problem.)

If the above numerical values are now substituted into Equation (7.3.18), we obtain the desired extremal:

$$u = 1.1947 - \sqrt{1.4651 - (t + 0.1928)^2}. \tag{7.3.22}$$

This function is shown plotted against t in Figure 7.3.

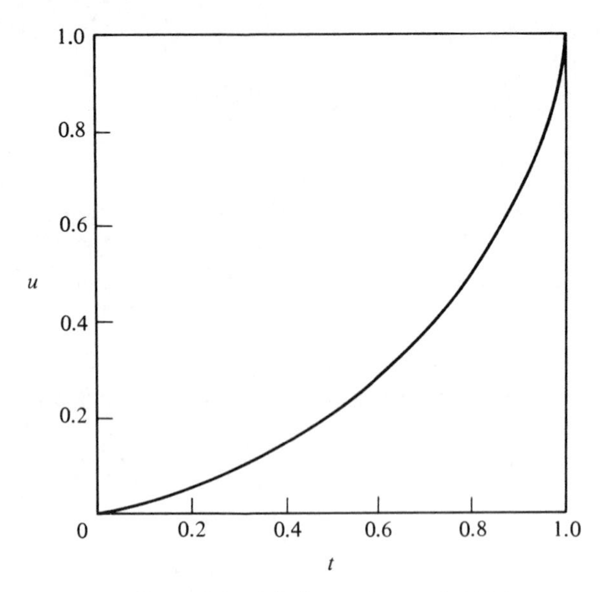

Figure 7.3. Solution to Example 7.3.1

It is straightforward to substitute the extremal into Equation (7.3.8) and carry out the integration, which results in

$$I_{min} = c_1 + \tfrac{1}{2}(1 - c_2)\sqrt{(\lambda^2 - c_2^2) + 2c_2 - 1} + \tfrac{1}{2}c_2\sqrt{(\lambda^2 - c_2^2)} - \tfrac{1}{2}S\lambda. \tag{7.3.23}$$

Evaluating the constants numerically, we obtain $I_{min} = 0.2944$.

Under special conditions, our problem can be considerably simplified. We have noted in our consideration of unconstrained minimization (Section 7.1) that when ϕ was expressed in terms of u and t only, the Euler–Lagrange conditions took a simple form. Similarly, if both ϕ and the integral constraint function ψ are expressed in terms of u and t only, then the necessary condition for an optimum becomes

$$\frac{\partial \phi}{\partial u} + \lambda \frac{\partial \psi}{\partial u} = 0. \tag{7.3.24}$$

The optimization of the integral I can then be treated as a succession of ordinary minimization problems for the values of t between t_0 and t_f.

In the absence of these special conditions we see from Example 7.3.1, just as we did from Example 7.1.1, that a great deal of cumbersome mathematical manipulation was required to solve this relatively simple problem. In particular, we had to obtain numerical solutions to three simultaneous, nonlinear algebraic equations, and then choose a set of roots that were consistent with the given problem. This is typical of problems of this type.

Furthermore, for many problems the integration cannot be carried out in closed form, thus compounding the difficulty considerably.

In Chapter 8 we shall develop an alternative procedure for solving isoperimetric problems. We shall see that this procedure, known as *dynamic programming*, has distinct computational advantages over variational calculus for solving problems of this type.

7.4 Differential Constraints

In the last section we discussed a constrained functional optimization problem that involved an invariant constraint; i.e., the constraining equation assumed a constant value. Let us now consider a more general constrained problem: optimization of an integral

$$\mathrm{I} = \int_{t_0}^{t_f} \phi(u_1, u_2, u_1', u_2', t)\, dt, \tag{7.4.1}$$

subject to a constraint of the form

$$G(u_1, u_2, u_1', u_2', t) = 0. \tag{7.4.2}$$

Notice that the constraint is now expressed as some implicit function of the independent variable t, rather than as a "fixed value" condition. A common example of this type of constraint is the differential equation

$$u_2' = \psi(u_1, u_2, t). \tag{7.4.3}$$

Problems of this type frequently arise in automatic-control theory.

In the above expressions, $u_1(t)$ and $u_2(t)$ are unknown extremizing functions. Frequently, in problems of this nature, only one initial condition is given; i.e., $u_2(t_0) = u_{20}$. We shall develop additional boundary restrictions as a consequence of the theory to be presented below. The functions $u_1(t)$ and $u_2(t)$ are assumed to have continuous second derivatives, as before. Note, however, that u_1 and u_2 are related by the constraining equation, and are therefore not independent.

Classical approach

As in Section 7.1, we represent any arbitrary continuous and differentiable functions that satisfy the initial conditions $\bar{u}_1(t)$ and $\bar{u}_2(t)$ in terms of the unknown extremals and the functions $\eta_1(t)$ and $\eta_2(t)$; i.e.,

$$\bar{u}_1(t) = u_1(t) + \epsilon_1 \eta_1(t), \tag{7.4.4}$$

$$\bar{u}_2(t) = u_2(t) + \epsilon_2 \eta_2(t). \tag{7.4.5}$$

The functions $\eta_1(t)$ and $\eta_2(t)$ are assumed to have continuous second derivatives, with $\eta_1(t_0) = \eta_2(t_0) = 0$. These functions are not independent, however, since $\epsilon_1\eta_1$ and $\epsilon_2\eta_2$ are related by Equation (7.4.2). Furthermore, η_1 and η_2 do not necessarily vanish at t_f, since we have not imposed any boundary conditions at this point.

Let us express the necessary condition for an extremum of Equation (7.4.1) in terms of the first variation. Thus

$$\delta \mathrm{I} = \epsilon_1 \int_{t_0}^{t_f} \left(\frac{\partial \phi}{\partial u_1} \eta_1 + \frac{\partial \phi}{\partial u_1'} \eta_1' \right) dt + \epsilon_1 \int_{t_0}^{t_f} \left(\frac{\partial \phi}{\partial u_2} \eta_2 + \frac{\partial \phi}{\partial u_2'} \eta_2' \right) \frac{d\epsilon_2}{d\epsilon_1}\, dt = 0, \tag{7.4.6}$$

where the development leading to this equation is analogous to that which led to Equation (7.1.11). From Equation (7.4.2), however, we can write

$$dG = \left(\frac{\partial G}{\partial u_1} \eta_1 + \frac{\partial G}{\partial u_1'} \eta_1' \right) d\epsilon_1 + \left(\frac{\partial G}{\partial u_2} \eta_2 + \frac{\partial G}{\partial u_2'} \eta_2' \right) d\epsilon_2 = 0. \tag{7.4.7}$$

Solving Equation (7.4.7) for $d\epsilon_2/d\epsilon_1$ and substituting into (7.4.6), we obtain

$$\begin{aligned} \delta \mathrm{I} = {} & \epsilon_1 \int_{t_0}^{t_f} \left(\frac{\partial \phi}{\partial u_1} \eta_1 + \frac{\partial \phi}{\partial u_1'} \eta_1' \right) dt \\ & - \epsilon_1 \int_{t_0}^{t_f} \frac{[(\partial \phi/\partial u_2)\eta_2 + (\partial \phi/\partial u_2')\eta_2']}{[(\partial G/\partial u_2)\eta_2 + (\partial G/\partial u_2')\eta_2']} \left(\frac{\partial G}{\partial u_1} \eta_1 + \frac{\partial G}{\partial u_1'} \eta_1' \right) dt = 0. \end{aligned} \tag{7.4.8}$$

Equation (7.4.8) is the necessary condition for an extremum of Equation (7.4.1), subject to the constraining equation (7.4.2).

Use of Lagrange multiplier

An alternative means of expressing the above necessary condition, which is actually more convenient than the procedure outlined, is to form the unconstrained, augmented integral

$$\mathrm{I} = \int_{t_0}^{t_f} (\phi + \lambda G)\, dt. \tag{7.4.9}$$

Note the similarity between Equations (7.4.9) and (7.4.5). In the present, more general case, however, we shall see that the Lagrange multiplier λ is a function of the independent variable t rather than a constant.

If we again think of ϕ and G as being dependent upon the *arbitrary* functions $\bar{u}_1(t)$ and $\bar{u}_2(t)$, where $\bar{u}_1$ and $\bar{u}_2$ are defined by Equations (7.4.4) and (7.4.5), then the necessary conditions for an extremum can be expressed as

$$\frac{\partial I}{\partial \epsilon_1} = \int_{t_0}^{t_f} \left\{ \frac{\partial \phi}{\partial u_1} \eta_1 + \frac{\partial \phi}{\partial u_1'} \eta_1' + \lambda\left(\frac{\partial G}{\partial u_1} \eta_1 + \frac{\partial G}{\partial u_1'} \eta_1' \right) \right\} dt = 0, \quad (7.4.10)$$

$$\frac{\partial I}{\partial \epsilon_2} = \int_{t_0}^{t_f} \left\{ \frac{\partial \phi}{\partial u_2} \eta_2 + \frac{\partial \phi}{\partial u_2'} \eta_2' + \lambda\left(\frac{\partial G}{\partial u_2} \eta_2 + \frac{\partial G}{\partial u_2'} \eta_2' \right) \right\} dt = 0. \quad (7.4.11)$$

Notice that we are now treating $u_1(t)$ and $u_2(t)$ as independent functions, since the augmented integral is assumed to be unconstrained.

Let us now choose a λ such that

$$\frac{\partial \phi}{\partial u_2} \eta_2 + \frac{\partial \phi}{\partial u_2'} \eta_2' + \lambda\left(\frac{\partial G}{\partial u_2} \eta_2 + \frac{\partial G}{\partial u_2'} \eta_2' \right) = 0 \quad (7.4.12)$$

for each t. Hence

$$\lambda = \lambda(t) = -\frac{[(\partial \phi/\partial u_2)\eta_2 + (\partial \phi/\partial u_2')\eta_2']}{[(\partial G/\partial u_2)\eta_2 + (\partial G/\partial u_2')\eta_2']}. \quad (7.4.13)$$

Substitution of this result into Equation (7.4.10) yields the necessary condition

$$\int_{t_0}^{t_f} \left(\frac{\partial \phi}{\partial u_1} \eta_1 + \frac{\partial \phi}{\partial u_1'} \eta_1' \right) dt$$
$$- \int_{t_0}^{t_f} \frac{[(\partial \phi/\partial u_2)\eta_2 + (\partial \phi/\partial u_2')\eta_2']}{[(\partial G/\partial u_2)\eta_2 + (\partial G/\partial u_2')\eta_2']} \left(\frac{\partial G}{\partial u_1} \eta_1 + \frac{\partial G}{\partial u_1'} \eta_1' \right) dt = 0, \quad (7.4.14)$$

which is equivalent to Equation (7.4.8). Thus we see that application of the necessary conditions to the unconstrained augmented integral is equivalent to an application of the necessary conditions to the original constrained problem. In the above formulation the Lagrangian function $\lambda(t)$ can be determined from the constraining equation (7.4.2).

Natural boundary conditions

Carrying the analysis one step farther, let us write Equations (7.4.10) and (7.4.11) as

$$\int_{t_0}^{t_f} \left\{ \left(\frac{\partial \phi}{\partial u_1} + \lambda \frac{\partial G}{\partial u_1} \right) \eta_1 + \left(\frac{\partial \phi}{\partial u_1'} + \lambda \frac{\partial G}{\partial u_1'} \right) \eta_1' \right\} dt = 0, \quad (7.4.15)$$

$$\int_{t_0}^{t_f} \left\{ \left(\frac{\partial \phi}{\partial u_2} + \lambda \frac{\partial G}{\partial u_2} \right) \eta_2 + \left(\frac{\partial \phi}{\partial u_2'} + \lambda \frac{\partial G}{\partial u_2'} \right) \eta_2' \right\} dt = 0. \quad (7.4.16)$$

Integrating the second term in each integral by parts, we obtain

$$\int_{t_0}^{t_f}\left(\frac{\partial\phi}{\partial u_1}+\lambda\frac{\partial G}{\partial u_1}\right)\eta_1\,dt+\left(\frac{\partial\phi}{\partial u_1'}+\lambda\frac{\partial G}{\partial u_1'}\right)\eta_1\Big|_{t_0}^{t_f}$$

$$-\int_{t_0}^{t_f}\frac{d}{dt}\left(\frac{\partial\phi}{\partial u_1'}+\lambda\frac{\partial G}{\partial u_1'}\right)\eta_1\,dt=0, \tag{7.4.17}$$

and

$$\int_{t_0}^{t_f}\left(\frac{\partial\phi}{\partial u_2}+\lambda\frac{\partial G}{\partial u_2}\right)\eta_2\,dt+\left(\frac{\partial\phi}{\partial u_2'}+\lambda\frac{\partial G}{\partial u_2'}\right)\eta_2\Big|_{t_0}^{t_f}$$

$$-\int_{t_0}^{t_f}\frac{d}{dt}\left(\frac{\partial\phi}{\partial u_2'}+\lambda\frac{\partial G}{\partial u_2'}\right)\eta_2\,dt=0. \tag{7.4.18}$$

Recall that $\eta_1(t_0)=\eta_2(t_0)=0$. Hence the second term in each of the above equations will vanish, providing

$$\frac{\partial\phi}{\partial u_1'}+\lambda\frac{\partial G}{\partial u_1'}\Big|_{t_f}=0 \tag{7.4.19}$$

and

$$\frac{\partial\phi}{\partial u_2'}+\lambda\frac{\partial G}{\partial u_2'}\Big|_{t_f}=0, \tag{7.4.20}$$

leaving simply

$$\int_{t_0}^{t_f}\left\{\left(\frac{\partial\phi}{\partial u_1}+\lambda\frac{\partial G}{\partial u_1}\right)-\frac{d}{dt}\left(\frac{\partial\phi}{\partial u_1'}+\lambda\frac{\partial G}{\partial u_1'}\right)\right\}\eta_1\,dt=0 \tag{7.4.21}$$

and

$$\int_{t_0}^{t_f}\left\{\left(\frac{\partial\phi}{\partial u_2}+\lambda\frac{\partial G}{\partial u_2}\right)-\frac{d}{dt}\left(\frac{\partial\phi}{\partial u_2'}+\lambda\frac{\partial G}{\partial u_2'}\right)\right\}\eta_2\,dt=0. \tag{7.4.22}$$

Equations (7.4.19)–(7.4.22) are an alternative expression of the necessary conditions for a constrained extremum. Observe that Equations (7.4.19) and (7.4.20) must apply in lieu of a given set of boundary conditions at t_f. Hence these expressions are frequently referred to as *natural boundary conditions.*

Euler–Lagrange equations

Recall that we have chosen $\lambda(t)$ such that the integrand in the necessary conditions will vanish for each t in the closed interval $t_0\le t\le t_f$. Therefore, we may rewrite the necessary conditions as

$$\frac{d}{dt}\left(\frac{\partial \phi}{\partial u_1'} + \lambda \frac{\partial G}{\partial u_1'}\right) - \left(\frac{\partial \phi}{\partial u_1} + \lambda \frac{\partial G}{\partial u_1}\right) = 0, \tag{7.4.23}$$

$$\frac{d}{dt}\left(\frac{\partial \phi}{\partial u_2'} + \lambda \frac{\partial G}{\partial u_2'}\right) - \left(\frac{\partial \phi}{\partial u_2} + \lambda \frac{\partial G}{\partial u_2}\right) = 0. \tag{7.4.24}$$

Equations (7.4.23) and (7.4.24) are the applicable Euler–Lagrange equations for this problem. Use of these equations also requires that the natural boundary conditions, (7.4.19) and (7.4.20), be satisfied. The unknown functions $u_1(t)$, $u_2(t)$, and $\lambda(t)$ can be determined for a particular problem through application of the Euler–Lagrange equations, the natural boundary conditions, and the original constraining equation. We shall see how to use these expressions in Example 7.4.1.

Before leaving this subject, let us summarize the procedure, at the same time generalizing to the case in which $u_1(t), u_2(t), \ldots, u_n(t)$ are the unknown functions that are subject to differential constraints of the form

$$\frac{du_i}{dt} = \psi(u_1, u_2, \ldots, u_n, x_1, x_2, \ldots, x_m, t), \qquad i = 1, 2, n, \tag{7.4.25}$$

and $x_1(t), x_2(t), \ldots, x_m(t)$ are the remaining unknown functions that are not constrained. We shall see the advantage of this new notation in the next few sections and in Chapter 8, where the functions $u_1, u_2, \ldots, u_n$ will be called *state variables* and the functions $x_1, x_2, \ldots, x_m$ will be referred to as *decision variables.*

Let us assume that $u_1, u_2, \ldots, u_n$ and $x_1, x_2, \ldots, x_m$ are continuous functions of t with continuous second derivatives. We can think of these functions as components of the n-dimensional vector $\mathbf{u}(t)$ and the m-dimensional vector $\mathbf{x}(t)$, where

$$\mathbf{u}(t) = \begin{bmatrix} u_1(t) \\ u_2(t) \\ \cdot \\ \cdot \\ \cdot \\ u_n(t) \end{bmatrix}, \qquad \mathbf{x}(t) = \begin{bmatrix} x_1(t) \\ x_2(t) \\ \cdot \\ \cdot \\ \cdot \\ x_m(t) \end{bmatrix}. \tag{7.4.26}$$

The problem is to determine $\mathbf{u}$ and $\mathbf{x}$ such that the integral

$$\mathrm{I} = \int_{t_0}^{t_f} \phi(\mathbf{u}, \mathbf{x}, \mathbf{u}', \mathbf{x}', t)\, dt \tag{7.4.27}$$

is extremized, subject to the constraint set

$$\mathbf{G}(\mathbf{u}, \mathbf{x}, \mathbf{u}', \mathbf{x}', t) = \mathbf{0}, \tag{7.4.28}$$

with $\mathbf{u}(t_0) = \mathbf{a}$. We shall assume that the vector $\mathbf{G}$ has n independent components (each of which may be a differential equation), so that $\mathbf{G}$ and $\mathbf{u}$ have the same dimensionality.

To solve this problem, we must write the governing Euler–Lagrange equations as

$$\frac{d}{dt}\left(\frac{\partial\phi}{\partial u_i'} + \sum_{j=1}^{n} \lambda_j \frac{\partial G_j}{\partial u_i'}\right) - \left(\frac{\partial\phi}{\partial u_i} + \sum_{j=1}^{n} \lambda_j \frac{\partial G_j}{\partial u_i}\right) = 0, \qquad i = 1, 2, \ldots, n, \tag{7.4.29}$$

$$\frac{d}{dt}\left(\frac{\partial\phi}{\partial x_k'} + \sum_{j=1}^{n} \lambda_j \frac{\partial G_j}{\partial x_k'}\right) - \left(\frac{\partial\phi}{\partial x_k} + \sum_{j=1}^{n} \lambda_j \frac{\partial G_j}{\partial x_k}\right) = 0, \qquad k = 1, 2, \ldots, m. \tag{7.4.30}$$

The natural boundary conditions that apply are

$$\left[\frac{\partial\phi}{\partial u_i'} + \sum_{j=1}^{n} \lambda_j \frac{\partial G_j}{\partial u_i'}\right]_{t=t_f} = 0, \qquad i = 1, 2, \ldots, n, \tag{7.4.31}$$

and

$$\left[\frac{\partial\phi}{\partial x_k'} + \sum_{j=1}^{n} \lambda_j \frac{\partial G_j}{\partial x_k'}\right]_{t=t_f} = 0, \qquad k = 1, 2, \ldots, m. \tag{7.4.32}$$

Equations (7.4.28)–(7.4.32), together with the initial conditions, completely specify the problem.

For the general problem formulation given above, it is noteworthy that a solution requires us to solve a total of $2n + m$ simultaneous equations, including differential equations, which may contain nonlinearities. Needless to say, this is not an easy task! Thus we see that the practical applicability of the above theory to real problems is largely limited to those problems in which u and x are low-dimensional vectors, and where ϕ and $G_j, j = 1, 2, \ldots, n$, are relatively simple functions.

The Euler–Lagrange equations which apply to each of the conditions discussed thus far are summarized in Table 7.1.

EXAMPLE 7.4.1

Determine the functions $u_1(t)$ and $u_2(t)$ that minimize the integral

$$\mathrm{I} = \int_0^1 (u_1'^2 + 2u_1 + u_2^2)\, dt, \tag{7.4.33}$$

subject to the constraining equation

$$G = u_2' + u_2 - u_1 = 0, \qquad 0 \le t \le 1, \tag{7.4.34}$$

with $u_2(0) = \frac{1}{2}$. Note that this problem requires only one initial condition, since only one derivative is present.

Application of the Euler–Lagrange equations, (7.4.23) and (7.4.24), yields

$$2(u_1 + 1) - \lambda = 0 \tag{7.4.35}$$

and

$$\lambda' - (2u_2 + \lambda) = 0. \tag{7.4.36}$$

We must also satisfy the natural boundary condition, (7.4.20), which results in simply

$$\lambda(1) = 0. \tag{7.4.37}$$

Observe that our problem involves two first-order differential equations, (7.4.34) and (7.4.36), and that we have two boundary conditions. We therefore have a fully defined problem, and the only remaining task is to integrate the differential equations.

Equations (7.4.34)–(7.4.36) can easily be combined to eliminate u_1 and λ, resulting in

$$u_2'' - 2u_2 = 1. \tag{7.4.38}$$

The solution to this linear differential equation is

$$u_2 = \cosh\sqrt{2}\,t + \frac{\sqrt{2}\,c}{2}\sinh\sqrt{2}\,t - \frac{1}{2}, \tag{7.4.39}$$

where c is an arbitrary constant of integration. Combining this expression with Equation (7.4.34) yields

$$u_1 = (1 + c)\cosh\sqrt{2}\,t + \sqrt{2}\left(1 + \frac{c}{2}\right)\sinh\sqrt{2}\,t - \frac{1}{2}. \tag{7.4.40}$$

Finally, Equations (7.4.35) and (7.4.40) can be combined to give

$$\lambda = 2(1 + c)\cosh\sqrt{2}\,t + 2\sqrt{2}\left(1 + \frac{c}{2}\right)\sinh\sqrt{2}\,t + 1. \tag{7.4.41}$$

Since we require that $\lambda(1) = 0$, we obtain

$$0 = 2(1 + c)\cosh\sqrt{2}\,t + 2\sqrt{2}\left(1 + \frac{c}{2}\right)\sinh\sqrt{2} + 1, \tag{7.4.42}$$

or

$$c = -\frac{1 + 2\cosh\sqrt{2} + 2\sqrt{2}\sinh\sqrt{2}}{2\cosh\sqrt{2} + \sqrt{2}\sinh\sqrt{2}} \tag{7.4.43}$$

$$= -1.52679. \tag{7.4.44}$$

Table 7.1 SUMMARY OF EULER–LAGRANGE EQUATIONS

Kind of Problem	*Objective Function*
Unconstrained	
Single parameter	$I = \int_{t_0}^{t_f} \phi(u, u', t)\, dt$ $I = \int_{t_0}^{t_f} \phi(u, u')\, dt$
Multi-parameter	$I = \int_{t_{10}}^{t_{1f}} \int_{t_{20}}^{t_{2f}} \cdots \int_{t_{n0}}^{t_{nf}} \phi(u, u', t_1, t_2, \ldots, t_n)\, dt_1\, dt_2 \ldots dt_n$
Multi-function	$I = \int_{t_0}^{t_f} \phi(u_1, u_2, \ldots, u_n, u'_1 \ldots u'_n, t)\, dt$
Constrained	
Single parameter Integral constraint	$I = \int_{t_0}^{t_f} \phi(u, u', t)\, dt$ $I = \int_{t_0}^{t_f} \phi(u, u')\, dt$ $I = \int_{t_0}^{t_f} \phi(u, t)\, dt$
Single parameter One differential constraint	$I = \int_{t_0}^{t_f} \phi(u_1, u'_1, t)\, dt$
Single parameter Many differential constraints (multi-function vectorial notation)	$I = \int_{t_0}^{t_f} \phi(\mathbf{u}, \boldsymbol{\theta}, \mathbf{u}', \boldsymbol{\theta}')\, dt$

Contraints	*Euler–Lagrange Equations*
	$\frac{d}{dt}\left(\frac{\partial\phi}{\partial u'}\right) - \frac{\partial\phi}{\partial u} = 0$ $u'\frac{\partial\phi}{\partial u} - \phi = c_1$
	$\frac{\partial}{\partial t_1}\left(\frac{\partial\phi}{\partial u_{t_1}}\right) + \frac{\partial}{\partial t_2}\left(\frac{\partial\phi}{\partial u_{t_2}}\right) + \cdots \frac{\partial}{\partial t_n}\left(\frac{\partial\phi}{\partial u_{t_n}}\right) - \frac{\partial\phi}{\partial u} = 0$ where $\partial u/\partial t_i = u_{t_i}$
	$\frac{d}{dt}\left(\frac{\partial\phi}{\partial u_i'}\right) + \frac{\partial\phi}{\partial u_i} = 0$ for $i = 1$ to n
$G = \int_{t_0}^{t_f} \psi(u, u'\ t)\, dt = 0$ $G = \int_{t_0}^{t_f} \psi(u, u')\, dt = 0$ $G = \int_{t_0}^{t_f} \psi(u, t)\, dt = 0$	$\frac{d}{dt}\left(\frac{\partial\phi}{\partial u'} + \lambda\frac{\partial\psi}{\partial u'}\right) - \left(\frac{\partial\phi}{\partial u} + \lambda\frac{\partial\psi}{\partial u}\right) = 0;\quad \lambda = \text{constant}$ $u'\frac{\partial}{\partial u'}(\phi + \lambda\psi) - (\phi + \lambda\psi) = c_1$ $\partial\phi/\partial u + \lambda\frac{\partial\psi}{\partial u} = 0$
$G(u_1, u_1'\ t) = 0$	$\frac{d}{dt}\left(\frac{\partial\phi}{\partial u_1'} + \lambda\frac{\partial G}{\partial u_1'}\right) - \left(\frac{\partial\phi}{\partial u_1} + \lambda\frac{\partial G}{\partial u_1}\right) = 0;\quad \lambda = f(t)$ + Nat. boundary cond. $\left[\frac{\partial\phi}{\partial u_1'} + \lambda\frac{\partial G}{\partial u_1'}\right]_{t_f} = 0$
$\mathbf{G(u, \theta, u'\ \theta'\ }t\mathbf{) = 0}$ Many constraints	$\frac{d}{dt}\left(\frac{\partial\phi}{\partial u_i'} + \sum_{j=1}^{n}\lambda_j\frac{\partial G_j}{\partial u_i'}\right) - \left(\frac{\partial\phi}{\partial u_i} + \sum_{j=1}^{n}\lambda_j\frac{\partial G_j}{\partial u_i}\right) = 0,$ $i = 1, 2 \ldots n$ $\frac{d}{dt}\left(\frac{\partial\phi}{\partial\theta_k'} + \sum_{j=1}^{n}\lambda_j\frac{\partial G_j}{\partial\theta_k'}\right) - \left(\frac{\partial\phi}{\partial\theta_k} + \sum_{j=1}^{n}\lambda_j\frac{\partial G_j}{\partial\theta_k}\right) = 0;$ $k = 1,2, \ldots m$ + Nat. Boundary Cond. $\left[\frac{\partial\phi}{\partial u_i'} + \sum_{j=1}^{n}\lambda_j\frac{\phi G_j}{\partial u_i'}\right]_{t=t_f} = 0$ $\left[\frac{\partial\phi}{\partial\theta_k'} + \sum_{j=1}^{n}\lambda_j\frac{\partial G_j}{\partial\theta_k'}\right]_{t=t_f} = 0;$ $\lambda = f(t)$

Hence we have obtained the desired solution:

$$u_1 = -0.52679 \cosh \sqrt{2}t + 0.33456 \sinh \sqrt{2}t - \tfrac{1}{2}, \tag{7.4.45}$$

$$u_2 = \cosh \sqrt{2}t - 1.07960 \sinh \sqrt{2}t - \tfrac{1}{2}, \tag{7.4.46}$$

$$\lambda = -1.05358 \cosh \sqrt{2}t + 0.66912 \sinh \sqrt{2}t + 1. \tag{7.4.47}$$

These three functions are shown as functions of t in Figure 7.4.

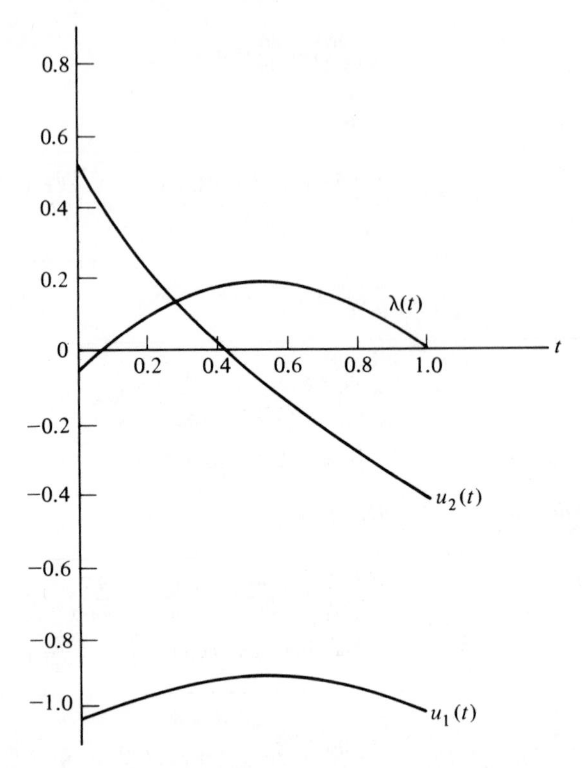

Figure 7.4. Curves Which Provide the Solution to Example 7.4.1

7.5 The Continuous Maximum Principle

The basic problem

In the last section we discussed the use of variational calculus for the minimization of the integral

$$\mathbf{I} = \int_{t_0}^{t_f} \phi(\mathbf{u}, \mathbf{x}, \mathbf{u}', \mathbf{x}', t)\, dt, \tag{7.5.1}$$

subject to a set of constraints of the form

$$\mathbf{G}(\mathbf{u}, \mathbf{x}, \mathbf{u}', \mathbf{x}', t) = \mathbf{0}, \tag{7.5.2}$$

where certain of the initial conditions are known. We have seen that problems of this type can be solved through the introduction of a set of Lagrange multipliers $\boldsymbol{\lambda}(t)$ and by subsequent application of the appropriate Euler–Lagrange equations and the applicable natural boundary conditions.

In this section we shall consider a somewhat different form of the above problem: Let us determine a decision vector $\mathbf{x}(t)$ that will minimize the integral

$$\mathrm{I} = \int_{t_0}^{t_f} \phi(\mathbf{u}, \mathbf{x})\, dt, \tag{7.5.3}$$

subject to the constraints

$$\frac{du_i}{dt} = g_i(\mathbf{u}, \mathbf{x}), \qquad i = 1, 2, \ldots, n. \tag{7.5.4}$$

As before, we shall assume that the state variables u_i, $i = 1, 2, \ldots, n$, and the decision variables x_k, $k = 1, 2, \ldots, m$, are continuous and differentiable, and that the functions ϕ and g_i, $i = 1, 2, \ldots, n$, are continuous and differentiable with respect to the dependent variables u_i and x_k. The state vector will be assumed to be known initially; i.e., $\mathbf{u}(t_0) = \mathbf{a}$.

Notice that this problem is somewhat more restrictive than the problem discussed in Section 7.4, since the functions ϕ and g_i, $i = 1, 2, \ldots, n$, are *autonomous* (i.e., they do not depend *explicitly* upon the independent variable t), and they do not contain the derivatives u_i' or x_k'. In spite of these restrictions, however, the above problem is sufficiently important to justify the following detailed development. Moreover, in later sections of this chapter we shall see that certain more general problems of this type can be recast into the above form.

Conditions for an optimum

Let us now apply the relevant necessary conditions of variational calculus to the above problem. For this case the Euler–Lagrange equations of Section 7.4 can easily be shown to reduce to

$$\frac{\partial \phi}{\partial u_i} - \sum_{j=1}^{n} \lambda_j \frac{\partial g_j}{\partial u_i} - \frac{d\lambda_i}{dt} = 0, \qquad i = 1, 2, \ldots, n, \tag{7.5.5}$$

and

$$\frac{\partial \phi}{\partial x_k} - \sum_{j=1}^{n} \lambda_j \frac{\partial g_j}{\partial x_k} = 0, \qquad k = 1, 2, \ldots, m. \tag{7.5.6}$$

Notice that Equation (7.5.6) takes on a simpler form than Equation (7.5.5); this is because the problem contains the derivatives of the state variables, u_i', but not the derivatives of the decision variables, x_j'.

The natural boundary conditions presented in Section 7.4 also take on a simpler form for this problem. Again, since u_i', but not x_k', appears in the problem, and since the integrand ϕ in the minimizing integral does not contain any derivatives of the dependent variables, the relevant natural boundary conditions reduce to

$$\lambda_i \frac{\partial}{\partial u_i'}\{u_i' - g_i(\mathbf{u}, \mathbf{x})\}\bigg|_{t_f} = 0, \qquad i = 1, 2, \ldots, n. \tag{7.5.7}$$

Furthermore, we see that g_i does not vary explicitly with u_i'. Hence the above expression reduces further, resulting in simply

$$\lambda_i(t_f) = 0, \qquad i = 1, 2, \ldots, n. \tag{7.5.8}$$

Finally, the original constraints and the given initial conditions must be combined with Equations (7.5.5), (7.5.6), and (7.5.8) for a complete statement of the necessary conditions for a particular problem.

Now let us examine this problem from a somewhat different point of view. First let us define a function $u_0(\mathbf{u}, \mathbf{x})$ such that

$$\frac{du_0}{dt} = \phi(\mathbf{u}, \mathbf{x}) = g_0(\mathbf{u}, \mathbf{x}). \tag{7.5.9}$$

(Note that the vector $\mathbf{u}$ contains $u_1, u_2, \ldots, u$, but not u_0.) Equation (7.5.9) can be integrated to yield

$$u_0(t_f) - u_0(t_0) = \int_{t_0}^{t_f} \frac{du_0}{dt}\, dt = \int_{t_0}^{t_f} g_0(\mathbf{u}, \mathbf{x})\, dt. \tag{7.5.10}$$

If we set $u_0(t_0) = 0$, then

$$u_0(t_f) = \mathrm{I} = \int_{t_0}^{t_f} g_0(\mathbf{u}, \mathbf{x})\, dt, \tag{7.5.11}$$

so that minimization of the given integral I is equivalent to minimization of $u_0(t_f)$. The problem then becomes

$$\underset{\mathbf{x}(t)}{\text{minimize}}\ u_0(t_f), \tag{7.5.12}$$

subject to

$$\frac{du_0}{dt} = g_0(\mathbf{u}, \mathbf{x}), \tag{7.5.13}$$

$$\frac{du_i}{dt} = g_i(\mathbf{u}, \mathbf{x}), \qquad i = 1, 2, \ldots, n, \tag{7.5.14}$$

with $u_0(t_0) = 0$ and $u_i(t_0) = a_i$, $i = 1, 2, \ldots, n$.

We now form a new function, called a *Hamiltonian function*, as follows:

$$H(\mathbf{u}, \mathbf{x}, \boldsymbol{\lambda}) = \sum_{j=0}^{n} \lambda_j(t) g_j(\mathbf{u}, \mathbf{x}). \tag{7.5.15}$$

Note that the vector λ is an $(n + 1)$-component vector given by

$$\boldsymbol{\lambda} = \begin{bmatrix} \lambda_0(t) \\ \lambda_1(t) \\ \lambda_2(t) \\ \cdot \\ \cdot \\ \cdot \\ \lambda_n(t) \end{bmatrix}, \tag{7.5.16}$$

whose components are simply Lagrange multipliers. However, since $g_0(\mathbf{u}, \mathbf{x})$ is actually an artificial constraint, we shall assign a particular final value to λ_0; i.e.,

$$\lambda_0(t_f) = -1. \tag{7.5.17}$$

The reason for this will soon become apparent.

Now let us write the following three equations, which involve certain partial derivatives of the Hamiltonian function:

$$\frac{dH}{du_i} = -\frac{d\lambda_i}{dt}, \qquad i = 0, 1, 2, \ldots, n, \tag{7.5.18}$$

$$\frac{\partial H}{\partial x_k} = 0, \qquad k = 1, 2, \ldots, m, \tag{7.5.19}$$

and

$$\frac{\partial H}{\partial \lambda_i} = \frac{du_i}{dt}, \qquad i = 0, 1, 2, \ldots, n. \tag{7.5.20}$$

To see the significance of these equations, let us combine them with Equation (7.5.15), which defines the Hamiltonian. Substitution of (7.5.15) into (7.5.18) yields

$$\frac{\partial H}{\partial u_0} = 0, \tag{7.5.21}$$

since the vector $\mathbf{u}$ does not contain u_0, and therefore none of the terms $g_0, g_1, \ldots, g_n$ depends upon u_0. Hence we see that

$$\frac{d\lambda_0}{dt} = 0, \tag{7.5.22}$$

which implies that λ_0 is a constant. Since we have specified that $\lambda_0(t_f)$ equal -1, we conclude that

$$\lambda_0(t) = -1, \qquad t_0 \leq t \leq t_f. \tag{7.5.23}$$

Also, we can write

$$\frac{\partial H}{\partial u_i} = \sum_{j=0}^{n} \lambda_j \frac{\partial g_j}{\partial u_i} = -\frac{d\lambda_i}{dt}, \qquad i = 1, 2, \ldots, n. \tag{7.5.24}$$

Since $\lambda_0 = -1$ and $g_0 = \phi$, we can express Equation (7.5.24) as

$$\frac{\partial \phi}{\partial u_i} - \sum_{j=1}^{n} \lambda_j \frac{\partial g_j}{\partial u_i} - \frac{d\lambda_i}{dt} = 0, \qquad i = 1, 2, \ldots, n, \tag{7.5.25}$$

which is one of the Euler–Lagrange equations [cf. Equation (7.5.5)].

Now let us substitute Equation (7.5.15) into Equation (7.5.19). This results in

$$\frac{\partial H}{\partial x_k} = \sum_{j=0}^{n} \lambda_j \frac{\partial g_j}{\partial x_k} = 0, \qquad k = 1, 2, \ldots, m, \tag{7.5.26}$$

which can be written as

$$\frac{\partial \phi}{\partial x_k} - \sum_{j=1}^{n} \lambda_j \frac{\partial g_j}{\partial x_k} = 0, \qquad k = 1, 2, \ldots, m. \tag{7.5.27}$$

This is the other Euler–Lagrange equation, (7.5.6).

Finally, substitution of (7.5.15) into (7.5.20) yields

$$\frac{\partial H}{\partial \lambda_i} = g_i = \frac{du_i}{dt}, \qquad i = 0, 1, \ldots, n, \tag{7.5.28}$$

which is simply the original set of constraints.

Thus we see that Equations (7.5.18)–(7.5.20) are concise expressions for the necessary conditions for an optimum written in terms of the Hamiltonian function. These expressions, together with the initial conditions

$$u_0(t_0) = 0, \tag{7.5.29}$$

$$u_i(t_0) = a_i, \qquad i = 1, 2, \ldots, n, \tag{7.5.30}$$

equation (7.5.23), and the natural boundary conditions

$$\lambda_i(t_f) = 0, \qquad i = 1, 2, \ldots, n, \tag{7.5.31}$$

enable us to obtain the desired optimum explicitly.

The maximum principle

We are now able to state the continuous maximum principle—*to obtain the set of decision variables $x_k(t)$, $k = 1, 2, \ldots, m$, that causes the objective function* (7.5.11) *to be minimized, subject to the constraints* (7.5.14); *introduce an additional set of variables $\lambda_i(t)$, $i = 0, 1, \ldots, n$, which satisfies* (7.5.23) *and* (7.5.31); *form a Hamiltonian function which satisfies* (7.5.15) *and* (7.5.18); *and maximize the Hamiltonian with respect to each of the $x_k(t)$.*

Although it is not obvious from the foregoing discussion, it happens to be more convenient to express the necessary conditions in terms of the Hamiltonian rather than in terms of the Euler–Lagrange equations. The simplicity of the Hamiltonian formulation will become evident in the following example.

EXAMPLE 7.5.1

Determine the functions $u_1(t)$ and $x(t)$ that minimize the integral

$$\mathrm{I} = \int_o^1 (x^2 + 2x + u_1^2)\, dt, \tag{7.5.32}$$

subject to

$$\frac{du_1}{dt} = x - u_1, \qquad 0 \le t \le 1, \tag{7.5.33}$$

and $u_1(0) = \frac{1}{2}$.

Notice that, except for a small change in notation, this problem is identical to Example 7.4.1. In the present example, however, we shall establish the necessary conditions for an optimum by means of the maximum principle.

We begin by defining the Hamiltonian:

$$H = \lambda_0 g_0 + \lambda_1 g_1 \tag{7.5.34}$$

$$= (-1)(x^2 + 2x + u_1^2) + \lambda_1(x - u_1). \tag{7.5.35}$$

From Equation (7.5.18) we can write

$$\frac{d\lambda_0}{dt} = 0. \tag{7.5.36}$$

We do not need this expression since we have already made use of the fact that $\lambda_0 = -1$ in the definition of the Hamiltonian. However, we can also utilize Equation (7.5.18) to write

$$\frac{d\lambda_1}{dt} = -\frac{\partial H}{\partial u_1}, \tag{7.5.37}$$

which results in

$$\frac{d\lambda_1}{dt} = 2u_1 + \lambda_1. \tag{7.5.38}$$

From Equation (7.5.19) we obtain

$$\frac{\partial H}{\partial x} = -2(x + 1) + \lambda_1 = 0. \tag{7.5.39}$$

Finally, Equation (7.5.20) yields

$$\frac{du_0}{dt} = \frac{\partial H}{\partial \lambda_0} = (x^2 + 2x + u_1^2), \tag{7.5.40}$$

and

$$\frac{du_1}{dt} = \frac{\partial H}{\partial \lambda_1} = x - u_1. \tag{7.5.41}$$

Notice that the desired minimum value of the objective function can be obtained by integrating Equation (7.5.40), with $u_0(0) = 0$, and then evaluating u_0 at $t = 1$ [cf. Equation (7.5.9) for the definition of u_0]. Observe also that Equation (7.5.41) is simply the original constraining equation.

The problem becomes completely specified once the boundary conditions are stated. We have already chosen $u_1(0) = \frac{1}{2}$. In addition, we have the condition

$$\lambda_1(1) = 0. \tag{7.5.42}$$

The solutions for $u_1(t)$, $x(t)$, and $\lambda_1(t)$ are obtained in a straightforward manner.

Analytical and graphical solutions to this problem are shown in Example 7.4.1. Note, however, that we now use the symbols u_1 and x for the variables, which were previously referred to as u_2 and u_1, respectively.

Before leaving this example, we note that the integration of the differential equations is straightforward only because these particular differential equations happen to be linear. If this were not the case, then a numerical integration procedure would probably be required. It should be understood that numerical integration of problems of this type (i.e., mixed boundary problems) can be difficult. This is particularly true of problems that involve a large number of state and decision variables.

7.6 Variants of the Problem

The maximum principle, although intended for variational-type problems with differential constraints, is not restricted solely to the particular problem described in the last section. In particular, it is possible to apply the maximum principle to problems that have different types of boundary conditions from those considered earlier; to problems that have additional terms in the constraining equations, including the independent variable t; or to problems that are constrained by higher-order differential equations. In each case it is necessary to modify the maximum principle somewhat by altering the boundary conditions or by introducing additional state variables. In the following paragraphs we shall examine each of these variants of our basic problem, and the maximum principle will be altered accordingly.

End conditions specified

If certain state variables are specified at t_f, then Equations (7.5.18)–(7.5.20) still apply, but the boundary conditions must be modified accordingly.

Recall that, as a variational problem, we required that the natural boundary conditions

$$\left[\frac{\partial \phi}{\partial u_i'} + \sum_{j=1}^{n} \lambda_j \frac{\partial G_j}{\partial u_i'}\right]_{t=t_f} = 0, \qquad i = 1, 2, \ldots, n \tag{7.6.1}$$

and

$$\left[\frac{\partial \phi}{\partial x_k'} + \sum_{j=1}^{n} \lambda_j \frac{\partial G_j}{\partial x_k'}\right]_{t=t_f} = 0, \qquad k = 1, 2, \ldots, m \tag{7.6.2}$$

be satisfied, where $G_j(\mathbf{u}, \mathbf{x}) = 0$ is a general expression of the jth constraining equation. For the particular problem described in Section 7.5, Equation (7.6.2) does not apply and Equation (7.6.1) reduces to simply

$$\lambda_i(t_f) = 0, \qquad i = 1, 2, \ldots, n. \tag{7.6.3}$$

Now if certain of the $u_i(t)$, say $u_{l1}(t), u_{l2}(t), \ldots, u_{lp}(t)$, take on the specified values $b_{l1}, b_{l2}, \ldots, b_{lp}$ at $t = t_f$, then the corresponding natural boundary conditions do not apply, and instead we have

$$u_i(t_f) = b_i, \qquad i = l1, l2, \ldots, lp. \tag{7.6.4}$$

Hence the boundary conditions that apply to this particular problem are

$$\lambda_i(t_f) = 0 \tag{7.6.5}$$

for those state variables not specified at t_f, and

$$u_i(t_f) = b_i \tag{7.6.6}$$

for those state variables whose values are fixed. Of course, the initial conditions, $\mathbf{u} = \mathbf{a}$, continue to apply as before.

Initial conditions not specified

If certain of the state variables do not have fixed initial values, then we must write corresponding natural boundary conditions that apply at $t = t_0$.

These conditions can be expressed in general as

$$\left[\frac{\partial \phi}{\partial u_i'} + \sum_{j=1}^{n} \lambda_j \frac{\partial G_j}{\partial u_i'}\right]_{t=t_0} = 0 \qquad i = l1, l2, \ldots, lp. \tag{7.6.7}$$

For the problem that we are considering, the above expression reduces to

$$\lambda_i(t_0) = 0, \qquad i = l1, l2, \ldots, lp.$$

Thus we have the initial conditions

$$u_0(t_0) = 0, \tag{7.6.8}$$

$$u_i(t_0) = a_i \tag{7.6.9}$$

for those state variables which take on fixed initial conditions, and

$$\lambda_i(t_0) = 0 \tag{7.6.10}$$

for the state variables which are initially unspecified.

Additional terms in constraining equations

Frequently the constraining equations contain an additional parameter or depend explicitly on the independent variable t. Problems of these types can easily be reduced to the form of the basic problem presented in Section 7.5 by introducing additional state variables in the correct manner. To see how this is done, let us first consider a set of constraining equations of the form

$$\frac{du_i}{dt}\, g_i(\mathbf{u}, \mathbf{x}, \alpha), \qquad i = 1, 2, \ldots, n, \tag{7.6.11}$$

where α is a scalar. Recall that $\mathbf{u}$ is an n-dimensional vector whose components are the state variables $u_1(t), u_2(t), \ldots, u_n(t)$. In addition, we have introduced the variable $u_0(t)$ as

$$u_0(t) = \int_{t_0}^{t} g_0(\mathbf{u}, \mathbf{x})\, dt; \tag{7.6.12}$$

hence we have $(n + 1)$ state variables in the problem. Now let us introduce an additional state variable, $u_{n+1}(t)$, as

$$u_{n+1}(t) = \alpha \tag{7.6.13}$$

or

$$\frac{du_{n+1}}{dt} = 0. \tag{7.6.14}$$

Hence we can write the constraining equations in the form

$$\frac{du_i}{dt} = g_i(\bar{\mathbf{u}}, \mathbf{x}), \qquad i = 0, 1, 2, \ldots, n + 1, \tag{7.6.15}$$

where $\bar{\mathbf{u}}$ is a vector whose components are the state variables $u_1, u_2, \ldots, u_n, u_{n+1}$.

Now let us consider the case in which the constraining equations vary explicitly with the independent variable; e.g.,

$$\frac{du_i}{dt} = g_i(\mathbf{u}, \mathbf{x}, t), \qquad i = 0, 1, 2, \ldots, n. \tag{7.6.16}$$

Problems of this kind are called *nonautonomous.*

As before, we introduce an additional state variable, $u_{n+1}(t)$, as

$$u_{n+1}(t) = t \tag{7.6.17}$$

or

$$\frac{du_{n+1}}{dt} = 1, \tag{7.6.18}$$

with $u_{n+1}(t_0) = t_0$. With this additional variable we can again express the constraint set by Equation (7.6.15), where $\bar{\mathbf{u}}$ is defined as before, except that $u_{n+1}(t)$ is now defined by (7.6.17).

Higher-order constraining equations

If one or more of the constraining equations are of higher order, e.g.,

$$\frac{d^2u_i}{dt^2} = g\left(\frac{du_i}{dt}, \mathbf{u}, \mathbf{x}\right), \tag{7.6.19}$$

with

$$u_i(t_0) = a_i \tag{7.6.20}$$

and

$$\left.\frac{du_i}{dt}\right|_{t_0} = c_i, \tag{7.6.21}$$

the problem can be reduced to the same form as the basic problem described in Section 7.5 as follows: Define

$$\frac{du_i}{dt} = u_{n+1}, \tag{7.6.22}$$

with

$$u_{n+1}(t_0) = a_{n+1} = c_i. \tag{7.6.23}$$

Equation (7.6.19) then becomes

$$\frac{du_{n+1}}{dt} = g(\bar{\mathbf{u}}, \mathbf{x}). \tag{7.6.24}$$

Therefore, we again have a set of $(n + 2)$ constraining equations, representing the behavior of the state variables $u_0, u_1, \ldots, u_n, u_{n+1}$. The vector $\bar{\mathbf{u}}$ is an $(n + 1)$-component vector, as before.

The above approach can be generalized to differential equations of the kth order, $k \geq 2$, by introducing $(k - 1)$ additional state variables. That is, given the constraint

$$\frac{d^{(k)}v_i}{dt^{(k)}} = g\left\{\frac{d^{(k-1)}u_i}{dt^{(k-1)}}, \frac{d^{(k-2)}u_i}{dt^{(k-2)}}, \ldots, \frac{du_i}{dt}, \mathbf{u}, \mathbf{x}\right\} \tag{7.6.25}$$

with known initial values for u_i and the first $(k - 1)$ derivatives, we can write

$$u_{n+1} = \frac{du_i}{dt}, \quad u_i(t_0) = a_i, \tag{7.6.26}$$

$$u_{n+2} = \frac{du_{n+1}}{dt}, \quad u_{n+1}(t_0) = a_{n+1}, \tag{7.6.27}$$

$$\vdots$$

$$u_{n+k-1} = \frac{du_{n+k-2}}{dt}, \quad u_{n+k-2}(t_0) = a_{n+k-2}, \tag{7.6.28}$$

$$\frac{du_{n+k-1}}{dt} = g(\bar{\mathbf{u}}, \mathbf{x}), \quad u_{n+k-1}(t_0) = a_{n+k-1}, \tag{7.6.29}$$

where $\bar{\mathbf{u}}$ contains the $(n + k)$ state variables $u_0, u_1, \ldots, u_n, u_{n+1}, \ldots, u_{n+k-1}$. Thus we have transformed a kth order initial value problem into a set of k first-order initial value problems.

There are other variants of the basic problem of Section 7.5 that can be transformed into the basic problem by the introduction of certain additional state variables. The reader who wishes to pursue this subject is referred to the text of Fan [1966].

EXAMPLE 7.6.1

Determine the trajectory $x(t)$ that will minimize the integral

$$\mathrm{I} = \int_0^1 (u^2 + x^2 + t^2)\, dt, \tag{7.6.30}$$

subject to

$$\frac{d^2u}{dt^2} + 2\frac{du}{dt} + u = x + \alpha, \tag{7.6.31}$$

where

$$u(0) = 0, \tag{7.6.32}$$

$$u(1) = 1, \tag{7.6.33}$$

and α is a positive constant.

We introduce the following notation:

$$u_1 = u, \tag{7.6.34}$$

$$u_2 = \frac{du_1}{dt}, \tag{7.6.35}$$

$$u_3 = t, \tag{7.6.36}$$

$$u_4 = \alpha, \tag{7.6.37}$$

$$u_0 = \int_0^1 g_0 \, dt = \int_0^1 (u_1^2 + u_3^2 + x^2) \, dt. \tag{7.6.38}$$

The corresponding constraining equations are

$$\frac{du_0}{dt} = g_0 = (u_1^2 + u_3^2 + x^2), \tag{7.6.39}$$

$$\frac{du_1}{dt} = g_1 = u_2, \tag{7.6.40}$$

$$\frac{du_2}{dt} = g_2 = -u_1 - 2u_2 + u_4 + x, \tag{7.6.41}$$

$$\frac{du_3}{dt} = g_3 = 1, \tag{7.6.42}$$

$$\frac{du_4}{dt} = g_4 = 0, \tag{7.6.43}$$

and the Hamiltonian becomes

$$H = -(u_1^2 + u_3^2 + x^2) + \lambda_1 u_2 + \lambda_2(-u_1 - 2u_2 + u_4 + x) + \lambda_3. \tag{7.6.44}$$

If we apply Equation (7.5.18), we obtain

$$\frac{d\lambda_1}{dt} = 2u_1 + \lambda_2, \tag{7.6.45}$$

$$\frac{d\lambda_2}{dt} = -\lambda_1 + 2\lambda_2, \tag{7.6.46}$$

$$\frac{d\lambda_3}{dt} = 2u_3, \tag{7.6.47}$$

$$\frac{d\lambda_4}{dt} = -\lambda_2. \tag{7.6.48}$$

Equation (7.5.19) yields

$$-2x + \lambda_2 = 0. \tag{7.6.49}$$

Thus we have a system of nine first-order differential equations and one algebraic equation, and the ten unknown functions u_0, u_1, u_2, u_3, u_4, λ_1, λ_2, λ_3, λ_4, x. We shall leave the manipulative details of the solution to the interested reader.

7.7 Behavior of the Hamiltonian Function

Let us again consider the problem of Section 7.5; i.e., minimize the integral

$$\mathrm{I} = \int_{t_0}^{t_f} g_0(\mathbf{u}, \mathbf{x})\, dt, \tag{7.7.1}$$

subject to

$$\frac{du_i}{dt} = g_i(\mathbf{u}, \mathbf{x}), \qquad i = 1, 2, \ldots, n, \tag{7.7.2}$$

and

$$\mathbf{u}(t_0) = \mathbf{a}.$$

Again, the variables u_i and x_k will be assumed continuous and differentiable functions of t, $t_0 \leq t \leq t_f$, and the functions $g_i(\mathbf{u}, \mathbf{x})$ are assumed to be continuous and differentiable with respect to the state and decision variables.

Constancy of the Hamiltonian

In this section we shall discuss the behavior of the Hamiltonian function $H(\mathbf{u}, \mathbf{x}, \lambda)$. To do this we shall make use of certain variational concepts, as developed previously.

We begin by examining the variation of the Hamiltonian function with the dependent variable t. Since

$$H = f\{\mathbf{u}(t), \mathbf{x}(t), \boldsymbol{\lambda}(t)\}, \tag{7.7.3}$$

we can write

$$\frac{dH}{dt} = \frac{\partial H}{\partial t} + \sum_{i=1}^{n} \frac{\partial H}{\partial u_i}\frac{du_i}{dt} + \sum_{k=1}^{m} \frac{\partial H}{\partial x_k}\frac{dx_k}{\partial t} + \frac{\partial H}{\partial \lambda_0}\frac{d\lambda_0}{dt} + \sum_{i=1}^{n} \frac{\partial H}{\partial \lambda_i}\frac{d\lambda_i}{dt}. \tag{7.7.4}$$

However, we observe that

$$\frac{\partial H}{\partial t} = 0 \tag{7.7.5}$$

because the Hamiltonian is autonomous, and that

$$\frac{d\lambda_0}{dt} = 0 \tag{7.7.6}$$

as a result of the constancy of λ_0. If we now substitute Equations (7.5.18)–(7.5.20) into the three remaining terms in Equation (7.7.4), we obtain

$$\frac{dH}{dt} = \sum_{i=1}^{n} \frac{\partial H}{\partial u_i}\frac{\partial H}{\partial \lambda_i} + 0 - \sum_{i=1}^{n} \frac{\partial H}{\partial \lambda_i}\frac{\partial H}{\partial u_i} \tag{7.7.7}$$

or

$$\frac{dH}{dt} = 0, \qquad t_0 \leq t \leq t_f. \tag{7.7.8}$$

Thus we see that *the Hamiltonian remains constant with respect to t* within the interval $t_0 \leq t \leq t_f$.

Suppose that we evaluate the Hamiltonian at $t = t_f$. Recall that, for the particular problem under consideration, we have $\lambda_0 = -1$ and $\lambda_i(t_f) = 0$, $i = 1, 2, \ldots, n$. Therefore, the Hamiltonian becomes

$$H\{\mathbf{u}(t_f), \mathbf{x}(t_f), \boldsymbol{\lambda}(t_f)\} = -g_0\{\mathbf{u}(t_f), \mathbf{x}(t_f)\}, \tag{7.7.9}$$

which is simply the integrand of the objective function, with the sign reversed, evaluated at t_f. Now since H remains constant over the entire interval $t_0 \leq t \leq t_f$, we conclude that

$$H\{\mathbf{u}(t), \mathbf{x}(t), \boldsymbol{\lambda}(t)\} = -g_0\{\mathbf{u}(t_f), \mathbf{x}(t_f)\}, \qquad t_0 \leq t \leq t_f. \tag{7.7.10}$$

If the end conditions are fixed, then the Hamiltonian will still be invariant with t, but its value will be different from that given above. We shall not pursue this topic; the interested reader is referred to the translation of Pontryagin et al. [1962].

Extremization of the Hamiltonian

Let us now examine the condition that the Hamiltonian be extremized with respect to $x_j(t)$, $j = 1, 2, \ldots, m$, when the object integral is minimized. Proceeding as in Section 7.1, we shall represent *any* set of continuous and differentiable decision variables $\bar{x}_k(t)$ in terms of the optimizing functions

$x_k(t)$ and the corresponding arbitrary functions $\psi_k(t)$; i.e.,

$$\bar{x}_k(t) = x_k(t) + \delta_k \psi_k(t), \tag{7.7.11}$$

where the δ_k are small scalars. Similarly, we can represent the corresponding state variables $\bar{u}_i(t)$ in terms of the optimal variables $u_i(t)$ and the arbitrary variables $\eta_i(t)$; i.e.,

$$\bar{u}_i(t) = u_i(t) + \epsilon_i \eta_i(t), \tag{7.7.12}$$

where the ϵ_i are small scalars. Since the initial values of the state variables are known, we shall require that

$$\eta_i(0) = 0, \qquad i = 0, 1, \ldots, n. \tag{7.7.13}$$

Recall that our objective is to minimize the integral represented by $u_0(t_f)$. Hence we can write

$$\int_{t_0}^{t_f} g_0(\bar{\mathbf{u}}, \bar{\mathbf{x}})\, dt - \int_{t_0}^{t_f} g_0(\mathbf{u}, \mathbf{x})\, dt \geq 0. \tag{7.7.14}$$

Since we can define $u_0(t)$ in terms of the differential equation

$$\frac{du_0}{dt} = g_0(\mathbf{u}, \mathbf{x}), \tag{7.7.15}$$

we have

$$g_0(\bar{\mathbf{u}}, \bar{\mathbf{x}}) - g_0(\mathbf{u}, \mathbf{x}) = \epsilon_0 \frac{d\eta_0}{dt}. \tag{7.7.16}$$

Combining this result with Equation (7.7.14) yields

$$\int_{t_0}^{t_f} \epsilon_0 \frac{d\eta_0}{dt}\, dt = \epsilon_0 \eta_0(t_f) \geq 0. \tag{7.7.17}$$

Now consider the quantity

$$\frac{d}{dt} \sum_{j=0}^{n} \epsilon_j \eta_j \lambda_j = \sum_{j=0}^{n} \lambda_j \epsilon_j \frac{d\eta_j}{dt} + \sum_{j=0}^{n} \epsilon_j \eta_j \frac{d\lambda_j}{dt}. \tag{7.7.18}$$

We shall express the quantity

$$\epsilon_j \frac{d\eta_j}{dt} = g_j(\bar{\mathbf{u}}, \bar{\mathbf{x}}) - g_j(\mathbf{u}, \mathbf{x}) \tag{7.7.19}$$

in terms of a first-order Taylor expansion, yielding

$$\epsilon_j \frac{d\eta_j}{dt} = \sum_{i=1}^{n} \epsilon_i \eta_i \frac{\partial g_j}{\partial u_i} + \sum_{k=1}^{m} \delta_k \psi_k \frac{\partial g_j}{\partial x_k}. \tag{7.7.20}$$

Also,

$$\frac{d\lambda_j}{dt} = -\frac{\partial H}{\partial u_j} = -\sum_{i=0}^{n} \lambda_i \frac{\partial g_i}{\partial u_j}. \tag{7.7.21}$$

Hence Equation (7.7.18) can be written

$$\frac{d}{dt} \sum_{j=0}^{n} \epsilon_j \eta_j \lambda_j = \sum_{j=0}^{n} \lambda_j \left\{ \sum_{i=1}^{n} \epsilon_i \eta_i \frac{\partial g_j}{\partial u_i} + \sum_{k=1}^{m} \delta_k \psi_k \frac{\partial g_j}{\partial x_k} \right\} - \sum_{j=0}^{n} \epsilon_j \eta_j \sum_{i=0}^{n} \lambda_i \frac{\partial g_i}{\partial u_j}. \tag{7.7.22}$$

When we interchange indices and cancel terms, Equation (7.7.22) simplifies to

$$\frac{d}{dt} \sum_{i=0}^{n} \epsilon_i \eta_i \lambda_i = \sum_{i=0}^{n} \lambda_i \sum_{k=1}^{m} \delta_k \psi_k \frac{\partial g_i}{\partial x_k}. \tag{7.7.23}$$

By integrating from t_0 to t_f, and remembering that $\eta_i(t_0) = 0$, $i = 0, 1, \ldots, n$, $\lambda_0 = -1$, and $\lambda_i(t_f) = 0$, $i = 1, 2, \ldots, n$, we obtain

$$\epsilon_0 \eta_0(t_f) = -\int_{t_0}^{t_f} \sum_{i=0}^{n} \lambda_i \sum_{k=1}^{m} \delta_k \psi_k \frac{\partial g_i}{\partial x_k}\, dt \tag{7.7.24}$$

$$= -\int_{t_0}^{t_f} \sum_{k=1}^{m} \delta_k \psi_k \sum_{i=0}^{n} \lambda_i \frac{\partial g_i}{\partial x_k}\, dt \tag{7.7.25}$$

$$= -\int_{t_0}^{t_f} \sum_{k=1}^{m} \delta_k \psi_k \frac{\partial H}{\partial x_k}\, dt. \tag{7.7.26}$$

In view of Equation (7.7.17), however, we can write

$$-\int_{t_0}^{t_f} \sum_{k=1}^{m} \delta_k \psi_k \frac{\partial H}{\partial x_k}\, dt \geq 0. \tag{7.7.27}$$

We see that Equation (7.7.27) implies that the Hamiltonian be a maximum with respect to the decision variables x_j, because

$$H(\mathbf{u}, \mathbf{x}, \boldsymbol{\lambda}) - H(\mathbf{u}, \bar{\mathbf{x}}, \boldsymbol{\lambda}) = -\sum_{k=1}^{m} \delta_k \psi_k \frac{\partial H}{\partial x_k} \geq 0, \tag{7.7.28}$$

where we have again utilized a first-order Taylor expansion. Therefore, we see that

$$H(\mathbf{u}, \mathbf{x}, \boldsymbol{\lambda}) \geq H(\mathbf{u}, \bar{\mathbf{x}}, \boldsymbol{\lambda}). \tag{7.7.29}$$

Thus we have shown that *the set of decision variables which causes the object integral to be minimized also maximizes the Hamiltonian for fixed* **u** *and* $\boldsymbol{\lambda}$. As mentioned earlier, this is the reason for referring to this approach as the maximum principle.

An alternative derivation of the maximum principle, which employs Green's functions, is given by Denn and Aris [1965]. For a rigorous treatment, however, the reader is referred to the translation of Pontryagin's work [1962].

7.8 Generalizations of the Basic Problem

So far in this chapter we have discussed the application of the maximum principle to problems that could have been attacked by the calculus of variations. The maximum principle is applicable to a broader class of problems, however, than the preceding discussion would indicate. This accounts for much of the interest shown in the maximum principle by analysts confronted with industrial optimization problems, particularly in the area of optimal control. In this section we shall discuss some of these more general problems, indicating how they can be solved by the maximum principle.

Final value of independent variable unspecified

In many problems the final values of the state variables may be known (in addition to the initial values), but the final value of the independent variable t_f may be unspecified. For example, consider the problem of finding $\mathbf{x}(t)$ such that

$$\mathrm{I} = \int_{t_0}^{t_f} g_0(\mathbf{u}, \mathbf{x})\, dt = u_0(t_f) \tag{7.8.1}$$

is minimized, subject to

$$\frac{d\mathbf{u}}{dt} = \mathbf{g}(\mathbf{u}, \mathbf{x}) \tag{7.8.2}$$

with

$$\mathbf{u}(t_0) = \mathbf{a}, \tag{7.8.3}$$

$$\mathbf{u}(t_f) = \mathbf{b}, \tag{7.8.4}$$

and t_f, the final value of the independent variable, to be determined. Under these conditions it can be shown that the Hamiltonian function vanishes at t_f (cf. Pontryagin et al. [1962]). Since we have established that the Hamiltonian is invariant with t, we conclude that

$$H(\mathbf{u}, \mathbf{x}, \boldsymbol{\lambda}) = 0, \qquad t_0 \leq t \leq t_f, \tag{7.8.5}$$

for this particular problem.* Thus Equation (7.8.5), together with Equations (7.5.18)–(7.5.20) and the boundary conditions, enables us to obtain the desired minimizing functions $\mathbf{x}(t)$, $\mathbf{u}(t)$, $\boldsymbol{\lambda}(t)$, and the end point t_f. We shall see a problem of this type in the examples which follow this section.

Constrained decision variables

Many problems require, in addition to the differential constraints which govern the state variables, that some relationship be satisfied among the decision variables; i.e.,

$$F(x_1, x_2, \ldots, x_m) \leq 0. \tag{7.8.6}$$

In particular, the magnitude of each decision variable frequently is upper bounded. Thus we would have a set of constraints of the form

$$|x_k| \leq M_k, \qquad k = 1, 2, \ldots, m. \tag{7.8.7}$$

Problems of this nature can also be solved through application of the maximum principle. In fact, the decision variables $x_k(t)$ need not be continuous functions of t, but can be piecewise continuous, as shown in Figure 7.5. For these more general conditions, however, it is necessary to modify our statement of the maximum principle. We shall present the maximum principle in its more general form and show how it may be used. The reader interested in the mathematical development is referred to Pontryagin et al. [1962].

The constraints which apply to the decision variables can be thought of as surfaces that bound the permissible $\mathbf{x}$ space. Thus Equation (7.8.6), when written as a strict inequality, applies at an interior point of the $\mathbf{x}$ space, whereas the bounding surface is represented by Equation (7.8.6) as an equality. Now if the functions $\mathbf{u}(t)$ and $\boldsymbol{\lambda}(t)$ are fixed, then the Hamiltonian, $H(\mathbf{u}, \mathbf{x}, \boldsymbol{\lambda})$, becomes a function only of $\mathbf{x}(t)$. Let us define a function $\mathfrak{M}(\mathbf{u}, \boldsymbol{\lambda})$ as

$$\max_{\mathbf{x}} H(\mathbf{u}, \mathbf{x}, \boldsymbol{\lambda}) = \mathfrak{M}(\mathbf{u}, \boldsymbol{\lambda}). \tag{7.8.8}$$

---

* Equation (7.8.5) is also valid for the corresponding problem with free right-end conditions.

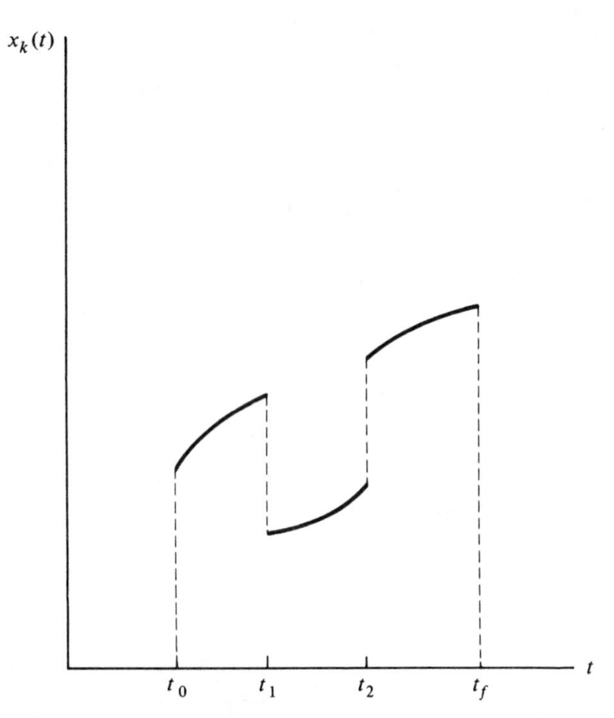

Figure 7.5. Piecewise-Continuous Decision Variable

If the maximizing value of $\mathbf{x}$ can be obtained by setting $\partial H/\partial x_k = 0$, $k = 1, 2, \ldots, m$, without violating any of the constraints expressed by (7.8.6) or (7.8.7), then the optimal policy falls within the permissible portion of the $\mathbf{x}$ space, and the maximum principle is unaltered from its previous form. If, however, the optimizing functions $x_k(t)$, obtained by setting $\partial H/\partial x_k = 0$, violate the constraints on the decision variables, then the necessary condition expressed by Equation (7.5.19) must be replaced by Equation (7.8.8).

In the latter case the function $\mathfrak{M}(\mathbf{u}, \boldsymbol{\lambda})$ is obtained by allowing one or more of the x_k to take on their limiting values. Thus if the constraining condition is expressed as a functional relationship, as given by Equation (7.8.6), then the constraint becomes a strict equality. In principle, (7.8.6) can be solved for one of the x_k in terms of the remaining x_k, and this expression for x_k can be substituted directly into the Hamiltonian. In fact, however, an explicit solution for x_k cannot always be obtained, thus complicating the computational procedure.

When the constraints that apply to the decision variables are simply upper bounds on the magnitude of the individual decision variables, as given by (7.8.7), then these bounding values can be substituted directly into the Hamiltonian and its associated necessary conditions. We shall see how this is done in an example at the end of this section.

Summary of the maximum principle

Before leaving this section, let us consolidate some of the material presented earlier by summarizing the maximum principle in its more general form.

We may state our problem as the determination of the set of piecewise continuous decision variables $x_1(t), x_2(t), \ldots, x_m(t)$ for which the integral

$$u_0(t_f) = \int g_0(\mathbf{u}, \mathbf{x})\, dt \tag{7.8.9}$$

is minimized, subject to the constraints

$$F(x_1, x_2, \ldots, x_m) \leq 0 \tag{7.8.10}$$

and

$$\frac{du_i}{dt} = g_i(\mathbf{u}, \mathbf{x}), \qquad i = 1, 2, \ldots, n, \tag{7.8.11}$$

or, in vector form,

$$\frac{d\mathbf{u}}{dt} = \mathbf{g}(\mathbf{u}, \mathbf{x}), \tag{7.8.12}$$

given a set of boundary conditions. We form the Hamiltonian function

$$H(\mathbf{u}, \mathbf{x}, \boldsymbol{\lambda}) = -g_0(\mathbf{u}, \mathbf{x}) + \sum_{i=1}^{n} \lambda_i g_i(\mathbf{u}, \mathbf{x}), \tag{7.8.13}$$

which satisfies the conditions

$$\frac{d\lambda_i}{dt} = -\frac{\partial H}{\partial u_i}, \qquad i = 1, 2, \ldots, n, \tag{7.8.14}$$

$$\frac{du_i}{dt} = \frac{\partial H}{\partial \lambda_i}, \qquad i = 1, 2, \ldots, n. \tag{7.8.15}$$

The $\lambda_i(t)$ are *adjoint variables* (Lagrange multipliers) that vary continuously within $t_0 \leq t \leq t_f$.

The maximum principle states that for the minimizing set of $x_k(t)$ the Hamiltonian must satisfy the conditions

$$\max_{\mathbf{x}} H(\mathbf{u}, \mathbf{x}, \boldsymbol{\lambda}) = \mathfrak{M}(\mathbf{u}, \boldsymbol{\lambda}) \tag{7.8.16}$$

and

$$\frac{\partial H}{\partial t} = 0, \qquad t_0 \leq t \leq t_f. \tag{7.8.17}$$

Here $\mathfrak{M}(\mathbf{u}, \boldsymbol{\lambda})$ is the least upper bound (supremum) of $H(\mathbf{u}, \mathbf{x}, \boldsymbol{\lambda})$ for fixed

u and **λ**. This function can be determined by setting

$$\frac{\partial H}{\partial x_k} = 0, \qquad k = 1, 2, \ldots, m, \tag{7.8.18}$$

providing the resulting set of $x_k(t)$ do not violate Equation (7.8.10).

When the state variables are specified on a boundary, such as $\mathbf{u}(t_0) = \mathbf{a}$, then there are no corresponding boundary conditions imposed on the adjoint variables $\boldsymbol{\lambda}$. If the state variables are unspecified, however, then the condition is imposed that $\boldsymbol{\lambda} = \mathbf{0}$ at the boundary. Finally, if one of the bounding values of the independent variable (i.e., t_0 or t_f) is not specified, then we invoke the additional requirement on the Hamiltonian that

$$H(\mathbf{u}, \mathbf{x}, \boldsymbol{\lambda}) = 0, \qquad t_0 \leq t \leq t_f. \tag{7.8.19}$$

Several variants of the above problem can be transformed into this given form by the techniques discussed in Section 7.6.

The reader should recognize that the maximum principle, like the Euler–Lagrange equations, represents a *necessary*, but *not a sufficient*, condition for an extremum. Thus we have no assurance that the optimal decision vector **x** will not *maximize* the object integral. Moreover, a problem that has nonlinear differential constraints may have several local minima, and we have no guarantee that a particular minimizing decision vector corresponds to the global optimum.

EXAMPLE 7.8.1

Let us consider the problem of a state variable $u(t)$ that decays in accordance with the expression

$$\frac{du}{dt} + u = x, \tag{7.8.20}$$

where $u(0) = 1$ and $u(t_f) = 0$. We wish to determine a decision variable $x(t)$ that will minimize the integral

$$I = \int_0^{t_f} (x^2 + 2x)\, dt, \tag{7.8.21}$$

where t_f is unspecified.

We form the Hamiltonian function

$$H = -g_0 + \lambda_1 g_1 \tag{7.8.22}$$

$$= -(x^2 + 2x) + \lambda_1(x - u). \tag{7.8.23}$$

By application of Equations (7.8.14), (7.8.15), and (7.8.16), we obtain

$$\frac{d\lambda_1}{dt} = -\frac{\partial H}{\partial u} = \lambda_1, \tag{7.8.24}$$

$$\frac{du}{dt} = \frac{\partial H}{\partial \lambda_1} = (x - u), \tag{7.8.25}$$

$$\frac{\partial H}{\partial x} = -2(x + 1) + \lambda_1 = 0. \tag{7.8.26}$$

Integration of Equation (7.8.24) yields

$$\lambda_1 = c_1 e^t \tag{7.8.27}$$

and substitution into (7.8.26) then gives

$$x = \tfrac{1}{2}c_1 e^t - 1. \tag{7.8.28}$$

Substitution of this result into (7.8.25) and then integrating leads to

$$u = c_2 e^{-t} + \tfrac{1}{4}c_1 e^t - 1. \tag{7.8.29}$$

We must still determine a set of values for c_1, c_2, and t_f. We first make use of Equation (7.8.19). If we substitute Equations (7.8.23), (7.8.27), (7.8.28), and (7.8.29) into this expression, we obtain

$$-(\tfrac{1}{2}c_1 e^t - 1)(\tfrac{1}{2}c_1 e^t + 1) + c_1 e^t(\tfrac{1}{2}c_1 e^t - 1 - c_2 e^{-t} - \tfrac{1}{4}c_1 e^t + 1) = 0, \tag{7.8.30}$$

which simplifies to

$$c_1 c_2 = 1. \tag{7.8.31}$$

Hence we can express $u(t)$ as

$$u = \frac{1}{4}c_1 e^t + \frac{1}{c_1}e^{-t} - 1. \tag{7.8.32}$$

Since $u(0) = 1$, we can write

$$\frac{1}{4}c_1 + \frac{1}{c_1} - 1 = 1 \tag{7.8.33}$$

or

$$c_1^2 - 8c_1 + 4 = 0. \tag{7.8.34}$$

The roots of this quadratic equation are easily found to be

$$c_1 = 4 \pm 2\sqrt{3}. \tag{7.8.35}$$

From our remaining boundary condition, we obtain

$$\frac{1}{4}c_1 e^{t_f} + \frac{1}{c_1}e^{-t_f} - 1 = 0. \tag{7.8.36}$$

If the above values for c_1 are substituted into this expression, we find that $t_f = 1.317$ when $c_1 = 4 - 2\sqrt{3}$, the other value for c_1 producing a negative value for t_f. We can now obtain c_2 from Equation (7.8.31). Substitution of these constants into Equations (7.8.27), (7.8.28), and (7.8.29) yields the final solutions:

$$\lambda_1 = 0.536e^t, \tag{7.8.37}$$

$$x = 0.268e^t - 1, \tag{7.8.38}$$

and

$$u = 1.866e^{-t} + 0.134e^t - 1, \tag{7.8.39}$$

for $0 \leq t \leq 1.317$. These functions are plotted against t in Figure 7.6.

EXAMPLE 7.8.2

Pontryagin et al. [1962] discuss the problem of minimizing the time for $u(t)$ to change from a given initial value to a prescribed final value in accordance with the expression

$$\frac{d^2u}{dt^2} = x, \qquad 0 \leq t \leq t_f, \tag{7.8.40}$$

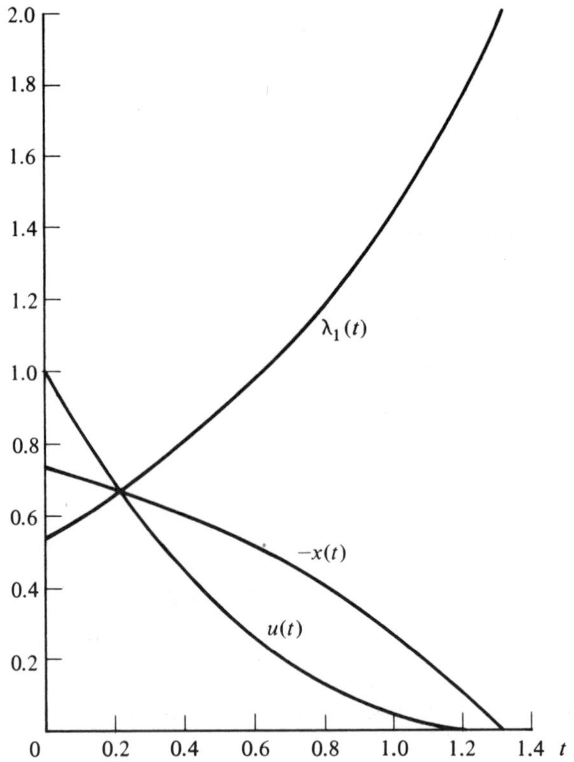

Figure 7.6. Functions Providing Solution to Example 7.8.1

where

$$|x(t)| \leq 1. \tag{7.8.41}$$

For example, suppose that $u(0) = du(0)/dt = 1$, and $u(t_f) = du(t_f)/dt = 0$. We wish to minimize the integral

$$\mathrm{I} = \int_0^{t_f} dt, \tag{7.8.42}$$

subject to the constraints given by Equations (7.8.40) and (7.8.41).

Let us introduce a new set of state variables defined by

$$\frac{du_0}{dt} = 1, \qquad u_0(0) = 0, \tag{7.8.43}$$

$$u_1 = u, \qquad u_1(0) = 1, \qquad u_1(t_f) = 0, \tag{7.8.44}$$

$$\frac{du_1}{dt} = u_2, \qquad u_2(0) = 1, \qquad u_2(t_f) = 0. \tag{7.8.45}$$

Equation (7.8.40) then becomes

$$\frac{du_2}{dt} = x, \qquad 0 \leq t \leq t_f. \tag{7.8.46}$$

We now define the Hamiltonian

$$H = -g_0 + \lambda_1 g_1 + \lambda_2 g_2 \tag{7.8.47}$$

$$= -1 + \lambda_1 u_2 + \lambda_2 x. \tag{7.8.48}$$

Application of Equations (7.8.14), (7.8.15), and (7.8.16) yields

$$\frac{d\lambda_1}{dt} = 0, \tag{7.8.49}$$

$$\frac{d\lambda_2}{dt} = -\lambda_1, \tag{7.8.50}$$

$$\frac{du_1}{dt} = u_2, \tag{7.8.51}$$

$$\frac{du_2}{dt} = x, \tag{7.8.52}$$

and

$$\begin{aligned} \mathfrak{M}(u_2, \lambda_1, \lambda_2) &= \max_x H(u_2, \lambda_1, \lambda_2, x) \\ &= -1 + \lambda_1 u_2 + |\lambda_2|. \end{aligned} \tag{7.8.53}$$

Thus we see that x must equal $+1$ when $\lambda_2 > 0$, and $x = -1$ when $\lambda_2 < 0$. Under these conditions, $x(t)$ will vary discontinuously with t while switching back and forth between ± 1. This kind of discontinuous switching of the decision variable is frequently referred to as *bang-bang control.*

If we integrate Equations (7.8.51) and (7.8.52), we obtain

(7.8.54) $$u_2 = xt + c_1,$$

(7.8.55) $$u_1 = \tfrac{1}{2}xt^2 + c_1 t + c_2, \qquad t > 0.$$

Similarly, integration of Equations (7.8.49) and (7.8.50) results in

(7.8.56) $$\lambda_1 = c_3,$$

(7.8.57) $$\lambda_2 = -c_3 t + c_4.$$

Since $\lambda_2(t)$ is linear with t, we see that it can change sign only once. Therefore, x will exhibit only one discontinuity within the interval $0 \leq t \leq t_f$. We shall refer to the time at which this discontinuity occurs as t^*.

Since the solutions for $u_1(t)$ and $u_2(t)$ depend explicitly upon $x(t)$, we must recognize that the constants of integration will be different for $0 \leq t \leq t^*$ and $t^* \leq t \leq t_f$. Hence we write equations (7.8.54) and (7.8.55) as

(7.8.58) $$u_1 = \tfrac{1}{2}xt^2 + c'_1 t + c'_2,$$

(7.8.59) $$u_2 = xt + c'_1,$$

for $t^* \leq t \leq t_f$. It is not necessary to proceed likewise for $\lambda_1(t)$ and $\lambda_2(t)$, since they do not depend explicitly upon $x(t)$.

An examination of Equations (7.8.54)–(7.8.59) reveals that six constants of integration must be determined. Also, t^* and t_f must be evaluated, and we must find out whether the initial value of x is positive or negative. We can write the following eight equations from the given boundary conditions, the requirement that $u_1(t)$ and $u_2(t)$ be continuous with time, and our general knowledge of the problem:

(7.8.60) $$u_1(0) = c_2 = 1,$$

(7.8.61) $$u_2(0) = c_1 = 1,$$

(7.8.62) $$u_1(t_f) = -\tfrac{1}{2}x_0 t_f^2 + c'_1 t_f + c'_2 = 0,$$

(7.8.63) $$u_2(t_f) = -x_0 t_f + c'_1 = 0,$$

(7.8.64) $$u_1(t^{*-}) - u_1(t^{*+}) - \tfrac{1}{2}x_0 t^{*2} + t^* + 1 + \tfrac{1}{2}x_0 t^{*2} - c'_1 t^* - c'_2 = 0,$$

(7.8.65) $$u_2(t^{*-}) - u_2(t^{*+}) = x_0 t^* + 1 + x_0 t^* - c'_1 = 0,$$

(7.8.66) $$H(t \leq t^*) = -1 + c_3 + c_4 x_0 = 0,$$

(7.8.67) $$\lambda_2(t^*) = -c_3 t^* + c_4 = 0.$$

In the above expressions we have let x_0 represent $x(t = 0)$, and we have made use of the fact that

(7.8.68) $$x(t^* \leq t \leq t_f) = -x_0.$$

If we combine Equations (7.8.62)–(7.8.65) to eliminate t^*, t_f, and c_2', we obtain

$$c_1' = -1 \pm \sqrt{2 - 4x_0}. \tag{7.8.69}$$

We see that x_0 must equal -1; otherwise c_1', and hence $u_2(t)$, will be complex for $t^* \leq t \leq t_f$. Equation (7.8.65) then yields

$$c_1' = 1 - 2t^*.$$

Since $t^* > 0$, we conclude that $c_1' < 1$, which requires that we select the negative value given by (7.8.69). Thus

$$c_1' = -1 - \sqrt{6} = -3.450. \tag{7.8.70}$$

It is then relatively straightforward to establish that $c_2' = 5.950$, $t^* = 2.225$, $t_f = 3.450$, $c_3 = -0.816$, and $c_4 = -1.816$. Thus we have solved the given minimum-time problem. The time-dependent functions $x(t)$, $u_1(t)$, $u_2(t)$, $\lambda_1(t)$, and $\lambda_2(t)$ are plotted against time in Figure 7.7.

Notice that $u_2(t)$ is continuous with t, but its slope exhibits a discontinuity at t^*. This, of course, is a direct consequence of Equation (7.8.52). Note also the

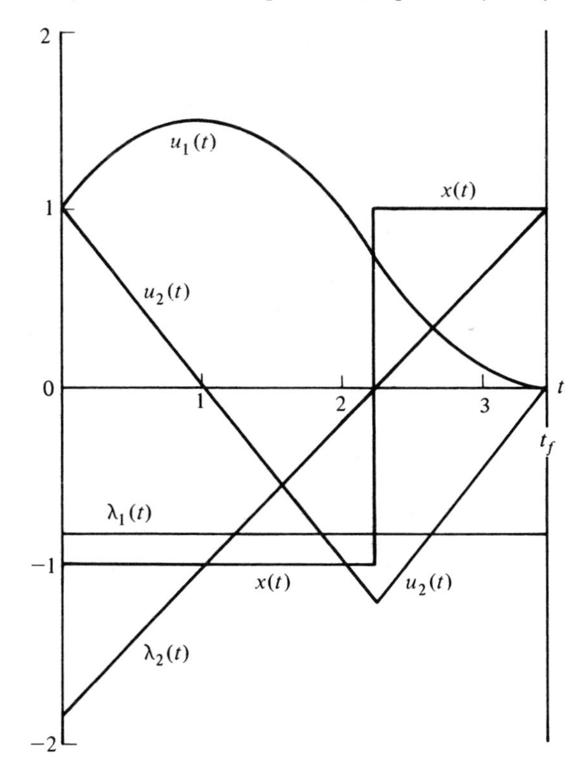

Figure 7.7. Functions Which Solve Example 7.8.2

difficulty of obtaining an explicit solution to this comparatively simple problem. The reader should also note that since this problem involves a constraint on the decision variable, $x(t)$, we could not have solved this problem using variational calculus. We were, however, able to obtain a solution using the maximum principle.

7.9 Direct Methods for Optimization of Functionals

The methods for functional optimization, which we have examined in the previous sections of this chapter, have been the so-called "indirect" methods. That is, we have determined the necessary conditions for an optimum and then attempted to solve the differential equations obtained from these conditions. These methods are analogous to solving ordinary optimization problems through the use of the necessary conditions for a stationary point obtained from the calculus. Just as we found for nonlinear algebraic optimization problems, the indirect methods are difficult and cumbersome. In ordinary optimization we abandoned the indirect methods and solved our problems by a number of "direct" numerical methods. Each of these methods required that we evaluate the objective function numerically and determine the minimum by direct comparison of function values. In many variational problems it is desirable to follow the same general procedure.

In Chapter 8 we shall indicate how dynamic programming may be used for optimization of functionals. Here we shall briefly indicate how nonlinear programming techniques may be used. Let us suppose that we are given an initial value, $u(t_0)$, and are asked to minimize the integral

$$\mathrm{I} = \int_{t_0}^{t_f} u(t)\,dt \tag{7.9.1}$$

subject to

$$u' = g(u, x), \tag{7.9.2}$$

where x is a decision variable under the control of the designer.

We shall replace this problem by a finite-difference approximation. We first break up our interval between t_0 and t_f into i adjacent subintervals, Δt_j, $j = 1, 2, \ldots i$. If these intervals are small enough, we can assume that in any interval, j, the values of u and x are approximately constant at u_j and x_j, respectively. Our objective function then is approximated by: Minimize

$$\mathrm{I} = \sum_{j=1}^{i} (\Delta t_j) u_j. \tag{7.9.3}$$

Further, we use our finite-difference approximation to write

$$u_{j+1} = u_j + (\Delta t_j)u'_j, \qquad j = 1, 2, \ldots, i - 1, \tag{7.9.4}$$

where

$$u'_j = g(u_j, x_j), \qquad j = 1, 2 \ldots i. \tag{7.9.5}$$

We see that we now have an algebraic objective function subject to algebraic, nonlinear constraints. We have thus approximated our functional minimization problem by an ordinary nonlinear programming problem. We can now use an appropriate nonlinear programming algorithm to determine a set of u_j and x_j which will solve our approximating problem.

In almost all instances it is possible to derive satisfactory finite-difference approximations for functional optimization problems. In view of the difficulty of variational approaches, it is generally desirable to consider nonlinear programming or dynamic programming methods when approximate numerical solutions are acceptable.

REFERENCES

Bolza, O., *Lectures on the Calculus of Variations*, 2nd ed., Chelsea, New York, 1904 (1st ed. copyright date).

Citron, S. J., *Elements of Optimal Control*, Holt, Rinehart and Winston, New York, 1969.

Courant, R., and D. Hilbert, *Methods of Mathematical Physics*, Vol. 1, Wiley-Interscience, New York, 1953.

Denn, M. M., and R. Aris, "An Elementary Derivation of the Maximum Principle," *A. I. Ch. E. Journal*, *11* (1965), 367.

Elsgolc, L. E., *Calculus of Variations*, Pergamon Press Ltd., 1961 (U.S. edition distributed by Addison-Wesley, Reading, Mass.).

Fan, L. T., *The Continuous Maximum Principle*, Wiley, New York, 1966.

Forsythe, A. R., *Calculus of Variations*, Dover, New York, 1960.

Lapidus, L., and R. Luus, *Optimal Control of Engineering Processes*, Ginn-Blaisdell, Waltham, Mass., 1967.

Merriam, C. W., *Optimization Theory and the Design of Feedback Control Systems*, McGraw-Hill, New York, 1964.

Pontryagin, L. S., V. G. Boltyanskii, R. V. Gamkrelidze, and E. F. Mishchenko, *The Mathematical Theory of Optimal Processes* (English translation by K. N. Trirogoff), Wiley-Interscience, New York, 1962.

PROBLEMS

7.1. Derive an equation for the function $u(t)$ that produces a surface of revolution about the t axis of minimum area. Assume that $0 \leq t \leq 1$, with $u(0) = 3$ and $u(1) = 2$. Verify that the solution does indeed minimize the area of the surface of revolution.

7.2. Determine the shape of the curve $u(t)$ that extremizes the integral

$$I = \int_0^1 (1 + t)[1 + (u')^2]^{1/2}\, dt,$$

with $u(0) = u_0$ and $u(1) = u_f$. Determine whether the extremum is a maximum or a minimum.

7.3. Set up the two-dimensional brachistochrone problem; i.e., we seek the functions $u(t)$, $v(t)$ that describe the height and depth, respectively, of a minimum-time path. Let t represent a horizontal coordinate, $0 \leq t \leq 1$, and let $u(0) = 0$, $v(0) = 1$, $u(1) = 1$, $v(1) = 0$. Define the coordinates as follows:

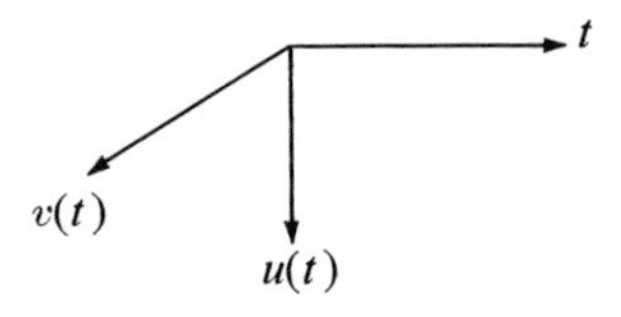

What problems are encountered in obtaining a closed-form solution which do not show up in the one-dimensional case?

7.4. Derive the necessary conditions for extremization of the integral

$$I = \int_{t_0}^{t_f} \phi(u, u', u'', t)\, dt.$$

What must be required concerning the continuity of ϕ and its derivatives? What must be stated about the boundary conditions?

7.5. Determine the shape of a curve $u(t)$, $0 \leq t \leq 2$, whose length S is equal to 4, which maximizes the area under the curve. Let $u(0) = 2$ and $u(2) = 1$. Show that the extremum is a maximum.

7.6. Determine the path of shortest distance between two points on the unit sphere. (This path is known as a *geodesic curve*.)

7.7. Minimize the integral

$$I = \int_0^2 u[1 + (u')^2]^{1/2}\, dt,$$

subject to

$$S = \int_0^2 [1 + (u')^2]^{1/2}\, dt = 2.8,$$

and $u(0) = u(2) = 1$.

The solution to this problem is known as a *catenary curve*. Physically, this solution can be obtained by minimizing the potential energy of a cable whose length is fixed.

7.8. The curves $u_1(t)$ and $u_2(t)$ that minimize the integral

$$I = \int_0^1 (u_1^2 + u_2^2 + t^2)\, dt,$$

subject to

$$\frac{d^2u_1}{dt^2} + 2\frac{du_1}{dt} + u_1 = u_2 + \alpha,$$

are to be determined, where $u_1(0) = 0$, $u_1(1) = 1$, and α is a positive constant. Express the necessary conditions in terms of explicit first-order differential equations with appropriate boundary conditions.

7.9. Obtain the functions $u_1(t)$ and $u_2(t)$ that minimize the integral

$$I = \int_0^{t_f} u_2(1 + 2u_2)\, dt,$$

subject to

$$\frac{du_1}{dt} + u_1 = u_2,$$

with $u_1(0) = 1$, $u_1(t_f) = 0$, and $t_f = 0.5$.

7.10. We wish to minimize the integral

$$I = \int_0^{t_f} (x^2 + u^2)\, dt,$$

where the state variable $u(t)$ is governed by the differential equation

$$\frac{du}{dt} = x - 2u, \qquad 0 \leq t \leq t_f.$$

Solve this problem by means of the continuous maximum principle, assuming that

(a) $u(0) = 1$, $t_f = 3$, and $u(3)$ is not specified.
(b) $u(0)$ is not specified, $t_f = 3$, and $u(3) = 1$.

(c) $u(0) = 1$, $t_f = 3$, and $u(3) = 0$.
(d) $u(0) = 1$, t_f is not specified, and $u(t_f) = 0$.
(e) $u(0) = 1$, t_f and $u(t_f)$ are not specified.
In each of the above examples show that the Hamiltonian is constant with respect to t.

7.11. Solve problem 7.9 by means of the continuous maximum principle.

7.12. Beginning with the Euler–Lagrange equations [(7.4.23), (7.4.24)] and the natural boundary conditions [(7.4.19), (7.4.20)], obtain equations (7.5.18)–(7.5.20), (7.5.23), and (7.5.31) for the problem: Minimize

$$I = \int_0^{t_f} \phi(u, x)\, dt,$$

subject to

$$\frac{du}{dt} = g(u, x), \qquad u(t_0) = a.$$

Assume that $u(t)$ and $x(t)$ are continuous and differentiable, and that ϕ and g are continuous and differentiable with respect to u and x.

7.13. Use the continuous maximum principle to minimize the time required for the state variable $u(t)$ to vary from some initial state $u(0) = u_0$, $u'(0) = u'_0$, to the final state $u(t_f) = 0$. The variation of u with time is governed by the differential equation

$$\frac{d^2u}{dt^2} + u = x(t),$$

where $x(t)$ is a decision function bounded by

$$-1 \leq x(t) \leq 1$$

(cf. Pontryagin et al., pp. 27–35 [1962]).
Suppose that the bounds were removed from the decision function. Can the problem be solved in closed form? What kind of extremum is obtained by setting $\partial H/\partial x = 0$?

7.14. Resolve problem 7.13, using the continuous maximum principle, for the case in which the state variable $u(t)$ is governed by

$$\frac{d^2u}{dt^2} + \alpha\frac{du}{dt} + \beta u = x(t),$$

$$u(0) = u_0, \qquad u'(0) = u'_0,$$

$$u(t_f) = u'(t_f) = 0,$$

$$\alpha > 0, \qquad \alpha^2 < 4\beta^2,$$

and

$$|x(t)| \leq \gamma.$$

Show that the Hamiltonian is constant with respect to t.

7.15. (a) Develop a suitable finite-difference approximation for problem 7.10.
(b) Solve the above by an appropriate mathematical programming technique for the boundary conditions of 7.10(a).

OPTIMIZATION OF STAGED SYSTEMS

This chapter is concerned with optimization problems that can be modeled as stagewise systems. The concepts of state and decision variables, which we introduced in the last chapter, will carry over to the staged systems that will now be discussed. In fact, some of the problems we shall encounter will be of the variational type. In general, however, we shall be concerned with a much broader class of problems than the variational problems discussed earlier. Thus our earlier study of the calculus of variations will provide certain background material for this chapter, just as our study of classical calculus provided a background for the material on linear and nonlinear programming.

Our main concern in this chapter will be with a technique known as *dynamic programming*. This is actually not a specific optimization technique, but a general solution strategy that applies to staged systems. In fact, dynamic programming can be applied not only to problems that describe physically staged systems, but also to variational problems and to a reasonably broad class of nonlinear programming problems. Moreover, the method can be used to solve problems that involve discrete decision variables, as we encountered in Chapter 6.

We shall begin this chapter with some general concepts that apply to

8

staged systems and to dynamic programming. Several different kinds of problems will then be solved from a dynamic programming point of view. Following this we shall derive the Euler–Lagrange equations of variational calculus by means of dynamic programming. Finally, we shall present a discrete version of the maximum principle that applies to stagewise systems. Thus we shall see some degree of unification between the calculus of variations, the maximum principle, and dynamic programming.

8.1 A Minimum-path Problem

Before becoming involved with the formalities of stagewise optimization techniques, we shall examine a simple trajectory problem that demonstrates the dynamic programming approach to problem solving.

Let us return to the minimum-path problem considered in Chapter 6. The problem requires that we find the shortest path, in terms of travel time, to go from point A to point G in Figure 8.1. Note that each of the nodes is lettered, and that the travel time between adjacent nodes is shown along the connecting path segments.

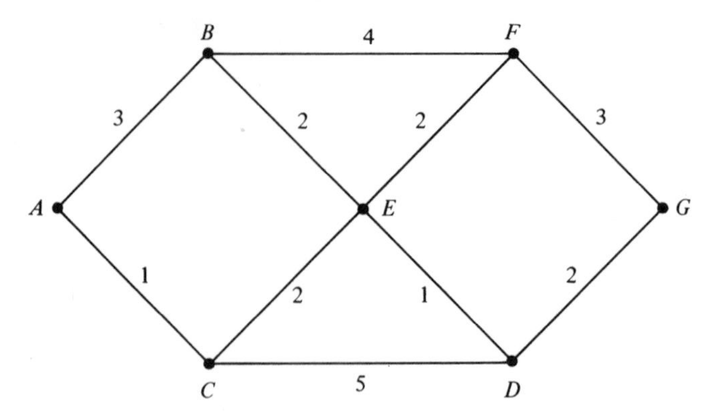

Figure 8.1. Minimum-Path Problem

Solution by exhaustive enumeration

One way to proceed with this problem, which is surely the most obvious, is simply to determine every possible path leading from point A to point G, compute the travel time for each of these paths, and then select from these the path that yields the desired minimum travel time. This is a "brute-force" approach. Because of the simplicity of the problem, however, it is not difficult to carry out this calculation. This method, known as *exhaustive enumeration*, results in the numerical values shown in Table 8.1. The optimal path is clearly $ACEDG$, and the minimum travel time is six time units.

Table 8.1. MINIMUM-PATH SOLUTION BY EXHAUSTIVE ENUMERATION

Path	*Driving Time*
$ABFG$	$t_{ABFG} = 3 + 4 + 3 = 10$
$ABEFG$	$t_{ABEFG} = 3 + 2 + 2 + 3 = 10$
$ABEDG$	$t_{ABEDG} = 3 + 2 + 1 + 2 = 8$
$ACEFG$	$t_{ACEFG} = 1 + 2 + 2 + 3 = 8$
$ACEDG$	$t_{ACEDG} = 1 + 2 + 1 + 2 = 6$
$ACDG$	$t_{ACDG} = 1 + 5 + 2 = 8$

Notice that in this problem we have implicitly assumed that we shall never travel backward from an intermediate nodal position. For example, path $ACDEFG$ would be inadmissible, since this would require us to backtrack from point D to point E. It is actually not necessary to impose this restriction on the problem; the more general case can be handled by the same method presented above, and also by the techniques to be discussed subsequently. By restricting the problem in this manner, however, we are able to minimize the computational effort and thereby keep the example as simple and as illustrative as possible.

Dynamic programming solution

Let us now reexamine the same problem from a different and less obvious point of view. Suppose, for the time being, that we wish to determine the minimum travel time from point A to point E. With the above restriction, which prevents backtracking, there are only two paths that lead to point E—from point B and from point C. Hence the minimum path from A to E will be the lesser of t_{ABE} and t_{ACE}. We can write this as

$$t_{A\to B} = \min \begin{Bmatrix} t_{ABE} \\ t_{ACE} \end{Bmatrix}, \tag{8.1.1}$$

where $t_{A\to E}$ denotes the minimum travel time from point A to point E.

We can carry the analysis a step farther and write

$$t_{ABE} = t_{BE} + t_{AB}. \tag{8.1.2}$$

Since there is only one path from A to B, this must be the minimum path from A to B, so that Equation (8.1.2) can be written

$$t_{ABE} = t_{BE} + t_{A\to B}. \tag{8.1.3}$$

Similarly,

$$t_{ACE} = t_{CE} + t_{A\to C}. \tag{8.1.4}$$

Substituting the last two expressions into Equation (8.1.1), we obtain

$$t_{A\to E} = \min \begin{Bmatrix} t_{BE} + t_{A\to B} \\ t_{CE} + t_{A\to C} \end{Bmatrix}. \tag{8.1.5}$$

The numerical evaluation of Equation (8.1.5) yields

$$t_{A\to E} = \min \begin{Bmatrix} 2 + 3 \\ 2 + 1 \end{Bmatrix} = 3_{(ACE)}. \tag{8.1.6}$$

Hence the minimum travel time from A to E is three time units, and the corresponding path is ACE. This is, of course, not the final solution to our problem. However, we shall see how this information can be used to obtain the desired minimum travel time $t_{A\to G}$.

Now let us consider point F. This point can be reached from two previous points—B and E. Therefore, an application of the above reasoning gives us

$$t_{A\to F} = \min\begin{Bmatrix} t_{BF} + t_{A\to B} \\ t_{EF} + t_{A\to E} \end{Bmatrix}. \tag{8.1.7}$$

The fact that there are several paths from point A to point E is of no concern to us now, since we have already determined the minimum path between these two points. Hence Equation (8.1.7) can be evaluated directly to yield

$$t_{A\to F} = \min\begin{Bmatrix} 4 + 3 \\ 2 + 3 \end{Bmatrix} = 5_{(ACEF)}. \tag{8.1.8}$$

Similarly, we can obtain $t_{A\to D}$ as follows:

$$t_{A\to D} = \min\begin{Bmatrix} t_{CD} + t_{A\to C} \\ t_{ED} + t_{A\to E} \end{Bmatrix} \tag{8.1.9}$$

or

$$t_{A\to D} = \min\begin{Bmatrix} 5 + 1 \\ 1 + 3 \end{Bmatrix} = 4_{(ACED)}. \tag{8.1.10}$$

Finally, we see that point G, the desired end point, can be reached only from point D or point F. Hence we can write

$$t_{A\to G} = \min\begin{Bmatrix} t_{FG} + t_{A\to F} \\ t_{DG} + t_{A\to D} \end{Bmatrix} \tag{8.1.11}$$

or

$$t_{A\to G} = \min\begin{Bmatrix} 3 + 5 \\ 2 + 4 \end{Bmatrix} = 6_{(ACEDG)}. \tag{8.1.12}$$

Thus we see that the minimum travel time is six units, along path $ACEDG$. This result agrees, of course, with that of our previous calculation.

Notice that the strategy we have employed is very different from the brute-force strategy of exhaustive enumeration. We have traversed our road map "from left to right," so to speak, by solving a sequence of partial minimization problems. Each calculation of the minimum time to travel a given partial path was used as a part of a subsequent calculation of a larger path. Moreover, any given path was later embedded within a subsequent path. Thus, by traversing *sequentially* from the starting point (node A) to the desired end point (node G), we obtained the desired optimal path. This is the dynamic programming strategy; we have, in fact, solved this problem precisely by means of dynamic programming.

At this point the reader might quite naturally ask what we have gained by the less obvious, dynamic programming strategy. The answer is a reduction in the required computational effort to solve the problem. If we return briefly to the earlier calculation, we see that six separate travel times had to be calculated. When employing dynamic programming, we had to calculate only four minimum travel times (viz., $t_{A \to E}$, $t_{A \to F}$, $t_{A \to D}$, and $t_{A \to G}$).

These differences may not seem significant for this problem, particularly since each computation in the dynamic programming problem required calculating the travel time for each of two paths and then determining the minimum of these. As the size of the problem increases, however, the computational advantages of dynamic programming over exhaustive enumeration become increasingly evident. We shall examine this point more closely in subsequent sections of this chapter.

Backward dynamic programming solution

Before leaving the simple minimum-path problem, which has already been solved by two different methods, let us again attempt a dynamic programming solution. This time, however, we shall work our way *backward* from point G, rather than *forward* from point A.

By proceeding as before, we can write

$$t_{E \to G} = \min \begin{Bmatrix} t_{EFG} \\ t_{EDG} \end{Bmatrix} \tag{8.1.13}$$

or

$$t_{E \to G} = \min \begin{Bmatrix} t_{EF} + t_{F \to G} \\ t_{ED} + t_{D \to G} \end{Bmatrix}. \tag{8.1.14}$$

This result can be expressed numerically as

$$t_{E \to G} = \min \begin{Bmatrix} 2 + 3 \\ 1 + 2 \end{Bmatrix} = 3_{(EDG)}. \tag{8.1.15}$$

Similarly,

$$t_{B \to G} = \min \begin{Bmatrix} t_{BF} + t_{F \to G} \\ t_{BE} + t_{E \to G} \end{Bmatrix} \tag{8.1.16}$$

$$= \min \begin{Bmatrix} 4 + 3 \\ 2 + 3 \end{Bmatrix} = 5_{(BEDG)} \tag{8.1.17}$$

and

$$t_{C\to G} = \min\begin{Bmatrix} t_{CD} + t_{D\to G} \\ t_{CE} + t_{E\to G} \end{Bmatrix} \tag{8.1.18}$$

$$= \min\begin{Bmatrix} 5 + 2 \\ 2 + 3 \end{Bmatrix} = 5_{(CEDG)}. \tag{8.1.19}$$

Finally, we have

$$t_{A\to G} = \min\begin{Bmatrix} t_{AB} + t_{B\to G} \\ t_{AC} + t_{C\to G} \end{Bmatrix} \tag{8.1.20}$$

$$= \min\begin{Bmatrix} 3 + 5 \\ 1 + 5 \end{Bmatrix} = 6_{(ACEDG)}. \tag{8.1.21}$$

Once again we have obtained the now-familiar solution, which tells us that a minimum travel time of six time units is obtained along path *ACEDG*.

The technique that we have just used to obtain our solution is very similar to that of the previous dynamic programming solution; the computational effort required for each problem is, in fact, identical. Thus we see that dynamic programming can be applied to this problem with equal efficiency by moving in either a forward or backward direction. There are many problems, however, in which only one end point is known, and the objective is to find the other end point such that some performance criterion is optimized. For problems of this nature the computation will always proceed along a path directed away from the known point. We shall encounter some more complex problems with various types of end-point conditions later in this chapter.

8.2 Dynamic Programming Fundamentals

To continue our discussion of dynamic programming on a more formal level, we must introduce certain definitions and basic ideas. These concepts provide a common basis for all dynamic programming problems. Furthermore, use of these ideas will offer us greater insight into the fundamental strategies that underlie all sequential decision processes.

Consider a series of events that occur in some successive order—such as a succession of discrete changes in the value of an objective function. Let us describe this series of events graphically, as in Figure 8.2. In this figure each of the squares represents a specific event—the transformation of the input value of u into an output value of u as a result of a decision x (e.g., u_i is transformed into u_{i-1} as a consequence of x_i in the ith square).

Each square in Figure 8.2 represents a *stage*. A stage may represent a physical quantity, such as some particular machine that is embedded within

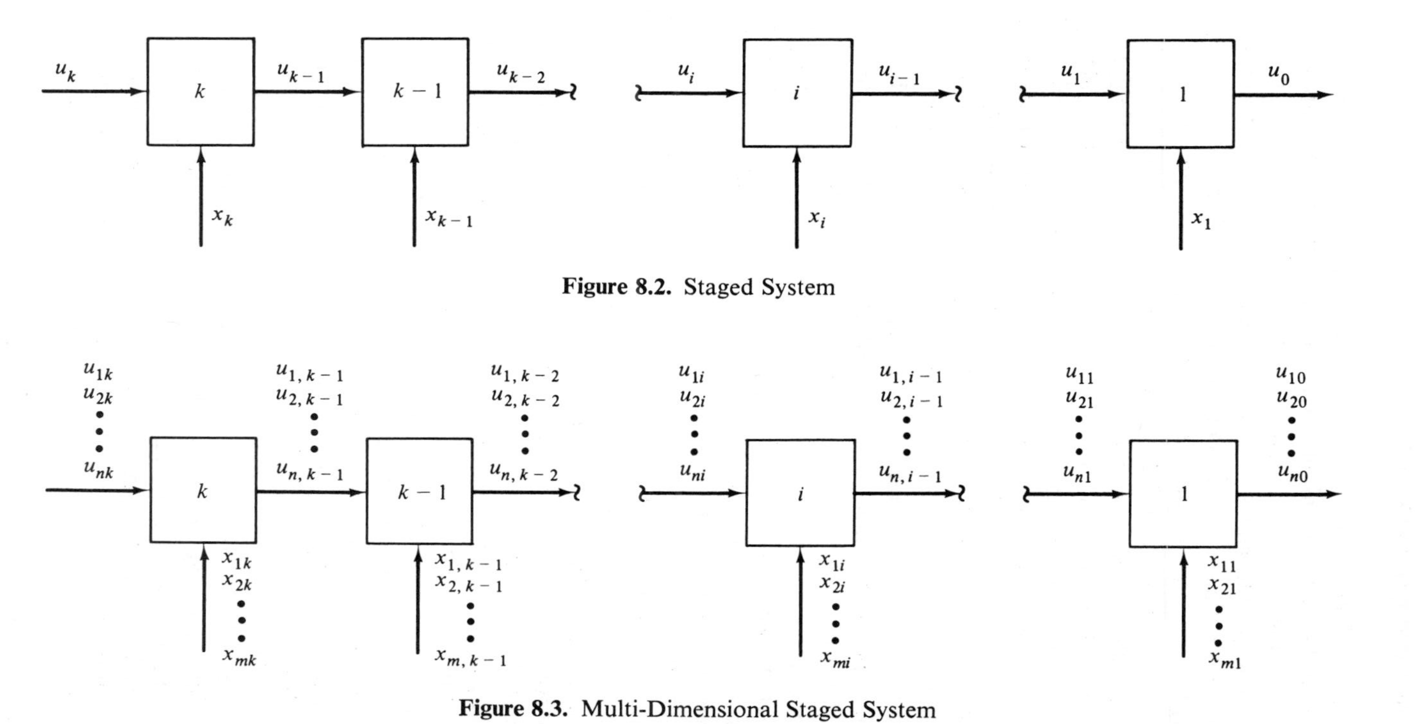

Figure 8.2. Staged System

Figure 8.3. Multi-Dimensional Staged System

a series of machines. More often, though, a stage represents some less tangible entity, such as a discrete unit of space or time. In the minimum-path problem discussed in the last section, nodes B and C make up one stage, node E constitutes a stage, and nodes D and F represent a stage. This example is not representative of a simple series of stages, however, because one can proceed directly from C to D or from B to F without going through point E.

If we return to Figure 8.2. it is seen that the u's which connect the stages represent different values of the *state* of the system. Accordingly, we refer to the variable u as simply a *state variable*. State variables undergo some change in value at each stage. Moreover, the system can be described at each stage by the corresponding values of the state variables. In our minimum-path problem (as in many problems) the state variable does not vary continuously between stages. Rather, the state of the system is given by a single letter, which refers to a particular node (a location) within the system.

The x_i's in Figure 8.2 represent parameters that determine the contribution of a particular stage to the overall objective function. These x_i's are called *decision variables*. In general the decision variables must be chosen so as to optimize some segment of the objective function at each stage. Thus it is the set of decision variables for all stages that must be determined in order to obtain the overall optimum for a system. A set of values for the decision variables is known as a *policy*.

Note that the above definitions are simply extensions of the definitions presented in Chapter 7. This becomes particularly evident if the continuous interval $t_0 \leq t \leq t_f$ is replaced by a sequence of finite increments and each increment is considered to be a stage.

In some problems the decision variables represent physically identifiable variables. For example, if a system consisted of a collection of machines operating in series, with material flowing from machine to machine, the decision variables might represent the individual settings for each machine (e.g., valve settings, temperatures, speeds, etc.). In other problems the decision variables are less obvious. In our minimum-path problem the decision variables are not continuously varying quantities but are the time subscripts that indicate which path to take between nodes. The policy for this problem is a particular set of nodes that defines a path from node A to node G. The *optimal policy* is, of course, that collection of nodes which defines the minimum-time path. The optimal policy need not necessarily be unique.

Returning to Figure 8.2 for a moment, notice that each stage has two input variables—u_i and x_i, and one output variable—u_{i-1}. Therefore, at each stage we require an equation that allows us to express the output variable as a function of the input variables. We can write such an equation in general form as

$$u_{i-1} = T_i(u_i, x_i). \tag{8.2.1}$$

This is known as a *stage-transformation* equation, since it represents the transformation of the state variable as it passes through the ith stage. In order to solve a specific problem using dynamic programming, we require an explicit transformation equation at each stage.

A dynamic programming problem need not be confined to one state variable and one decision variable. Figure 8.3 illustrates the more general case of n state variables and m decision variables per stage. Problems of this type can be solved by dynamic programming, providing they have the proper serial structure and at each stage an explicit stage-transformation equation can be written for each of the state variables.

As a practical matter, however, the computational effort required to solve a problem by dynamic programming increases exponentially as the number of state variables increases. In fact, it is generally not possible to solve a problem having more than three state variables, even on the largest computers, because of core storage limitations and excessive computing time requirements.

The situation is somewhat more encouraging with regard to decision variables. It is possible to consider several decision variables at each stage, although it may be necessary to determine the value of these variables simultaneously in order to obtain an optimal value for the objective function. Thus we may be faced with the need to solve a nonlinear programming problem with respect to the decision variables at each stage. Moreover, the optimal decision policy at each stage must be stored with similar information for all other stages in order to obtain a final optimum for a given system. Hence the storage requirements that are imposed upon a computer can easily become excessive.

From the particular structure of Figures 8.2 and 8.3 we infer that not all problems can be solved by dynamic programming. To use dynamic programming, we must first have a problem that can be expressed in terms of a serial structure. Also we must be able to identify an individual *return function* for each stage. Such a function will necessarily be dependent upon the inputs to the corresponding stage; hence we shall refer to the return function for the ith stage as $r_i(u_i, x_i)$. Finally, we must relate the *overall objective function* $y(u_k, x_k, x_{k-1}, \ldots, x_1)$ to the individual return functions $r_i(u_i, x_i)$, $i = k, k-1, \ldots, 1$. Usually, although not always, the overall objective function will be expressed in terms of the *sum* of the individual returns; i.e.,

$$y(u_k, x_k, x_{k-1}, \ldots, x_1) = r_k(u_k, x_k) + r_{k-1}(u_{k-1}, x_{k-1}) + \cdots + r_1(u_1, x_1). \tag{8.2.2}$$

Notice that each return function depends upon the decision variables corresponding only to that particular stage. We shall see that this is a most important property in structuring a dynamic programming problem.

The object of the game, of course, is to choose the particular value of all decision variables that will cause the objective function to be maximized (or minimized). If we denote this maximum (minimum) value for k stages as $y_k(u_k)$, then we can write

$$y_k(u_k) = \max_{x_k, x_{k-1}, \ldots, x_1}(\min)\{y(u_k, x_k, x_{k-1}, \ldots, x_1)\} \tag{8.2.3}$$

or

$$y_k(u_k) = \max_{x_k, x_{k-1}, \ldots, x_1}(\min)\{r_k(u_k, x_k) + r_{k-1}(u_{k-1}, x_{k-1}) + \cdots + r_1(u_1, x_1)\}. \tag{8.2.4}$$

In our minimum-path problem, the return functions are simply the travel times between adjacent nodes, e.g., t_{AB}. The overall objective function is the time required to travel from point A to point G. The desired optimal value is simply the minimum time required to travel from A to G.

8.3 Generalized Mathematical Formulation

In this section we shall develop a formal mathematical expression for the basic strategy that underlies all dynamic programming problems. We shall then see that this formalism easily reduces to the simple expressions employed in our minimum-travel-time problem.

Serial decomposition

To begin our discussion, consider a two-state system, as shown in Figure 8.4. Suppose that the output state of the system, u_0, is fixed, and we wish to determine the values of x_1 and x_2 that will cause the objective function to be maximized for a given value of u_2 (i.e., a given initial stage). Hence we can write

$$y_2(u_2) = \max_{x_2, x_1}\{r_2(u_2, x_2) + r_1(u_1, x_1)\}. \tag{8.3.1}$$

Notice that the objective function is expressed as the sum of the individual stage returns, and that each return function depends only upon the decision variable for the corresponding stage. However, the input state to stage 1 is equivalent to the output state from stage 2; thus the value of the state variable u_1 will depend upon the choice of the decision variable x_2.

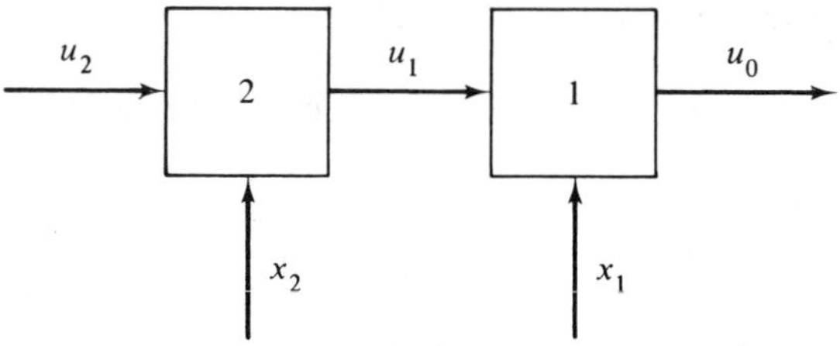

Figure 8.4. Two-Stage System

Since the choice of x_1 affects only $r_1(u_1, x_1)$, we can rewrite equation (8.3.1) as

$$y_2(u_2) = \max_{x_2} \{r_2(u_2, x_2) + \max_{x_1} [r_1(u_1, x_1)]\}. \tag{8.3.2}$$

Observe that Equation (8.3.2) is, *in general*, a more restrictive expression than Equation (8.3.1). That is, if it were not for the particular structure of this problem, we would write

$$\max_{x_2} \{r_2(u_2, x_2) + \max_{x_1} [r_1(u_1, x_1)]\} \leq \max_{x_2, x_1} \{r_2(u_2, x_2) + r_1(u_1, x_1)\}. \tag{8.3.3}$$

The condition which allows Equation (8.3.3) to become a strict equality is the fact that $r_2(u_2, x_2)$ is independent of x_1. If this were not true, then the functional equations that we are about to develop would not be valid.

Since we have moved the maximization with respect to x_1 inside the brackets and applied it only to r_1, it is natural to see whether or not we can apply the maximization with respect to x_2 only to the return function r_2. In other words, can we rewrite Equation (8.3.2) in the form

$$y_2(u_2) = \max_{x_2} [r_2(u_2, x_2)] + \max_{x_1} [r_1(u_1, x_1)]? \tag{8.3.4}$$

If this were the case, then the determination of the overall maximum would be most elementary, since we would only have to maximize each individual return and add up the results.

Unfortunately, things are not this simple. The reason is that the choice of x_2 affects the value of u_1, which in turn will affect the optimal value of $r_1(u_1, x_1)$. Thus r_1 depends upon x_2 as well as on x_1; however, r_2 depends only upon x_2. The explanation for this seemingly contradictory state of affairs is the direction of flow of the state variable: left to right. So the output of r_2 becomes the input to r_1.

Hence we *cannot* decompose our overall maximization problem to the extent indicated by Equation (8.3.4). The correct expression, which we repeat below, is Equation (8.3.2):

$$y_2(u_2) = \max_{x_2} \{r_2(u_2, x_2) + \max_{x_1} [r_1(u_1, x_1)]\}. \tag{8.3.2}$$

Now let us refer to

$$\max_{x_1} [r_1(u_1, x_1)]$$

as simply $y_1(u_1)$. The significance of this seemingly unimportant change in notation will soon become apparent. Equation (8.3.2) then becomes

$$y_2(u_2) = \max_{x_2} \{r_2(u_2, x_2) + y_1(u_1)\}. \tag{8.3.5}$$

This is the desired general formulation for a two-stage system. The meaning of this equation is that the overall maximum is equal to the *greatest* sum of (1) the return from stage 2 (which determines u_1), and (2) the *maximum* return from stage 1 (given the correct value for u_1).

Let us now consider a three-stage system, as represented in Figure 8.5. Notice that the structure of this problem is identical to our two-stage problem, the only difference being the addition of one additional stage. Again considering a maximization problem, we can write the objective function as

$$y_3(u_3) = \max_{x_3, x_2, x_1} \{r_3(u_3, x_3) + r_2(u_2, x_2) + r_1(u_1, x_1)\}. \tag{8.3.6}$$

Since, as before, the decision variable x_1 affects only stage 1, we can write Equation (8.3.6) as

$$y_3(u_3) = \max_{x_3, x_2} \{r_3(u_3, x_3) + r_2(u_2, x_2) + \max_{x_1} [r_1(u_1, x_1)]\}, \tag{8.3.7}$$

or

$$y_3(u_3) = \max_{x_3, x_2} \{r_3(u_3, x_3) + r_2(u_2, x_2) + y_1(u_1)\}. \tag{8.3.8}$$

Now let us examine the dependence of Equation (8.3.8) on the decision variable x_2. Clearly, $r_3(u_3, x_3)$ is unaffected by x_3, whereas $r_2(u_2, x_2)$ is obviously dependent upon x_2. Furthermore, the choice of x_2 will affect the output from stage 2, which is u_1; hence $y_1(u_1)$ is also dependent upon the

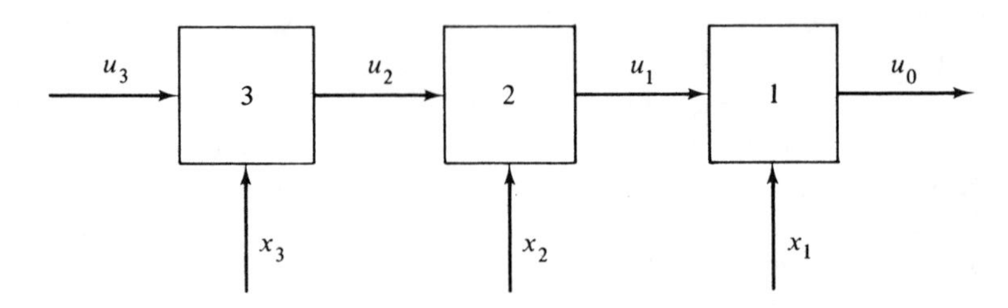

Figure 8.5. Three-Stage System

value of the decision variable x_2. When can therefore restrict our attention to the last two stages with respect to x_2, and Equation (8.3.8) becomes

$$y_3(u_3) = \max_{x_3} = \{r_3, (u_3, x_3) + \max_{x_2}[r_2, (u_2, x_2) + y_1(u_1)]\} \quad (8.3.9)$$

or

$$y_3(u_3) = \max_{x_3}\{r_3(u_3, x_3) + y_2(u_2)\}. \quad (8.3.10)$$

The similarity between Equations (8.3.5) and (8.3.10) is noteworthy.

The meaning of Equation (8.3.10) is as follows: Given an input value for the state variable (i.e., given a value for u_3), and given a fixed value for the output variable u_0, the maximum return resulting from the transformation of u_3 into u_0 is equal to the *greatest* sum of (1) the return from stage 3 (which determines u_2), and (2) the *maximum* return from the remaining two stages, given a value for u_2 that corrresponds to the outlet value from stage 3.

We could easily extend this train of thought to a four-stage system, then a five-stage system, and so on. The logic is unchanged, although the development becomes more lengthy. In any event, we would merely confirm that the objective function can be expressed in a manner analogous to Equation (8.3.10); i.e.,

$$y_4(u_4) = \max_{x_4} \{r_4(u_4, x_4) + y_3(u_3)\}, \quad (8.3.11)$$

$$y_5(u_5) = \max_{x_5} \{r_5(u_5, x_5) + y_4(u_4)\}, \quad (8.3.12)$$

and so on. The entire development can be generalized by writing the objective function for a k-stage system as

$$y_k(u_k) = \max_{x_k}(\min) \{r_k(u_k, x_k) + y_{k-1}(u_{k-1})\}. \quad (8.3.13)$$

The only restriction on the validity of Equation (8.3.13) is the particular serial-like structure of the problem as shown in Figure 8.2. At each stage, however, we shall require a stage-transformation equation to calculate u_{k-1} as a function of u_k and x_k.

Notice that the above equations are entirely general in nature; i.e., the equations are valid for any set of returns and any objective function, as long as the problem has the correct structure. Hence the result may be applied to a variety of problems.

Validity of the decomposition procedure

The question of what kinds of problems can be decomposed into a sequence of individual-stage optimization problems has received some attention (cf. Mitten [1964]). Functional equations that reflect the dynamic programming strategy can be written for certain kinds of problems other than problems with additive returns. A sufficient, though not necessary, condition is that the overall objective function be *separable* in terms of the individual returns, and that the k-stage objective function $y[r_k(u_k, x_k), r_{k-1}(u_{k-1}, x_{k-1}), \ldots, r_1(u_1, x_1)]$ be a *monotonically nondecreasing* function of $r_{k-1}(u_{k-1}, x_{k-1}), r_{k-2}(u_{k-2}, x_{k-2}), \ldots, r_1(u_1, x_1)$. These conditions, which we shall assume to be satisfied in the remainder of this chapter, are discussed in considerable detail by Nemhauser [1966].

EXAMPLE 8.3.1

As a specific example of the application of the functional equation concept to a particular problem, let us once again return to the minimum-path problem of Section 8.1. For simplicity we shall begin with only a part of the original problem—the determination of the minimum travel time from point E to point G, as shown in Figure 8.6. This may be thought of as a two-stage system, as represented by the dashed lines in the figure. Recall that each node represents a particular state of the system, and the state is transformed (i.e., we travel from one node to an adjacent node) within each stage. The decision variables indicate the path we take through the network.

Let us refer to the minimum travel time from point E to point G as $y_2(u_2)$, where u_2 is the state of the system at the start of the two-stage traversal. In other words, u_2 refers to a particular point—node E. Thus we can write the objective function as simply $y_2(E)$. In terms of a functional equation, we can write

$$y_2(E) = \min \{r_2(E, x_2) + y_1(u_1, x_1)\}, \tag{8.3.14}$$

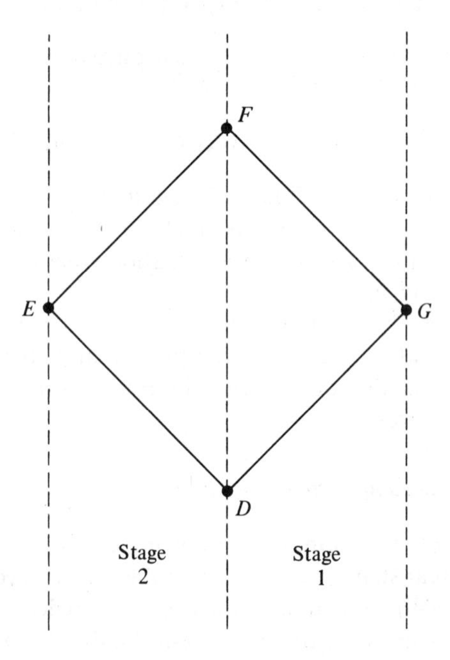

Figure 8.6. Relationship Between Path Nodes and Stages

where x_2 and x_1 refer to the paths leading from E to u_1 and from u_1 to G, respectively. Note that u_1 represents either node D or node F.

By consideration of both alternatives (i.e., that x_2 leads us to $u_1 = F$ and x_1 leads us from F to G, or that x_2 leads to $u_1 = D$ and x_1 leads from D to G), we can rewrite equation (8.3.14) as

$$y_2(E) = \min \begin{Bmatrix} r_2(E, E \longrightarrow F) + y_1(F, F \longrightarrow G) \\ r_2(E, E \longrightarrow D) + y_1(D, D \longrightarrow G) \end{Bmatrix}. \tag{8.3.15}$$

However, the return function r_2 is simply the minimum travel time in stage 2, so that

$$r_2(E, E \longrightarrow F) = t_{EF} \tag{8.3.16}$$

and

$$r_2(E, E \longrightarrow D) = t_{ED}, \tag{8.3.17}$$

in terms of the notation used in Section 8.1. Similarly, y_1 can be expressed as

$$y_1(F, F \longrightarrow G) = t_{F \to G} \tag{8.3.18}$$

and

$$y_1(D, D \longrightarrow G) = t_{D \to G}. \tag{8.3.19}$$

Hence our functional equation can be written

$$y_2(E) = \min \begin{Bmatrix} t_{EF} + t_{F \to G} \\ t_{ED} + t_{D \to G} \end{Bmatrix}. \tag{8.3.20}$$

Since $y_2(E)$ is simply $t_{E \to G}$, we have

$$t_{E \to G} = \min \begin{Bmatrix} t_{EF} + t_{F \to G} \\ t_{ED} + t_{D \to G} \end{Bmatrix}, \tag{8.3.21}$$

which is simply Equation (8.1.14). Thus we see that our functional equation easily reduces to the minimum-time expressions of Section 8.1, providing each of the terms in the functional equation is interpreted properly. It should be noted, however, that the correct interpretation of the terms in the functional equation is not obvious, and such interpretation usually requires considerable care.

We cannot extend the functional-equation concept to the complete minimum-path problem as it was originally developed. This is because of the structure of the original problem. However, we can modify our representation of the problem slightly, without altering the actual problem, so as to have a serial-like structure to which we can apply our functional equation.

Suppose, for example, that we add two additional nodes to the network, one between B and F, which we shall call B', and the other, called C', between points C and D. This representation of the problem is shown in Figure 8.7. Notice that the distances from B to F and from C to D have been arbitrarily divided between the respective nodes. We now have a properly structured problem, so that the functional

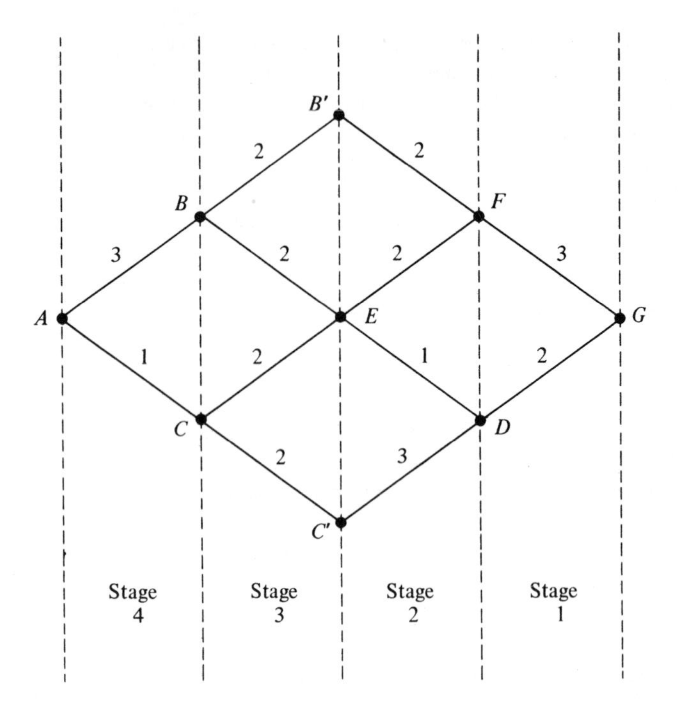

Figure 8.7. Minimum-Path Problem as a Four-Stage System

equation can be applied, resulting in a solution which reduces to that given in Section 8.1.

It would at this point be instructive for the reader to work out the entire minimum-path problem using the functional notation developed earlier. We shall apply the functional-equation concept to other kinds of dynamic programming problems in subsequent sections of this chapter.

8.4 The Principle of Optimality

In developing a functional equation to represent the dynamic programming strategy, we have seen that the determination of an optimal policy from stage k through stage 1 *requires previous knowledge* of the optimal policy from stage $k - 1$ through stage 1, given the state of the system on entering stage $k - 1$. Generally, we must determine a $(k - 1)$-stage optimal policy for *each of several values* of u_{k-1}, the correct value of u_{k-1} not being known until the k-stage optimization is carried out. Thus we see that, given the state of the system entering the $(k - 1)$st stage (u_{k-1}), *we must determine the optimal policy for the remainder of the system* (*stage* k $-$ 1 *through stage* 1) *without regard for the decisions to be made in the earlier part of the system* (*stage* n *through stage* k). The object is to build up the optimal policies for successively increasing numbers of stages in such a manner that the state variables remain consistent; i.e., the stage of the system leaving the kth stage must always be equivalent to the state of the system when entering the $(k - 1)$st stage. This idea is the basic strategy of dynamic programming.

Richard Bellman, the creator of the dynamic programming algorithm, refers to the basic strategy as the *principle of optimality*. To quote Bellman [1957], "An optimal policy has the property that whatever the initial state and initial decision are, the remaining decisions must constitute an optimal policy with regard to the state resulting from the first decision." In other words, given a value for x_k and for u_k (referring back to Figure 8.2), the policy for the remaining stages ($k - 1$ through 1) must be optimal with respect to u_{k-1} (where u_{k-1} is determined from the values of x_k and u_k by means of the stage-transformation equations). Moreover, the optimality with respect to the remaining $(k - 1)$ stages must hold regardless of the particular values of u_k and x_k that may be given. Hence the need to obtain a $(k - 1)$-stage optimal policy for each of several values of u_{k-1}.

The basic concept represented by the principle of optimality is simple but tricky, and its complete mastery is very elusive. We shall therefore present some additional examples of problems solved by dynamic programming, thus clarifying the use of functional equations and the meaning of the principle of optimality.

8.5 Discrete Decision Models—An Optimal Automobile Replacement Policy

As an example of the use of functional equations as well as an interesting application of the principle of optimality, let us consider a problem that involves a set of discrete yes–no decisions—that of determining when to trade in an automobile for a shiny new model. Although there may be as many opinions on this subject as there are automobile owners, we shall see that dynamic programming allows one to make certain rational decisions regarding an optimal policy.

Optimal policy for a finite time period

Let us examine 15 consecutive time intervals of 1 year each; hence a total period of 15 years. At the beginning of each time interval we shall make one simple decision—whether to replace the car with a newer one, or to keep the car for at least 1 more year. Based upon a sequence of such decisions, we shall construct a keep–replace stragy for the entire 15-year period.

Table 8.2. AUTOMOBILE ECONOMIC DATA

t	C	T	E
0	3000	2600	600
1	2250	1950	600
2	1800	1500	650
3	1350	1100	725
4	1000	800	800
5	750	550	900
6	600	400	1000
7	500	300	1200
8	400	200	1400
9	300	100	1600
10	200	0	1800
11	200	0	2000
12	200	0	2000
13	200	0	2000
14	200	0	2000
15	200	0	2000

C = cost of car t years old
T = trade-in value of car t years old
E = operating and maintenance expense of car t years old

The present example will be based upon replacement of an existing automobile with a newer model of age τ. Automobile costs as a function of age are given in Table 8.2. Also shown are the trade-in value of an originally new car with time, and the annual operating expense as a function of time. Notice that a car more than 9 years old has no trade-in value; but a used car, in good condition, will always cost at least \$200. Furthermore, the annual operating expense, which includes maintenance and major repairs as well as gas, oil, and insurance, becomes very high as the car ages. This is to account for the possibility that the car may suffer a total breakdown, requiring an enormous expenditure to be put back into reasonably good condition.

The age of the "new" car is determined by the value of the parameter τ. By maintaining $\tau = 0$ throughout the problem, we shall be considering trading for a brand-new car. If $\tau = 2$, we are considering a trade for a 2-year-old car, and so on. This parameter will be held constant throughout any given problem.

Let us now proceed to define the objective function. Suppose that we wish to minimize the aggregate of operating expense, depreciation, and purchase cost over a period of several years. In any year if we should choose to keep a car of age t for 1 more year, then the cost can be expressed as the sum of the operating and depreciation expenses. Thus

$$r_i(t) = E(t) + D(t), \tag{8.5.1}$$

where the depreciation can be expressed as the difference in trade-in values for successive years; i.e.,

$$D(t) = T(t) - T(t+1), \quad t > \tau. \tag{8.5.2}$$

An exception to this rule should be made in the first year of ownership, for the depreciation in this case is the purchase price less the trade-in value 1 year later. Hence

$$D(\tau) = C(\tau) - T(\tau + 1). \tag{8.5.3}$$

Suppose now that we decide to trade in a car of age t for a newer car of age τ (where τ remains fixed throughout the problem). The total cost incurred during that year will consist of the net purchase cost and the operating and depreciation expenses of the newer car. Hence

$$r_i(t) = C(\tau) - T(t) + E(\tau) + D(\tau). \tag{8.5.4}$$

The overall objective function can be expressed as $y_i(t)$, the minimum cost associated with following an optimal policy for a car t years old over i remaining time intervals (years). Hence we can write the following functional equation for $y_i(t)$:

$$y_i(t) = \min \{r_i(t) + y_{i-1}(t+1)\}. \tag{8.5.5}$$

By use of Equations (8.5.1) and (8.5.4), we can write Equation (8.5.5) more explicitly as

$$y_i(t) = \min \begin{Bmatrix} E(t) + D(t) + y_{i-1}(t+1) \\ C(\tau) - T(t) + E(\tau) + D(\tau) + y_{i-1}(\tau + 1) \end{Bmatrix}, \tag{8.5.6}$$

where we shall choose whichever line yields the lesser value. Selection of the top line indicates a decision to keep the old car for at least 1 more year, whereas selection of the bottom line represents a decision to replace the old car with a newer car of age τ.

To begin the actual computation, let us consider the problem of when to trade for a brand-new car; i.e., $\tau = 0$. If we consider only a one-stage problem, the objective function can be written

$$y_1(t) = \min \begin{Bmatrix} E(t) + D(t) \\ C(0) - T(t) + E(0) + D(0) \end{Bmatrix}. \tag{8.5.7}$$

Since the state of the system (i.e., the age of the old car) is unknown when entering the last stage, we must obtain an optimal one-stage policy for each of several initial states. For $t = 0$, we have

$$y_1(0) = \min \begin{Bmatrix} E(0) + D(0) \\ C(0) - T(0) + E(0) + D(0) \end{Bmatrix}, \tag{8.5.8}$$

which becomes (cf. Table 8.2)

$$y_1(0) = \min \begin{Bmatrix} 600 + 1050 \\ 3000 - 2600 + 600 + 1050 \end{Bmatrix} = 1650(K). \tag{8.5.9}$$

Since the top line resulted in a lower value than the bottom line, we see that the optimal policy is to keep the car for the remaining year; hence the (K) following the cost.

When $t = 1$, we can write

$$y_1(1) = \min \begin{Bmatrix} E(1) + D(1) \\ C(0) - T(1) + E(0) + D(0) \end{Bmatrix}, \tag{8.5.10}$$

which takes on the value

$$y_1(1) = \min \begin{Bmatrix} 600 + 450 \\ 3000 - 1950 + 600 + 1050 \end{Bmatrix} = 1050(K). \tag{8.5.11}$$

Again, it is cheaper to keep the 1-year-old car for another year than to trade it in on a new one.

If we continue to evaluate $y_1(t)$ for $t = 2, 3, \ldots, 15$, we obtain the following table:

t	$y_1(t)$	t	$y_1(t)$
0	1650 (K)	8	1500 (K)
1	1050 (K)	9	1700 (K)
2	1050 (K)	10	1800 (K)
3	1025 (K)	11	2000 (K)
4	1050 (K)	12	2000 (K)
5	1050 (K)	13	2000 (K)
6	1100 (K)	14	2000 (K)
7	1300 (K)	15	2000 (K)

Notice that in all cases the decision is to keep rather than replace.

Now let us consider a two-stage problem, maintaining a fixed value of $\tau = 0$. Our functional equation is

$$y_2(t) = \min \begin{Bmatrix} E(t) + D(t) + y_1(t+1) \\ C(0) - T(t) + E(0) + D(0) + y_1(1) \end{Bmatrix}. \tag{8.5.12}$$

When $t = 0$ we have

$$y_2(0) = \min \begin{Bmatrix} 600 + 1050 + 1050 \\ 3000 - 2600 + 600 + 1050 + 1050 \end{Bmatrix} = 2700(K). \tag{8.5.13}$$

Similarly, when $t = 1$,

$$y_2(1) = \min \begin{Bmatrix} 600 + 450 + 1050 \\ 3000 - 1950 + 600 + 1050 + 1050 \end{Bmatrix} = 2100(K), \tag{8.5.14}$$

and when $t = 2$,

$$y_2(2) = \min \begin{Bmatrix} 650 + 400 + 1025 \\ 3000 - 1500 + 600 + 1050 + 1050 \end{Bmatrix} = 2075(K), \tag{8.5.15}$$

and so on.

Again summarizing the optimal results for $t = 0, 1, \ldots, 15$, we obtain the following table:

t	$y_2(t)$	t	$y_2(t)$
0	2700 (K)	8	3200 (K)
1	2100 (K)	9	3500 (K)
2	2075 (K)	10	3800 (K)
3	2075 (K)	11	4000 (K)
4	2100 (K)	12	4000 (K)
5	2150 (K)	13	4000 (K)
6	2400 (K)	14	4000 (K)
7	2800 (K)		

Notice that we have had to use data from the previous table in order to construct the present table.

Let us now generate additional such tables for $i = 3, 4, \ldots, 15$, and combine all the result into one large table. For example, the values of $y_1(t)$ that follow Equation (8.5.11) have been entered into the first row of Table 8.3; the values of $y_2(t)$ that follow Equation (8.5.15) appear in the second row

Table 8.3. SOLUTION OF AUTOMOBILE REPLACEMENT PROBLEM FOR $\tau = 0$

	T = 0.	1.	2.	3.	4.	5.	6.	7.	8.	9.	10.	11.	12.	13.	14.
1	1650.	1050.	1050.	1025.	1050.	1050.	1100.	1300.	1500.	1700.	1800.	2000.	2000.	2000.	2000.
2	2700.	2100.	2075.	2075.	2100.	2150.	2400.	2800.	3200.	3500.	3800.	4000.	4000.	4000.	4000.
3	3750.	3125.	3125.	3125.	3200.	3450.	3900.	4500.	5000.	5500.	5800.	6000.	6000.	6000.	4000.
4	4775.	4175.	4175.	4225.	4500.	4950.	5600.	6300.	7000.	7500.	−7775.	−7775.	−7775.	−7775.	−7775.
5	5825.	5225.	5275.	5525.	6000.	6650.	7400.	8300.	−8625.	−8725.	−8825.	−8825.	−8825.	−8825.	−8825.
6	6875.	6325.	6575.	7025.	7700.	8450.	9400.	−9575.	−9675.	−9775.	−9875.	−9875.	−9875.	−9875.	−9875.
7	7975.	7625.	8075.	8725.	9500.	−10425.	−10575.	−10675.	−10775.	−10875.	−10975.	−10975.	−10975.	−10975.	−10975.
8	9275.	9125.	9775.	10525.	−11475.	−11725.	−11875.	−11975.	−12075.	−12175.	−12275.	−12275.	−12275.	−12275.	−12275.
9	10775.	10825.	11575.	12500.	12775.	12925.	13075.	13375.	−13575.	−13675.	−13775.	−13775.	−13775.	−13775.	−13775.
10	12475.	12625.	13550.	13800.	13975.	14125.	14475.	14875.	15175.	−15375.	−15475.	−15475.	−15475.	−15475.	−15475.
11	14275.	14600.	14850.	15000.	15175.	15525.	15975.	16475.	16875.	−17175.	−17275.	−17275.	−17275.	−17275.	−17275.
12	16250.	15900.	16050.	16200.	16575.	17025.	17575.	18175.	18675.	18975.	19075.	−19250.	−19250.	−19250.	−19250.
13	17550.	17100.	17250.	17600.	18075.	18625.	19275.	19975.	−20350.	−20450.	−20550.	−20550.	−20550.	−20550.	−20550.
14	18750.	18300.	18650.	19100.	19675.	20325.	21075.	−21450.	−21550.	−21650.	−21750.	−21750.	−21750.	−21750.	−21750.
15	19950.	19700.	20150.	20700.	21375.	22125.	−22550.	−22650.	−22750.	−22850.	−22950.	−22950.	−22950.	−22950.	−22950.

of Table 8.3; and so on. We see that each row in Table 8.3 is dependent upon the previous row because of the dependence of $y_i(t)$ on either $y_{i-1}(t+1)$ or $y_{i-1}(\tau+1)$.

Each decison is a "keep" for $t \geq 0$, $i = 1, 2$, and 3. When $i = 4$, the decision is to "keep" for $t \leq 9$, but to "replace" when $t \geq 10$. In other words, for $i = 4$, the optimal policy is to keep a car less than 10 years old for 4 additional years, but to trade a car which is at least 10 years old for a new one.

We have indicated the "replace" decision by prefixing a minus sign to $y_i(t)$ in Table 8.3. For each i the keep–replace decisions have been separated by the heavy line that zigzags through Table 8.3. The reader should at this point reproduce at least some of the numerical entries in Table 8.3 so that he becomes familiar with the generation of such a table.

Determining the optimal policy

Once Table 8.3 has been generated we are by no means finished, since the use of the table requires considerable care and insight into the meaning of each numerical entry. Let us first concentrate on the determination of an optimal replacement policy for the stated 15-year period. We shall see that this information can be extracted from Table 8.3 quite easily. Later, however, we shall see that Table 8.3 can be used to obtain a more general result, which is independent of the length of the time period, by expending only a little more effort.

Suppose that we now consider beginning a 15-year time period with a brand-new car (i.e., $i = 15$, $t = 0$). We see that the optimal policy is to keep the new car for at least 1 more year, which moves us diagonally up the chart to $i = 14$, $t = 1$. Now we have a 1-year-old car, with 14 more years to consider. Again we keep the car for at least another year, and so on, until we have kept the car for 7 years. After 7 years, we have a car 7 years old, and we have 8 more years to consider. If we follow our way diagonally up Table 8.3, we see that we enter the replace region when $i = 8$ and $t = 7$, so that we trade our car at the end of the seventh year in order to follow an optimal policy. After the eighth year, we shall again have a 1-year-old car, so that $i = 7$ and $t = 1$. We continue to follow our way diagonally upward, arriving at a keep decision at $i = 1$, $t = 7$.

The cost of following this optimal policy can be determined for any number of stages up to and including 15. For example, the cost for the entire 15-year period is $19,950, or $1330 per year. The last 10 years will cost $14,125, or $1412.50 per year, and the last 5 years will cost $5525, or $1105 per year. The *first* 5 years will cost $19,950 − $14,125 = $5825, or $1165 per year, and the *first* 10 years will cost $19,950 − $5525 = $14,425, or $1442.50 each year. The cost for the *second* 5 years is $14,125 − $5525 = $8600, which is equivalent to $1720 per year. Notice that the second 5-year period appears most expensive; this is because the one-and-only purchase is made during this time.

Optimal replacement cycle for an infinite time period

Having analyzed the given 15-year period, let us now turn our attention to a more general question—that of determining an optimal trade-in *cycle* over an infinite time span. We shall approximate infinite time with a 15-year period. The 15-year time period is not unique, however, and we could consider a lesser time period by using only a part of Table 8.3, or a greater time period by adding more entries to the table.

If the line separating the "keep" from the "replace" decisions were vertical rather than zigzaged, this would be easy to do. Presumably, this line would become vertical if we were to consider a large enough number of stages. Given simply the 15-stage data of Table 8.3, however, we can still determine an approximate optimal cycle time as follows.

We have seen previously that the optimal policy, beginning a 15-year period with a new car, is to keep the car for 7 years and then trade it in. The cost of this 7-year cycle is the trade-in cost, \$3000 − \$300 = \$2700, plus the operating and depreciation cost, \$19,950 − \$11,975. This totals to \$10,675, or \$1525 per year.

Now consider a 14-year time period. Beginning with a new car (i.e., $i = 14$, $t = 0$), we see that the optimal policy is to keep the car for only 6 years. The cost will be the sum of the trade-in cost, \$3000 − \$400 = \$2600, and the operating and depreciation expense, \$18,750 − \$11,875 = \$6875. This adds up to \$9475, which is equivalent to \$1579.17 per year—clearly more expensive than the 7-year cycle.

If we consider a 13-year time period, the optimal cycle is to keep for 5 years and then replace. The cost of this policy is \$2450 (for trade-in) + \$5825 (for operation and depreciation), or \$8275. This figure is equivalent to \$1655 per year, which exceeds both the 6- and 7-year cycles.

Finally, consider a 12-year time period. We now keep for only 4 years, which yields a trade-in cost of \$2200 and an operation and depreciation cost of \$4775. On an annual basis this results in a total cost of \$1743.75—the highest figure we have yet obtained.

We cannot consider a time period less than 12 years because such a time period does not involve a replacement decision. The tentative conclusion, then, is that the optimal policy *for a new car* is to keep the car for 7 years and then trade. We assume that this policy would be borne out more consistently if we considered a longer time span (i.e., if $i < 15$).

Variation of problem parameters

Table 8.4 shows the solution of the automobile replacement problem for $\tau = 2$; i.e., given a 2-year-old car, how long should we keep the car before

Table 8.4. SOLUTION OF AUTOMOBILE REPLACEMENT PROBLEM FOR $\tau = 2$

	T = 2.	3.	4.	5.	6.	7.	8.	9.	10.	11.	12.	13.	14.	15.
1	1350.	1025.	1050.	1050.	1100.	1300.	1500.	1700.	1800.	2000.	2000.	2000.	2000.	2000.
2	2375.	2075.	2100.	2150.	2400.	2800.	3200.	3500.	3800.	4000.	4000.	4000.	4000.	2000.
3	3425.	3125.	3200.	3450.	3900.	4500.	5000.	−5125.	−5225.	−5225.	−5225.	−5225.	−5225.	−5225.
4	4475.	4225.	4500.	4950.	5600.	−5975.	−6075.	−6175.	−6275.	−6275.	−6275.	−6275.	−6275.	−6275.
5	5575.	5525.	6000.	6650.	−6975.	−7075.	−7175.	−7275.	−7375.	−7375.	−7375.	−7375.	−7375.	−7375.
6	6875.	7025.	7700.	8025.	8175.	−8375.	−8475.	−8575.	−8675.	−8675.	−8675.	−8675.	−8675.	−8675.
7	8375.	8725.	9075.	9225.	9475.	9775.	−9975.	−10075.	−10175.	−10175.	−10175.	−10175.	−10175.	−10175.
8	10075.	10100.	10275.	10525.	10875.	11275.	11575.	−11775.	−11875.	−11875.	−11875.	−11875.	−11875.	−11875.
9	11450.	11300.	11575.	11925.	12375.	12875.	−13050.	−13150.	−13250.	−13250.	−13250.	−13250.	−13250.	−13250.
10	12650.	12600.	12975.	13425.	13975.	−14150.	−14250.	−14350.	−14450.	−14450.	−14450.	−14450.	−14450.	−14450.
11	13950.	14000.	14475.	15025.	15250.	−15450.	−15550.	−15650.	−15750.	−15750.	−15750.	−15750.	−15750.	−15750.
12	15350.	15500.	16075.	16300.	16550.	−16850.	−16950.	−17050.	−17150.	−17150.	−17150.	−17150.	−17150.	−17150.
13	16850.	17100.	17350.	17600.	17950.	18250.	−18450.	−18550.	−18650.	−18650.	−18650.	−18650.	−18650.	−18650.
14	18450.	18375.	18650.	19000.	19350.	19750.	−20050.	−20150.	−20250.	−20250.	−20250.	−20250.	−20250.	−20250.
15	19725.	19675.	20050.	20400.	20850.	−21225.	−21325.	−21425.	−21525.	−21525.	−21525.	−21525.	−21525.	−21525.

trading it in on another 2-year-old car? Again we see that the line which separates the "keep" from the "replace" decision zigzags down the table, so that there is no clear-cut policy. However, we can see that a 5-year cycle is indicated if we begin a 15-year, 11-year, or 10-year time period with a 2-year-old car. A 7-year cycle results when $i = 14$, a 6-year cycle for $i = 13$ or $i = 12$, and a 4-year cycle for $i = 9$. The total and yearly costs for each cycle are given below:

Length of Cycle	*Total Cost*	*Yearly Cost*
4	$ 5,875	$1468.75
5	7,075	1415.00
6	8,475	1412.50
7	10,075	1439.29

Clearly, the optimal cycle for a 2-year-old car is 6 years. In other words, given a 2-year-old car, keep it until it is 8 years old, and then trade for another 2-year-old car.

Notice that the yearly cost of this policy ($1412.50) is less than the cost of following an optimal policy for a new car ($1525). Thus a trade-in for a 2-year-old car makes more sense economically, than a trade-in for a new car. Observe also that the yearly cost of a 5-year cycle is almost the same as the optimal 6-year cycle, so that the determination of a precise trade-in policy is not crucial in this case.

Now let us again resolve the problem, assuming a trade-in for a *4-year-old* car. The results of this study are shown in Table 8.5. Notice that in this case the optimal keep–replace line stabilizes rapidly, indicating a 3-year cycle. The optimal policy, then, given a 4-year-old car, is to keep the car for 3 years and then trade it on another 4-year-old car. The cost of this policy is $4100 for trade-in plus operating and maintenance expenses, or $1366.67 yearly. This policy is somewhat better than the optimal policy for a 2-year-old car. It would appear from these results that consideration should be given to the cost of an optimal policy when trading for a 3-year-old car and a 5-year-old car. We leave these computations as exercises for the reader.

Finally, the reader is cautioned against taking the results of this study too seriously. The mathematical model that we have considered is an over-simplification of reality. Not included in the model are such significant factors as the earning power of cash reserves and the change in car values with time because of inflation. Also neglected are quantitative expressions for value judgments, such as the value, in dollars, of the greater dependability or the greater personal satisfaction that may be associated with new-car ownership. While introduction of such additional considerations would make the problem considerably more complicated, the basic idea of constructing a sequence of keep–replace decisions would be unchanged.

Table 8.5. SOLUTION OF AUTOMOBILE REPLACEMENT PROBLEM FOR $\tau = 4$

	T = 4.	5.	6.	7.	8.	9.	10.	11.	12.	13.	14.	15.
1	1250.	1050.	1100.	1300.	1500.	1700.	1800.	2000.	2000.	2000.	2000.	2000.
2	2300.	2150.	2400.	2800.	−3100.	−3200.	−3300.	−3300.	−3300.	−3300.	−3300.	−3300.
3	3400.	3450.	3900.	−4100.	−4200.	−4300.	−4400.	−4400.	−4400.	−4400.	−4400.	−4400.
4	4700.	4950.	5200.	−5400.	−5500.	−5600.	−5700.	−5700.	−5700.	−5700.	−5700.	−5700.
5	6200.	6250.	6500.	6800.	−7000.	−7100.	−7200.	−7200.	−7200.	−7200.	−7200.	−7200.
6	7500.	7550.	7900.	−8200.	−8300.	−8400.	−8500.	−8500.	−8500.	−8500.	−8500.	−8500.
7	8800.	8950.	9300.	−9500.	−9600.	−9700.	−9800.	−9800.	−9800.	−9800.	−9800.	−9800.
8	10200.	10350.	10600.	−10900.	−11000.	−11100.	−11200.	−11200.	−11200.	−11200.	−11200.	−11200.
9	11600.	11650.	12000.	−12300.	−12400.	−12500.	−12600.	−12600.	−12600.	−12600.	−12600.	−12600.
10	12900.	13050.	13400.	−13600.	−13700.	−13800.	−13900.	−13900.	−13900.	−13900.	−13900.	−13900.
11	14300.	14450.	14700.	−15000.	−15100.	−15200.	−15300.	−15300.	−15300.	−15300.	−15300.	−15300.
12	15700.	15750.	16100.	−16400.	−16500.	−16600.	−16700.	−16700.	−16700.	−16700.	−16700.	−16700.
13	17000.	17150.	17500.	−17700.	−17800.	−17900.	−18000.	−18000.	−18000.	−18000.	−18000.	−18000.
14	18400.	18550.	18800.	−19100.	−19200.	−19300.	−19400.	−19400.	−19400.	−19400.	−19400.	−19400.
15	19800.	19850.	20200.	−20500.	−20600.	−20700.	−20800.	−20800.	−20800.	−20800.	−20800.	−20800.

A somewhat more realistic automobile-replacement problem, involving a probabilistic model, has been solved by Howard [1960] and reported by Bellman and Dreyfus [1962]. Aside from the use of probabilities, it is interesting to see that Howard approaches the automobile replacement problem from a standpoint of maximizing value rather than minimizing cost.

8.6 Nonlinear Continuous Models

Many problems in nonlinear programming can be transformed into stagewise optimization problems and solved using dynamic programming. As a rule dynamic programming is less effective in solving such problems than the techniques presented in Chapter 5 or, if the problem is particularly simple, the methods of Chapter 2. Nevertheless, it is very instructive to see how a nonlinear programming problem can be solved by means of stagewise decomposition.

Consider a multidimensional minimization problem with one inequality constraint. Suppose that both the objective function and the constraint are written in terms of a sum of one-dimensional functions; i.e.: Minimize

$$y(\mathbf{X}) = f_1(x_1) + f_2(x_2) + \cdots + f_n(x_n), \tag{8.6.1}$$

subject to

$$g(\mathbf{X}) = g_1(x_1) + g_2(x_2) + \cdots + g_n(x_n) \geq b, \tag{8.6.2}$$

where b is some specified constant. We can, if we wish, express the constraining equation as an equality rather than an inequality—this will not alter the strategy we are about to develop. Moreover, we can bound the independent (decision) variables; e.g., $\mathbf{X} \geq \mathbf{0}$.

Introduction of state variables

To solve this problem by means of dynamic programming, we must first introduce an appropriate set of state variables. This can be accomplished as follows: Let

$$g_1(x_1) = u_1 - u_0, \tag{8.6.3}$$

$$g_2(x_2) = u_2 - u_1, \tag{8.6.4}$$

$$g_3(x_3) = u_3 - u_2, \tag{8.6.5}$$

$$\vdots$$

$$g_n(x_n) = u_n - u_{n-1}. \tag{8.6.6}$$

Note that Equations (8.6.3)–(8.6.6) can easily be rearranged to yield a set of n stage-transformation equations.

Introduction of the state variables causes the problem to be cast into the form of a staged system, as shown in Figure 8.2. By imposing the conditions

$$u_0 = 0, \tag{8.6.7}$$

$$u_n \geq b, \tag{8.6.8}$$

we see that Equations (8.6.3)–(8.6.6) can be added to yield Equation (8.6.2), our original constraint condition. Hence our strategy has been to introduce a set of state variables in such a manner that the given constraint can be expressed in terms of a set of stage-transformation equations.

Overall objective function

To decompose the problem into a sequence of partial optimization problems (each of which will be embedded in a larger problem), we must combine the stagewise returns with the stage-transformation equations. Thus we obtain

$$y_1(u_1) = \min_{x_1} \{f_1(x_1)\} = \min_{x_1} \{f_1(u_1)\},$$

$$y_2(u_2) = \min_{x_2} \{f_2(x_2) + y_1(u_1)\} \tag{8.6.9}$$

$$= \min_{x_2} \{f_2(x) + y_1[u_2 - g_2(x_2)]\}, \tag{8.6.10}$$

$$y_3(u_3) = \min_{x_3} \{f_3(x_3) + y_2(u_2)\}$$

$$= \min_{x_3} \{f_3(x_3) + y_2[u_3 - g_3(x_3)]\}, \tag{8.6.11}$$

$$\vdots$$

$$y_n(u_n) = \min_{x_n} \{f_n(x_n) + y_{n-1}(u_{n-1})\}$$

$$= \min_{x_n} \{f_n(x_n) + y_{n-1}[u_n - g_n(x_n)]\}. \tag{8.6.12}$$

Notice that each of the above expressions has been combined with a transformation equation in such a manner that the right-hand side of the ith expression [i.e., $y_i(u_i)$] is a function only of x_i and u_i.

The problem can now be solved by dynamic programming. Either a set of tabular constructions can be used, as in the automobile-replacement problem, or an analytical procedure, as described in Example 8.6.1. Before proceeding with a detailed solution, however, let us first consider some extensions of the general approach.

Multiplicative problems

The procedure can be applied equally well to decomposable problems that are expressed in terms of products of functions rather than sums. Suppose, for example, that Equation (8.6.2) is replaced by

$$g(\mathbf{X}) = g_1(x_1)g_2(x_2)\ldots g_n(x_n) \geq b. \tag{8.6.13}$$

The state variables would then be introduced as

$$g_1(x_1) = \frac{u_1}{u_0}, \tag{8.6.14}$$

$$g_2(x_2) = \frac{u_2}{u_1}, \tag{8.6.15}$$

$$\vdots$$

$$g_n(x_n) = \frac{u_n}{u_{n-1}}. \tag{8.6.16}$$

By setting $u_0 = 1$ and writing

$$u_n \geq b, \tag{8.6.17}$$

we would then be able to express the original constraint in terms of the desired set of stage-transformation equations.

Multiple constraints and Lagrange multipliers

The procedure also applies when several constraints are present. Now, however, we must introduce a new set of state variables for each constraint. Since dynamic programming quickly becomes ineffective with increasing dimensionality, we see that this approach is impractical.

An alternative that is sometimes useful is to eliminate one or more constraints through the introduction of Lagrange multipliers. For example, consider the problem: Minimize

$$y(\mathbf{X}) = f_1(x_1) + f_2(x_2) + \cdots + f_n(x_n), \tag{8.6.18}$$

subject to

$$g_1(\mathbf{X}) = g_{11}(x_1) + g_{12}(x_2) + \cdots + g_{1n}(x_n) - b_1 = 0, \tag{8.6.19}$$

$$g_2(\mathbf{X}) = g_{21}(x_1) + g_{22}(x_2) + \cdots + g_{2n}(x_n) - b_2 = 0. \tag{8.6.20}$$

Rather than introduce two sets of state variables, we can form a Lagranian function that incorporates one of the constraints. Hence the problem becomes: Minimize

$$z(\mathbf{X}, \lambda) = f_1(x_1) + f_2(x_2) + \cdots + f_n(x_n) + \lambda[g_{11}(x_1) + g_{12}(x_2) + \cdots + g_{1n}(x_n) - b_1], \qquad (8.6.21)$$

subject to

$$g_2(\mathbf{X}) = g_{21}(x_1) + g_{22}(x_2) + \cdots + g_{2n}(x_n) - b_2 = 0. \qquad (8.6.22)$$

The procedure is to assume a value for λ and solve the problem (with only one set of state variables) using dynamic programming. Once an optimal solution has been found, say $\mathbf{X}^*$, the decision variables are substituted into the expression for $g_1(\mathbf{X})$. If the correct value for λ was selected, then $g_1(\mathbf{X}^*)$ will equal zero. As a rule, however, the correct value for λ will not have been selected, and $g_1(\mathbf{X}^*)$ will take on some value other than zero. The value for λ is then adjusted and the entire problem resolved. This is continued until an optimal solution is obtained for a value of λ that causes $g_1(\mathbf{X}^*)$ to equal zero.

For many problems it is possible to plot $g_1(\mathbf{X}^*)$ versus λ, as shown in Figure 8.8. It is then very simple to obtain a value of λ, say λ^*, that causes $g_1(\mathbf{X}^*)$ to vanish. The problem can then be resolved, using this value of λ^*.

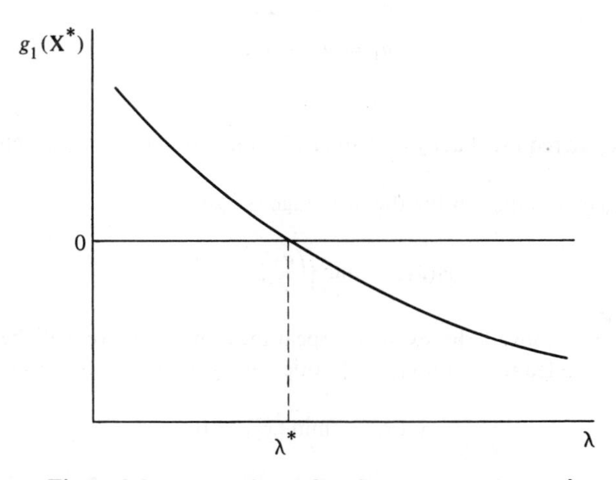

Figure 8.8. Constraint a Continuous Function of λ

The difficulty with this procedure is that $g_1(\mathbf{X}^*)$ may not be a continuous function of λ. Thus for some problems we may obtain a plot of $g_1(\mathbf{X}^*)$ versus λ as shown in Figure 8.9. Under these circumstances the entire procedure is invalid.

EXAMPLE 8.6.1

Solve the following problem by means of dynamic programming: Minimize

$$y = x_1^2 + x_2^2 + x_3^2, \qquad (8.6.23)$$

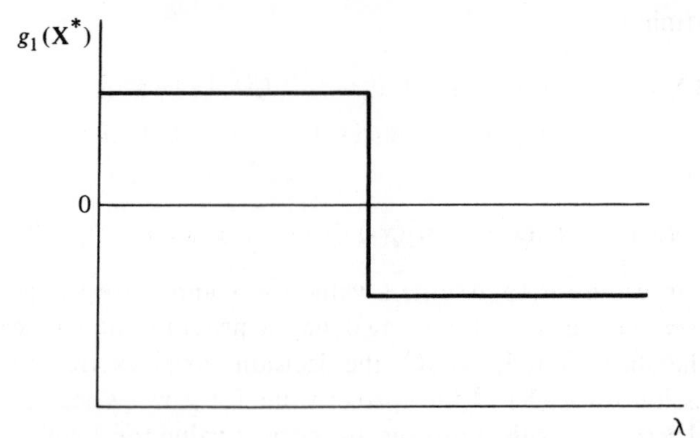

Figure 8.9. Constraint a Discontinuous Function of λ

subject to

$$g = 2x_1 + x_2 + 3x_3 \geq 14. \tag{8.6.24}$$

The appropriate state variables, and the corresponding state transformations, are obtained from Equations (8.6.3)–(8.6.6) as

$$u_0 = u_1 - 2x_1, \tag{8.6.25}$$

$$u_1 = u_2 - x_2, \tag{8.6.26}$$

$$u_2 = u_3 - 3x_3. \tag{8.6.27}$$

Furthermore, we require that $u_0 = 0$ and $u_3 \geq 14$, in order that the given constraint be satisfied.

The objective function for the last stage (stage 1) is

$$y_1(u_1) = \min_{x_1}\left\{\left(\frac{u_1}{2}\right)^2\right\} = \frac{u_1^2}{4}. \tag{8.6.28}$$

Since we have removed the explicit appearance of x_1, there will be no actual minimization needed to obtain $y_1(u_1)$. [Notice that this expression *cannot* be written as

$$y_1(u_1) = \min_{x_1}\{x_1^2\} = 0$$

because in this form the right-hand side is not a function of u_1, as is required.]

The objective function for the last two stages (stages 2 and 1) is

$$\begin{aligned} y_2(u_2) &= \min_{x_2}\{x_2^2 + \tfrac{1}{4}u_1^2\} \\ &= \min_{x_2}\{x_2^2 + \tfrac{1}{4}(u_2 - x_2)^2\}. \end{aligned} \tag{8.6.29}$$

It is easy to see that $y_2(u_2)$ will be minimized for $x_2 = u_2/5$. Thus we can write

$$y_2(u_2) = \left\{\left(\frac{u_2}{5}\right)^2 + \frac{1}{4}\left(\frac{4u_2}{5}\right)^2\right\} = \frac{u_2^2}{5}. \tag{8.6.30}$$

Finally, the overall objective function can be written

$$\begin{aligned} y_3(u_3) &= \min_{x_3} \{x_3^2 + \tfrac{1}{5}u_2^2\} \\ &= \min_{x_3} \{x_3^2 + \tfrac{1}{5}(u_3 - 3x_3)^2\}. \end{aligned} \tag{8.6.31}$$

This function will be minimized for $x_3 = 3u_3/14$. Therefore, $y_3(u_3)$ becomes

$$y_3(u_3) = \left\{\left(\frac{3u_3}{14}\right)^2 + \frac{1}{5}\left(\frac{5u_3}{14}\right)^2\right\} = \frac{u_3^2}{14}. \tag{8.6.32}$$

Since $u_3 \geq 14$, we see that $y_3(u_3)$ will take on its smallest value if we set $u_3 = 14$. Hence we obtain $y_3(u_3) = 14$, $x_3 = 3$, $u_2 = u_3 - 3x_3 = 5$, $x_2 = 1$, $u_1 = u_2 - x_2 = 4$, and $x_1 = 2$. Thus the final solution is $y_{\min} = 14$ at $x_1 = 2$, $x_2 = 1$, $x_3 = 3$.

This solution can easily be verified by defining an appropriate Lagrangian function and proceeding as in Chapter 2. In fact, such a procedure would offer a more direct approach to problems of this nature. We again remark that solving this problem by dynamic programming has been for illustrative purposes only.

8.7 Dynamic Programming and the Optimization of Functionals

In Chapter 7 we discussed the use of analytical methods for solving problems that involve the optimization of functionals. We saw that the use of these methods for the extremization of relatively simple integrals is quite arduous. Often it is not possible to obtain a solution to a problem in closed form. Moreover, solutions that are obtainable are frequently nonlinear algebraic equations, requiring the use of some numerical method to determine the final answer. Thus we concluded that analytical methods are of limited value in obtaining numerical solutions to many functional optimization problems. Let us now examine the use of dynamic programming to solve such problems.

Functional decomposition

Suppose, for example, that we wish to determine $u(\xi)$ such that the integral

$$\mathrm{I} = \int_t^{t_f} \phi(u, u', \xi)\, d\xi \tag{8.7.1}$$

is minimized, given an initial value $u(t)$ and a final value $u(t_f)$. We shall assume that $u(t)$ is a continuous and differentiable function of t, and that $\phi(u, u', \xi)$ is piecewise continuous over the interval $y \leq \xi \leq t_f$. We shall also assume that u', the derivative of u with respect to ξ, is also continuous over the above interval. Letting $y(u, t)$ be the desired minimum, we can write

$$y(u, t) = \min_{u(\xi)} \int_t^{t_f} \phi(u, u', \xi)\, d\xi. \tag{8.7.2}$$

If the function $\phi(u, u', \xi)$ is continuous over the subinterval $t \leq \xi \leq t + \Delta t$, then Equation (8.7.2) can be rewritten as

$$y(u, t) = \min_{u(\xi)} \left\{ \int_t^{t+\Delta t} \phi(u, u', \xi)\, d\xi + \int_{t+\Delta t}^{t_f} \phi(u, u', \xi)\, d\xi \right\}. \tag{8.7.3}$$

Applying the law of the mean for integrals, we can write

$$\int_t^{t+\Delta t} \phi(u, u', \xi)\, d\xi = \phi(u, u', \xi^*)\Delta t = r(u, u', t), \qquad t \leq \xi^* \leq t + \Delta t. \tag{8.7.4}$$

Hence we can write

$$y(u, t) = \min_{u(\xi)} \left\{ r(u, u', t) + \int_{t+\Delta t}^{t_f} \phi(u, u', \xi)\, d\xi \right\}. \tag{8.7.5}$$

Finite-difference approximation

Now let us break up the interval $t + \Delta t \leq \xi \leq t_f$ into i adjacent subintervals Δt_j, $j = 1, 2, \ldots, i$, with $\phi(u, u', \xi)$ assumed to be continuous within each subinterval. We shall refer to the *last* subinterval (i.e., $t_f - \Delta t \leq \xi \leq t_f$) as subinterval 1, the *second-last* subinterval as subinterval 2, and so on, through subinterval i (cf. Figure 8.10). Let u_i represent the value of

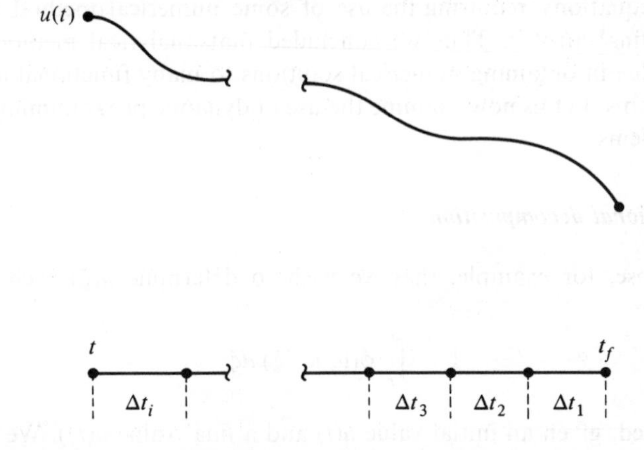

Figure 8.10. Stagewise Representation of a Continuous Extremizing Function

u entering the ith increment. Equation (8.7.5) can then be rewritten as

$$y_i(u_i, t_i) = \min_{\substack{u_{i-1} \\ u_{i-2} \\ \vdots \\ u_1}} \left\{ r_i(u_i, u'_i, t_i) + \sum_{j=1}^{i-1} r_j(u_j, u'_j, t_j) \right\}. \tag{8.7.6}$$

So far we have not specified a stage transformation between u_j and u'_j. One such relationship, which is simply a finite-difference approximation for u'_j, is

$$u_j = u_{j-1} - u'_j \Delta t_j. \tag{8.7.7}$$

Thus we can think of u_j as being a state variable that undergoes a change in stage j as determined by the choice of the decision variable u'_j. Hence we can write Equation (8.7.6) as

$$y_i(u_i, t_i) = \min_{\substack{u'_i \\ u'_{i-1} \\ \vdots \\ u'_1}} \left\{ r_i(u_i, u'_i, t_i) + \sum_{j=1}^{i-1} r_j(u_j, u'_j, t_j) \right\}. \tag{8.7.8}$$

Notice, however, that the choice of u'_1 affects only r_1; the choice of u'_2 affects r_2 and r_1; u'_3 affects r_3, r_2, and r_1; and so on. We can therefore decompose the determination of the optimal u'_j, and Equation (8.7.8) becomes

$$y_i(u_i, t_i) = \min_{u_i'} \left\{ r_i(u_i, u'_i, t_i) + \min_{\substack{u'_{i-1} \\ u'_{i-2} \\ \vdots \\ u'_1}} \sum_{j=1}^{i-1} r_j(u_j, u'_j, t_j) \right\}. \tag{8.7.9}$$

The summation term is simply $y_{i-1}(u_{i-1}, t_{i-1})$, however, so that we can express Equation (8.7.9) as

$$y_i(u_i, t_i) = \min_{u_i'} \{ r_i(u_i, u'_i, t_i) + y_{i-1}(u_{i-1}, t_{i-1}) \}. \tag{8.7.10}$$

This, of course, is our now-familiar functional equation upon which the dynamic programming strategy is based. Thus we see that a problem involving optimization of a functional can be recast into the form of a dynamic programming problem.

In the following example we shall make use of dynamic programming to solve a specific functional minimization problem. The computational aspects of the dynamic programming approach will be pointed out in detail.

EXAMPLE 8.7.1

In Chapter 7 we solved the brachistochrone problem through the use of the calculus of variations. Now let us apply dynamic programming to the solution of the same problem.

Recall that the problem is to obtain the trajectory $u(t)$ which allows a bead to slide down a frictionless wire, under the force of gravity, in minimum time. Mathematically,

$$y = \min_{u(t)} \frac{1}{\sqrt{2g}} \int_{t_0}^{t_f} \sqrt{\frac{1 + (u')^2}{u}}\, dt, \tag{8.7.11}$$

where $u(t)$ is measured vertically downward and $u(t_0)$ and $u(t_f)$ are known. In Chapter 7 we obtained the analytical solution to this problem:

$$u = \frac{1 - \cos \tau}{2c_1^2}, \tag{8.7.12}$$

$$t = c_2 + \frac{\tau - \sin \tau}{2c_1^2}. \tag{8.7.13}$$

If we let $u(0) = 0$ and $u(1) = 1$, we obtain numerical values for the constants of $c_1^2 = 0.87272$ and $c_2 = 0$, and the time of descent of $\tau_f = 2.4120$.

To reformulate the problem for solution by dynamic programming, let us follow the procedure outlined in the last section by writing

$$y(u, t) = \min_{u(\xi)} \left\{ \frac{1}{\sqrt{2g}} \int_{t}^{t\Delta + t} \sqrt{\frac{1 + (u')^2}{u}}\, d\xi + \frac{1}{\sqrt{2g}} \int_{t+\Delta t}^{t_f} \sqrt{\frac{1 + (u')^2}{u}}\, d\xi \right\}. \tag{8.7.14}$$

Within the interval $t \leq \xi \leq t + \Delta t$, let us approximate $u(\xi)$ as

$$u = a + bt, \tag{8.7.15}$$

with

$$a_i = \frac{u_i t_{i-1} - u_{i-1} t_i}{\Delta t}, \tag{8.7.16}$$

$$b_i = \frac{u_{i-1} - u_i}{\Delta t} \tag{8.7.17}$$

where

$$t_i = t, \tag{8.7.18}$$

$$t_{i-1} = t + \Delta t, \tag{8.7.19}$$

$$u_i = u(t_i), \tag{8.7.20}$$

$$u_{i-1} = u(t_{i-1}). \tag{8.7.21}$$

If $u'(\xi)$ is assumed to remain constant within the interval $t_i \leq \xi \leq t_{i-1}$, then the first integral in Equation (8.7.14) becomes

$$r_i(u_i, u'_i, t_i) = \sqrt{\frac{1 + (u'_i)^2}{2g}} \int_{t_i}^{t_{i-1}} \frac{d\xi}{\sqrt{a_i + b_i\xi}}$$

$$= \frac{2}{b_i} \sqrt{\frac{1 + (u'_i)^2}{2g}} \left\{\sqrt{a_i + b_i t_{i-1}} - \sqrt{a_i + b_i t_i}\right\} \tag{8.7.22}$$

and the minimum integral can be expressed as the functional equation

$$y_i(u_i, t_i) = \min_{u'_i} \left\{\frac{2}{b_i}\sqrt{\frac{1 + (u'_i)^2}{2g}}\left[\sqrt{a_i + b_i t_{i-1}} - \sqrt{a_i + b_i t_i}\right] + y_{i-1}(u_{i-1}, t_{i-1})\right\}, \tag{8.7.23}$$

where u and u' are related by

$$u_{i-1} = u_i + u'_i \, \Delta t. \tag{8.7.24}$$

This problem has been solved numerically on a digital computer using values of $\Delta t = 0.10$ and $g = 0.50$. Again the boundary conditions $u(0) = 0$ and $u(1) = 1$ were invoked. Within each interval the decision variable u'_i was considered to be nonnegative, and it was allowed to take on numerical values that are discrete multiples of 0.10. The value of each state variable u_i was restricted to lie within the interval $t_i \leq u_i \leq 1$ for computational efficiency. In addition, the u_i were restricted to taking on numerical values that are multiples of 0.01; this caused the u_i and u'_i to be consistent with Equation (8.7.24), and a numerical interpolation scheme was not needed.

The results of the computation are shown in Tables 8.6–8.16. Table 8.6 shows the results obtained for the first stage ($0.9 \leq t \leq 1.0$). In this stage there is no actual minimization performed. The value of $u_0 = u(1.0)$ is known to be 1.0. Thus, given a value of $u_1 = u(0.90)$, the corresponding value of u'_1 can be readily determined from Equation (8.7.24), and a value of y_1 can be computed from the expression

$$y_1(u_1, 0.9) = \frac{2}{b_1}\sqrt{\frac{1 + (u'_1)^2}{2g}}\left[\sqrt{a_1 + b_1} - \sqrt{a_1 + 0.9b_1}\right] \tag{8.7.25}$$

The values of u_1, u'_1 and $y(u_1, 0.9)$ are shown in Table 8.6.

For stages 2 through 9 (Tables 8.7–8.14) the value of u_{i-1} is not known in advance. Therefore, we choose a value of u_i and then find the value of u'_i [and, from Equation (8.7.24), the corresponding value of u_{i-1}], which causes $y_i(u_i, t_i)$ to be a minimum. The value of $y_i(u_i, t_i)$ is obtained from Equation (8.7.23). The optimal values of u'_i and $y_i(u_i, t_i)$ that correspond to different choices of u_i are shown in Tables 8.7–8.14 for stages 2 through 9, respectively.

The value of u_i for the tenth stage (i.e., u_{10}) is known to be equal to zero. Hence we must find the value of u'_{10} that causes $y_{10}(u_{10}, 0)$ to be minimized. The results of this computation are shown in Table 8.15.

Finally, we must make use of these tabular results to obtain the desired optimal trajectory. From Table 8.15 we see that the correct value of $u'_{10} = u(1.0 \leq t \leq 0.9)$ is 3.1. Knowing that $u_{10} = u(t = 0) = 0$, we can determine $u_9 = u(t = 0.9)$ as

Table 8.6. BRACHISTOCHRONE PROBLEM

STAGE 1: X = 0.90

U(J)	U'(I, J)	Y(I, J)
0.9900	0.1000	0.1008
0.9800	0.2000	0.1025
0.9700	0.3000	0.1052
0.9600	0.4000	0.1088
0.9500	0.5000	0.1132
0.9400	0.6000	0.1184
0.9300	0.7000	0.1243
0.9200	0.8000	0.1307
0.9100	0.9000	0.1377
0.9000	1.0000	0.1451

Table 8.7

STAGE 2: X = 0.80

U(J)	U'(I, J)	Y(I, J)	U(J)	U'(I, J)	Y(I, J)
0.9900	0.0	0.2013	0.8900	0.6000	0.2348
0.9800	0.1000	0.2020	0.8800	0.6000	0.2407
0.9700	0.2000	0.2038	0.8700	0.7000	0.2467
0.9600	0.2000	0.2060	0.8600	0.7000	0.2533
0.9500	0.3000	0.2088	0.8500	0.8000	0.2600
0.9400	0.3000	0.2120	0.8400	0.8000	0.2673
0.9300	0.4000	0.2157	0.8300	0.9000	0.2746
0.9200	0.4000	0.2199	0.8200	0.9000	0.2824
0.9100	0.5000	0.2244	0.8100	1.0000	0.2902
0.9000	0.5000	0.2295	0.8000	1.1000	0.2985

Table 8.8

STAGE 3: X = 0.70

U(J)	U'(I, J)	Y(I, J)	U(J)	U'(I, J)	Y(I, J)
0.9900	0.0	0.3018	0.8400	0.6000	0.3545
0.9800	0.1000	0.3025	0.8300	0.6000	0.3606
0.9700	0.1000	0.3038	0.8200	0.7000	0.3669
0.9600	0.2000	0.3056	0.8100	0.7000	0.3735
0.9500	0.2000	0.3079	0.8000	0.7000	0.3804
0.9400	0.3000	0.3106	0.7900	0.8000	0.3874
0.9300	0.3000	0.3134	0.7800	0.8000	0.3948
0.9200	0.3000	0.3168	0.7700	0.8000	0.4024
0.9100	0.4000	0.3205	0.7600	0.9000	0.4100
0.9000	0.4000	0.3243	0.7500	0.9000	0.4182
0.8900	0.4000	0.3286	0.7400	1.0000	0.4264
0.8800	0.5000	0.3332	0.7300	1.0000	0.4348
0.8700	0.5000	0.3381	0.7200	1.1000	0.4435
0.8600	0.5000	0.3433	0.7100	1.1000	0.4525
0.8500	0.6000	0.3488	0.7000	1.1000	0.4614

Table 8.9

STAGE 4:	X = 0.60				
U(J)	U'(I, J)	Y(I, J)	U(J)	U'(I, J)	Y(I, J)
0.9900	0.0	0.4023	0.7900	0.6000	0.4776
0.9800	0.1000	0.4030	0.7800	0.7000	0.4840
0.9700	0.1000	0.4043	0.7700	0.7000	0.4906
0.9600	0.2000	0.4061	0.7600	0.7000	0.4975
0.9500	0.2000	0.4079	0.7500	0.7000	0.5047
0.9400	0.2000	0.4102	0.7400	0.8000	0.5119
0.9300	0.3000	0.4130	0.7300	0.8000	0.5195
0.9200	0.3000	0.4158	0.7200	0.8000	0.5273
0.9100	0.3000	0.4192	0.7100	0.9000	0.5353
0.9000	0.3000	0.4226	0.7000	0.9000	0.5433
0.8900	0.4000	0.4264	0.6900	0.9000	0.5518
0.8800	0.4000	0.4303	0.6800	1.0000	0.5604
0.8700	0.4000	0.4346	0.6700	1.0000	0.5691
0.8600	0.4000	0.4391	0.6600	1.0000	0.5780
0.8500	0.5000	0.4439	0.6500	1.1000	0.5872
0.8400	0.5000	0.4488	0.6400	1.1000	0.5967
0.8300	0.5000	0.4542	0.6300	1.1000	0.6062
0.8200	0.5000	0.4597	0.6200	1.1000	0.6159
0.8100	0.6000	0.4653	0.6100	1.2000	0.6258
0.8000	0.6000	0.4713	0.6000	1.2000	0.6360

Table 8.10

STAGE 5:	X = 0.50				
U(J)	U'(I, J)	Y(I, J)	U(J)	U'(I. J)	Y(I, J)
0.9900	0.0	0.5028	0.7400	0.7000	0.6040
0.9800	0.1000	0.5035	0.7300	0.7000	0.6109
0.9700	0.1000	0.5048	0.7200	0.7000	0.6181
0.9600	0.2000	0.5066	0.7100	0.7000	0.6255
0.9500	0.2000	0.5084	0.7000	0.8000	0.6329
0.9400	0.2000	0.5107	0.6900	0.8000	0.6406
0.9300	0.2000	0.5131	0.6800	0.8000	0.6485
0.9200	0.3000	0.5159	0.6700	0.8000	0.6567
0.9100	0.3000	0.5187	0.6600	0.9000	0.6650
0.9000	0.3000	0.5221	0.6500	0.9000	0.6734
0.8900	0.3000	0.5256	0.6400	0.9000	0.6821
0.8800	0.4000	0.5294	0.6300	1.0000	0.6911
0.8700	0.4000	0.5333	0.6200	1.0000	0.7002
0.8600	0.4000	0.5374	0.6100	1.0000	0.7095
0.8500	0.4000	0.5418	0.6000	1.0000	0.7188
0.8400	0.4000	0.5464	0.5900	1.1000	0.7286
0.8300	0.5000	0.5512	0.5800	1.1000	0.7385
0.8200	0.5000	0.5563	0.5700	1.1000	0.7486
0.8100	0.5000	0.5615	0.5600	1.1000	0.7589
0.8000	0.5000	0.5670	0.5500	1.1000	0.7693
0.7900	0.6000	0.5727	0.5400	1.2000	0.7799
0.7800	0.6000	0.5784	0.5300	1.2000	0.7909
0.7700	0.6000	0.5846	0.5200	1.3000	0.8020
0.7600	0.6000	0.5910	0.5100	1.3000	0.8133
0.7500	0.6000	0.5974	0.5000	1.3000	0.8248

Table 8.11

		STAGE 6:	X = 0.40		
U(J)	U′(I, J)	Y(I, J)	U(J)	U′(I, J)	Y(I, J)
0.9900	0.0	0.6033	0.6900	0.7000	0.7344
0.9800	0.1000	0.6040	0.6800	0.7000	0.7418
0.9700	0.1000	0.6053	0.6700	0.8000	0.7495
0.9600	0.2000	0.6071	0.6600	0.8000	0.7572
0.9500	0.2000	0.6089	0.6500	0.8000	0.7651
0.9400	0.2000	0.6112	0.6400	0.8000	0.7735
0.9300	0.2000	0.6136	0.6300	0.9000	0.7819
0.9200	0.3000	0.6164	0.6200	0.9000	0.7905
0.9100	0.3000	0.6193	0.6100	0.9000	0.7993
0.9000	0.3000	0.6222	0.6000	0.9000	0.8082
0.8900	0.3000	0.6256	0.5900	0.9000	0.8175
0.8800	0.3000	0.6291	0.5800	1.0000	0.8268
0.8700	0.4000	0.6329	0.5700	1.0000	0.8365
0.8600	0.4000	0.6369	0.5600	1.0000	0.8462
0.8500	0.4000	0.6411	0.5500	1.0000	0.8561
0.8400	0.4000	0.6455	0.5400	1.1000	0.8663
0.8300	0.4000	0.6502	0.5300	1.1000	0.8767
0.8200	0.4000	0.6549	0.5200	1.1000	0.8874
0.8100	0.5000	0.6598	0.5100	1.1000	0.8982
0.8000	0.5000	0.6649	0.5000	1.2000	0.9092
0.7900	0.5000	0.6703	0.4900	1.2000	0.9204
0.7800	0.5000	0.6758	0.4800	1.2000	0.9317
0.7700	0.6000	0.6816	0.4700	1.2000	0.9435
0.7600	0.6000	0.6875	0.4600	1.3000	0.9554
0.7500	0.6000	0.6936	0.4500	1.3000	0.9675
0.7400	0.6000	0.6999	0.4400	1.3000	0.9799
0.7300	0.6000	0.7065	0.4300	1.3000	0.9925
0.7200	0.6000	0.7131	0.4200	1.4000	1.0053
0.7100	0.7000	0.7199	0.4100	1.4000	1.0183
0.7000	0.7000	0.7270	0.4000	1.4000	1.0316

Table 8.12

		STAGE 7:	X = 0.30		
U(J)	U′(I, J)	Y(I, J)	U(J)	U′(I, J)	Y(I, J)
0.9900	0.0	0.7038	0.8600	0.4000	0.7370
0.9800	0.1000	0.7045	0.8500	0.4000	0.7411
0.9700	0.1000	0.7058	0.8400	0.4000	0.7453
0.9600	0.2000	0.7076	0.8300	0.4000	0.7497
0.9500	0.2000	0.7094	0.8200	0.4000	0.7545
0.9400	0.2000	0.7117	0.8100	0.4000	0.7593
0.9300	0.2000	0.7141	0.8000	0.5000	0.7642
0.9200	0.3000	0.7169	0.7900	0.5000	0.7694
0.9100	0.3000	0.7198	0.7800	0.5000	0.7748
0.9000	0.3000	0.7227	0.7700	0.5000	0.7803
0.8900	0.3000	0.7261	0.7600	0.5000	0.7860
0.8800	0.3000	0.7296	0.7500	0.6000	0.7918
0.8700	0.3000	0.7332	0.7400	0.6000	0.7979

Table 8.12. (Continued)

		STAGE 7:	X = 0.30		
U(J)	U′(I, J)	Y(I, J)	U(J)	U′(I, J)	Y(I, J)
0.7300	0.6000	0.8041	0.5100	1.0000	0.9884
0.7200	0.6000	0.8105	0.5000	1.1000	0.9991
0.7100	0.6000	0.8172	0.4900	1.1000	1.0098
0.7000	0.6000	0.8240	0.4800	1.1000	1.0210
0.6900	0.7000	0.8309	0.4700	1.1000	1.0323
0.6800	0.7000	0.8380	0.4600	1.2000	1.0438
0.6700	0.7000	0.8453	0.4500	1.2000	1.0556
0.6600	0.7000	0.8529	0.4400	1.2000	1.0675
0.6500	0.7000	0.8607	0.4300	1.2000	1.0797
0.6400	0.8000	0.8685	0.4200	1.3000	1.0922
0.6300	0.8000	0.8764	0.4100	1.3000	1.1048
0.6200	0.8000	0.8847	0.4000	1.3000	1.1178
0.6100	0.8000	0.8933	0.3900	1.4000	1.1311
0.6000	0.8000	0.9020	0.3800	1.4000	1.1446
0.5900	0.9000	0.9108	0.3700	1.4000	1.1584
0.5800	0.9000	0.9197	0.3600	1.4000	1.1725
0.5700	0.9000	0.9288	0.3500	1.4000	1.1868
0.5600	0.9000	0.9382	0.3400	1.5000	1.2014
0.5500	1.0000	0.9479	0.3300	1.5000	1.2162
0.5400	1.0000	0.9577	0.3200	1.6000	1.2316
0.5300	1.0000	0.9678	0.3100	1.6000	1.2472
0.5200	1.0000	0.9780	0.3000	1.6000	1.2632

Table 8.13

		STAGE 8:	X = 0.20		
U(J)	U′(I, J)	Y(I, J)	U(J)	U′(I, J)	Y(I, J)
0.9900	0.0	0.8043	0.7800	0.5000	0.8744
0.9800	0.1000	0.8050	0.7700	0.5000	0.8799
0.9700	0.1000	0.8063	0.7600	0.5000	0.8855
0.9600	0.2000	0.8081	0.7500	0.5000	0.8912
0.9500	0.2000	0.8099	0.7400	0.6000	0.8971
0.9400	0.2000	0.8122	0.7300	0.6000	0.9032
0.9300	0.2000	0.8146	0.7200	0.6000	0.9095
0.9200	0.3000	0.8174	0.7100	0.6000	0.9159
0.9100	0.3000	0.8203	0.7000	0.6000	0.9225
0.9000	0.3000	0.8232	0.6900	0.6000	0.9293
0.8900	0.3000	0.8266	0.6800	0.7000	0.9362
0.8800	0.3000	0.8301	0.6700	0.7000	0.9433
0.8700	0.3000	0.8337	0.6600	0.7000	0.9506
0.8600	0.4000	0.8375	0.6500	0.7000	0.9581
0.8500	0.4000	0.8416	0.6400	0.7000	0.9658
0.8400	0.4000	0.8458	0.6300	0.8000	0.9737
0.8300	0.4000	0.8500	0.6200	0.8000	0.9817
0.8200	0.4000	0.8546	0.6100	0.8000	0.9898
0.8100	0.4000	0.8593	0.6000	0.8000	0.9981
0.8000	0.5000	0.8642	0.5900	0.8000	1.0067
0.7900	0.5000	0.8691	0.5800	0.9000	1.0156

Table 8.13. (Continued)

STAGE 8:			X = 0.20		
U(J)	U′(I, J)	Y(I, J)	U(J)	U′(I, J)	Y(I, J)
0.5700	0.9000	1.0246	0.3800	1.3000	1.2350
0.5600	0.9000	1.0337	0.3700	1.3000	1.2484
0.5500	0.9000	1.0430	0.3600	1.3000	1.2621
0.5400	0.9000	1.0524	0.3500	1.4000	1.2762
0.5300	0.9000	1.0623	0.3400	1.4000	1.2907
0.5200	1.0000	1.0722	0.3300	1.4000	1.3054
0.5100	1.0000	1.0824	0.3200	1.4000	1.3204
0.5000	1.0000	1.0929	0.3100	1.5000	1.3358
0.4900	1.0000	1.1034	0.3000	1.5000	1.3515
0.4800	1.0000	1.1142	0.2900	1.5000	1.3675
0.4700	1.0000	1.1252	0.2800	1.6000	1.3840
0.4600	1.1000	1.1363	0.2700	1.6000	1.4007
0.4500	1.1000	1.1477	0.2600	1.6000	1.4180
0.4400	1.1000	1.1595	0.2500	1.7000	1.4357
0.4300	1.2000	1.1715	0.2400	1.7000	1.4539
0.4200	1.2000	1.1836	0.2300	1.7000	1.4725
0.4100	1.2000	1.1961	0.2200	1.8000	1.4917
0.4000	1.2000	1.2088	0.2100	1.8000	1.5115
0.3900	1.2000	1.2218	0.2000	1.9000	1.5318

Table 8.14

STAGE 9:			X = 0.10		
U(J)	U′(I, J)	Y(I, J)	U(J)	U′(I, J)	Y(I, J)
0.9900	0.0	0.9048	0.7600	0.5000	0.9855
0.9800	0.1000	0.9055	0.7500	0.5000	0.9912
0.9700	0.1000	0.9068	0.7400	0.5000	0.9970
0.9600	0.2000	0.9086	0.7300	0.6000	1.0029
0.9500	0.2000	0.9104	0.7200	0.6000	1.0091
0.9400	0.2000	0.9127	0.7100	0.6000	1.0155
0.9300	0.2000	0.9151	0.7000	0.6000	1.0220
0.9200	0.3000	0.9179	0.6900	0.6000	1.0286
0.9100	0.3000	0.9208	0.6800	0.6000	1.0355
0.9000	0.3000	0.9237	0.6700	0.7000	1.0425
0.8900	0.3000	0.9271	0.6600	0.7000	1.0496
0.8800	0.3000	0.9306	0.6500	0.7000	1.0570
0.8700	0.3000	0.9342	0.6400	0.7000	1.0646
0.8600	0.4000	0.9380	0.6300	0.7000	1.0722
0.8500	0.4000	0.9421	0.6200	0.7000	1.0802
0.8400	0.4000	0.9463	0.6100	0.8000	1.0882
0.8300	0.4000	0.9505	0.6000	0.8000	1.0964
0.8200	0.4000	0.9551	0.5900	0.8000	1.1047
0.8100	0.4000	0.9598	0.5800	0.8000	1.1133
0.8000	0.5000	0.9647	0.5700	0.8000	1.1221
0.7900	0.5000	0.9696	0.5600	0.9000	1.1311
0.7800	0.5000	0.9746	0.5500	0.9000	1.1404
0.7700	0.5000	0.9800	0.5400	0.9000	1.1498

Table 8.14. (Continued)

STAGE 9:			X = 0.10		
U(J)	U′(I, J)	Y(I, J)	U(J)	U′(I, J)	Y(I, J)
0.5300	0.9000	1.1593	0.3100	1.4000	1.4280
0.5200	0.9000	1.1690	0.3000	1.5000	1.4436
0.5100	0.9000	1.1789	0.2900	1.5000	1.4595
0.5000	1.0000	1.1890	0.2800	1.5000	1.4757
0.4900	1.0000	1.1994	0.2700	1.5000	1.4924
0.4800	1.0000	1.2101	0.2600	1.6000	1.5095
0.4700	1.0000	1.2209	0.2500	1.6000	1.5270
0.4600	1.0000	1.2320	0.2400	1.6000	1.5451
0.4500	1.1000	1.2432	0.2300	1.7000	1.5636
0.4400	1.1000	1.2547	0.2200	1.7000	1.5825
0.4300	1.1000	1.2663	0.2100	1.7000	1.6020
0.4200	1.1000	1.2783	0.2000	1.8000	1.6221
0.4100	1.2000	1.2906	0.1900	1.8000	1.6428
0.4000	1.2000	1.3030	0.1800	1.8000	1.6642
0.3900	1.2000	1.3158	0.1700	1.9000	1.6863
0.3800	1.2000	1.3289	0.1600	1.9000	1.7093
0.3700	1.2000	1.3422	0.1500	2.0000	1.7331
0.3600	1.3000	1.3557	0.1400	2.0000	1.7578
0.3500	1.3000	1.3696	0.1300	2.1000	1.7836
0.3400	1.3000	1.3837	0.1200	2.1000	1.8105
0.3300	1.3000	1.3981	0.1100	2.2000	1.8388
0.3200	1.4000	1.4129	0.1000	2.2000	1.8685

Table 8.15

STAGE 10:	X = 0.0	
U(J)	U′ (I, J)	Y(I, J)
0.0	3.1000	2.5981

$u_9 = u_{10} + u'_{10}\ \Delta t = 0 + (3.1)(0.1) = 0.31$. From Table 8.14 we see that the value of u'_9 corresponding to $u_9 = 0.31$ is $u'_9 = 1.4$. Hence $u_8 = u_9 + u'_9\ \Delta t = 0.31 + (1.4)(0.1) = 0.45$. From Table 8.13 we obtain $u_7 = u_8 + u'_8\ \Delta t = 0.45 + (1.1)(0.1) = 0.56$, and so on.

Table 8.16. OPTIMAL TRAJECTORY OF BRACHISTOCHRONE DETERMINED BY DYNAMIC PROGRAMMING

i	t	u(t)	i	t	u(t)
10	0	0.00	4	0.6	0.80
9	0.1	0.31	3	0.7	0.86
8	0.2	0.45	2	0.8	0.91
7	0.3	0.56	1	0.9	0.96
6	0.4	0.65	0	1.0	1.00
5	0.5	0.73			

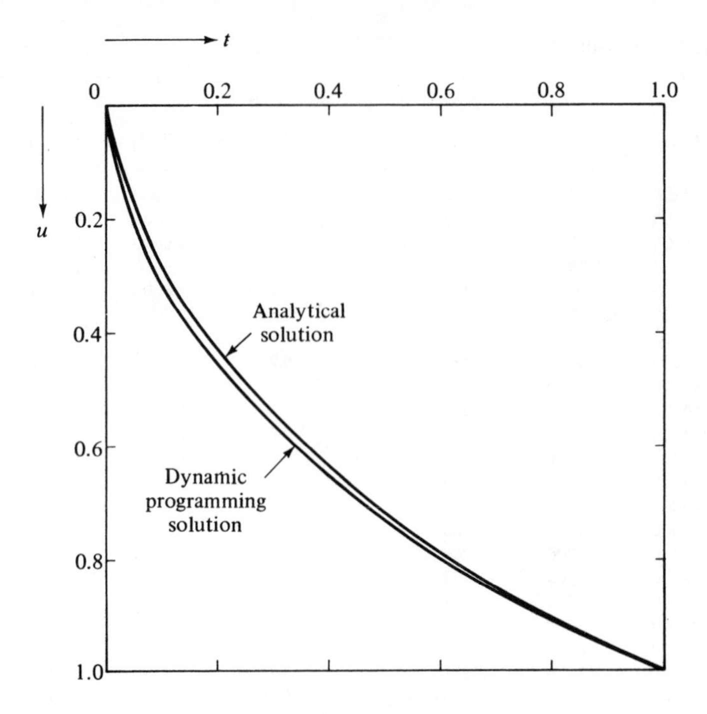

Figure 8.11. The Brachistochrone

The complete optimal trajectory, obtained in this manner, is shown in Table 8.16 and is plotted in Figure 8.11. The correct analytical solution is also shown in Figure 8.11. Notice that the two solutions are similar, though not identical. The accuracy of the dynamic programming solution can be improved by either decreasing Δt (and increasing the number of stages), or by employing a better approximation for $u(t)$ than that given by Equation (8.7.15).

From Table 8.15 we can also see that the computed minimum descent time, $y_{10}(0, 0)$, is equal to 2.5981. This compares reasonably well with the value of 2.4120 determined analytically.

8.8 Dynamic Programming and Variational Calculus

In the last section we used dynamic programming to solve a problem that was formerly approached by means of the calculus of variations. It is natural, then, to expect that there is a relationship between the two methods, in spite of their apparent differences. In this section we shall use dynamic programming to obtain the Euler–Lagrange equation. As we saw in Chapter 7, this equation represents a necessary condition that must be satisfied by all variational problems.

Recall that we wish to extremize the integral

$$I = \int_{t}^{t_f} \phi(u, u', \xi)\, d\xi. \tag{8.8.1}$$

We shall not consider constrained problems in this section, although the same development could be applied to the augmented integral of a problem in which equality constraints were present. Moreover, we shall again restrict our attention to a minimization problem, realizing that the same analysis applies to maximization problems.

Functional decomposition

From Equations (8.7.10) and (8.7.4), we can write

$$y_i(u_i, t_i) = \min_{u'} \int_{t_i}^{t_f} \phi(u, u', \xi)\, d\xi = \min_{u_i'} \{\phi_i(u_i, u_i', t_i)\Delta t + y_{i-1}(u_{i-1}, t_{i-1})\}. \tag{8.8.2}$$

Equation (8.8.2) is the functional-equation representation of the original problem. The dynamic programming solution is constructed by using this equation recursively.

Euler–Lagrange equations

Let us now expand the last term in Equation (8.8.2) in a Taylor series, retaining only the first-order terms. We have

$$\begin{aligned} y_{i-1}(u_{i-1}, t_{i-1}) &= y_i(u_i, t_i) + \frac{\partial y}{\partial u_i}\frac{du_i}{dt_i}\Delta t + \frac{\partial y}{\partial t_i}\Delta t \\ &= y_i(u_i, t_i) + u_i'\frac{\partial y}{\partial u_i}\Delta t + \frac{\partial y}{\partial t_i}\Delta t. \end{aligned} \tag{8.8.3}$$

Substituting into Equation (8.8.2),

$$y_i(u_i, t_i) = \min_{u_i'} \left\{\phi_i(u_i, u_i', t_i)\,\Delta t + y_i(u_i, t_i) + u_i'\frac{\partial y}{\partial u_i}\Delta t + \frac{\partial y}{\partial t_i}\Delta t\right\}. \tag{8.8.4}$$

If $y_i(u_i, t_i)$ is subtracted from both sides, we have, dropping subscripts,

$$\min_{u'} \left\{\phi + u'\frac{\partial y}{\partial u} + \frac{\partial y}{\partial t}\right\} = 0. \tag{8.8.5}$$

The necessary condition for Equation (8.8.5) to be minimized is

$$\frac{\partial}{\partial u'}\left\{\phi + u'\frac{\partial y}{\partial u} + \frac{\partial y}{\partial t}\right\} = 0 \tag{8.8.6}$$

or

$$\frac{\partial \phi}{\partial u'} + \frac{\partial y}{\partial u} = 0. \tag{8.8.7}$$

For the correct minimizing function $u(t)$, Equation (8.8.5) can be written simply as

$$\phi + u'\frac{\partial y}{\partial u} + \frac{\partial y}{\partial u} = 0. \tag{8.8.8}$$

We now wish to eliminate the partial derivatives of y from Equations (8.8.7) and (8.8.8). Let us first take the total derivative of (8.8.7) with respect to t:

$$\frac{d}{dt}\frac{\partial \phi}{\partial u'} + u'\frac{\partial^2 y}{\partial u^2} + \frac{\partial}{\partial t}\frac{\partial y}{\partial u} = 0. \tag{8.8.9}$$

Now we obtain the partial derivative of (8.8.8) with respect to u:

$$\frac{\partial \phi}{\partial u} + u'\frac{\partial^2 y}{\partial u^2} + \frac{\partial}{\partial u}\frac{\partial y}{\partial t} = 0. \tag{8.8.10}$$

Substracting (8.8.10) from (8.8.9), we have

$$\frac{d}{dt}\frac{\partial \phi}{\partial u'} - \frac{\partial \phi}{\partial u} = 0. \tag{8.8.11}$$

Equation (8.8.11) is the Euler–Lagrange equation, which we derived using classical considerations in Section 7.1.

It is noteworthy that the Euler–Lagrange equation represents a *global* necessary condition for an extremum; i.e., Equation (8.8.11) is a differential equation that must be satisfied by the function $\phi(u, u', t)$ of the minimizing function $u(t)$ *simultaneously for all t* within the interval of integration. The functional equations of dynamic programming, on the other hand, represent a condition that applies *locally*. This condition relates the return of a given stage to the optimal returns of a sequence of adjacent stages. Thus we are able to examine the same problem from two very different viewpoints.

8.9 The Discrete Maximum Principle

Another approach to the optimization of staged systems involves use of a discrete form of the maximum principle. Many stagewise optimization problems are approached from this point of view. We shall see, however, that the maximum principle applies to discrete problems only in a modified and weakened form unless the problem meets certain special requirements. We shall also point out some computational difficulties associated with the discrete maximum principle.

The discrete problem statement

Let us begin with a discrete form of the basic problem discussed in Section 7.5. Let the region of integration $t_0 \leq t \leq t_f$ be broken up into N discrete subregions $\Delta t^{(\alpha)}$, with

$$t_f - t_0 = \sum_{\alpha=1}^{N} \Delta t^{(\alpha)}, \tag{8.9.1}$$

where

$$\begin{aligned}
\Delta(t)^{(1)} &= t^{(1)} - t_0, \\
&\vdots \\
\Delta t^{(\alpha)} &= t^{(\alpha)} - t^{(\alpha-1)}, \\
&\vdots \\
\Delta t^{(N)} &= t_f - t^{(N-1)}.
\end{aligned} \tag{8.9.2}$$

The objective function, which we wish to minimize, can be represented as

$$u_0^{(N)} = \sum_{\alpha=1}^{N} g_0^{(\alpha)}[u_1^{(\alpha-1)}, u_2^{(\alpha-1)}, \ldots, u_n^{(\alpha-1)}, x_1^{(\alpha)}, x_2^{(\alpha)}, \ldots, x_m^{(\alpha)}]\Delta t^{(\alpha)}, \tag{8.9.3}$$

where $u_0^{(\alpha)}$, $u_0^{(\alpha-1)}$, and $g_0^{(\alpha)}$ are related by

$$u_0^{(\alpha)} = u_0^{(\alpha-1)} + g_0^{(\alpha)}\Delta t^{(\alpha)}, \tag{8.9.4}$$

and $u_0^{(0)} = 0$. Thus we simply have a finite-difference representation of Equation (7.5.11).

Similarly, we represent the constraining differential equations in finite-difference form as

$$u_i^{(\alpha)} = u_i^{(\alpha-1)} + g_i^{(\alpha)}[u_1^{(\alpha-1)}, u_2^{(\alpha-1)}, \ldots, u_n^{(\alpha-1)}, x_1^{(\alpha)}, x_2^{(\alpha)}, \ldots, x_m^{(\alpha)}]\Delta t^{(\alpha)},$$
$$i = 1, 2, \ldots, n, \tag{8.9.5}$$

where the initial values of the state variables are known; i.e., $u_i^{(0)} = a_i$, $i = 1, 2, \ldots, n$. We seek the set of values for the decision variables $x_k^{(\alpha)}$; $k = 1, 2, \ldots, m$; $\alpha = 1, 2, \ldots, N$, which causes the objective function to be minimized. As before, we shall assume that the $g_i^{(\alpha)}$ have continuous partial derivatives with respect to the state and decision variables.

Lagrange multiplier formulation

Since we have replaced all differential relationships with algebraic approximations, we have essentially transformed our problem into a nonlinear programming problem. Therefore, we can introduce a set of n Lagrange multipliers for each α (i.e., at each stage) and proceed to establish a stationary point for the Lagrangian function

$$z = u_0^{(N)} + \sum_{\alpha=1}^{N} \sum_{j=1}^{n} \lambda_j^{(\alpha)}[u_j^{(\alpha-1)} + g_j^{(\alpha)}\Delta t^{(\alpha)} - u_j^{(\alpha)}]. \tag{8.9.6}$$

Let us now rewrite Equations (8.9.4) and (8.9.5) as

$$u_i^{(\alpha)} = u_i^{(\alpha-1)} + h_i^{(\alpha)}[u_1^{(\alpha-1)}, u_2^{(\alpha-1)}, \ldots, u_n^{(\alpha-1)}, x_1^{(\alpha)}, x_2^{(\alpha)}, \ldots, x_m^{(\alpha)}], \quad i = 0, 1, 2, \ldots, n, \tag{8.9.7}$$

or simply

$$u_i^{(\alpha)} = T_i^{(\alpha)}[u_0^{(\alpha-1)}, u_1^{(\alpha-1)}, \ldots, u_n^{(\alpha-1)}, x_1^{(\alpha)}, x_2^{(\alpha)}, \ldots, x_m^{(\alpha)}], \quad i = 0, 1, 2, \ldots, n. \tag{8.9.8}$$

Notice that this small change in notation not only allows us to write the given problem more concisely, but also lets us view the problem in more general terms. We are now no longer confined to a finite-difference representation of the integral minimization problem with differential constraints. Rather, we are now concerned with *any* discrete, staged system, as shown in Figure 8.12. The $T_i^{(\alpha)}$ can be thought of as a set of stage transformations that transform the state variables entering stage α, $u_i^{(\alpha-1)}$, into the state variables leaving stage α, $u_i^{(\alpha)}$, as a result of the choice of the decision variables $x_k^{(\alpha)}$. Thus we are now concerned with applying the discrete maximum principle to the same class of problems that we attacked by dynamic programming. (Note, however, that we are now numbering the stages *forward*, i.e., from left to right, and that we have changed notation somewhat.)

Introducing our new terminology, the Lagrangian function becomes

$$z = \sum_{\alpha=1}^{N} h_0^{(\alpha)} + \sum_{\alpha=1}^{N} \sum_{j=1}^{n} \lambda_j^{(\alpha)}[T_j^{(\alpha)} - u_j^{(\alpha)}]. \tag{8.9.9}$$

With the application of the necessary conditions for an extremization of the Lagrangian, as presented in Chapter 2, we obtain

$$\frac{\partial z}{\partial u_i^{(\alpha-1)}} = -\lambda_i^{(\alpha-1)} + \frac{\partial h_0^{(\alpha)}}{\partial u_i^{(\alpha-1)}} + \sum_{j=1}^{n} \lambda_j^{(\alpha)} \frac{\partial T_j^{(\alpha)}}{\partial u_i^{(\alpha-1)}} = 0, \quad i = 1, 2, \ldots, n, \quad \alpha = 1, 2, \ldots, N, \tag{8.9.10}$$

$$\frac{\partial z}{\partial u_i^{(N)}} = -\lambda_i^{(N)} = 0, \qquad i = 1, 2, \ldots, n, \tag{8.9.11}$$

$$\frac{\partial z}{\partial \lambda_i^{(\alpha)}} = T_i^{(\alpha)} - u_i^{(\alpha)} = 0, \qquad i = 1, 2, \ldots, n, \tag{8.9.12}$$

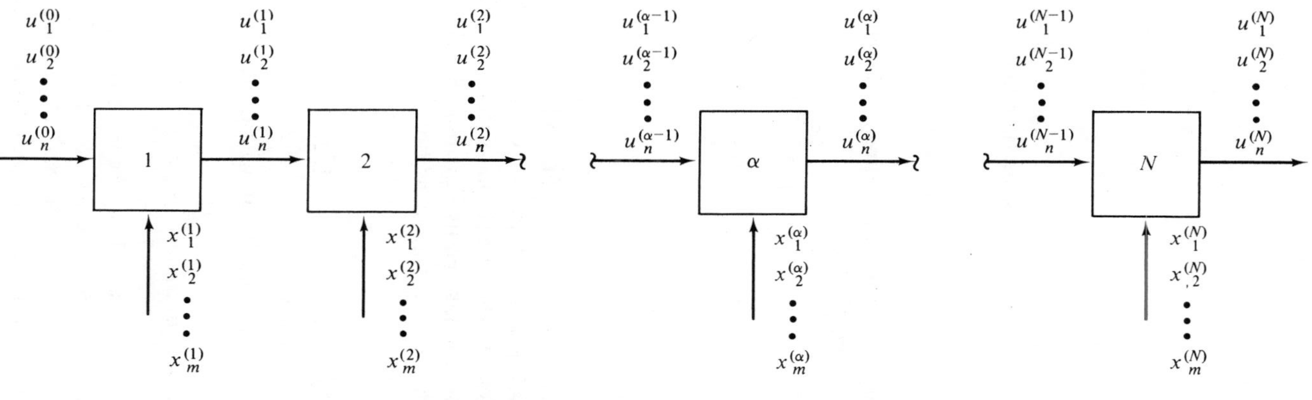

Figure 8.12. Multi-Dimensional Staged System

and

$$\frac{\partial z}{\partial x_k^{(\alpha)}} = \frac{\partial h_0^{(\alpha)}}{\partial x_k^{(\alpha)}} + \sum_{j=1}^{n} \lambda_j^{(\alpha)} \frac{\partial T_j^{(\alpha)}}{\partial x_k^{(\alpha)}} = 0,$$

(8.9.13) $$k = 1, 2, \ldots, m, \quad \alpha = 1, 2, \ldots, N.$$

The discrete Hamiltonian

An alternative means of establishing the same set of necessary conditions is to define a discrete, stagewise Hamiltonian function $H^{(\alpha)}$, where

(8.9.14) $$H^{(\alpha)} = \sum_{j=0}^{n} \lambda_j^{(\alpha)} T_j^{(\alpha)},$$

the $T_j^{(\alpha)}$ are the transformation functions defined earlier, the $\lambda_j^{(\alpha)}$ are unknown constants that satisfy the conditions

(8.9.15) $$\lambda_0^{(\alpha)} = 1, \qquad \alpha = 1, 2, \ldots, N,$$

(8.9.16) $$\lambda_i^{(N)} = 0, \qquad i = 1, 2, \ldots, n,$$

and

(8.9.17) $$\frac{\partial H^{(\alpha)}}{\partial u_i^{(\alpha-1)}} = \lambda_i^{(\alpha-1)}, \qquad i = 1, 2, \ldots, n, \qquad \alpha = 1, 2, \ldots, N.$$

By combining Equations (8.9.14), (8.9.15), and (8.9.17), we easily obtain Equation (8.9.10), which is our first necessary condition. Equations (8.9.11) and (8.9.16) are identical, so that the second necessary condition is satisfied. The third necessary condition, given by Equation (8.9.12), is simply a definition of $T_i^{(\alpha)}$. Finally, the last necessary condition, as represented by Equation (8.9.13), is obtained if we render $H^{(\alpha)}$ stationary with respect to $x_k^{(\alpha)}$; i.e.,

(8.9.18) $$\frac{\partial H^{(\alpha)}}{\partial x_k^{(\alpha)}} = 0, \qquad k = 1, 2, \ldots, m,$$

providing the resulting values for $x_k^{(\alpha)}$ fall within their allowable range.

The discrete maximum principle

From the foregoing we see that our discrete "maximum principle" requires that *the set of decision variables* $x_k^{(\alpha)}$, $k = 1, 2, \ldots, m$; $\alpha = 1, 2, \ldots, N$, *which causes the objective function (8.9.3) to take on a stationary value, subject to the constraints (8.9.8,) be obtained as follows: introduce an additional set of constants* $\lambda_i^{(\alpha)}$ $i = 1, 2, \ldots, n$; $\alpha = 0, 1, \ldots, N-1, N$, *which satisfy (8.9.15) and (8.9.16); form a stagewise Hamiltonian that satisfies (8.9.14) and (8.9.17); and solve for the values of the decision variables that cause the Hamiltonian to take on a stationary value at each stage.*

Weak and strong forms

It is important to recognize several significant points in connection with the above statement. First, we are considering only a necessary condition for a minimum of $u_0^{(N)}$, not a sufficient condition. In this respect the discrete maximum principle resembles the continuous maximum principle. In the continuous case, however, *maximization* of the Hamiltonian with respect to the decision variables is necessary in order to *minimize* the objective function. On the other hand, the discrete case simply states that a *stationary value* of the Hamiltonian with respect to the decision variables corresponds to a *stationary value* of the objective function. The latter case is sometimes referred to as the *weak form* of the discrete maximum principle.

There is also a *strong form* of the discrete maximum principle which applies to problems that are convex with respect to the decision and state variables, and to problems that involve only one state variable. The strong form states, essentially, that a necessary condition for minimization of the objective function is that the stagewise Hamiltonian be *minimized* with respect to the decision variables. (Note the distinction between the strong discrete maximum principle, which is a min–min principle, and the continuous maximum principle, which is a min–max principle. Some authors approach the continuous case from a min–min standpoint, by reversing the sign of $\lambda_0(t)$. We, however, have presented the continuous maximum principle in the same form presented by Pontryagin et al. [1962].)

The subtleties that distinguish the weak and strong forms of the discrete maximum principle had not been recognized by many earlier investigators (cf. Katz [1962] and Fan and Wang [1964]), although Rozonoer [1959] realized that the strong form of the discrete maximum principle could produce erroneous results if a problem were not linear in $u_i^{(\alpha)}$ and $x_k^{(\alpha)}$. Horn and Jackson [1965] and Jackson and Horn [1965] were able to distinguish between the strong and weak forms of the discrete maximum principle on the basis of second-derivative information. The role of convexity, and later directional convexity (a weaker and thus more general condition than convexity), as a requirement for the strong form of the discrete maximum principle were established by Halkin [1964], Holtzman [1966], and Holtzman and Halkin [1966]. Denn and Aris [1965] tabulate several special cases in which the strong form of the discrete maximum principle is either a necessary condition or a necessary and sufficient condition for an optimum.

Variants of the problem

There are several variations of the basic problem to which the discrete maximum principle applies. The procedure for applying the discrete maximum principle to such problems is directly analogous to that for the continuous maximum principle under the same conditions. That is, by altering the boundary conditions or by introducing additional variables we can consider discrete, stagewise problems for which the end values of the state variables are specified, the initial values of the state variables are unspecified, or there appear additional terms in the constraining equations. We have discussed these cases in Section 7.6. Other variants of the basic problem are considered by Katz [1962] and by Fan and Wang [1964].

When the decision variables at each stage must satisfy an inequality constraint of the form

$$F^{(\alpha)}[x_1^{(\alpha)}, x_2^{(\alpha)}, \ldots, x_m^{(\alpha)}] \leq 0, \tag{8.9.19}$$

then the strong form of the discrete maximum principle will apply if the minimizing set of the $x_k^{(\alpha)}$ lies on a boundary in the $\mathbf{x}$ space. However, if the values of the decision variables that cause the stagewise Hamiltonian to take on a stationary point should fall within the $\mathbf{x}$ space, then the strong form of the discrete maximum principle applies only to the special cases cited earlier.

Considerable interest has been shown in applying the discrete maximum principle to large, complex systems that contain staged elements, but also contain converging or diverging chains or recycle loops. Systems of this nature are particularly common in the chemical-process industries. Katz [1962], Fan and Wang [1964], and Denn and Aris [1965] have all considered the application of the discrete maximum principle to such systems. We shall not pursue this area any further as it is a subject unto itself.

Computational algorithm

Let us now return to the basic problem that we used to introduce the discrete maximum principle earlier in this section. Having established a set of equations that defines the discrete form of the maximum principle, we seek a computational algorithm that will allow us to implement the theoretical development. This is not a particularly easy task, since we are required to solve a set of mixed-boundary difference equations. Algorithms that attempt to solve problems of this type characteristically encounter convergence problems.

We begin by restating Equation (8.9.17) in a somewhat different form. If we combine Equations (8.9.14) and (8.9.17), we easily obtain

$$\lambda_i^{(\alpha-1)} = \frac{\partial}{\partial u_i^{(\alpha-1)}} \sum_{j=0}^{n} \lambda_j^{(\alpha)} T_j^{(\alpha)} \tag{8.9.20}$$

or

$$\lambda_i^{(\alpha-1)} = \sum_{j=0}^{n} \lambda_j^{(\alpha)} \frac{\partial T_i^{(\alpha)}}{\partial u_i^{(\alpha-1)}}, \quad i = 1, 2, \ldots, n,$$
$$\alpha = 1, 2, \ldots, N. \tag{8.9.21}$$

This expression is frequently referred to as the *adjoint equation*. Since the final values of the adjoint variables, i.e., $\lambda_i^{(N)}$, are known, it is natural to think of Equation (8.9.21) as a backward recursive relationship among the $\lambda_i^{(\alpha)}$.

With the forward recursive equations that transform the state variables, i.e., Equation (8.9.8) and the backward recursive adjoint equations, the following computational algorithm suggests itself:

1. Choose an initial set of decision variables; i.e., select a set of numerical values for $x_k^{(\alpha)}$, $k = 1, 2, \ldots, m$; $\alpha = 1, 2, \ldots, N$.
2. Using these estimates for the decision variables and the initial values of the state variables, generate a set of state variables $u_i^{(\alpha)}$, $i = 1, 2, \ldots, n$; $\alpha = 1, 2, \ldots, N$, using Equation (8.9.8). Note that this computation is carried out in the direction of increasing α.
3. The estimated values for the decision variables and the state variables are now used, together with the final values of the adjoint variables, to generate a set of adjoint variables $\lambda_i^{(\alpha)}$, $i = 1, 2, \ldots, n$; $\alpha = N - 1, N - 2, \ldots, 0$, by means of Equation (8.9.21). In contrast to step 2, we see that this calculation is performed in the direction of decreasing α.
4. A new set of decision variables is obtained by rendering the stagewise Hamiltonian stationary, with the state variables and the adjoint variables retaining their most recent values. We then return to step 2 and obtain a new set of state variables and a new set of adjoint variables.
5. This iterative procedure is continued until successive sets of the state variables have converged to within some preassigned tolerance, ϵ.

In the event that a satisfactory set of initial values for the $x_k^{(\alpha)}$ cannot be estimated on physical grounds, then we could proceed by starting directly with step 2, choosing a set of $x_k^{(\alpha)}$ for each α is such a manner that $T_0^{(\alpha)}$ is minimized. The rationale behind this strategy is the fact that our overall objective is the minimization of the function $u_0^{(N)}$, which is related to the sequence of $T_0^{(\alpha)}$. This suggestion is due to Katz [1962].

Another variation of the algorithm which Katz [1962] has suggested concerns the possibility of extrapolation to an improved set of decision variables. Specifically, if $x_k^{(\alpha)*}$ is the old set of decision variables and $x_k^{(\alpha)**}$ is the new set obtained from step 4 above, then a new estimate for the decision variables can be obtained as

$$x_k^{(\alpha)} = x_k^{(\alpha)*} + \mu[x_k^{(\alpha)**} - x_k^{(\alpha)*}], \quad \mu > 0, \quad k = 1, 2, \ldots, m,$$
$$\alpha = 1, 2, \ldots, N. \tag{8.9.22}$$

When $0 < \mu < 1$, we are essentially interpolating between the two sets of estimates, whereas $\mu > 1$ corresponds to extrapolation.

The above algorithm was first suggested by Katz [1962]. It is not the only means of implementation of the discrete maximum principle. It is probably the best-known computational algorithm, however, because of its simplicity and its straightforward nature. The method has been shown by Denn [1965] to be convergent for systems having certain structural features, the most notable being linear, straight-chain systems. Unfortunately, the method may not converge for problems that lack Denn's restrictive conditions.

The above algorithm is applied to a specific problem in Example 8.9.1. Several more applications, including applications in the aerospace and chemical-process areas, are shown by Fan and Wang [1964].

We shall not present additional discussion of the computational aspects of the discrete maximum principle. In addition to the examples provided by Fan and Wang [1964], the interested reader should consult Lapidus and Luus [1967]. The latter authors present a very thorough discussion of computational algorithms for the solution of nonlinear, multidimensional control problems (cf. Chapter 4 of the cited reference), including a discussion of iterative algorithms for solution via the discrete maximum principle. Finally, Gurel and Lapidus [1968] present a very clever approach to the steady-state optimization of discrete stagewise systems. They apply the continuous maximum principle to such systems, allowing the problem to evolve in time until an asymptotic solution is reached. This avoids the difficulties that are attendant to the weak form of the discrete maximum principle. One drawback, however, is an increase in the number of decision and state variables for a given problem.

EXAMPLE 8.9.1

Resolve Example 7.5.1 using the discrete maximum principle. Let the interval $0 \leq t \leq 1$ be broken up into ten equally spaced intervals, so that $\Delta t = 0.10$.

In finite-difference form, the problem is to determine numerical values for $x^{(\alpha)}$ and $u_1^{(\alpha)}$ that minimize the quantity

$$u_0^{(N)} = \sum_{\alpha=1}^{10} 0.10[x^{(\alpha)^2} + 2x^{(\alpha)} + u_1^{(\alpha-1)^2}], \tag{8.9.23}$$

subject to the constraining equations

$$u_0^{(\alpha)} = u_0^{(\alpha-1)} + 0.10[x^{(\alpha)^2} + 2x^{(\alpha)} + u_1^{(\alpha-1)^2}] = T_0^{(\alpha)}, \tag{8.9.24}$$

$$u_1^{(\alpha)} = u_1^{(\alpha-1)} + 0.10[x^{(\alpha)} + u_1^{(\alpha-1)}] = T_1^{(\alpha)}, \tag{8.9.25}$$

and the initial conditions $u_0^{(0)} = 0$, $u_1^{(0)} = \frac{1}{2}$.

To solve this problem, we introduce the stagewise Hamiltonian function

$$H^{(\alpha)} = \lambda_0^{(\alpha)} T_0^{(\alpha)} + \lambda_1^{(\alpha)} T_1^{(\alpha)}, \tag{8.9.26}$$

where

$$\lambda_0^{(\alpha)} = 1, \qquad \alpha = 1, 2, \ldots, 10, \tag{8.9.27}$$

and

$$\lambda_1^{(10)} = 0. \tag{8.9.28}$$

If we substitute Equations (8.9.24), (8.9.25), and (8.9.27) into Equation (8.9.26), we obtain the following explicit representation for the stagewise Hamiltonian:

$$H^{(\alpha)} = \{u_0^{(\alpha-1)} + 0.10[x^{(\alpha)2} + 2x^{(\alpha)} + u_1^{(\alpha-1)2}]\} + \lambda_1^{(\alpha)}\{u_1^{(\alpha-1)} + 0.10[x^{(\alpha)} - u_1^{(\alpha-1)}]\},$$
$$\alpha = 1, 2, \ldots, 10. \tag{8.9.29}$$

In order that $H^{(\alpha)}$ be rendered stationary with respect to $x^{(\alpha)}$, it is necessary that

$$\frac{\partial H^{(\alpha)}}{\partial x^{(\alpha)}} = 0, \tag{8.9.30}$$

which results in

$$x^{(\alpha)} = -[\tfrac{1}{2}\lambda_1^{(\alpha)} + 1], \qquad \alpha = 1, 2, \ldots, 10. \tag{8.9.31}$$

[It is easy to see that $H^{(\alpha)}$ will be *minimized* with respect to $x^{(\alpha)}$, since $\partial^2 H^{(\alpha)}/\partial x^{(\alpha)2} > 0$]. The stagewise Hamiltonian must also satisfy Equation (8.9.21), which yields

$$\lambda_1^{(\alpha-1)} = 0.20u_1^{(\alpha-1)} + 0.90\lambda_1^{(\alpha)}, \qquad \alpha = 2, 3, \ldots, 10. \tag{8.9.32}$$

We now have the necessary equations to develop a computational algorithm. Clearly, we would simply like to integrate in the direction of increasing t. That is, we would prefer to evaluate each of the unknown functions for $\alpha = 1$, then repeat the procedure for $\alpha = 2$, $\alpha = 3, \ldots$, until we have reached the last stage, designated by $\alpha = 10$. This cannot be done, however, because u_1 is specified as an *initial* condition [$u_1^{(0)}$ is known], whereas λ_1 is given as a *final* condition [$\lambda_1^{(10)}$ is specified]. Hence we must use an iterative procedure, as outlined in the last section.

To begin the computation, we assume a set of values for $x^{(1)}, x^{(2)}, \ldots, x^{(10)}$. Let us allow each of these functions to equal zero. We then use Equation (8.9.25) to compute a set of $u_1^{(\alpha)}$, $\alpha = 1, 2, \ldots, 10$. This is followed by a "backward" calculation of $\lambda_1^{(\alpha)}$ using Equation (8.9.32); i.e., we evaluate $\lambda_1^{(\alpha)}$ for $\alpha = 9, 8, \ldots, 1, 0$. We then calculate a new set of $x^{(\alpha)}$ by means of Equation (8.9.31). If the new set of $x^{(\alpha)}$ is sufficiently close to the old set, then we can calculate $u_0^{(\alpha)}$, $\alpha = 1, 2, \ldots, 10$, using Equation (8.9.23), and we have our desired final answer. If the new and the old $x^{(\alpha)}$'s do not agree, however, we use the most recent set of $x^{(\alpha)}$ and repeat the entire procedure, starting with the determination of a new set of $u_1^{(\alpha)}$, $\alpha = 1, 2, \ldots, 10$. Since we have a quadratic objective function and two state variables, we have no assurance a priori that the method will converge (cf. Denn [1965]).

The computation was carried out with a digital computer, and convergence was obtained to within four significant figures in nine iterations. The numerical results obtained during each iteration are shown in Tables 8.17–8.25, respectively. [Notice that a set of values for $u_0^{(\alpha)}$ is obtained in each iteration. This is not necessary, since the $u_0^{(\alpha)}$ do not enter into the iterative computation. On the other hand, it does allow us to see how rapidly we attain convergence of the objective function.]

Table 8.17. SAMPLE PROBLEM VIA DISCRETE MAXIMUM PRINCIPLE
ITERATION NUMBER 1

Alpha	Assumed X	U_1	Lambda	Calculated X	U_0
0	0.0	0.500000E 00	0.462328E 00	−0.684279E−78	0.0
1	0.0	0.450000E 00	0.402586E 00	−0.120129E 01	−0.756981E−01
2	0.0	0.405000E 00	0.347318E 00	−0.117366E 01	−0.156280E 00
3	0.0	0.364500E 00	0.295909E 00	−0.114795E 01	−0.240805E 00
4	0.0	0.328050E 00	0.247788E 00	−0.112389E 01	−0.328508E 00
5	0.0	0.295245E 00	0.202420E 00	−0.110121E 01	−0.418767E 00
6	0.0	0.265720E 00	0.159301E 00	−0.107965E 01	−0.511072E 00
7	0.0	0.239148E 00	0.117953E 00	−0.105898E 01	−0.605005E 00
8	0.0	0.215234E 00	0.779145E−01	−0.103896E 01	−0.700220E 00
9	0.0	0.193710E 00	0.387420E−01	−0.101937E 01	−0.796430E 00
10	0.0	0.174339E 00	0.0	−0.100000E 01	−0.893391E 00

Table 8.18. SAMPLE PROBLEM VIA DISCRETE MAXIMUM PRINCIPLE
ITERATION NUMBER 2

Alpha	Assumed X	U_1	Lambda	Calculated X	U_0
0	0.0	0.500000E 00	0.323917E−01	−0.684279E−78	0.0
1	−0.120129E 01	0.329871E 00	−0.751203E−01	−0.962440E 00	−0.889773E−01
2	−0.117366E 01	0.179518E 00	−0.156772E 00	−0.921614E 00	−0.185140E 00
3	−0.114795E 01	0.467709E−01	−0.214084E 00	−0.892958E 00	−0.283776E 00
4	−0.112389E 01	−0.702954E−01	−0.248264E 00	−0.875868E 00	−0.381741E 00
5	−0.110121E 01	−0.173387E 00	−0.260228E 00	−0.869886E 00	−0.477041E 00
6	−0.107965E 01	−0.264013E 00	−0.250612E 00	−0.874694E 00	−0.568501E 00
7	−0.105898E 01	−0.343509E 00	−0.219788E 00	−0.890106E 00	−0.655493E 00
8	−0.103896E 01	−0.413054E 00	−0.167874E 00	−0.916063E 00	−0.737727E 00
9	−0.101937E 01	−0.473685E 00	−0.947371E−01	−0.952631E 00	−0.815065E 00
10	−0.100000E 01	−0.526317E 00	0.0	−0.100000E 01	−0.887364E 00

Table 8.19. SAMPLE PROBLEM VIA DISCRETE MAXIMUM PRINCIPLE
ITERATION NUMBER 3

Alpha	Assumed X	U_1	Lambda	Calculated X	U_0
0	0.0	0.500000E 00	0.118825E 00	−0.684279E−78	0.0
1	−0.962440E 00	0.353756E 00	0.209165E−01	−0.101046E 01	−0.874747E−01
2	−0.921614E 00	0.226219E 00	−0.553718E−01	−0.972314E 00	−0.182280E 00
3	−0.892958E 00	0.114302E 00	−0.111795E 00	−0.944102E 00	−0.280661E 00
4	−0.875868E 00	0.152847E−01	−0.149617E 00	−0.925191E 00	−0.380078E 00
5	−0.869886E 00	−0.732323E−01	−0.169638E 00	−0.915181E 00	−0.478822E 00
6	−0.874694E 00	−0.153378E 00	−0.172213E 00	−0.913893E 00	−0.575728E 00
7	−0.890106E 00	−0.227051E 00	−0.157264E 00	−0.921368E 00	−0.669955E 00
8	−0.916063E 00	−0.295952E 00	−0.124282E 00	−0.937859E 00	−0.760810E 00
9	−0.952631E 00	−0.361620E 00	−0.723240E−01	−0.963838E 00	−0.847602E 00
10	−0.100000E 01	−0.425458E 00	0.0	−0.100000E 01	−0.929500E 00

Table 8.20. SAMPLE PROBLEM VIA DISCRETE MAXIMUM PRINCIPLE
ITERATION NUMBER 4

Alpha	Assumed X	U_1	Lambda	Calculated X	U_0
0	0.0	0.500000E 00	0.101719E 00	−0.684279E−78	0.0
1	−0.101046E 01	0.348954E 00	0.191009E−02	−0.100095E 01	−0.878230E−01
2	−0.972314E 00	0.216828E 00	−0.754231E−01	−0.962288E 00	−0.182979E 00
3	−0.944102E 00	0.100735E 00	−0.131987E 00	−0.934006E 00	−0.281529E 00
4	−0.925191E 00	−0.185788E−02	−0.169038E 00	−0.915481E 00	−0.380814E 00
5	−0.915181E 00	−0.931901E−01	−0.187408E 00	−0.906296E 00	−0.479068E 00
6	−0.913893E 00	−0.175260E 00	−0.187522E 00	−0.906239E 00	−0.575117E 00
7	−0.921368E 00	−0.249871E 00	−0.169411E 00	−0.915295E 00	−0.668156E 00
8	−0.937859E 00	−0.318670E 00	−0.132707E 00	−0.933646E 00	−0.757560E 00
9	−0.963838E 00	−0.383187E 00	−0.766373E−01	−0.961681E 00	−0.842730E 00
10	−0.100000E 01	−0.444868E 00	0.0	−0.100000E 01	−0.922939E 00

Table 8.21. SAMPLE PROBLEM VIA DISCRETE MAXIMUM PRINCIPLE
ITERATION NUMBER 5

Alpha	Assumed X	U_1	Lambda	Calculated X	U_0
0	0.0	0.500000E 00	0.105088E 00	−0.684279E−78	0.0
1	−0.100095E 01	0.349905E 00	0.565392E−02	−0.100283E 01	−0.877559E−01
2	−0.962288E 00	0.218685E 00	−0.714744E−01	−0.964263E 00	−0.182846E 00
3	−0.934006E 00	0.103416E 00	−0.128013E 00	−0.935994E 00	−0.281366E 00
4	−0.915481E 00	0.152677E−02	−0.165218E 00	−0.917391E 00	−0.380684E 00
5	−0.906296E 00	−0.892555E−01	−0.183915E 00	−0.908043E 00	−0.479041E 00
6	−0.906239E 00	−0.170954E 00	−0.184515E 00	−0.907742E 00	−0.575267E 00
7	−0.915295E 00	−0.245388E 00	−0.167027E 00	−0.916486E 00	−0.668548E 00
8	−0.933646E 00	−0.314214E 00	−0.131055E 00	−0.934472E 00	−0.758246E 00
9	−0.961681E 00	−0.378960E 00	−0.757920E−01	−0.962104E 00	−0.843741E 00
10	−0.100000E 01	−0.441064E 00	0.0	−0.100000E 01	−0.924287E 00

Table 8.22. SAMPLE PROBLEM VIA DISCRETE MAXIMUM PRINCIPLE
ITERATION NUMBER 6

Alpha	Assumed X	U_1	Lambda	Calculated X	U_0
0	0.0	0.500000E 00	0.104425E 00	−0.684279E−78	0.0
1	−0.100283E 01	0.349717E 00	0.491720E−02	−0.100246E 01	−0.877692E−01
2	−0.964263E 00	0.218319E 00	−0.722514E−01	−0.963874E 00	−0.182872E 00
3	−0.935994E 00	0.102888E 00	−0.128795E 00	−0.935603E 00	−0.281399E 00
4	−0.917391E 00	0.860512E−03	−0.165969E 00	−0.917015E 00	−0.380710E 00
5	−0.908043E 00	−0.900297E−01	−0.184602E 00	−0.907699E 00	−0.479047E 00
6	−0.907742E 00	−0.171801E 00	−0.185107E 00	−0.907447E 00	−0.575239E 00
7	−0.916486E 00	−0.246269E 00	−0.167496E 00	−0.916252E 00	−0.668473E 00
8	−0.934472E 00	−0.315090E 00	−0.131380E 00	−0.934310E 00	−0.758113E 00
9	−0.962104E 00	−0.379791E 00	−0.759582E−01	−0.962021E 00	−0.843545E 00
10	−0.100000E 01	−0.441812E 00	0.0	−0.100000E 01	−0.924025E 00

Table 8.23. SAMPLE PROBLEM VIA DISCRETE MAXIMUM PRINCIPLE
ITERATION NUMBER 7

Alpha	Assumed X	U_1	Lambda	Calculated X	U_0
0	0.0	0.500000E 00	0.104556E 00	−0.684279E−78	0.0
1	−0.100246E 01	0.349754E 00	0.506210E−02	−0.100253E 01	−0.877666E−01
2	−0.963874E 00	0.218391E 00	−0.720986E−01	−0.963951E 00	−0.182867E 00
3	−0.935603E 00	0.102992E 00	−0.128641E 00	−0.935679E 00	−0.281393E 00
4	−0.917015E 00	0.991523E−03	−0.165822E 00	−0.917089E 00	−0.380705E 00
5	−0.907699E 00	−0.898775E−01	−0.184467E 00	−0.907767E 00	−0.479046E 00
6	−0.907447E 00	−0.171634E 00	−0.184990E 00	−0.907505E 00	−0.575245E 00
7	−0.916252E 00	−0.246096E 00	−0.167404E 00	−0.916298E 00	−0.668488E 00
8	−0.934310E 00	−0.314917E 00	−0.131316E 00	−0.934342E 00	−0.758139E 00
9	−0.962021F 00	−0.379628E 00	−0.759255E−01	−0.962037E 00	−0.843583E 00
10	−0.100000E 01	−0.441665E 00	0.0	−0.100000E 01	−0.924076E 00

Table 8.24. SAMPLE PROBLEM VIA DISCRETE MAXIMUM PRINCIPLE
ITERATION NUMBER 8

Alpha	Assumed X	U_1	Lambda	Calculated X	U_0
0	0.0	0.500000E 00	0.104530E 00	−0.684279E−78	0.0
1	−0.100253E 01	0.349747E 00	0.503385E−02	−0.100252E 01	−0.877671E−01
2	−0.963951E 00	0.218377E 00	−0.721284E−01	−0.963936E 00	−0.182868E 00
3	−0.935679E 00	0.102972E 00	−0.128671E 00	−0.935664E 00	−0.281394E 00
4	−0.917089E 00	0.965893E−03	−0.165851E 00	−0.917075E 00	−0.380706E 00
5	−0.907767E 00	−0.899073E−01	−0.184493E 00	−0.907753E 00	−0.479047E 00
6	−0.907505E 00	−0.171667E 00	−0.185013E 00	−0.907493E 00	−0.575244E 00
7	−0.916298E 00	−0.246130E 00	−0.167422E 00	−0.916289E 00	−0.668485E 00
8	−0.934342E 00	−0.314951E 00	−0.131329E 00	−0.934336E 00	−0.758134E 00
9	−0.962037E 00	−0.379660E 00	−0.759319E−01	−0.962034E 00	−0.843576E 00
10	−0.100000E 01	−0.441694E 00	0.0	−0.100000E 01	−0.924066E 00

Table 8.25. SAMPLE PROBLEM VIA DISCRETE MAXIMUM PRINCIPLE
ITERATION NUMBER 9

Alpha	Assumed X	U_1	Lambda	Calculated X	U_0
0	0.0	0.500000E 00	0.104535E 00	−0.684279E−78	0.0
1	−0.100252E 01	0.349748E 00	0.503927E−02	−0.100252E 01	−0.877669E−01
2	−0.963936E 00	0.218380E 00	−0.721226E−01	−0.963939E 00	−0.182868E 00
3	−0.935664E 00	0.102976E 00	−0.128665E 00	−0.935667E 00	−0.281393E 00
4	−0.917075E 00	0.970840E−03	−0.165845E 00	−0.917078E 00	−0.380706E 00
5	−0.907753E 00	−0.899015E−01	−0.184488E 00	−0.907756E 00	−0.479046E 00
6	−0.907493E 00	−0.171661E 00	−0.185009E 00	−0.907496E 00	−0.575244E 00
7	−0.916289E 00	−0.246123E 00	−0.167418E 00	−0.916291E 00	−0.668485E 00
8	−0.934336E 00	−0.314945E 00	−0.131326E 00	−0.934337E 00	−0.758135E 00
9	−0.962034E 00	−0.379653E 00	−0.759307E−01	−0.962035E 00	−0.843577E 00
10	−0.100000E 01	−0.441688E 00	0.0	−0.100000E 01	−0.924068E 00

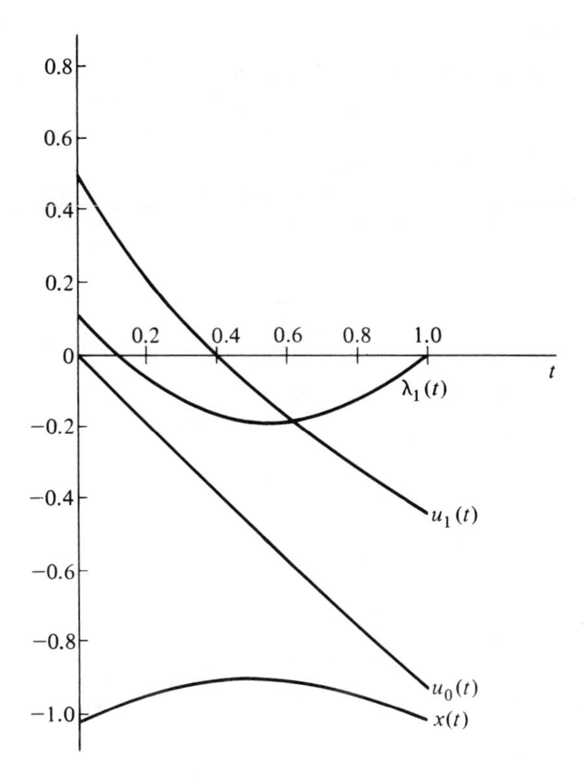

Figure 8.13. Solution to Example 8.9.1

Figure 8.13 shows a plot of the results obtained in the last iteration. In preparing this figure, the functions $u_0^{(\alpha)}$, $u_1^{(\alpha)}$, and $\lambda_1^{(\alpha)}$ were taken to correspond to integer multiples of $t = 0.10$ [i.e., $u_1^{(1)}$ applies at $t = 0.10$, $u_1^{(2)}$ at $t = 0.20$, etc.], whereas the $x^{(\alpha)}$ were assumed to apply at the center of an increment [$x^{(1)}$ applies at $t = 0.05$, $x^{(2)}$ at $t = 0.15$, etc.]. By plotting the results in this manner, we obtain a set of curves that resembles the analytical solutions shown in Figure 7.4 (cf. Example 7.5.1).

The only noticeable difference between the solutions shown in Figures 7.4 and 8.13 concerns the function $\lambda_1(t)$, which is of opposite sign in the two figures. This is to be expected, since the discrete maximum principle is actually a *minimum* principle, in its strong form, as discussed in the last section. Recall that the reason for this apparent discrepancy is that we defined $\lambda_0^{(\alpha)}$ to be equal to $+1$ for all α in the discrete form of the maximum principle. In contrast, we set $\lambda_0(t) = -1$ in our development of the continuous maximum principle.

Another difference between Figures 7.4 and 8.13 concerns the two curves for $u_1(t)$ [labeled $u_2(t)$ in Figure 7.4], which are not quite the same. This merely reflects the fact that the ten-stage approximation to the interval $0 \leq t \leq 1$ is relatively coarse. The agreement between the two curves can be improved simply by increasing the number of stages within the given interval. This would, however, increase the amount of computation required to solve the discrete form of the problem.

EXAMPLE 8.9.2

An interesting example of a two-stage, discrete system is presented by Horn and Jackson [1965]. This problem demonstrates clearly that the strong form of the discrete maximum principle can produce erroneous results when the constraining equations or the adjoint equations are not linear.

Specifically, we are given the two-stage system shown in Figure 8.14. We wish to determine numerical values for the decision variables $x^{(1)}$ and $x^{(2)}$, such that the function

$$u_0^{(2)} = [1 - 2x^{(1)} - \tfrac{1}{2}x^{(1)^2}] + [u_1^{(1)^2} + x^{(2)^2}] \tag{8.9.33}$$

is minimized. The state variables are transformed by the expressions

$$u_0^{(1)} = u_0^{(0)} + [1 - 2x^{(1)} - \tfrac{1}{2}x^{(1)^2}] = T_0^{(1)}, \tag{8.9.34}$$

$$u_0^{(2)} = u_0^{(1)} + u_1^{(1)^2} + x^{(2)^2} = T_0^{(2)}, \tag{8.9.35}$$

$$u_1^{(1)} = u_1^{(0)} + x^{(1)} = T_1^{(1)}, \tag{8.9.36}$$

$$u_1^{(2)} = \text{arbitrary}.$$

We are also given the initial conditions $u_0^{(0)} = 0$ and $u_1^{(0)} = 1$.

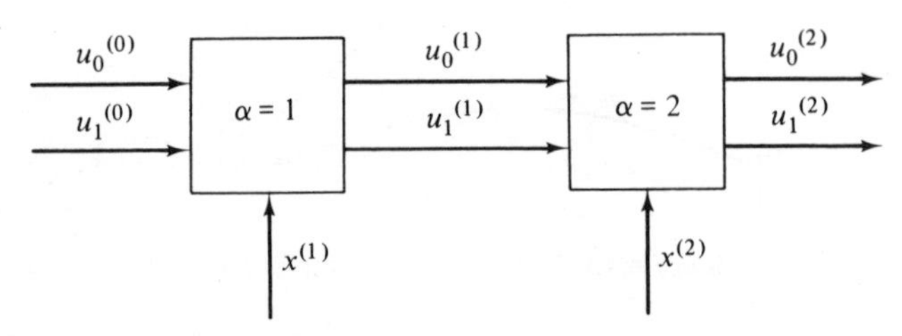

Figure 8.14. Description of Example 8.9.2

Owing to the simplicity of the problem, let us first obtain a solution by direct substitution. If we combine Equations (8.9.33) and (8.9.36), we obtain

$$u_0^{(2)} = u_1^{(0)^2} + 1 + \tfrac{1}{2}x^{(1)^2} + x^{(2)^2}. \tag{8.9.37}$$

Inclusion of the known initial condition then yields the following problem:

$$\underset{x^{(1)},\, x^{(2)}}{\text{minimize}}\; u_0^{(2)} = 2 + \tfrac{1}{2}x^{(1)^2} + x^{(2)^2}. \tag{8.9.38}$$

Clearly, this function will be minimized for $x^{(1)} = x^{(2)} = 0$, resulting in the solution $u_0^{(1)} = 1$, $u_1^{(1)} = 1$, and $u_0^{(2)} = 2$. Thus the desired minimum value of the objective function is equal to 2, obtained by setting the decision variables equal to zero.

Now let us apply the discrete maximum principle to this problem. We define a Hamiltonian function for each stage as

$$H^{(1)} = [1 - 2x^{(1)} - \tfrac{1}{2}x^{(1)^2}] + \lambda_1^{(1)}[1 + x^{(1)}], \tag{8.9.39}$$

$$H^{(2)} = [u_0^{(1)} + u_1^{(1)^2} + x^{(2)^2}]. \tag{8.9.40}$$

In addition, Equation (8.9.21) must be used to relate $\lambda_1^{(1)}$ to the known final conditions. This yields

$$\lambda_1^{(1)} = 2\lambda_0^{(2)}\frac{\partial T_0^{(2)}}{\partial u_1^{(1)}} + \lambda_1^{(2)}\frac{\partial T_1^{(2)}}{\partial u_1^{(1)}}. \tag{8.9.41}$$

Since $\lambda_0^{(2)} = 1$ and $\lambda_1^{(2)} = 0$, the above expression becomes simply

$$\lambda_1^{(1)} = 2\frac{\partial T_0^{(2)}}{\partial u_1^{(1)}} = 2u_1^{(1)}. \tag{8.9.42}$$

To obtain a solution by means of Katz's iterative algorithm, we must assume a pair of values for $x^{(1)}$ and $x^{(2)}$, solve Equation (8.9.36) for $u_1^{(1)}$, then obtain $\lambda_1^{(1)}$ from Equation (8.9.42), and, finally, compute new values for $x^{(1)}$ and $x^{(2)}$ by rendering $H^{(1)}$ and $H^{(2)}$ stationary. [We would seek to *minimize* $H^{(1)}$ and $H^{(2)}$ if the strong form of the discrete maximum principle were to apply.] Let us examine this last part of the algorithm. To render $H^{(1)}$ stationary, we require that

$$\frac{\partial H^{(1)}}{\partial x^{(1)}} = -2 - x^{(1)} + \lambda^{(1)} = 0, \tag{8.9.43}$$

or

$$x^{(1)} = \lambda^{(1)} - 2. \tag{8.9.44}$$

Taking the second derivative, however, we see that

$$\frac{\partial^2 H^{(1)}}{\partial x^{(1)2}} = -1, \tag{8.9.45}$$

which indicates that the value of $x^{(1)}$ obtained from Equation (8.9.44) actually *maximizes* $H^{(1)}$ with respect to $x^{(1)}$. Thus we see that the strong form of the discrete maximum principle does not apply. Rather, we minimize the desired objective function, $u_2^{(0)}$, by maximizing the appropriate stagewise Hamiltonians with respect to the decision variables. This is, of course, entirely consistent with the discrete maximum principle in its weak form.

8.10 Comparison of the Methods

In this chapter we have seen several unique advantages and disadvantages associated with the use of dynamic programming. On the plus side, dynamic programming proceeds by decomposing a problem into a sequence of simpler problems. If applied properly, the method always yields a global optimum. Moreover, we can obtain various suboptimal policies, or optimal policies for different initial conditions, from a given set of dynamic programming arrays. The method is particularly well suited to stagewise problems that involve either discrete decision variables or stochastic decision variables (we shall discuss problems of the latter type in Chapter 9).

On the other hand, use of dynamic programming is restricted to stagewise problems of low dimensionality. The method requires that large arrays of information be generated and stored, and it may be necessary to perform tedious interpolations to extract required information from the arrays. Furthermore, it is quite difficult to write a general-purpose computer code to implement the dynamic programming strategy.

For the reasons cited above, dynamic programming is usually not a preferred solution technique for those problems which can be attacked in another manner. In particular, stagewise problems which involve continuous variables and functions that are continuous and differentiable may best be approached using nonlinear programming techniques (as, for example, the methods discussed in Chapter 5). Generally speaking, these methods do not suffer from the storage and dimensionality restrictions of dynamic programming, and they can be programmed for a computer in general form.

The discrete maximum principle offers still another approach to the solution of stagewise problems. This method is not subject to storage and dimensionality restrictions. However, the introduction of the adjoint variables does increase the dimensionality of the solution space. Furthermore, use of the maximum principle requires solution of a mixed-boundary-type problem. Problems of this nature are often beset with severe convergence difficulties. Even if a solution is obtained using the maximum principle, we have no assurance that the desired optimum has been located, since the maximum principle represents only a necessary condition for an optimum. Finally, use of the discrete maximum principle is restricted to problems in which the stage-transformation equations have continuous partial derivatives with respect to the state and decision variables.

REFERENCES

Bellman, R. E., *Dynamic Programming*, Princeton University Press, Princeton, N.J. 1957.

———, and S. E. Dreyfus, *Applied Dynamic Programming*, Princeton University Press, Princeton, N.J., 1962.

Denn, M. M., "Convergence of a Method of Successive Approximations in the Theory of Optimal Processes " *Ind. Eng. Chem. Fundamentals*, *4* (1965), 231.

———, and R. Aris, "Second Order Variational Equations and the Strong Maximum Principle," *Chem. Eng. Sci.*, *20* (1965), 373.

Fan, L. T., and C. S. Wang, *The Discrete Maximum Principle*, Wiley, New York, 1964.

Gurel, O. and L. Lapidus, "The Maximum Principle and Discrete Systems," *Ind. Eng. Chem. Fundamentals*, *7* (1968), 617.

Halkin, H., "Optimal Control for Systems Described by Difference Equations," Chapter 4 of *Advances in Control Systems*, C. T. Leondes, ed., Vol. 1, Academic Press, New York, 1964.

Holtzman, J. M., "On the Maximum Principle for Nonlinear Discrete-Time Systems," *IEEE Trans. Auto. Control, 11* (1966), 273.

———, and H. Halkin, "Directional Convexity and the Maximum Principle for Discrete Systems," *J. SIAM Control, 4* (1966), 263.

Horn, F., and R. Jackson, "Correspondence on the Discrete Maximum Principle," *Ind. Eng. Chem. Fundamentals, 4* (1965), 110.

Howard, R. A., *Dynamic Programming and Markov Processes*, Wiley, New York, 1960.

Jackson, R., and F. Horn, "On Discrete Analogues of Pontryagin's Maximum Principle," *Intern. J. Control, 1* (1965), 389.

Katz, S., "Best Operating Points for Staged Systems," *Ind. Eng. Chem. Fundamentals, 4* (1962), 226.

Lapidus, L., and R. Luus, *Optimal Control of Engineering Processes*, Ginn-Blaisdell, Waltham, Mass., 1967.

Mitten, L. G., "Composition Principles for Synthesis of Optimal Multistage Processes," *Operations Res., 12* (1964), 610–19.

Nemhauser, G. L., *Introduction to Dynamic Programming*, Wiley, New York, 1966.

Pontryagin, L. S., V. G. Boltyanskii, R. V. Gamkrelidze, and E. F. Mischenko, *The Mathematical Theory of Optimal Processes* (English translation, K. N. Trirogoff), Wiley-Interscience, New York, 1962.

Roberts, S. M., *Dynamic Programming in Chemical Engineering and Process Control*, Academic Press, New York, 1964.

Rozonoer, L. I., "The Maximum Principle of L. S. Pontryagin in Optimal System Theory," Part III, *Automation & Remote Control, 20* (1959), 1517.

PROBLEMS

8.1. In the network shown, determine the minimum path from point A to point F using dynamic programming. Assume all routes are one-way; i.e., "backward" moves will not be allowed.

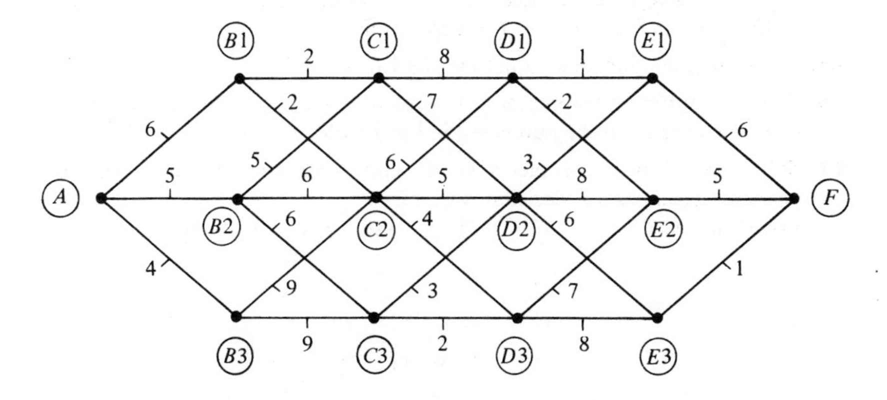

8.2. In the network shown, find the right-hand nodal point (i.e., point *I*1, *I*3, *I*5, *I*7, or *I*9) that is closest to point *A*1. Use dynamic programming and assume that "backward" moves are not allowed.

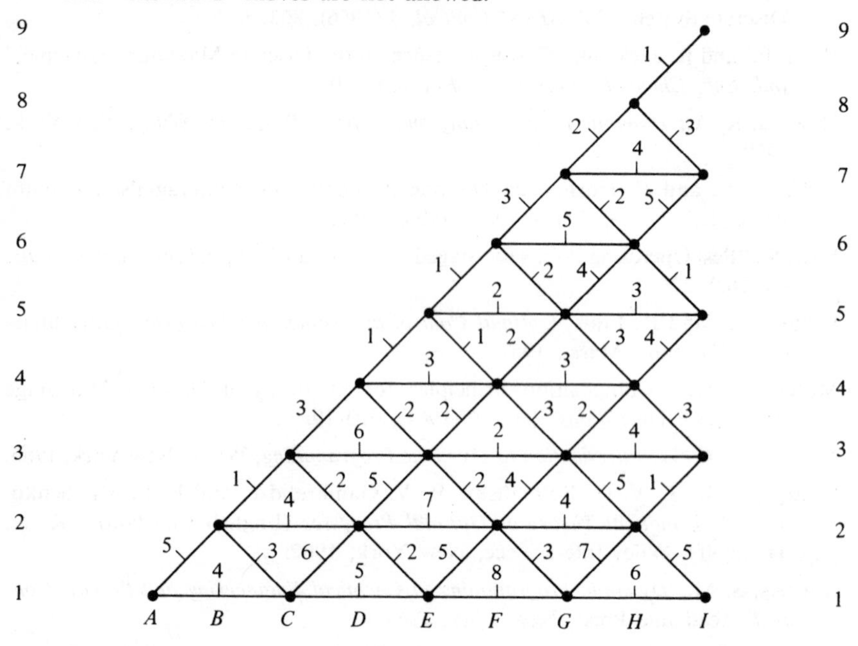

8.3. Can problem 8.2 be solved by dynamic programming if "backward" moves are allowed? Explain.

8.4. Suppose that the overall objective function for some particular problem is expressed in terms of the product of individual stage returns; i.e.,

$$y_k(u_k) = \max_{x_k, x_{k-1}, \ldots, x_1}(\min) \left\{ \prod_{j=1}^{k} r_j(u_j, x_j) \right\}.$$

If this problem has a serial structure, as represented by Figure 8.2, then the objective function can be decomposed, resulting in an expression analogous to Equation (8.3.13). Derive this expression.

8.5. Prove by induction that Equation (8.3.13) is valid.

8.6. Solve the automobile-replacement problem described in Section 8.5 for the case when the car to be purchased is 3 years old ($\tau = 3$).

8.7. Solve the following problems by dynamic programming. Treat the independent variables as decision variables.

(a) Minimize $y = (x_1 - 2)^2 + 2(x_2 - 1)^2 + (x_3 - 3)^2$, subject to

$$2x_1 + x_2 + 2x_3 \geq 4,$$
$$x_1^2 + 2x_2^2 + 3x_3^2 \geq 48,$$
$$x_1, x_2, x_3 \geq 0.$$

(b) Maximize $y = x_1 x_2 x_3$, subject to

$$x_1 + 2x_2 + x_3 = 6,$$
$$x_1, x_2, x_3 \geq 0.$$

(c) Minimize $y = x_1 + 2x_2 + x_3$, subject to

$$x_1 x_2 x_3 = 4,$$
$$x_1, x_2, x_3 \geq 0.$$

8.8. Solve problem 2.7 by means of dynamic programming.

8.9. Obtain a numerical dynamic programming solution to the problem presented in Example 8.6.1. Solve by constructing three tables, one for each stage, as shown in Section 8.5. The optimal policy is obtained by retracing a path through the tables, as with the automobile-replacement problem in Section 8.5.

8.10. A small manufacturer of specialty items anticipates a large but temporary increase in his product demand over the next 5 years. He is considering expanding his manufacturing facilities by either choosing expansion plan A or expansion plan B at some undetermined time in the future. He may also choose not to expand his facilities at all. An optimal policy is sought that would determine which expansion plan should be adopted (if any), and when the expansion should be undertaken.

The estimated yearly profits obtainable with each of the three physical facilities (states) for each year (stage) are given in the table in thousands of dollars.

Stage (year)	*State* →	*I (present facilities)*	*II (expansion A)*	*III (expansion B)*
1		100	120	120
2		100	150	190
3		100	150	250
4		100	150	230
5		100	130	130

The costs of expanding from state I to state II or state III are \$110,000 and \$250,000, respectively. The objective is to maximize the present worth of the total profit; hence, operating profit at year i is discounted back for i years. Expansion costs are assumed to be incurred at the beginning of year i and are therefore discounted for $(i - 1)$ years.

Let $u_{i,j}$ be the maximum discounted net profit for the i-stage problem terminating in state j; let p_{ij} be the yearly operating profit for stage i and state j; and let d_i be the discount factor: $d_i = 1/(1 + r)^i$, where r, the discount rate, is 0.10. Write the functional equations to solve this problem by dynamic programming. Proceed by counting forward in time and by writing the governing equations for each state.

8.11. Using the functional equations of problem 8.10, determine y_{ij} for $i = 1, 2, 3, 4, 5$ and $j = \text{I, II, III}$. Show that the optimal policy is to expand from state I to state III at the beginning of stage 2, and that the policy yields a net discounted total profit of $446,300. Since $y_{4,\text{II}}$ and $y_{4,\text{III}}$ are both greater than $y_{4,\text{I}}$, is it necessary to evaluate $y_{5,\text{I}}$?

8.12. The personnel director of a firm engaged in cyclical work has estimated the manpower requirements for each of the three cycles in a typical year. He assumes that the company will incur an "employee imbalance" cost which is proportional to the square of the difference between the actual and desired employment levels during each cyclical period. Furthermore, an additional cost, equal to the square of the difference between employment levels in succeeding periods, is imposed to account for the cost of acquiring, training, and releasing employees.

The personnel director's estimate of the firm's manpower requirements for each cycle are given in the table.

Cycle	*A*	*B*	*C*	*A*
Estimated manpower	100	200	160	100
Cost proportionality constant	1	2	2	1

Assuming that a level of 100 is maintained during period A, determine the employment level for each period such that the total yearly cost is minimized.

8.13. A clever young member of the jet set has decided to analyze the way she spends her time during a typical evening out in the social world. In particular, if she has 5 hours available, she is interested in determining how she can best divide her time among three social activities. She has constructed the following table of personal satisfaction versus hours spent in each activity:

	Number of Hours					
Activity	*0*	*1*	*2*	*3*	*4*	*5*
A	0	5	13	17	20	26
B	0	3	11	16	23	28
C	0	6	14	19	24	30

Use dynamic programming to show that she will achieve a maximum personal satisfaction level of 32 by engaging in activity A for 2 hours and in activity C for 3 hours.

8.14. A supplier maintains an inventory of some particular commodity in a warehouse whose capacity is 50,000 units. The supply and demand for this commodity are seasonal, and the monthly purchase cost and sale price per unit for a typical year are shown in the table.

Month	*Purchase Cost* p_i	*Sale Price* s_i
Jan.	20	40
Feb.	16	36
Mar.	13	30
Apr.	10	24
May	12	20
June	15	17
July	18	15
Aug.	22	18
Sept.	26	22
Oct.	30	27
Nov.	28	32
Dec.	24	37

If the total inventory is 40,000 units on Jan. 1, find an optimal buy–sell strategy that will maximize the yearly profit.

In deriving the dynamic programming functional equations, let

w_i = inventory at start of the ith month (counting backward),
$y_i(x_i)$ = maximum profit for the i remaining months,
u_i = number of units bought in the ith month, and
v_i = number of units sold in the ith month.

Write all the constraints carefully, and show that they form a convex polyhedron for each month. Using this fact and the linearity of the objective function, devise a simple search scheme for determining the optimal values of u_i and v_i.

8.15. Solve problem 7.2 using dynamic programming. (Replace the integral with a trapezoidal rule approximation, with $\Delta t = 0.10$.)

8.16. Solve problem 7.3 using dynamic programming. [*Hint:* Consider $u(t)$ and $v(t)$ to both be state variables.]

8.17. Find the function $u(t)$, $0 \leq t \leq 1$, that causes the area under the curve to be maximized. Assume that $u(0) = u(1) = 0$, and that $u(t)$ is bounded between 0 and 1 for all t. Let the length of the curve be equal to 2. (*Hint:* See Example 7.3.1.)

8.18. Solve problem 7.9, using dynamic programming, for the special case in which $t_f = 1$.

8.19. A servomechanism is being designed that will control the movement of a hydraulic valve on a jet airliner. The output of the servomechanism is a connecting rod that controls the quantity of flow through the valve. The position of this connecting rod, $u(t)$, is governed by the differential equation

$$\frac{d^2u}{dt^2} = x(t),$$

where t represents time and $x(t)$ is an electrical actuating signal (e.g., a voltage), subject to the restrictions

$$-1 \leq x(t) \leq 1.$$

Determine by dynamic programming the variation in the actuating signal with time such that the time required for $u(t)$ to travel from $u(0) = 0$ to $u(t_f) = 1$ will be minimized. Assume that

$$\left.\frac{du}{dt}\right|_{t=0} = 1.$$

8.20. Show by dynamic programming that two necessary conditions for extremization of the integral

$$I = \int_{t_0}^{t_f} \phi(u, v, u', v', t)\, dt$$

are that

$$\frac{d}{dt}\frac{\partial \phi}{\partial u'} - \frac{\partial \phi}{\partial u} = 0$$

and

$$\frac{d}{dt}\frac{\partial \phi}{\partial v'} - \frac{\partial \phi}{\partial v} = 0,$$

providing u and v are continuous, differentiable functions of t, with $u(t_0)$, $v(t_0)$, $u(t_f)$ and $v(t_f)$ known. These are the *Euler–Lagrange equations* of Section 7.2.

8.21. Solve problems 8.7(a), (b), and (c) using the discrete maximum principle.

8.22. Solve problems 7.10(a), (b), and (c) numerically by means of the discrete maximum principle. Let $\Delta t = 0.50$. Compare this approach with that of introducing Lagrange multipliers, viewing the problem as one in nonlinear programming.

8.23. Solve problem 8.12 by means of the discrete maximum principle. Which form of the maximum principle applies to this particular problem? Compare your results with the dynamic programming solution.

8.24. Solve problem 8.13 by means of the discrete maximum principle. Compare this solution with that obtained using dynamic programming.

8.25. Solve problem 8.14 using the discrete maximum principle. Again compare this solution with the dynamic programming solution.

OPTIMIZATION UNDER UNCERTAINTY AND RISK

It has been assumed in the previous chapters that we live in a deterministic world. Our objective function and all constraint parameters were considered to be known with certainty. We all recognize that in many instances this is not the case. For example, suppose that we wish to optimize the design of a manufacturing plant which is to be built some time hence. We would attempt to determine the equipment sizes and operating conditions that would provide minimum costs. On the basis of known current equipment prices and operating costs, we could find the proper balance between capital investment and operating expenses. However, current costs may no longer be applicable when the plant is finally constructed. In addition, we may expect further changes in operating cost levels over the life of the plant. We can attempt to estimate the costs we shall encounter in the future based on past cost behavior and then optimize our design using these estimated costs. However, we recognize that any estimate of future costs cannot be precise. At best, all we can say is that there is some probability that a particular cost will be above a given level or fall below a given level. The possible variations in cost may be so large that it is imprudent to ignore them. We must then abandon the idea that our decision is to be made under deterministic conditions.

9

We can consider the various cost levels that we may obtain as representing various states of the world. When we know what the various possible states of the world are, and have some means of estimating the probability of these states occurring, then we are faced with a decision under *risk*. If, however, we know what the various possible states of the world are, but are unable to estimate the probability that any state will occur, we are faced with a decision under *uncertainty*.

Uncertain situations may arise when we have no past information on which to base predictions or when past information cannot be assumed to be a reliable guide. This latter possibility will arise in competitive situations. Suppose, for example, that two large manufacturing organizations control the production of a given commodity. If we assume that the share of the market each receives is determined by their relative advertising expenditure, past information can tell company I what fraction of the market it will receive with known expenditures by companies I and II. However, it is unlikely that this past information will enable company I to estimate what advertising expenditures company II will make for the next year. Company I would normally have no knowledge of the decision-making process of company II. Thus company I knows what the various states of the world may be (fraction of the market it will receive for given expenditures by I and II), but cannot associate any probabilities with these states. Company I is thus required to make a decision under uncertainty.

In the deterministic models presented in earlier chapters, we did not concern ourselves with the formulation of our objective function. It was assumed that the decision maker had a clear idea of what he wished to do; e.g., minimize costs, maximize profits, or maximize some design property. In the presence of significant risk or uncertainty it is not clear what the real costs, profits, or design properties will be. The selection of a proper objective function is therefore not obvious. In this chapter our primary concern will be the development of a mathematical model that can account for risk or uncertainty and that will provide an appropriate objective function. We shall also concern ourselves with the proper formulation of constraints under risky conditions.

9.1 Optimization Under Uncertainty

Let us return to the example of companies I and II competing for a given market. We shall suppose that advertising agencies are willing to carry out campaigns with $250,000, $500,000, or $750,000 expenditures. With an expenditure of $750,000, full coverage of all the media is obtained, and expenditures beyond this level produce no further change. Thus each organization has three possibilities for action. We shall call each of these possibilities a *strategy*. For each combination of the strategies of companies I and II, we may estimate, on the basis of past experience, the share of the market to be lost or gained by each organization. A set of such hypothetical estimates is shown in Table 9.1. The values shown are the gains that will be made by

Table 9.1. HYPOTHETICAL EFFECT OF ADVERTISING EXPENDITURE ON MARKET SHARE FOR ORGANIZATION I

	Player II Strategies		
Player I Strategies	1 = *$250,000 campaign*	2 = *$500,000 campaign*	3 = *$750,000 campaign*
1 = *$250,000* campaign	0%	−10%	−20%
2 = *$500,000 campaign*	+10%	0%	−10%
3 = *$750,000 campaign*	+20%	+10%	0%

organization I. Organization I will obviously seek to maximize the gains shown. Since the gains of organization I are the losses of organization II, organization II will seek to minimize organization I's gains.

The situation we have just discussed has many points of similarity to a simple gambling game. Consider the simple game of odd–even. In this game, each player extends one or two fingers from a closed fist. The players' actions occur simultaneously so that neither has advance knowledge of the other's play. Player II pays player I one unit if the total number of fingers extended is even, and player I pays player II one unit if they are odd. We represent this information in the simple matrix of Table 9.2. The values in the matrix are the *pay offs* received by player I for all strategy combinations, and therefore the matrix is called the *payoff matrix*. Player I seeks to maximize his payoffs and is therefore called the *maximizing* player. Since the gains made by player I are losses to player II, he seeks to minimize the losses and is the *minimizing* player. We observe that the algebraic sum of the gains by player I and losses by player II is zero. Such a game is therefore called a *zero-sum* game.

Table 9.2. PAYOFF MATRIX FOR PLAYER I IN GAME OF ODD–EVEN

	Player II Strategies	
Player I Strategies	1 = *One finger*	2 = *Two fingers*
1 = *One finger*	+1	−1
2 = *Two fingers*	−1	+1

If we compare Tables 9.1 and 9.2, we observe the similarities. The possible advertising strategies are analogous to the player strategies. The market shares gained by organization I are entirely analogous to the payoffs received by player I in our simple game. The algebraic sum of the market gains and losses will be zero, just as the algebraic sum of the payoffs was in our game. We can thus think of our competitive market situation in terms of a simple game whose payoffs are market-share losses and gains.

The analogy between games and competitive economic behavior was first delineated by Von Neumann and Morgenstern in their *Theory of Games and Economic Behavior* [1953]. The game analogy leads to very useful insights, and we shall therefore consider all such competitive situations as games. For the sake of simplicity, we shall confine ourselves to situations in which there are only two players and the game is a zero-sum game.

The payoff matrices shown in Tables 9.1 and 9.2 are particularly simple. In the situations we have considered, the options open to both players are identical, and the payoff matrix is symmetric. This need not be the case. In most cases, the alternatives open to player I will be entirely different than those open to player II. Consider the payoff matrix shown in Table 9.3. The different options open to the two players have led to an unsymmetric payoff matrix. The game is still a zero-sum game, since the payoffs to player I (the values shown in the matrix) are the losses of player II.

Table 9.3. UNSYMMETRIC ZERO-SUM PAYOFF MATRIX

	Player II		
Player I	(*d*)	(*e*)	(*f*)
(*a*)	3	7	1
(*b*)	6	5	2
(*c*)	4	3	7

Optimum pure strategies

We have not yet considered how to establish guidelines for strategy selection. Let us return to our market-share game (Table 9.1) and examine the various strategies. If player I selects strategy 1, the worst thing that can happen to him is that he loses 20 per cent of the market. With strategy 2, the worst possibility is that he loses 10 per cent of the market; with strategy 3, he loses 0 per cent at worst. If player I were very cautious, he would want to protect himself from the worst possible occurrence He therefore would choose that strategy which maximized his minimum payoff, strategy 3. The cautious man thus examines all the strategies, determines the minimum possible payoff for each, and then selects that strategy which would yield the maximum of these minima. This principle, which was first stated by Wald [1950], is called the *maximin* principle.

We can state our result more formally if we represent our payoff matrix by

$$\mathbf{A} = \begin{bmatrix} a_{11} a_{12} \ldots a_{1n} \\ \text{-----------} \\ a_{m1} a_{m2} \ldots a_{mn} \end{bmatrix}, \tag{9.1.1}$$

where n and m are (3) in our example. Player I first determines the minimum in each row, since the minimum represents the worst that can happen to him with the given strategy:

$$\begin{aligned} \min_j a_{1j} &= a_{13} = -20\%, \\ \min_j a_{2j} &= a_{23} = -10\%, \\ \min_j a_{3j} &= a_{33} = \quad 0\%. \end{aligned} \tag{9.1.2}$$

He then selects the maximum of these minima:

$$\max_i \min_j a_{ij} = a_{33} = 0\%. \tag{9.1.3}$$

Now let us examine our market-share game from the standpoint of player II. Since the gains of player I are his losses, the maxima of each column represent the worst possible occurrences. He therefore examines

$$\begin{aligned} \max_i a_{i1} &= a_{31} = 20\%, \\ \max_i a_{i2} &= a_{32} = 10\%, \\ \max_i a_{i3} &= a_{33} = \ \ 0\%, \end{aligned} \tag{9.1.4}$$

and then will minimize his losses by selecting the minimum of these maxima:

$$\min_j \max_i a_{ij} = a_{33} = 0\%. \tag{9.1.5}$$

We observe that

$$\min_j \max_i a_{ij} = \max_i \min_j a_{ij} = v \tag{9.1.6}$$

in this game. We call v the *value* of the game. By choosing his third pure strategy, player I is certain of winning at least v. Any departure from this strategy lowers his opportunity for winning at least v. Similarly, player II can prevent player I from winning more than v by using his third strategy. Any departure from this strategy will give player I an opportunity to win more than v. We therefore say that the two strategies are in *equilibrium*. When the two players use their equilibrium strategy, player I will win v, the value of the game.

In the payoff matrix of Table 9.1, element a_{33} is both the minimum of its row and the maximum of its column. Such an element is called a *saddle point*. When element a_{lk} is a saddle point, then the optimal pure strategies are l and k, and $v = a_{lk}$. When we examine the payoff matrix of Table 9.2, we see that it does not contain a saddle point. In the odd–even game, no two pure strategies are in equilibrium.

Mixed strategies

In games, as well as in economic competition, the play is usually repeated many times. Thus, in our odd–even game, the players are really concerned with their relative position after a number of plays. If both players are cautious, each will try to minimize the other's gain. It is obvious that if player I always uses his first strategy, player II has merely to use his second strategy to be assured of winning. Similarly, if player I always chooses his second strategy, player I will be frustrated by player II choosing his first strategy. It is clear that consistent selection of a single strategy by player I is not desirable in this game. Our observations about player I apply equally well to player II, and therefore both players will play the game by using both their first and second strategies. When a player does this, we say that he is employing a *mixed strategy*. In contrast, when a player consistently uses the same strategy at every play, we say that he is using a *pure strategy*.

We shall represent the mixed strategy of player I by the row vector **X**, where

$$\mathbf{X} = [x_1, x_2, \ldots, x_m] \tag{9.1.7}$$

and x_i represents that fraction of the time player I will choose strategy i. Similarly, we represent the mixed strategy of player II by the column vector **W**, where

$$\mathbf{W} = \begin{bmatrix} w_1 \\ w_2 \\ \cdot \\ \cdot \\ \cdot \\ w_n \end{bmatrix} \tag{9.1.8}$$

and w_j represents the fraction of the time player II chooses strategy j. On the average, player I may then expect to receive from each play $E(\mathbf{X}, \mathbf{W})$, which is given by

$$E(\mathbf{X}, \mathbf{W}) = \mathbf{XAW} = \sum_{i=1}^{m} \sum_{j=1}^{n} x_i a_{ij} w_j. \tag{9.1.9}$$

Von Neumann and Morgenstern [1953] have shown that, for every two-person zero-sum game, there exists a set of *optimal strategies* $\mathbf{X}_o$ and $\mathbf{W}_o$. These strategies will guarantee that player I will win at least v and that player II will lose no more than v. We may express their results as

$$E(\mathbf{X}_o, \mathbf{W}) \geq v, \tag{9.1.10}$$

$$E(\mathbf{X}, \mathbf{W}_o) \leq v. \tag{9.1.11}$$

Equations (9.1.10) and (9.1.11) imply

$$E(\mathbf{X}_o, \mathbf{W}_o) = v. \tag{9.1.12}$$

Thus if both players use their optimal strategy, it leads to a predetermined result—a payoff of v. We again call v the *value* of the game.

For player I, the set of optimal strategies represents maximization of his payoff over his own strategies with the assumption that his efforts will be minimized by player II. Hence for player I

$$E(\mathbf{X}_o, \mathbf{W}_o) = \max_{\mathbf{X}} \min_{\mathbf{W}} \{E(\mathbf{X}, \mathbf{W})\}. \tag{9.1.13}$$

Similarly, we have for player II

$$E(\mathbf{X}_o, \mathbf{W}_o) = \min_{\mathbf{W}} \max_{\mathbf{X}} \{E(\mathbf{X}, \mathbf{W})\}. \tag{9.1.14}$$

Therefore, we may write

$$\max_{\mathbf{X}} \min_{\mathbf{W}} \{E(\mathbf{X}, \mathbf{W})\} = \min_{\mathbf{W}} \max_{\mathbf{X}} \{E(\mathbf{X}, \mathbf{W})\} = v. \tag{9.1.15}$$

This is the well known min–max statement of Von Neumann and Morgenstern [1953].

Games and linear programming

We now have the problem of determining the optimum mixed strategies, $\mathbf{X}_o$, $\mathbf{W}_o$, for players I and II. Since we know from Equations (9.1.10) and (9.1.11) that $E(\mathbf{X}_o, \mathbf{W}) \geq v$ and $E(\mathbf{X}, \mathbf{W}_o) \leq v$, we can state our problem as that of finding a number v such that these inequalities hold. We shall show that we can state our problem as a linear programming problem. To do so, we shall make use of the following definitional theorem: A solution to a matrix game is a pair of mixed strategies $(\mathbf{X}_o, \mathbf{W}_o)$ and a real number v such that

$$E(\mathbf{X}_o, w_j) \geq v \qquad \text{for pure strategies } j = 1, 2, \ldots, m, \tag{9.1.16}$$

$$E(x_i, \mathbf{W}_o) \leq v \qquad \text{for pure strategies } i = 1, 2, \ldots, m. \tag{9.1.17}$$

Let us use the foregoing theorem to determine the optimum mixed strategy for player I. Assume that player I adopts an optimal mixed strategy, $\mathbf{X}_o = (x_1, x_2, \ldots, x_m)$, and player II adopts the pure strategy where $w_1 = 1$ and all other $w_j = 0$. Equation (9.1.16) tells us that

$$a_{11}x_1 + a_{21}x_2 + \cdots + a_{m1}x_m \geq v. \tag{9.1.18}$$

We can write a similar equation for any pure strategy chosen by player II. We therefore have the set of inequalities

$$\begin{aligned} a_{11}x_1 + a_{21}x_2 + \cdots + a_{m1}x_m &\geq v, \\ a_{12}x_1 + a_{22}x_2 + \cdots + a_{m2}x_m &\geq v, \\ &\cdots\cdots \\ a_{1n}x_1 + a_{2n}x_2 + \cdots + a_{mn}x_m &\geq v. \end{aligned} \tag{9.1.19}$$

Furthermore, we know that

$$x_1 + x_2 + \cdots + x_m = 1 \tag{9.1.20}$$

and

$$x_i \geq 0 \qquad \text{for } i = 1, 2, \ldots, m. \tag{9.1.21}$$

It is obvious that our problem would have the format of a linear programming problem if we provided an objective function. We can do so by recognizing that every element of **A**, our payoff matrix, can be made positive by addition of a constant to all the a_{ij}. This will not change the value of $\mathbf{X}_o$ but will increase v by the constant added and thus assure that $v > 0$. We shall therefore assume that $v > 0$ and divide (9.1.19) and (9.1.20) by v. Also, let

$$x_i' = \frac{x_i}{v}. \tag{9.1.22}$$

Then from Equation (9.1.20) we have

$$x_1' + x_2' + \cdots + x_m' = \frac{1}{v}. \tag{9.1.23}$$

We wish to choose a vector **X** that will maximize v, or minimize $1/v$. We therefore let our objective function be: Minimize

$$y = x_1' + x_2' + \cdots + x_m' \tag{9.1.24}$$

and require that the constraints

$$\begin{aligned} a_{11}x_1' + a_{21}x_2' + \cdots + a_{m1}x_m' &\geq 1, \\ a_{12}x_1' + a_{22}x_2' + \cdots + a_{m2}x_m' &\geq 1, \\ &\cdots\cdots\cdots \\ a_{1n}x_1' + a_{2n}x_2' + \cdots + a_{mn}x_m' &\geq 1, \\ x_i' \geq 0 \qquad \text{for } i = 1, 2, \ldots, n& \end{aligned} \tag{9.1.25}$$

be met. Solution of this linear programming problem provides the optimum mixed strategy for player I.

Now let us use Equation (9.1.17) to determine the optimum mixed strategy for player II, i.e., the strategy that minimizes v. As before, we assume player II adopts an optimal mixed strategy $\mathbf{W}_o$ and that player I adopts the pure strategy where $x_1 = 1$ and all other $x_i = 0$. Then, from Equation (9.1.11),

$$a_{11}w_1 + a_{12}w_2 + \cdots + a_{1n}w_n \leq v. \tag{9.1.26}$$

We may write similar equations for any other pure strategy choice by player

I, and therefore we obtain the set of inequalities

$$a_{i1}w_1 + a_{i2}w_2 + \cdots + a_{in}w_n \leq v, \qquad i = 1, 2, m. \tag{9.1.27}$$

We also have

$$w_1 + w_2 + \cdots + w_n = 1 \tag{9.1.28}$$

and

$$w_i \geq 0 \qquad \text{for } j = 1, 2, \ldots, n. \tag{9.1.29}$$

We may proceed just as we did previously and divide all expressions by v. If we let

$$w'_j = \frac{w_j}{v}, \tag{9.1.30}$$

then

$$\sum_{j=1}^{n} w'_j = \frac{1}{v}. \tag{9.1.31}$$

Since player II wishes to minimize v, he desires to maximize $1/v$, We therefore state our problem as: Maximize

$$y = w'_1 + w'_2 + \cdots + w'_n, \tag{9.1.32}$$

subject to

$$\begin{aligned}
&a_{11}w'_1 + a_{12}w'_2 + \cdots + a_{1n}w'_n \leq 1,\\
&a_{21}w'_1 + a_{22}w'_2 + \cdots + a_{2n}w'_n \leq 1,\\
&\text{-----------------------------------}\\
&a_{m1}w'_1 + a_{m2}w'_2 + \cdots + a_{mn}w'_n \leq 1,\\
&w'_j = 0 \qquad \text{for } j = 1, 2, \ldots, m.
\end{aligned} \tag{9.1.33}$$

If we compare this problem statement with the one developed for obtaining $\mathbf{X}_o$, we see immediately that one problem is the dual of the other. It will be recalled from Chapter 4 that when we solve the primal problem we obtain the dual solution as the $(y_j - c_j)$ elements for the slack vectors. We can thus obtain both $\mathbf{X}_o$ and $\mathbf{W}_o$ by solution of a single linear programming problem.

EXAMPLE 9.1.1

Let us illustrate our procedure by determining the optimum mixed strategy for player II in the simple game of odd–even. Since the matrix of Table 9.2 contains negative quantities, we shall add the constant 2 to each a_{ij}. Addition of this constant (2) will increase v by two units. We then have

$$\mathbf{A} = \begin{bmatrix} 3 & 1 \\ 1 & 3 \end{bmatrix}, \tag{9.1.34}$$

which leads to the dual problem: Maximize

$$y = w'_1 + w'_2, \tag{9.1.35}$$

subject to

$$3w'_1 + w'_2 \leq 1, \quad (9.1.36)$$

$$w'_1 + 3w'_2 \leq 1. \quad (9.1.37)$$

After appropriate slack variables w'_3 and w'_4 are introduced, we have

$$3w'_1 + w'_2 + w'_3 = 1, \quad (9.1.38)$$

$$w'_1 + 3w'_2 + w'_4 = 1. \quad (9.1.39)$$

The initial simplex tableau is shown in Table 9.4 and the final tableau in Table 9.5.

Table 9.4. INITIAL SIMPLEX TABLEAU FOR DUAL OF ODD–EVEN GAME

	c		*1*	*1*	*0*	*0*
Basis		$\mathbf{P}_0$	$\mathbf{P}_1$	$\mathbf{P}_2$	$\mathbf{P}_3$	$\mathbf{P}_4$
$\mathbf{P}_3$	0	1	3	1	1	0
$\mathbf{P}_4$	0	1	1	3	0	1
$(y_j - c_j)$			−1	−1	0	0

Table 9.5. FINAL SIMPLEX TABLEAU FOR DUAL OF ODD–EVEN GAME

	c		*1*	*1*	*0*	*0*
Basis		$\mathbf{P}_0$	$\mathbf{P}_1$	$\mathbf{P}_2$	$\mathbf{P}_3$	$\mathbf{P}_4$
$\mathbf{P}_1$	1	$\frac{1}{4}$	1	0	$\frac{3}{8}$	$-\frac{1}{8}$
$\mathbf{P}_2$	1	$\frac{1}{4}$	0	1	$-\frac{1}{8}$	$\frac{3}{8}$
$(y_j - c_j)$		$\frac{1}{2}$	0	0	$\frac{1}{4}$	$\frac{1}{4}$

We see that

$$w'_1 = \frac{1}{4}, \quad w'_2 = \frac{1}{4} \quad (9.1.40)$$

and

$$\frac{1}{v} = w'_1 + w'_2 = \frac{1}{2}. \quad (9.1.41)$$

Since $w_i = vw'_i$, our final result is

$$w_1 = \frac{1}{2}, \quad w_2 = \frac{1}{2}. \quad (9.1.42)$$

We also see that the value of this game is two units. We have noted previously that this value is two units greater than that of the original game, since we increased each member of the payoff matrix by two units. The value of the original game is therefore zero.

By examination of the $(y_j - c_j)$ row, we obtain the solution to the primal in the columns corresponding to the slack variables as

$$x_1' = \frac{1}{4}, \qquad x_2' = \frac{1}{4} \tag{9.1.43}$$

or

$$x_1 = \frac{1}{2} \qquad x_2 = \frac{1}{2}. \tag{9.1.44}$$

Since the payoff matrix is symmetric, this is what we would expect. Obtaining the solution to the primal in this manner is slightly easier than solving the primal directly. There is no unit matrix present in the statement of the primal problem, and an artificial basis would have to be introduced.

In the simple example we chose for illustration, one could determine the optimum strategies for the two players without the formation of a linear programming problem. However, when confronted with an unsymmetric payoff matrix, such as that of Table 9.3, this would not be true. The optimum mixed strategy for either player would not use each of the pure strategies equally. Furthermore, the strategies for the two players would not be the same. It is in such situations that the linear programming approach becomes highly useful.

Although we have confined our discussion to two-person zero-sum games, it is well to remember that many other situations are possible. For example, we may have more than two players. We call such games *n-person games*. A situation of this nature could arise in the competition between several large corporations. We may also have *non-zero-sum games*, where the losses of one player are not necessarily the gains of another. Non-zero-sum and *n*-person games are considerably more complicated than those we have discussed. In fact, the complications are such that general agreement on what constitutes a solution to these game types is still lacking.

9.2 Probability and Risk

Probability

Risky situations are, perhaps, more common than uncertain ones. We observed that in a risky situation we had some idea of the probability with which events might occur. To proceed further, we need to clarify what we mean by "probability."

If we were to toss a coin, there are only two possible results: a head or a tail. We say that the possible results are *mutually exclusive* since any given result precludes the possibility of the other. With a fair coin, it is no more likely that we shall obtain a head than a tail on a given toss. We thus would expect that, on the average, we would obtain a head in half the tosses and a tail in half the tosses. We therefore say that the probability of a head on a given toss is $\frac{1}{2}$ and that the probability of a tail is $\frac{1}{2}$. Similarly, if we were tossing a single die, with the numbers 1 through 6 on its faces, the probability of any given number being thrown (a success) on a particular toss is $\frac{1}{6}$. The probability of not tossing this given number (failing) is $\frac{5}{6}$.

We may ask what the probability is of tossing a die three times and obtaining a given number twice. One way this could occur would be to obtain the given number two times in a row and to fail the third time. The probability of throwing the desired number on the first toss is $\frac{1}{6}$, succeeding on the second toss is $\frac{1}{6}$, and failing on the third is $\frac{5}{6}$, The individual events are independent, and we obtain the overall probability as the product $\frac{1}{6} \times \frac{1}{6} \times \frac{5}{6}$. In more general terms, if the probability of a success is p and the probability of failure q, then the probability of obtaining x consecutive successes followed by $(n - x)$ failures is

$$\overbrace{(p)(p)\dots(p)}^{x} \quad \overbrace{(q)(q)\dots(q)}^{n-x} = p^x q^{n-x}, \tag{9.2.1}$$

where n is the total number of trials. In our example,

$$p = \tfrac{1}{2}, \qquad q = \tfrac{1}{2}, \qquad n = 3, \qquad \text{and } x = 2.$$

When the number of successes x can vary randomly from trial to trial, x is called a *random variable.*

There are other ways in which we could obtain two successes and one failure, e.g., one failure followed by two successes, or a success followed by a failure and a success. Thus the total probability of tossing a given number twice out of three trials is given by $3 \times \frac{1}{6} \times \frac{1}{6} \times \frac{5}{6}$. In general, the number of ways we can obtain x successes and $(n - x)$ failures is given by

$$N(x) = \frac{n!}{x!(n - x)!}. \tag{9.2.2}$$

The probability of achieving x successes and $(n - x)$ failures is then the number of ways this can occur, multiplied by the probability that a given permutation occurs; i.e.,

$$f(x) = \frac{n!}{x!(n - x)!} p^x q^{n-x}. \tag{9.2.3}$$

Frequency functions

The function $f(x)$ yields the probability that the random variable x (in our example, the number of successes) will assume any particular value. This function is called the *frequency function* of x.* The particular frequency function of Equation (9.2.3) is called the *binomial frequency function.* It holds whenever we are dealing with mutually exclusive, discrete events.

In any random process it is certain that one of the possible alternatives must occur. Therefore, if we sum the probabilities for all possibilities, we must obtain 1. That is,

$$\sum_{x=1}^{n} f(x) = 1. \tag{9.2.4}$$

This is shown graphically in Figure 9.1 for the example of the numbers of successes in three tosses of a die. Note that (9.2.4) requires that the sum of the areas of the bars be 1. The area of a given bar is the probability that x

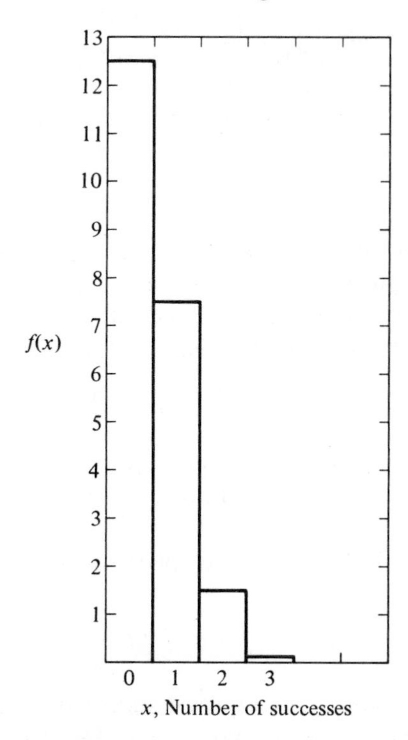

Figure 9.1. Frequency Function for Number of Successes in Three Tosses of a Die

* The frequency function is often referred to as the *probability density function.*

will have the given value. The probability that x will be greater than 1 equals the sum of the bar areas for $x = 2$ and $x = 3$. We call a graph of this type, in which the areas represent relative frequencies, a *histogram.*

We characterize a frequency function in terms of two quantities: its *mean* and *variance.* We define the mean μ of the frequency function of a discrete variable as

$$\mu = \sum_{x=1}^{n} x f(x). \tag{9.2.5}$$

For the binomial distribution,

$$\mu = \sum_{x=1}^{n} (x) \frac{n!}{x![n - x]!} p^x q^{n-x}. \tag{9.2.6}$$

This may be rewritten as

$$\mu = np\left[\sum_{x=1}^{n} \frac{[n - 1]!}{[x - 1]![n - x]!} p^{x-1} q^{n-x}\right]. \tag{9.2.7}$$

The summation in brackets has a value of 1 and, therefore, for the binomial distribution,

$$\mu = np. \tag{9.2.8}$$

Now if the probability of a success in any given trial is p, our probable number of successes in two trials would simply be $(p + p)$. Similarly, in n trials the probable, or most likely, number of successes would be np. We then say that $E(x)$, the *expected* (most likely) value of x, is given by

$$E(x) = np. \tag{9.2.9}$$

We see that the mean is identical with our expected value. Hence we interpret the mean of a frequency function as being the expected value of our random variable x.

We define the variance, σ^2, of a frequency function for a discrete variable as

$$\sigma^2 = \sum_{x=1}^{n} [x - \mu]^2 f(x). \tag{9.2.10}$$

The variance is a measure of the expected scatter of the variables about the mean. The greater σ^2, the greater the scatter. The quantity σ is called the *standard deviation.* For a binomial distribution, we can show that

$$\sigma = \sqrt{\sigma^2} = \sqrt{npq}. \tag{9.2.11}$$

We have been discussing discrete variables, i.e., variables that can assume only certain distinct values. However, in the more usual situation this is not the case. Suppose that we were measuring the height of all persons residing in a given community. The heights we measure could have any value within the range of heights for humans. Such variables are called *continuous* variables. Other examples of continuous variables are weights, velocities, prices, temperatures, incomes, etc. When such a quantitative characteristic is measured for a large group, it is found that the distribution of measured values can be described with the aid of a frequency function. If we divide the quantity being measured into intervals, we can determine the number of measurements in the given interval (e.g., number of people between 60 and 64 inches in height). If we divide the numbers in each interval, n_i, by the total number of measurements made, N, the quotient n_i/N represents the relative frequency, f_i, with which the given range occurs. Just as we did for a discrete variable in Figure 9.1, we may represent the results graphically by the histogram shown in Figure 9.2. Again, the area of the rectangles represents the relative frequencies f_i. If the interval width is h(h is 4 inches in our example), then let the height of rectangle i be $f_i/h = n_i/Nh$. Since the total area under the curve is 1, the area of rectangle i represents the probability that a measurement made at random will lie within that interval.

The number of intervals used to construct the histogram of Figure 9.2 was purely arbitrary. We could have used a considerably larger number of intervals, as shown in Figure 9.3. We preserve the property that the area of each rectangle represents the relative frequency that x will lie within the cor-

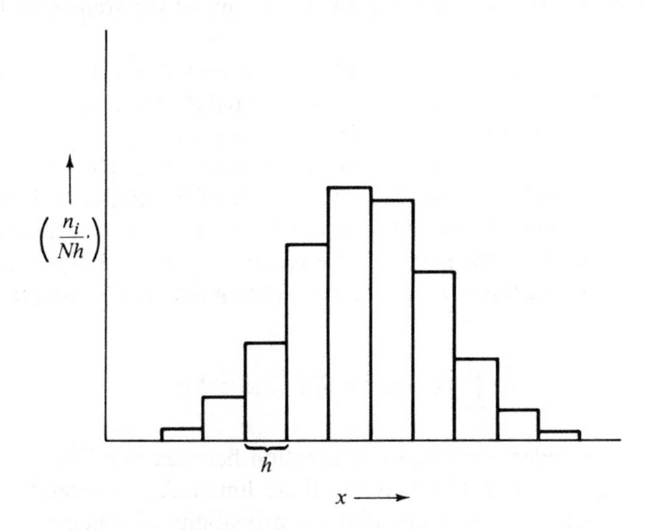

Figure 9.2. Typical Histogram Representing Experimental Data

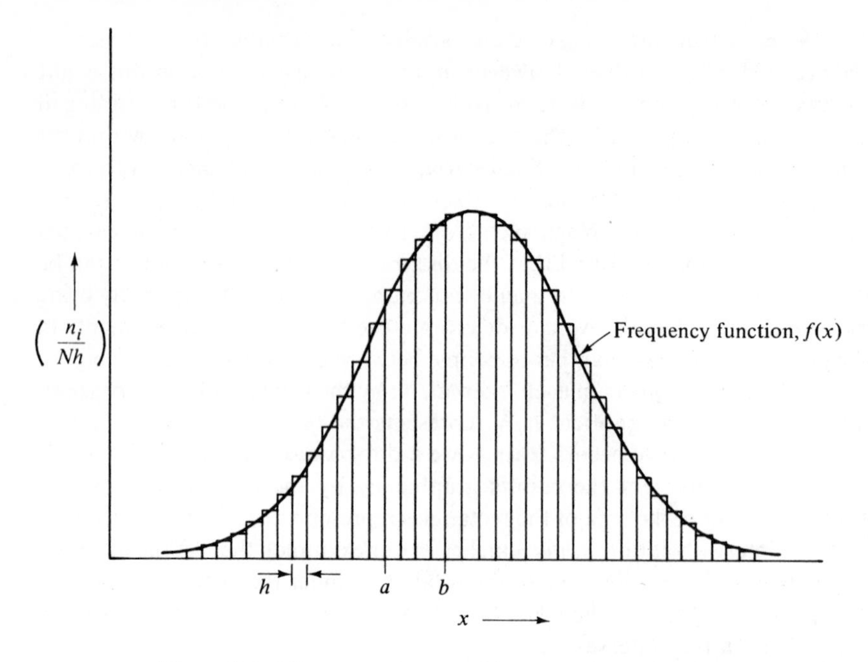

Figure 9.3. Histogram Approaching a Frequency Function

responding interval. As we increase the number of intervals chosen, the histogram approaches a smooth curve. The smooth curve that would be obtained with an infinite number of intervals is the graph of the frequency function $f(x)$.

If we sum the areas of the adjacent rectangles between a and b in the histogram of Figure 9.3, the total area must equal the relative frequency with which x will lie between a and b. That is, the probability p that x will lie between a and b equals the sum of the area of the rectangles. This equality will continue to hold as we increase the number of rectangles in the interval. The equality remains valid when we increase the number of rectangles indefinitely. As we do that, the height of the rectangles approaches $f(x)$ and their width approaches dx. Thus, in the limit, we obtain our area by integration and have

$$\int_a^b f(x)\,dx = P\{a < x < b\}. \tag{9.2.12}$$

Hence the area under our frequency function between two limits represents the probability that x will lie between those limits. As we extend the limits toward $+\infty$ and $-\infty$, we know that the probability of x being within that range approaches 1. Therefore, we may write

$$\int_{-\infty}^{\infty} f(x)\, dx = 1. \tag{9.2.13}$$

We also know that $f(x)$ cannot be negative, so that

$$f(x) \geq 0. \tag{9.2.14}$$

Any function that possesses the properties of Equations (9.2.12), (9.2.13), and (9.2.14) can be used as a frequency function for a continuous random variable.

Very often we shall be concerned with the probability that the random variable x will be less than a given value b. This is readily obtained from

$$P\{x \leq b\} = \int_{-\infty}^{b} f(x)\, dx. \tag{9.2.15}$$

This question is so common that to provide the answer we define a new function, the *distribution function,** $F(x)$, such that

$$F(x) = \int_{-\infty}^{x} f(s)\, ds, \tag{9.2.16}$$

where s is simply a dummy variable. When values of the distribution function are available, these can also be used to find the probability that x will lie in a given interval. The probability that x will lie between a and b is simply $[F(b) - F(a)]$.

A frequency function of a continuous random variable is also characterized by its mean and variance. We define the mean and variance in the same manner as for a discrete random variable, but replace the summation by an integral. We have

$$\mu = \int_{-\infty}^{\infty} x f(x)\, dx, \tag{9.2.17}$$

$$\sigma^2 = \int_{-\infty}^{\infty} [x - \mu]^2 f(x)\, dx, \tag{9.2.18}$$

$$\sigma = \left[\int_{-\infty}^{\infty} [x - \mu]^2 f(x)\, dx\right]^{1/2}. \tag{9.2.19}$$

We again interpret the mean μ as the most likely value of our random variable and the variance σ^2 as a measure of the expected scatter about the mean.

---

* Some authors refer to the frequency function $f(x)$ as the *distribution function* and the function $F(x)$ as the *cumulative distribution function.*

There have been many frequency functions devised to represent the distribution of continuous random variables. However, the most frequently occurring distribution is the *normal* or *Gaussian* distribution. This distribution is represented by

(9.2.20) $$f(x) = \frac{1}{\sigma\sqrt{2\pi}} e^{-[1/2][(x-\mu)/\sigma]^2},$$

where μ and σ are the mean and the standard deviation, respectively. It is found that when the variation in a quantity is caused by the sum of a large number of small, independent effects, the normal distribution is a very close approximation for the observed distribution. The frequency function for such diverse quantities as the height of persons in a given population, the diameter of metal rods from a production line, the demand per day for a given product, can usually be well represented by the normal distribution. In Figure 9.4 two normal distributions with identical means are shown. It is readily seen that the distribution with the greater σ represents a situation in which there is a wider spread in the observed values. It will also be seen that the distribution is symmetric about the mean; therefore, there is as great a probability of a given observation being above the mean as below the mean.

It is often desirable to express x in terms of the number of standard deviations by which it departs from the mean. That is,

(9.2.21) $$x = \mu + t\sigma,$$

where t is the number of standard deviations from the mean. We call t the standard normal variate and, by rewritting (9.2.21), we express it as

(9.2.22) $$t = \frac{x - \mu}{\sigma}.$$

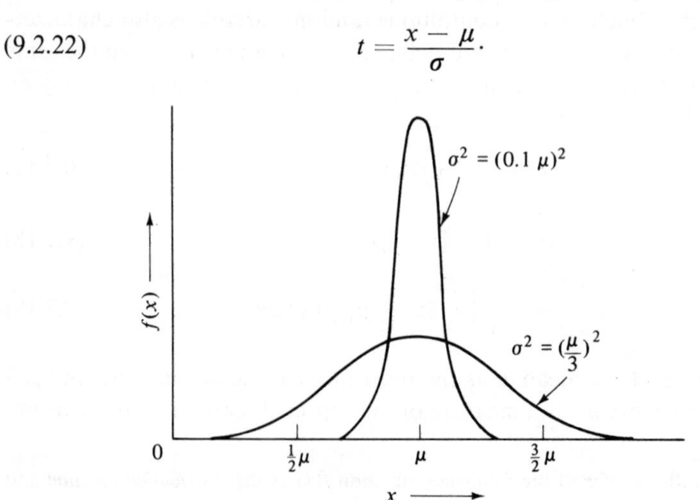

Figure 9.4. Effect of the Variance on the Normal Distribution

We may rewrite (9.2.20) in terms of t to obtain

$$f(t) = f(x)\frac{dx}{dt} = \frac{1}{\sqrt{2\pi}}e^{-t^2/2}. \tag{9.2.23}$$

The probability that x will have a value less than b can now be rewritten as

$$P\{x \leq b\} = P\left\{t \leq \frac{b-\mu}{\sigma}\right\} = F\left(\frac{b-\mu}{\sigma}\right) = \int_{-\infty}^{[(b-\mu)/\sigma]} \frac{1}{\sqrt{2\pi}}e^{-t^2/2}\,dt. \tag{9.2.24}$$

Equation (9.2.24) tells us that the probability of observing a deviation from the mean as large as t standard deviations is the same as for any normally distributed variable. Hence any time we specify t we can determine the probability that x will be greater or less than $(\mu + t\sigma)$.

Since the integral of Equation (9.2.24) cannot be evaluated readily, tables of $F(t)$ are provided by most texts on statistics. For such tables, as well as for a more comprehensive treatment of the subject of probability and statistics, the reader is referred to standard texts such as Feller [1968] or Guttman, Wilks, and Hunter [1971].

Now that we have examined the mathematical description of random variables, we shall consider how such variables can affect optimization problems. In this process we shall need the concepts we have just developed.

Frequency functions and design under risk

Optimization under risk implies that one or more of the quantities affecting the objective function will be a random variable. By our definition of risk, we must be able to estimate the probability that any such random quantity will fall within a given range. We can then describe the behavior of our random variable by means of an appropriate frequency function. It is possible to formulate and, under certain conditions, to solve optimization problems that contain random variables. This is generally accomplished by incorporating characteristics of the frequency functions, such as the mean and standard deviation, into the mathematical model. Optimization problems of this type are called *stochastic* programming problems.

When our objective function contains randomly varying quantities, we cannot assign a definite value to the objective function. The value of the objective function is itself a randomly varying quantity, which can be described by a frequency function $f(y)$. We can, however, seek to maximize (minimize) the most probable (expected) value of our objective function. If y represents the value of the objective function without consideration of risk, then upon consideration of risk, the objective becomes: Maximize (minimize)

$$E(y) = \mu_y = \int_{-\infty}^{\infty} y f(y)\, dy. \tag{9.2.25}$$

An estimate of the probable loss in the objective function based upon values less than the most probable value is given by

$$\int_{-\infty}^{\mu_y} [\mu_y - y] f(y)\, dy. \tag{9.2.26}$$

This is exactly counterbalanced by the possible benefit resulting from values of the objective function in excess of the most probable value,

$$\int_{\mu_y}^{\infty} [y - \mu_y] f(y)\, dy. \tag{9.2.27}$$

Random variables may also appear in our constraints. When this occurs we must also introduce into the constraints the characteristics of the frequency functions describing these variables. In the succeeding sections we shall examine several means of accomplishing this.

9.3 Risk Elements Only in the Objective Function

Random variables are most commonly introduced into optimization problems through the objective function. The objective function usually is some economic index, which is determined by such items as prices, costs, and demands. Such quantities can generally only be estimated, and often they must be regarded as having appreciable variability. The quantities in the constraint equations may, however, be much better known, and often they can be considered as deterministic.

As an illustration of such a situation, let us consider the selection of the optimum number of products to be produced by some hypothetical plant in the coming year. We are told that the cost C of amortization and operation of the plant is

$$C = k_1 + k_2 x, \tag{9.3.1}$$

where k_1 and k_2 are constants and x is the number of products produced. The maximum number of products that can be produced is W. Therefore, we have as a constraint

$$0 \leq x \leq W. \tag{9.3.2}$$

Plant management has set the sale price of a product unit as p. The demand for the product, D, is a random variable with a frequency function $f(D)$. If the demand for the product is less than the number produced, the unsold products will be obsolete and have no value. For any given demand, the profit y will be

$$y = pD - [k_1 + k_2 x] \tag{9.3.3}$$

if $D < x$. Otherwise,

$$y = px - [k_1 + k_2 x]. \tag{9.3.4}$$

Since we have no control over D, our profit will be a random variable.

Suppose that we wish to determine the number of products which will maximize our most probable (expected) profit. The expected profit for underproduction conditions ($x \leq D$) is given by

$$E(D \geq x) = \int_x^\infty \{px - [k_1 + k_2 x]\} f(D)\, dD. \tag{9.3.5}$$

Similarly, the expected profit for overproduction conditions ($x \geq D$) can be written

$$E(D \leq x) = \int_{-\infty}^x \{pD - [k_1 + k_2 x]\} f(D)\, dD. \tag{9.3.6}$$

The total expected profit is the sum of the above two expressions, which is

$$\begin{aligned} E(y) = {} & p \int_{-\infty}^x D f(D)\, dD + px \int_x^\infty f(D)\, dD \\ & - [k_1 + k_2 x] \left\{ \int_{-\infty}^x f(D)\, dD + \int_x^\infty f(D)\, dD \right\}. \end{aligned} \tag{9.3.7}$$

Since

$$\int_{-\infty}^\infty f(D)\, dD = 1, \tag{9.3.8}$$

however, the above expression simplifies to

$$E(y) = p \int_{-\infty}^x D f(D)\, dD + px \int_x^\infty f(D)\, dD - [k_1 + k_2 x]. \tag{9.3.9}$$

We cannot have negative demands, and therefore the first integral in equation (9.3.9) is really evaluated over the range of 0 to x. However, any realistic demand distribution must be such that a negligible area is under the tail of the frequency function extending from 0 to $-\infty$. Essentially the same numerical result is obtained whether the lower limit on the integral is taken as 0 or $-\infty$.

The solution to our problem is obtained by maximizing (9.3.9) subject to the linear constraint of (9.3.2). Introduction of risk has considerably complicated our problem, since our objective function has obviously become nonlinear. Problems of this nature, in which the demand followed a normal distribution, have been considered by Dantzig [1955]. He showed that, if the constraints are linear and if the objective function would have been linear without the introduction of risk, the problem is a nonlinear concave (convex, with an objective function to be minimized) programming problem. We may thus obtain the global optimum by use of an appropriate nonlinear programming technique.

If we had not considered the elements of risk and simply assumed the demand for products to be the expected demand $E(D)$, we would have concluded that we should produce exactly $E(D)$ products. This would have yielded a lower value of $E(y)$ and could very well have exposed us to an unacceptably high probability of being overstocked.

If our problem, prior to the consideration of the random nature of some of the quantities, had been nonlinear, our stochastic programming problem would not necessarily have been concave (convex). A solution could still be obtained by an appropriate nonlinear programming method, but we would have no assurance that the result obtained was the global optimum.

We have previously observed that there is no simple analytical expression for the normal distribution function. Hence evaluation of our objective function [Equation (9.3.9)] requires the use of the tabular values of normal frequency and distribution functions. This does not affect the convexity of the problem, but does increase the computational effort required. One means for reducing this effort is to approximate the normal frequency function by one which can be integrated analytically. The Weibull distribution is one which may be employed for this purpose. Dubey [1966] has shown that the normal frequency and distribution functions can be approximated very closely by Weibull functions of the form

$$f(t) = 1.001[0.9013 + 0.27786t]^{2.60232} \exp\{-[0.9013 + 0.27186t]^{3.60232}\}, \quad (9.3.10)$$

$$F(t) = 1 - \exp\{-[0.9013 + 0.27786t]^{3.602326}\}, \quad (9.3.11)$$

providing $t \leq 3.243$.

9.4 Chance-constrained Programming

Sources of variability

There are many quantities which enter a problem that may be random variables. We have already noted that factors which affect our objective function—such as costs, prices, and demands—are likely to be random variables.

There are, however, other variable factors. For example:

1. *Decision variables.* Although a decision variable may be specified by the designer, its actual value may be different than that specified due to variability in materials, inaccuracy of measurement, etc. For example, a flow rate may be specified by a plant designer after optimizing the system, but in actual plant operation the flow cannot be determined precisely. If the measurement errors were significant, the flow rate would have to be considered as a random variable having as its mean the value specified by the designer.

2. *Technological coefficients.* These quantities may be coefficients in various relationships (e.g., reaction-rate constants, thermal conductivity, etc.), or they may be quantities that under different circumstances would be a decision variable but are specified for a particular problem.

3. *Constraint variability.* The relationship between a quantity being constrained and the variables affecting it may be imperfectly understood. For example, a chemical process may require that, for a given temperature difference Δt, the rate of heat transfer per unit area from material 1 to material 2 be no less than some quantity q. The relationship between the rate of heat transfer and the temperature difference is

$$q = UA\Delta t, \tag{9.4.1}$$

where A is the area available for heat transfer and U is a coefficient of proportionality. The coefficient U may be obtained from a statistical correlation

$$U = \psi(V, \nu, \rho), \tag{9.4.2}$$

where V represents a fluid velocity, ν viscosity, and ρ density. Actual measured values of U can be found with deviations from 10 to 15 per cent of the predicted values, due to the effects of other variables (such as surface roughness, impurities, etc.) not included in the correlation. Weisman [1968] has suggested that such situations be treated by introducing a new random variable s_i, where

$$U = s_i\psi(V, \nu, \rho). \tag{9.4.3}$$

The new random variable s_i would have a mean of 1 and frequency function appropriate to the correlation considered. The desired frequency function can be determined by computing the ratio of the predicted value to the observed value for each of the data points. A histogram, similar to that shown in Figure 9.3, but with the ratio of predicted to observed value replacing x, is then constructed. An appropriate frequency function is then fitted to the histogram.

After examining all the sources of variability and determining their magnitude, it is possible to determine what consideration must be given to their variability during the optimization process. If our analysis shows that these effects are small everywhere, we can consider our problem as deterministic. If we conclude that the variability is negligible everywhere except in the objective function, the method of the previous section is applicable. If the variability in both objective function and constraints is appreciable, we must seek a revised procedure.

Probabilistic constraints

When the variations in technological coefficients, decision variables, and constraint correlations are significant, consideration of constraints in a deterministic manner is no longer meaningful. An inequality such as

$$g_i(\mathbf{X}) \leq b_i \tag{9.4.4}$$

has little meaning when we cannot assign a definite value to **X**. We can never be completely certain that $g_i(\mathbf{X})$ will not exceed b_i, but we can set conditions such that there is a very low probability of this occurring. If it is desired to keep the probability of satisfactory behavior above some high level, say k_i, we may replace the deterministic constraint of (9.4.4) by

$$P\{g_i(\mathbf{X}) \leq b_i\} \geq k_i, \tag{9.4.5}$$

where $P\{g_i(\mathbf{X}) \leq b_i\}$ represents that probability that b_i will not be exceeded. Our optimization problem now becomes the maximization (minimization) of the expected value of our objective function $E(y)$, subject to a series of constraints having the form of (9.4.5). That is: Maximize or Minimize

$$E(y), \tag{9.4.6}$$

subject to

$$P\{g_i(\mathbf{X}) \leq b_i\} \geq k_i, \qquad i = 1, 2, \ldots, m. \tag{9.4.7}$$

This approach to optimization under risk was originated by Charnes, Cooper, and Symonds [1958]; they call this *chance-constrained programming* (cf. Charnes and Cooper [1959]).

Evaluation of probabilities is time consuming, and therefore Charnes et al. sought a way to avoid this during the optimization procedure. It will be recalled that we can express a random variable in terms of its mean and standard deviation [Equation (9.2.21)]. Since $g_i(\mathbf{X})$ can itself be considered a random variable when the x_j are random variables, we have

$$g_i(\mathbf{X}) = \mu g_i(\mathbf{X}) + t_i \sigma_{g_i(\mathbf{X})}, \tag{9.4.8}$$

where $\sigma_{(g_i(\mathbf{X}))}$ is the standard deviation of $g_i(\mathbf{X})$. If we also make use of the fact that $\mu g_i(\mathbf{X}) = E(g_i(\mathbf{X}))$, we can rewrite Equation (9.4.7) as

$$P\{E(g_i(\mathbf{X}) + t_i \sigma_{g_i(\mathbf{X})} \leq b_i\} \geq k_i, \qquad i = 1, 2, \ldots, m. \tag{9.4.9}$$

We have seen that if we specify t_i, the probability of the occurrence is also specified. Hence specifying t_i is equivalent to specifying k_i. That is, when we express our frequency function in terms of the standard deviate t_i,

$$k_i = F(t_i). \tag{9.4.10}$$

In view of this, if t_i is specified in accordance with (9.4.10), we can replace Equation (9.4.9) by

$$E(g_i(\mathbf{X})) + t_i\sigma_{(g_i(\mathbf{X}))} \leq b_i. \tag{9.4.11}$$

Equation (9.4.11) is called the certainty equivalent of (9.4.5). If our restraint had been $P\{g_i(\mathbf{X}) \geq b_i\} \geq k_i$, then instead of (9.4.11) we would write

$$E(g_i(\mathbf{X})) - t_i\sigma_{(g_i(\mathbf{X}))} \geq b_i. \tag{9.4.12}$$

To make use of (9.4.11), we must be able to describe $g_i(\mathbf{X})$ in terms of a frequency function. We shall limit our consideration to those conditions where all the randomly varying components are normally distributed. If $g_i(\mathbf{X})$ is a linear function of the decision variables x_j and technological coefficients u_j, then

$$g_i(\mathbf{X}) = a_{i1}x_1 + a_{i2}x_2 + \cdots + a_{in}x_n + d_{i1}u_1 + d_{i2}u_2 + \cdots + d_{is}u_s + K. \tag{9.4.13}$$

If both the x_j and the u_j can vary randomly, and K, the a_{ij}, and the d_{ij} are constants, then

$$E(g_i(\mathbf{X})) = \sum_{j=1}^{n} a_{ij}E(x_j) + \sum_{j=1}^{s} d_{ij}E(u_j) + K. \tag{9.4.14}$$

If the behavior of each of the random variables is independent of the behavior of the other, then it may be shown that

$$\sigma^2_{(g_i(\mathbf{X}))} = \sum_{j=1}^{n} [a_{ij}]^2\sigma^2_{(x_j)} + \sum_{j=1}^{s} [d_{ij}]^2\sigma^2_{(u_j)}. \tag{9.4.15}$$

Furthermore, if all the x_j and u_j are normally distributed, $g_i(\mathbf{X})$ is normally distributed.

In the more general case, we do not restrict the form of $g_i(\mathbf{X})$, i.e.,

$$g_i(\mathbf{X}) = g_i(x_1, x_2, \ldots, x_n, u_1, u_2, \ldots, u_s), \tag{9.4.16}$$

but require that the random variables be independent and that the σ^2/μ be small for each u and x. We can then approximate $g_i(\mathbf{X})$ by a Taylor series. If we ignore higher-order terms, we have

$$
\begin{aligned}
g_i(\mathbf{X}) \simeq\ & g_i(E(x_1), E(x_2), \ldots, E(x_n), E(u_1), E(u_2), \ldots, E(u_s)) \\
& + \frac{\partial g_i(\mathbf{X})}{\partial x_1}[x_1 - E(x_1)] + \frac{\partial g_i(\mathbf{X})}{\partial x_2}[x_2 - E(x_2)] + \cdots \\
& + \frac{\partial g_i(\mathbf{X})}{\partial x_n}[x_n - E(x_n)] + \frac{\partial g_i(\mathbf{X})}{\partial u_1}[u_1 - E(u_1)] \\
& + \frac{\partial g_i(\mathbf{X})}{\partial u_2}[u_2 - E(u_2)] + \cdots + \frac{\partial g_i(\mathbf{X})}{\partial u_s}[u_s - E(u_s)].
\end{aligned}
\tag{9.4.17}
$$

To determine the expected value of this expression, we consider the $E(x_j)$ and $E(u_j)$ as constants. The expected value of a constant is that constant. Also,

$$
E[x_j - E(x_j)] = E(x_j) - E(x_j) = 0. \tag{9.4.18}
$$

We then conclude that

$$
E(g_i(\mathbf{X})) \simeq g_i(E(x_1), E(x_2), \ldots, E(x_n), E(u_1), E(u_2), \ldots, E(u_m)). \tag{9.4.19}
$$

To obtain the variance of $g_i(\mathbf{X})$, we observe that we have a linear relationship and therefore can make use of Equation (9.4.15). We then obtain

$$
\begin{aligned}
\sigma^2[g_i(\mathbf{X})] =\ & \left(\frac{\partial g_i}{\partial x_1}\right)^2 \sigma^2_{(x_1)} + \left(\frac{\partial g_i}{\partial x_2}\right)^2 \sigma^2_{(x_2)} + \cdots \\
& + \left(\frac{\partial g_i}{\partial x_n}\right)^2 \sigma^2_{(x_n)} + \left(\frac{\partial g_i}{\partial u_1}\right)^2 \sigma^2_{(u_1)} \\
& + \left(\frac{\partial g_i}{\partial u_2}\right)^2 \sigma^2_{(u_2)} + \cdots + \left(\frac{\partial g_i}{\partial u_s}\right)^2 \sigma^2_{(u_s)}.
\end{aligned}
\tag{9.4.20}
$$

Furthermore, if the foregoing provisions apply and all random quantities are normally distributed, $g_i(\mathbf{X})$ will approximately be normally distributed if computation of $g_i(\mathbf{X})$ involves no divisions by variables that are zero or close to zero. Note that any s_i, expressing constraint variability, are treated just as the u_i. We are now able to estimate all the quantities required for the statement of our constraints for a number of common situations.

EXAMPLE 9.4.1. Stock Portfolio Selection under Risk

Let us consider the situation wherein an investor wishes to invest, for a specified period, a fixed amount distributed among n different stocks. The investor's objective is to maximize the final value of his portfolio, but he wishes to protect himself from the possibility of severe losses. He therefore will select stocks in a manner such that there is only a low probability, say 5 per cent, of any losses occurring. We wish to determine what fraction of his investment should be placed in each stock.

We shall assume that the investor has available an estimate of the prices of the n stocks at the end of the interval. Furthermore, we shall assume that for each stock this estimate is described by a normal distribution having a mean μ_i and a variance σ_i^2. If x_i is the fraction of funds to be invested in stock i, then the objective function is: Maximize

$$E(y) = \sum_{i=1}^{n} \left[\frac{\mu_i}{C_i}\right](x_i), \tag{9.4.21}$$

where C_i is the known initial cost of stock i. Our loss constraint is

$$P\left\{\left[\sum_{i=1}^{n} \left(\frac{\mu_i}{C_i}\right)x_i\right] \geq 1.0\right\} \geq 0.95. \tag{9.4.22}$$

We may write the certainty equivalent of this constraint by

$$\sum_{i=1}^{n} \left[\frac{\mu_i}{C_i}\right]x_i - t\left[\sum_{i=1}^{n} \left(\frac{x_i}{C_i}\right)^2 \sigma_i^2\right]^{1/2} \geq 1.0. \tag{9.4.23}$$

The value of t is determined by setting $k_i = 0.95$ in Equation (9.4.10). By consulting a table of the normal distribution function we find that $F(t) = 0.95$ when $t = 1.645$. Maximization of (9.4.21) subject to the constraint of (9.4.23) is a concave nonlinear programming problem, which we can solve by a variety of methods. It will be observed that both the objective function and constraint are separable functions, and therefore separable programming would be applicable here. More complex chance-constrained problems than our illustrative example would probably not be convex or concave, and we would not be certain that the solution obtained was the global optimum.

As a simple numerical example, let us assume that our investment portfolio is to consist only of government bonds having a known appreciation and a common-stock mutual fund. We shall assume that the ratio of final value to cost for the bonds is 1.5, and is estimated at 1.8 for the stocks. The standard deviation of the final value of the mutual fund is estimated at 80 per cent of its initial cost. If x_1 represents the fraction invested in the mutual fund, the objective function becomes

$$E(y) = 1.8x_1 + 1.5[1 - x_1], \tag{9.4.24}$$

and the constraint is

$$1.8x_1 + 1.5[1 - x_1] - 1.645[0 + 0.64x_1^2]^{1/2} \geq 1.0. \tag{9.4.25}$$

Our constraint can only be met when $x_1 \leq 0.49$. Since the objective function increases monotonically with x_1, we conclude that 49 per cent of the investment should be in the mutual fund.

9.5 Risk Minimization by Failure Penalties

Costs of exceeding a scarce resource

Chance-constrained programming requires that probability limits be established for each constraint. However, no guidance is furnished to help establish these limits. Limits that are unduly restrictive may unnecessarily increase costs; limits that are too loose may lead to undesirable risks. Ferguson and Dantzig [1956] suggested an alternative approach when dealing with constraints arising from utilization of a scarce resource.* They concluded that if a limit was exceeded, an additional cost was encountered. For example, in a deterministic situation we might require that our total production be set so that the demands for raw material can be met from inventory. In a real situation our randomly varying production may exceed our available inventory. When this happens, an additional cost is added either by the good will lost by not filling the order or by extra costs that may be encountered in obtaining the needed raw material. We may consider this additional cost per unit, π, as a unit penalty. If our production is x and the available inventory is b, then the expected additional cost (total penalty) for exceeding our inventory is

$$\pi \int_b^\infty [x - b] f(x)\, dx. \tag{9.5.1}$$

We would add this penalty to our objective function. Now if the optimal value of x turns out to be large, then the probability of exceeding the available inventory is low. Thus we shall automatically keep the probability of exceeding the available inventory at a reasonably low level, since any optimization technique employed will choose a large production level and therefore avoid high penalty costs. If we were to increase the unit penalty π, a lower probability of exceeding the inventory would be selected. For every unit penalty cost there will be some optimum probability of exceeding the inventory. Thus whenever we require that there be a given probability of not exceeding a particular resource, it is equivalent to assigning some cost penalty. However, when we arbitrarily select probability limits, the equivalent cost penalty may be highly unrealistic.

Failure penalties

Hadley [1964] has observed that this approach can be generalized so that any constraints which require that the probability of an event be less than

---

* Ferguson and Dantzig [1956] referred to their approach as *programming under uncertainty*. However, in light of the present use of the term "uncertainty," this name is unfortunate.

or equal to some value can be replaced by assigning a cost to the occurrence of the undesired event. We can consider the condition under which we exceed a given constraint as a failure, and the additional expected cost encountered as a failure penalty. Our probable total failure cost is the product of the probability of failure $P(F_i)$, and the expected failure cost $E(C_i)$. We thus replace a problem that is in the form of (9.4.6) and (9.4.7) by: Minimize

$$E(y) + \sum_{i=1}^{m} P(F_i) \cdot E(C_i), \tag{9.5.2}$$

where $E(y)$ represents the expected value of the sum of all costs except those arising from failures. If realistic failure costs have been assigned, the resulting failure levels will be low.

In the example in which we wished to avoid exceeding a given raw-material inventory, the costs of failure depended on the degree by which we exceeded our constraint. If we were to put our failure costs, given by Equation (9.5.1), in the form of (9.5.2), we would have

$$P(F_i) = \int_b^\infty f(x)\,dx, \tag{9.5.3}$$

$$E(C_i) = \frac{\pi \int_b^\infty [x - b] f(x)\,dx}{\int_b^\infty f(x)\,dx}. \tag{9.5.4}$$

The formulation of Equation (9.5.2) would appear awkward in this case. However, as we shall see, there are many situations in which full failure cost is encountered as soon as our constraint is exceeded. In these cases, Equation (9.5.2) is particularly useful.

We may illustrate the use of the failure-probability concept by consideration of the activities of a process-equipment manufacturer. We shall assume that he has accepted a fixed-price order for a piece of apparatus that will transfer a given amount of heat, q, between two flowing streams. The cost of the apparatus may be assumed to be directly proportional to the available heat-transfer area, A. The manufacturer therefore desires to provide the smallest area that will be satisfactory.

The relationship between the heat transferred and process conditions was discussed in Section 9.4. We had that the total heat transfer q was given by

$$q = UA\,\Delta t, \tag{9.5.5}$$

where U was a coefficient of proportionality and Δt the temperature difference between the two streams. We shall assume that the temperature difference, flows, and physical properties of the flowing streams are estabilshed by the process, but that as before, an allowance must be made for uncertainties in the correlation. We must therefore treat U as a random variable.

If the apparatus fails to transfer the required amount of heat, the manufacturer will have to replace the unit sold by a larger one, as well as lose his customer's good will. Also, the small unit that is returned will have only scrap value. If π represents the sum of all the foregoing failure costs, the probable cost of failure is then

$$\pi \cdot P\{UA\Delta t \leq q\}. \tag{9.5.6}$$

We then solve our problem by selecting that value of A which will: Minimize

$$CA + \pi \cdot P\{UA\Delta t \leq q\}, \tag{9.5.7}$$

where C is the cost per unit heat transfer area. We may rewrite our objective function as: Minimize

$$CA + \pi P\left\{U \leq \frac{q}{A\Delta t}\right\}. \tag{9.5.8}$$

Since U is the only random variable in the expression, we can evaluate our objective function from

$$CA + \pi \int_0^{[q/A\Delta t]} f(U)\, dU. \tag{9.5.9}$$

In this simple example we must solve a convex, nonlinear optimization problem. The frequency function for $f(U)$ can be determined by the methods described in the previous section. In more complex problems which may be realistically expected, the objective functions are not likely to be convex, nor are the frequency functions determined in a straightforward manner.

9.6 Risk Programming

Utility functions

Although the failure-penalty concept can be widely applied, there are some problems in which it does not appear to be of assistance. For example, in the stock-portfolio problem considered in Example 9.4.1, it is difficult to assign a meaningful penalty when there is a high probability of losing our investment. Nothing we have yet said will explain a willingness to accept a lower expected profit in return for increased investment safety. Yet we all know that a prudent investor will behave in this manner. The explanation for this behavior lies in the value we place on money.

Implicit in our previous assumption, that we wish to maximize our profits from an investment, is the assumption that our value for money is linear with the quantity of money. If this is so, an investment for which there is the same probability of gaining \$20,000 as of losing \$20,000 would have an expected value of zero. However, for most individuals this is not so—the investment would have a negative value. That is, although most people would like to gain \$20,000, they would be much more concerned about losing \$20,000. Most individuals place a higher value on the \$20,000 they may have than on the \$20,000 they may gain—their value for money is nonlinear with money.

Usually, an individual values money in the manner shown in Figure 9.5, where the ordinate indicates the individual's value, or *utility*, for money, and the abscissa indicates the total worth of the individual. As an individual's total worth increases, his value for each additional unit of wealth decreases.

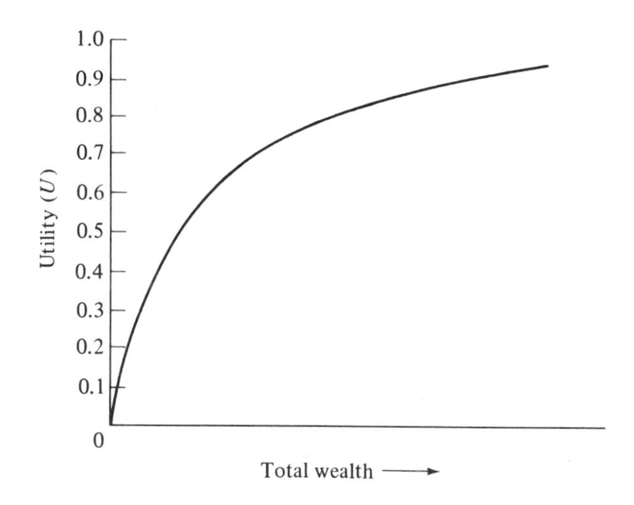

Figure 9.5. Typical Utility Function for Money

Obviously, an individual with a total worth of $\$10^6$ will be far less concerned about a loss of \$5000 than an individual whose total worth is \$20,000. The scale for value is arbitrary, but it is almost always assumed that one's value varies from 0 to 1. This normalized value is referred to as an individual's *utility* for money. A relationship between this normalized value and wealth, such as shown in Figure 9.5, is called a *utility function.* Such utility functions have validity for corporate enterprises as well as individuals. The utility function for a large enterprise will usually have an entirely different scale than that of an individual's utility function (cf. Figure 9.6). In both cases, however, the utility function will appear approximately linear when the considered change in worth is small with respect to the present worth. Thus, when the possible changes in net worth are small, we are justified in maximizing ex-

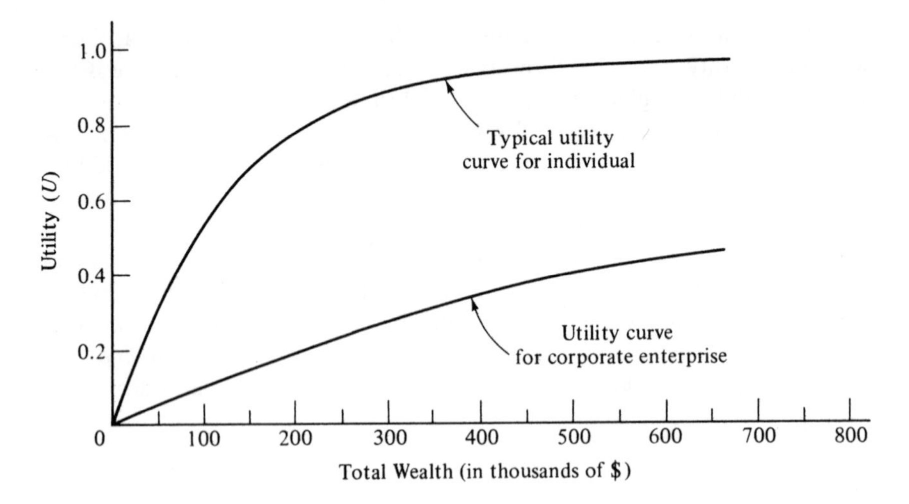

Figure 9.6. Comparison of Individual and Business Enterprise Utilities

pected profit. However, when the possible changes in net worth are large, we should maximize our expected utility.

Freund [1956] has observed that utility functions having the general form shown in Figure 9.5 can be closely approximated by

$$U = 1 - e^{-ar}, \tag{9.6.1}$$

where U represents utility, r represents total worth, and a is a constant chosen to fit the particular utility function. Let us suppose that as the result of a particular investment we expect our net worth to be $E(r)$. Furthermore, let us assume that the profits will be normally distributed with variance $\sigma^2_{(r)}$. Then

$$f(r) = \frac{e^{-[r-E(r)]^2/[2\sigma^2_{(r)}]}}{\sigma_{(r)}\sqrt{2\pi}}. \tag{9.6.2}$$

Our expected utility $E(U)$ would then be

$$E(U) = \int_{-\infty}^{\infty} f(r)[1 - e^{-ar}]\, dr. \tag{9.6.3}$$

The integration may be shown to yield

$$E(U) = 1 - e^{-[aE(r)-(a^2)[\sigma^2_{(r)}]/2]}. \tag{9.6.4}$$

If we were selecting among investments, we would make our selections such that $E(U)$ were maximized. It is obvious that maximization of $E(U)$ is attained by minimizing the exponent. This is equivalent to maximizing

$$E(r) - \frac{[a\sigma_{(r)}^2]}{2}. \tag{9.6.5}$$

Thus to maximize our expected utility, we not only have to consider the expected values of our new net worth, but its possible variance. When the variance in expected worth is large, there is a significant possibility of a loss rather than a profit. We therefore try to maximize our expected worth while keeping the possible profit fluctuations low. The relative emphasis placed on profit and risk [risk being indicated by $\sigma_{(r)}$] is determined by the constant a.

EXAMPLE 9.6.1

Let us now reexamine the investment-portfolio problem described earlier. If we assume that the investor has a utility function of the form given by (9.6.1), his objective function will follow Equation (9.6.5). If the investor's net worth, r_1, and his initial investment are identical, we have for an investment of r_1 distributed among n items: Maximize

$$r_1\left[\sum_{i=1}^{n}\left(\frac{\mu_i}{C_i}\right)x_i\right] - r_1^2\frac{a}{2}\left[\sum_{i=1}^{n}\left(\frac{x_i}{C_i}\right)^2\sigma_i^2\right], \tag{9.6.6}$$

where x_i represents the fraction invested in item i, μ_i the expected value of item i at the end of the investment period, C_i the cost of item i, and σ_i^2 the variance in the final value of item i. Consider the simple example presented in Example 9.4.1, in which an investment is to be divided between bonds with a known return and a mutual fund. We shall see that a judicious choice for the parameter a in Equation (9.6.6) will eliminate the need for a constraining equation. With use of the expected values and variances assumed earlier, we have that $E(r)$ is

$$E(r) = 1.8r_1x_1 + 1.5r_1[1 - x_1] \tag{9.6.7}$$

and $\sigma_{(r_1)}^2$ is calculated to be $0.64x_1^2r_1^2$. Our objective function is then: Maximize

$$E(U) = E(r) - \frac{a}{2}\sigma_{(r)}^2 = 1.8x_1r_1 - 1.5[1 - x_1]r_1 - \frac{a}{2}[0.64x_1^2r_1^2]. \tag{9.6.8}$$

The value of x_1 will be 1 when $a = 0$ and will decrease as a is increased. We obtain $x_1 = 0.49$, which is identical to the solution obtained with the probabilistic constraint of Section 9.4, if we let $a = 0.94/r_1$. Any constraint limiting the probability that the final worth will fall below some value is equivalent to selecting a value of a for the current problem.

In many problems we shall be concerned with minimizing costs rather than maximizing profits. If our final net worth is $(r_1 - C)$, where r_1 is our known initial worth and C represents our costs, then our expected utility is given by

$$E(U) = E(r) - \frac{a\sigma^2_{(r)}}{2} = E(r_1 - C) - \frac{a}{2}\sigma^2_{(r_1 - C)}, \tag{9.6.9}$$

providing our costs are normally distributed. Since r_1 is a constant, the variance of $(r_1 - C)$ is simply $\sigma^2_{(C)}$, and therefore

$$E(U) = r_1 - E(C) - \frac{a\sigma^2_{(C)}}{2}. \tag{9.6.10}$$

In order to maximize $E(U)$ we need to minimize

$$E(C) + \frac{a\sigma^2_{(C)}}{2}. \tag{9.6.11}$$

A satisfactory strategy thus requires not only low costs but also low cost variability.

Effect of constraints

We may deal with constraints as failure-probability limits. In that event, our problem would be a variant of chance-constrained programming. Alternatively, we may eliminate explicit constraints and add failure penalties to the objective function. This procedure has been called *risk programming* (cf. Weisman [1968]). The advantages to be gained by this latter procedure have been discussed in the preceding section.

In the risk-programming approach, our expected total cost is given by

$$E(C) = E(y) + \sum_{i=1}^{m} P(F_i)E(C_i), \tag{9.6.12}$$

where $E(y)$ is the sum of our usual costs and $\sum_{i=1}^{m} P(F_i)E(C_i)$ is the sum of our probable failure costs. When the probable failure costs are significant, the total expected cost will not necessarily be normally distributed, even when our usual costs are normally distributed. In that event, we cannot use (9.6.9) as the objective function. Under these conditions, Weisman and Holzman [1972] suggest that the total utility be determined by considering the utilities of all the possible states. If $E(U_i)$ designates the expected utility of the state which will exist if the i'th failure occurs, the $[P(F_i)\,E(U_i)]$ represents the expected contribution of the i'th failure mode to the total utility. Further, since the $P(F_i)$ will be low, the probability of two simultaneous failures is so low as to be negligible. Therefore, if $E(U_0)$ represents the expected utility of the state which will exist if no failure occurs, then $[1 - \sum_i P(F_i)]\,E(U_0)$ represents the expected contribution of the no failure state to the total utility.

We then have

$$E(U) = [(1 - \sum_i P(F_i)]E(U_0) + \sum_i P(F_i)E(U_i). \tag{9.6.13}$$

If the costs for each of the possible states are normally distributed and Equation (9.6.1) holds, we may write Equation (9.6.13) as

$$\begin{aligned} E(U) = [1 - \sum_i P(F_i)]\left\{1 - e^{-ar_1} \exp\left[aE(y) + \frac{a^2}{2}\sigma^2_{(y)}\right]\right\} \\ + \sum_i P(F_i)\left\{1 - e^{-ar_1} \exp\left[aE(C'_i) + \frac{a^2}{2}\sigma^2_{(C'_i)}\right]\right\}, \end{aligned} \tag{9.6.14}$$

where

$$C'_i = E(y) + C_i. \tag{9.6.15}$$

Since r_1 is a constant, our objective function can be simplified and written as: Minimize

$$\begin{aligned} [1 - \sum_i P(F_i)] \exp\left[aE(y) + \frac{a^2}{2}\sigma^2_{(y)}\right] \\ + \sum_i P(F_i) \exp\left[aE(C'_i) + \frac{a^2}{2}\sigma^2_{(C'_i)}\right]. \end{aligned} \tag{9.6.16}$$

In order to minimize this objective function, Weisman and Holzman conducted a numerical search. False optima were avoided by using penalties to minimize the importance of the failure costs during the early stages of the search.

9.7 Sequential Decisions Under Risk

Solutions via dynamic programming

The optimizations under risk we have previously considered have all been single-stage problems. However, we know that there are many situations in which we must make a sequence of decisions. The classic example of such a situation is the inventory problem faced by most manufacturers and wholesalers. The demand for their product is not a constant with respect to time but will vary from period to period (e.g., from month to month). Let us consider the situation in which a manufacturer's normal production capacity is limited, and he would have to work at overtime rates to produce the required demand during the peak months. It may, therefore, be cheaper for him to produce more than is required during periods of low demand and carry the excess production in inventory for later sale. Whenever items are in inventory, he incurs a carrying cost. We shall assume the cost is directly proportional to the number of items in his inventory. Thus at the beginning of each period the manufacturer must decide how much to produce in order to maximize his return.

The manufacturer's decision at period k is influenced by both his past decisions (which account for his present inventory) and future expectations. We thus have a sequence of related decisions that must be adjusted to give an overall minimum. Such a problem can be put in a form suitable for solution by dynamic programming. If the total time period considered by the manufacturer is divided into intervals (e.g., months), each of these intervals may then be considered to be a stage in a dynamic programming problem.

In accordance with our previous dynamic programming nomenclature (cf. Figure 8.2), we designate the final month considered as stage 1 and the first month considered as stage n. Our decision variables x_k are then the quantities to be produced during each month, and our state variables u_k are the inventories at hand at the beginning of each month. The demand for the product during any month, D_k, will temporarily be assumed as known.

All orders not filled during any given month will be assumed lost. The monthly sales are then equal to the demand, providing the stock on hand plus the monthly production is equal to or greater than demand. Monthly income is therefore given by

$$S \cdot \min(D_k, u_k + x_k), \tag{9.7.1}$$

where S is the unit price of the product. If normal and overtime production costs are linear with output, the total monthly production and inventory costs are given by

$$C_1 x_k + C_2 \max(x_k - q, 0) + C_3 u_k, \tag{9.7.2}$$

where C_1, C_2, C_3 are unit costs, and q is the maximum production before overtime premium is imposed. The return function for any stage is then

$$r_k(u_k, x_k) = S \cdot \min(D_k, u_k + x_k) - C_1 x_k - C_2 \max(x_k - q, 0) - C_3 u_k \tag{9.7.3}$$

and our overall objective function is simply the sum of these return functions:

$$y_n = \max_{x_n, x_{n-1}, \ldots, x_1} \{r_n(u_n, x_n) + r_{n-1}(u_{n-1}, x_{n-1}) + \cdots + r_1(u_1, x_1)\}. \tag{9.7.4}$$

To evaluate the return function at any stage, we need to determine the inventory on hand. This can be expressed in terms of the inventory at the previous stage. With our reverse numbering procedure, we have

$$u_k + x_k - D_k = u_{k-1}. \tag{9.7.5}$$

When we solve for u_k and prohibit negative inventories, we get

$$u_k = \min(u_{k-1} - x_k + D_k, 0) \qquad \text{for } k = 1, 2, 3, \ldots, (n-1), \tag{9.7.6}$$

where u_n, the initial inventory, is a specified constant. For simplicity we shall assume that the u_k can only be one of a set of finite levels. We now solve the problem by making use of the recursive relationship (cf. Section 8.3) that for the first k stages

$$y_k(u_k) = \max_{x_k}\{r_k(u_k, x_k) + y_{k-1}(u_{k-1})\}. \tag{9.7.7}$$

Starting with the first stage, we determine the maximum y_1 for each of the allowable inventory levels. Since the objective function has no meaning for anything prior to the first stage, the value of y_0 is taken as zero. We then proceed to the second stage and determine the maximum y_2 for each of the possible inventory levels at stage 2 and each of the y_1 previously determined. We proceed in this manner until the final stage is reached. There we shall select the maximum value of y_n that meets the restrictions on the initial inventory level u_n. That is, our optimal solution must be based on the manufacturer's present inventory.

Now let us consider a somewhat more realistic situation. We are aware that, in most cases, we shall not know the demand exactly during some future period. We may be able to estimate an expected demand, but this estimate may have wide variability. It is usually more realistic to consider such a future demand as a random variable. If the demand were a discrete random variable, which could have one of l possible values, the expected demand $E(D_k)$ for period k would be given by

$$E(D_k) = D_{k1}P_{k1} + D_{k2}P_{k2} + \cdots + D_{kl}P_{kl}, \tag{9.7.8}$$

where P_{kj} is the probability that the demand for period k will be D_{kj}. Actually, such a situation would be very unusual and we would expect the demand for each period to be represented by a continuous random variable. The demand behavior in any period can then be described by some frequency function of D_k and its equivalent distribution function $F(D_k)$. However, such continuous functions are difficult to handle directly in dynamic programming. It is more convenient to assume that Equation (9.7.8) applies and to define P_{kj} by

$$P_{kj} = F\left(D_{kj} + \frac{h}{2}\right) - F\left(D_{kj} - \frac{h}{2}\right), \qquad j = 2, 3, \ldots, (l-1), \tag{9.7.9}$$

$$P_{k1} = F\left(D_{k1} + \frac{h}{2}\right), \tag{9.7.10}$$

$$P_{kl} = 1 - F\left(D_{kl} - \frac{h}{2}\right), \tag{9.7.11}$$

where h is simply the interval size chosen, i.e., $(D_{kl} - D_{k1})/l$. Since we are lumping the range of demands above D_{kl} into one interval, and the entire range of demands below D_{k1} into another single interval, it is desirable that we choose values of D_{k1} and D_{kl} such that P_{k1} and P_{kl} are low.

If we knew that the demand at stage k was D_{kj}, we would compute the return for stage k, for a given u_k and x_k, from Equation (9.7.3), using D_{kj} in place of D_k. However, there is only a probability P_{kj} that the demand will actually be D_{kj}. We can say that the probable return due to demand D_{kj} is given by $P_{kj} r_{kj}(u_k, x_k, D_{kj})$, where

$$r_{kj}(u_k, x_k, D_{kj}) = \{S \cdot \min(D_{kj}, u_k + x_k) - C_1 x_k - C_2 \max(x_k - q, 0) - C_3 u_k\}. \tag{9.7.12}$$

We may make the same statement for every demand level. Our total probable return (expected return) is then obtained by summing over all the possible demand levels. That is, for a given u_k and x_k,

$$E(r_k) = \sum_{j=1}^{l} P_{kj}\{S \cdot \min(D_{kj}, u_k + x_k) - C_1 x_k - C_2 \max(x_k - q, 0) - C_3 u_k\}. \tag{9.7.13}$$

We see from Equation (9.7.5) that any set of x_k, u_k, and D_{kj} fix u_{k-1}. Since D_{kj} may be at any one of several different levels, u_{k-1} may be at several levels. The probability that u_{k-1} will be at a particular level equals the probability that the corresponding D_{kj} will occur. Therefore, to obtain the expected return from stage $(k - 1)$ we must weight the returns in accordance with the probability that demand D_{kj} occurs. For a given u_k and given choices of x_k and x_{k-1} we have

$$E(r_{k-1}) = \sum_{j=1}^{l} P_{kj}\left[\sum_{j=1}^{l} P_{(k-1),j} r_{(k-1),j}(u_{k-1}, x_{k-1}, D_{(k-1),j})\right]. \tag{9.7.14}$$

Obviously, the expected returns from each of the preceding stages must be obtained by weighting the returns in accordance with the probabilities that the corresponding demand levels will occur.

We may obtain $E[y(u_k)]$, the expected value of $y_k(u_k)$, by maximizing the sum of the expected returns. That is,

$$E[y_k(u_k)] = \max_{x_k, x_{k-1}, \ldots, x_1} \left\{ \sum_{j=1}^{l} P_{kj}\left[r_{kj}(u_k, x_k, D_{kj}) \right.\right.$$
$$+ \sum_{j=1}^{l} P_{(k-1),j}\{r_{k-1,j}(u_{k-1}, x_{k-1}, D_{(k-1),j}) + \cdots$$
$$\left.\left. + \sum_{j=1}^{l} P_{1j} r_{1j}(u_1, x_1, D_{1j}) \cdots \} \right] \right\}. \tag{9.7.15}$$

Just as we simplified the earlier expression for $y_k(u_k)$(cf. Section 8.3), we can simplify the above to

$$E[y_k(u_k)] = \max_{x_k} \left\{ \sum_{j=1}^{l} P_{kj}\big(r_{kj}(u_k, x_k, D_{kj}) + E[y_{k-1}(u_{k-1})]\big) \right\}. \tag{9.7.16}$$

If our utility for money may be considered linear with quantity of money, then $E(y_n)$, the expected value of y_n, is the appropriate objective function for us to choose. We now solve our problem in the same manner previously used for the solution of deterministic problems, but use $E[(y_k(u_k)]$ in place of $y_k(u_k)$.

One significant difference between the results for a varying and known demand must be recognized. With a known demand, once the problem is solved and the decision for the first month obtained, the decisions for each succeeding month are set. This is not the case when demand is a random variable. The decisions for the second and succeeding months were based on the demand distribution for the first month and, therefore, would not be the ones chosen if we happened to know precisely the demand for the first month. However, when it is time to make the decision on production for the second month, we will know precisely what the demand was during the first month. It would be foolish to ignore this information and proceed as if we were not aware of it. We therefore redetermine the optimum production for the second month by reconsidering the $(n - 1)$-stage problem with the inventory at the end of the first month as the input condition that must be met at stage $(n - 1)$. Similarly, when the second, third, etc., months arrive, we redetermine our optimum x_k based on the appropriately reduced number of stages and observed inputs. Although we must redetermine the x_k, it is not necessary to perform additional computations. The values of the x_k corresponding to the observed inventory levels are embedded in the original solution. To determine the optimum x_n, we had to determine the optimum x_k for all the allowable u_k levels at stages $(n - 1)$, $(n - 2)$, etc. We need only consult our original solution in order to select the desired x_k for any observed level of u_k. Therefore, the dynamic programming solution may be considered to supply us with a production policy that is a function of the observed inventory levels.*

Although we have limited our consideration to a particular inventory problem, the dynamic programming approach is applicable to many types of sequential decisions that must be made under risk. In all cases we must consider that the solution has provided a policy which is a function of the random variables. The specific decisions, after the first, must be determined when the actual state of the system is observed prior to each stage of the problem.

The limitations of dynamic programming, described in Chapter 8, apply equally well here. We are limited to problems that involve only a few state variables and do not have numerous constraints. Furthermore, the problem must have a serial-like structure. If these conditions are satisfied, however, we can consider models that contain nonlinearities, discrete variables, and random variables. Thus the dynamic programming approach is very powerful for a certain class of problems.

Sequential decision problems as nonlinear programming problems

In Chapter 8 we observed that an n-stage dynamic programming problem, involving a single decision variable at each stage, could be replaced by a single nonlinear programming problem in n variables. Because of the limitations of dynamic programming, the nonlinear programming approach must be used when the number of state variables, decision variables, or constraints is large.

We can, if we wish, state the inventory problem of the previous section as a nonlinear programming problem. In establishing an appropriate objective function, the discrete formulation of the demand probability distribution is not convenient, and we shall, therefore, go back to the original continuous demand distribution. Let us first consider underproduction conditions at stage k; i.e., $u_k + x_k \leq D_k$, assuming u_k is known. In the manner of Section 9.3, we have for a given u_k,

$$E[r_k(u_k + x_k \leq D_k]$$
$$= \int_{u_k+x_k}^{\infty} [Su_k + Sx_k - C_1x_k - C_2 \max(x_k - q, 0) - C_3u_k]f(D_k)\,dD_k. \tag{9.7.17}$$

However, the value of u_k cannot be established absolutely since u_k is a func-

* In some instances the optimizer may designate the planning period as a specified time period (i.e., 1 year) that always begins at the present month. When information for the present month is obtained, a new policy is developed for the 12 months following. Here the dynamic programming problem would have to be resolved at the beginning of each month.

tion of the demands that are stochastic variables. We shall therefore consider u_k to be a stochastic variable. When we do so, we must consider all possible values of u_k; therefore, we write our expected return as

$$\begin{aligned} &E[r_k(u_k + x \leq D_k)] \\ &= \int_0^\infty \int_{u_k + x_k}^\infty [Su_k + Sx_k - C_1 x_k - C_2 \max(x_k - q, 0) \\ &- C_3 u_k] f(D_k) f(u_k) d(D_k) d(u_k). \end{aligned} \tag{9.7.18}$$

We may write a similar expression for overproduction conditions:

$$\begin{aligned} &E[r_k(u_k + x_k \geq D_k)] \\ &= \int_0^\infty \int_{-\infty}^{u_k + x_k} [SD_k - C_1 x_k - C_2 \max(x_k - q, 0) - C_3 u_k] \\ &\times f(D_k) f(u_k) d(D_k) d(u_k). \end{aligned} \tag{9.7.19}$$

If we sum these expressions over the n stages, we find the objective function to be

$$\begin{aligned} \max E(y) = \sum_{k=1}^{n} \Big\{ &\int_0^\infty \int_{u_k + x_k}^\infty [Su_k + Sx_k] f(D_k) f(u_k) d(D_k) d(u_k) \\ &+ \int_0^\infty \int_0^{u_k + x_k} [SD_k] f(D_k) f(u_k) d(D_k) d(u_k) \\ &- C_1 x_k - C_2 \max(x_k - q, 0) - C_3 E(u_k). \end{aligned} \tag{9.7.20}$$

To make use of the above expression, we must obtain an expression for $f(u_k)$. We may do so in terms of $E(u_k)$ and $\sigma_{(u_k)}$. We may obtain $E(u_k)$ by rewriting Equation (9.7.6) as

$$E(u_k) = \min[E(u_{k-1} - x_k + D_k), 0]. \tag{9.7.21}$$

If we were to assume that the probability of $u_k = 0$ is small, then we could write (9.7.21) in terms of D_k and x_k only; i.e.,

$$E(u_k) \simeq \sum_{j=1}^{k} D_j - \sum_{j=1}^{k} x_j. \tag{9.7.22}$$

We shall use this approximation to estimate the variance of u_k. Since (9.7.22) is linear, we know from Equation (9.4.15) that

$$\sigma_{(u_k)}^2 = \sum_{j=1}^{k} \sigma_{D_j}^2 \tag{9.7.23}$$

as the x_j are not random variables. If all the D_j are normally distributed,

u_k will also be normally distributed. Hence

$$f(u_k) = \frac{1}{\sigma_{(u_k)}\sqrt{2\pi}} \exp\left[-\frac{1}{2}\left(\frac{u_k - E(u_k)}{\sigma_{(u_k)}}\right)^2\right]. \tag{9.7.24}$$

By substituting Equations (9.7.21) and (9.7.24) into Equation (9.7.20), we now can evaluate our objective function for any set of $x_1, x_2, \ldots, x_n$ selected. We may therefore regard the problem as an unconstrained minimization problem in n variables and may solve it by one of the methods of Chapter 3.

Solution of sequential decision problems by chance-constrained programming

The probability integrals that appear in (9.7.20) cannot be evaluated analytically when the demands are normally distributed. We see that in this case the dynamic programming approach appears convenient. However, this is not always the case. We have previously noted the difficulties dynamic programming encounters when there are numerous constraints. When some of the constraints imposed are stochastic in nature, the difficulties in the use of dynamic programming are increased still further. For the case of stochastic constraints, we shall examine an alternative—the consideration of sequential decisions under risk as a chance-constrained programming problem.

In Section 9.4 we considered the proper selection of an investment portfolio. At the time we assumed that the investor made a single initial investment and held it for some specified period. Investors are not likely to do this under real conditions. In most instances they will have an income surplus, which they can invest at intervals; also, they may wish to modify their investments in the future. Such an investor is really faced with a sequence of decisions under risk.

EXAMPLE 9.7.1

Let us consider a simple problem of this type, originally studied by Näslund [1967]. An investor has the objective of earning as much money as possible over a fixed number of years. At each year he will have additional funds available for investment. The investor is to divide his available funds between a common-stock mutual fund and a cash reserve. He wishes to protect himself against losses by requiring that losses beyond a given amount L, in any period, have a probability of occurrence of less than α.

Näslund concludes that the average yearly advance of stock prices can be represented by the random variable v_j, where

$$v_j = \frac{[C_j - C_{j-1}]}{C_{j-1}} \tag{9.7.25}$$

and C_j is the price, or cost, of the mutual fund at the end of the year j. Furthermore, it appears that v_j is approximately normally distributed. The expected value of v_j indicates the expected fractional increase in mutual-fund price. If x_j were invested at the beginning of year j, the increase in value over year j would be $x_j E(v_j)$. Hence, to maximize our earnings over n periods, our objective function is: Maximize

$$\sum_{j=1}^{n} x_j E(v_j). \tag{9.7.26}$$

When a loss occurs, the magnitude is given by $(x_j v_j)$. Therefore, the limitation on the maximum loss during any one year is easily written as

$$P[x_j v_j \leq L] \leq \alpha \qquad \text{for } j = 1, 2, \ldots, n, \tag{9.7.27}$$

where the maximum allowable annual loss L is a negative quantity. If the quantity w_i is the accumulated capital that the investor has available for investment in the mutual fund at the beginning of the ith period, apart from returns on earlier fund investments, the investor's total worth is w_i plus investment appreciation; i.e.,

$$\text{total worth} = w_i + \sum_{j=1}^{i-1} x_j v_j. \tag{9.7.28}$$

There must be a very low probability, say η, that the investment program call for an investment x_i in excess of the total worth at period i. We therefore require that

$$P\left[w_i + \sum_{j=1}^{i-1} x_j v_j \leq x_i\right] \leq \eta. \tag{9.7.29}$$

Since v_i is considered to follow a normal distribution, we can readily replace equation sets (9.7.27) and (9.7.29) by their deterministic equivalents. That is,

$$x_i[E(v_i) - K_1 \sigma_{(v_i)}] \leq L \qquad \text{for } i = 1, 2, \ldots, n, \tag{9.7.30}$$

and

$$w_i + \left[\sum_{j=1}^{i-1} x_j E(v_j)\right] - K_2\left[\sum_{j=1}^{i-1} x_j^2 \sigma_{(v_j)}^2\right]^{1/2} \leq x_i \qquad \text{for } i = 1, 2, \ldots, n. \tag{9.7.31}$$

Since the sum of normally distributed variables is itself normally distributed, $\sum_i x_i v_i$ is normally distributed. Therefore, the constants K_1 and K_2 are both determined from tables of normal distribution so as to ensure that the respective probability limits, α and η, are observed.

To obtain the required x_i, we must solve a nonlinear programming problem in n decision variables and $(2n)$ constraints. The problem is convex and can be solved by several of the methods described in Chapter 5. The value of $x_2, x_3, \ldots, x_n$ obtained will not be acted upon. After the first year the actual value of v_1 will be available. An improved value of x_2 can then be obtained by resolving the problem for the $(n - 1)$ remaining years using the observed v_1. In contrast to the dynamic programming approach, the policies to be followed for the different values of v_1

usually are not contained within the original solution. We must therefore resolve our problem at the beginning of each time interval.

The failure-penalty and risk-programming approaches may, of course, also be adapted to determination of sequential decisions under risk. The basic approach is the same as that outlined for our chance-constrained programming approach. Each time interval is examined and the contribution to the objective function is expressed in terms of the decision variables for that time interval. The optimization is then accomplished by an appropriate nonlinear programming technique.

When the decision variables in a sequential problem under risk are constrained to be discrete, it is usually desirable to use the dynamic programming approach. Dynamic programming readily provides discrete solutions, whereas they are attainable only with difficulty using nonlinear programming. We would have to consider the discrete problem in terms of an integer programming problem and attempt to solve it by coupling a branch-and-bound algorithm to a nonlinear programming technique. This is not desirable since, with nonlinear problems, the branch-and-bound methods generally require large amounts of computing effort. Furthermore, unless we are certain that the problem is convex (concave), the branch-and-bound procedure may only provide a local optimum.

9.8 Status of Optimization Under Risk and Uncertainty

When we concluded our examination of nonlinear programming techniques in Chapter 5, we observed that there was no one procedure that was clearly superior in all situations. When nondeterministic problems are considered, the situation is even more ambiguous. Not only are there a number of ways to solve a problem once it is formulated, but there are usually several ways of formulating a problem.

In optimization under uncertainty, the formulation of Von Neumann and Morgenstern is generally accepted for two-person situations. Linear programming provides us with an efficient and rapid means of obtaining an optimum solution. However, as we observed earlier, in n-person situations no general agreement has yet been reached as to the proper problem formulation.

The most useful procedure for optimization under conditions of risk depends upon the particular problem. In general, we shall be maximizing or minimizing the expected value of some quantity. When the quantity is cost or monetary return, we must consider whether our value for money is approximately linear with money over the range involved. When the possible variation in cost or return is large with respect to net worth, a linear relationship is a poor assumption. An objective function in terms of utility of money is preferable under such circumstances.

Deterministic constraints have little meaning under conditions of risk. If it is possible to assign a cost penalty to the violation of a constraint, it is desirable to do so. The probable cost of violating this constraint can then be included in the objective function as a *failure penalty.* Use of the failure-penalty concept will generally avoid solutions that are unnecessarily conservative.

In some situations it is extremely difficult to assign a cost penalty when a constraint is violated. For example, a businessman wishing to maximize the return from a new venture may conclude that his ability to obtain bank credit when needed will be impaired unless the ratio of sales to investment is above a certain level. It may be very difficult to assign a cost to the impairment of a credit rating and therefore easier to treat such a limitation by *chance-constrained programming.* It should be observed that in any one problem one may encounter limitations that can be treated by assigning a failure penalty and other limitations where this is difficult to do. In such a situation one may formulate the problem as a chance-constrained program for some constraints but include failure penalties for the remaining constraints in the objective function. Such a combination approach is often desirable.

When formulated as a chance-constrained program, we may obtain a solution by one of the nonlinear programming techniques described in Chapter 5. If the failure-penalty approach is used, we have an unconstrained optimization, which can be treated by one of the methods discussed in Chapter 3. If the problem requires a sequence of decisions, both of these approaches provide information on the first decision only. After the results of the initial decision have been observed, the problem must be resolved to determine the next decision. For sequential decisions under risk, it is desirable to consider dynamic programming, since the initial dynamic programming solution provides the policy to be followed for all the events possible. Dynamic programming can be a powerful tool for the solution of sequential decision problems, providing the usual limitations on dynamic programming are observed. The difficulties introduced by constraints can often be circumvented by replacing these constraints by failure penalties appearing in the objective function.

9.9 Quantitative Decision Making

In a sense, the chief contribution of optimization theory may be considered the extension of our ability to make rational, quantitative decisions in areas where only intuitive judgments were possible previously. Initially, the extension of our rational decision-making ability was quite limited. However, it has taken only a relatively short time for optimization theory to advance from the position of solving only deterministic linear problems to the provision of solutions for complex, nonlinear stochastic problems. This advance, coupled with the rapid expansion in digital-computer technology, has enabled the solution of successively more realistic models of real-world problems. Since it is now generally recognized that much of the world is not completely deterministic, we may expect increased emphasis on solutions of problems under risk as still more realistic models are developed. We may, of course, also expect a continued growth of optimization theory, which will lead to more powerful algorithms and the solution of many problems now considered too complex.

The use of formal optimization techniques has been largely limited to a few areas of endeavor. Application of these techniques to science and engineering, public administration, and the social sciences has barely begun. While more widespread application of optimization procedures is unlikely to achieve the best of all possible worlds, extensive use of optimization methods can help us progress to a more rational allocation of this planet's limited resources.

REFERENCES

Charnes, A. and W. W. Cooper, "Chance-Constrained Programming," *Management Sci., 6* (1959), 73.

Charnes, A., W. W. Cooper, and G. H. Symonds, "Cost Horizons and Certainty Equivalents: An Approach to Statistical Programming of Heating Oil," *Management Sci., 4* (1958), 235.

Dantzig, G. B., "Linear Programming Under Uncertainty," *Management Sci., 1* (1955), 197.

Dubey, S. D., "Normal and Weibull Distributions," *Naval Res. Logistics Quart., 14* (1966), 69.

Feller, W., *An Introduction to Probability Theory and its Application, Vol. 1*, 3rd ed., Wiley, New York, 1968.

Ferguson, A. R., and G. B. Dantzig, "Allocation of Aircraft to Routes—Examples of Linear Programming of Uncertain Demand," *Management Sci., 3* (1956), 45.

Freund, R. J., "Introduction of Risk into a Programming Model," *Econometrica, 24* (1956), 253.

Guttman, I., S. S. Wilks, and J. S. Hunter, *Introductory Engineering Statistics*, 2nd ed., Wiley, New York, 1971.

Hadley, G., *Nonlinear and Dynamic Programming*, Addison-Wesley, Reading, Mass., 1964.

Näslund, B., *Decisions Under Risk*, Economic Research Institute, Stockholm School of Economics, Stockholm, Sweden, 1967.

Von Neumann, J., and O. Morgenstern, *Theory of Games and Economic Behavior*, Princeton University Press, Princeton, N.J., 1953.

Wald, A., *Statistical Decision Functions*, Wiley, New York, 1950.

Weisman, J., *Engineering Design Optimization under Conditions of Risk*, Ph. D thesis, University of Pittsburgh, Pittsburgh, Pa., 1968.

Weisman, J., and A. G. Holzman, "Optimal Process System Design Under Risk," *Ind. Eng. Chem. Process Design and Devel.*, *11* (1972), 386.

PROBLEMS

9.1. A two-person zero-sum game has the following payoff matrix:

Player I Strategies	*Player II Strategies*				
	1	*2*	*3*	*4*	*5*
1	1	3	2	0	1
2	2	4	0	5	1
3	3	6	2	1	0
4	4	3	2	1	0
5	5	1	0	2	5

The values in the matrix are payoffs to player I. Determine the strategies each player should follow.

9.2. Show that solution of the following problem yields a solution to the two-person zero-sum game: Maximize

$$a_{11}x_1 + a_{21}x_2 + \cdots + a_{m1}x_m - x_{m+1},$$

subject to

$$(a_{12} - a_{11})x_1 + (a_{22} - a_{21})x_2 + \cdots + (a_{m2} - a_{m1})x_m + x_{m+1} - x_{m+2} = 0,$$

$$(a_{13} - a_{11})x_1 + (a_{23} - a_{21})x_2 + \cdots + (a_{m3} - a_{m1})x_m + x_{m+1} - x_{m+3} = 0,$$

$$\cdots\cdots\cdots\cdots\cdots\cdots\cdots\cdots$$

$$(a_{1n} - a_{11})x_1 - (a_{2n} - a_{21})x_2 + \cdots + (a_{mn} - a_{m1})x_m + x_{m+1} - x_{m+n} = 0,$$

where the a_{ij} are payoff matrix values and the x_{m+1} to x_{m+n} represent surplus variables.

9.3. In a two-person zero-sum game a given strategy of player 1 (say strategy k) has the property that

$$a_{kj} < a_{ij} \qquad \text{for } j = 1, 2, \ldots m, \quad i = 1, 2, \ldots, k-1, k+1, \ldots, m.$$

Show that the same solution to the game is obtained whether or not this strategy is included in the payoff matrix.

9.4. The owner of a small restaurant specializing in omelettes has reason to believe that some of the eggs he has on hand have spoiled. He has no idea, however, of the probability that a given egg is bad. He therefore decides that instead of breaking all the eggs for an omelette into a single large bowl, and possibly losing the entire bowl because of one bad egg, he will break each egg into a separate bowl. He will only add the egg to the larger bowl if examination shows the egg to be unspoiled. The owner finds that this procedure is time consuming and that his costs have risen by the equivalent of $\frac{5}{8}$ egg for each egg examined separately. He therefore has second thoughts about his decision and wonders whether, for some size omelettes, it is worthwhile to inspect each egg separately. Consider this problem in terms of a series of games, and determine whether eggs should be examined individually when making a 1-egg, 2-egg, 3-egg, 4-egg, or 5-egg omelette. The owner considers as negligible the chance that more than 1 egg in a given omelette will be bad.

9.5. The Weibull distribution function is given by

$$F(x) = 1 - e^{-[(x-a/\theta)^m]},$$

where θ, m, and a are specified constants.

(a) Show that the corresponding frequency function is given by

$$f(x) = \frac{m}{\theta}(x - a)^{m-1}e^{-[(x-a)^m/\theta]}.$$

(b) If $m = 3.259$, show that the mean, $\mu(x) = a + 0.89640\theta^{1/3.259}$.

9.6. If x_1 and x_2 are stochastic quantities which have variances that are small with respect to their means, show that the variance of their product, x_1x_2, is given by

$$\sigma^2(x_1x_2) = (\mu_{x_1})^2\sigma^2_{x_2} + (\mu_{x_2})^2\sigma^2_{x_1}.$$

9.7. A manufacturer finds that the mean demand for his product now equals the present capacity of his plant. He believes that the increase in the expressed annual demand will be given by

$$[E(\Delta D)] = 5000t,$$

where t is time in years from the present. The manufactuter has reason to believe that the increased demand is normally distributed with a $\sigma = 0.3E(\Delta D)$.

The manufacturer expects that the selling price of his product will remain stable at \$50/unit and that his operating costs will remain at \$10/unit. Annual capital costs are estimated at \$30 for each unit of plant capacity.

(a) If a 10-year planning horizon is considered, what size plant addition should be built?

(b) How does this compare with the addition that would have been built had demand been deterministic?

9.8. The manufacturer of animal feeds has at his disposal barley, oats, sesame flakes, and peanut meal. He finds that these materials vary from batch to batch. His samples have the properties shown in the table.

Material	*Protein Content (%)*	*σ Protein Content*	*Fat Content (%)*	*σ Fat Content*	*Cost (\$/ton)*
Barley	12.0	0.53	2.3	0.1	24.50
Oats	11.9	0.44	5.6	0.2	26.70
Seasame flakes	41.8	4.5	11.1	0.5	39.00
Peanut meal	52.1	0.79	1.3	0.1	40.50

To ensure healthy animals, the manufacturer desires that there be no more than a 5 per cent probability that his product wil contain less than 21 per cent protein or 5 per cent fat. What is the optimal composition of the feed? How does this compare with the solution that would have been obtained under deterministic conditions? [Cf. Van de Panne and Popp, *Management Sci.*, *9*, (1962), 405.]

9.9. (a) Show that when a linear programming problem of the form: Minimize

$$y = \mathbf{C}'\mathbf{X},$$

subject to

$$\mathbf{A}\mathbf{X} \geq \mathbf{B},$$
$$\mathbf{A}_0\mathbf{X} = \mathbf{B}_0,$$

is transformed into a chance-constrained problem, because the components of **A** are stochastic quantities having mean values a_{ij}, we obtain: Minimize

$$y = \mathbf{C}'\mathbf{X},$$

subject to

$$\mathbf{A}_0\mathbf{X} = \mathbf{B}_0,$$
$$\mathbf{A}\mathbf{X} - k(\mathbf{X}'\mathbf{V}\mathbf{X})^{1/2} \geq \mathbf{B}.$$

Here k is a positive scalar and **V** is a matrix having nonzero elements only on the diagonal.

(b) Show that this chance-constrained problem is a convex programming problem. [*Hint:* Assume that $\mathbf{A}\mathbf{X} - k(\mathbf{X}'\mathbf{V}\mathbf{X})$ is strictly convex and show, by using the basic definition of strict convexity, that this leads to

$(\mathbf{X}_2)'\mathbf{V}\mathbf{X}_1 < ((\mathbf{X}_1)'\mathbf{V}\mathbf{X}_1(\mathbf{X}_1)'\mathbf{V}\mathbf{X}_2)^{1/2}$. Transform this latter expression so that the Schwarz inequality can be applied.]

9.10. A real-estate investor is considering the relative merits of two investments. He expects that his investment in area A will be worth 170 per cent of its present value in 5 years. However, he recognizes that his investment is risky and concludes that there is a 10 per cent probability that his investment will not appreciate at all. He can also invest in area B, where he expects a 100 per cent increase in value, but concludes that there is a 35 per cent probability that his investment will not appreciate.

If he can accumulate an additional $20,000 through his investments, he will have enough capital to become one of the partners in large real-estate firm. He believes that this partnership is worth $150,000 to him. However, he cannot afford to lose more than $35,000 through his investments and desires that there be no more than a 0.2 per cent probability that this will occur. How much money should he invest in each of the two areas? You may assume that the final real-estate prices are normally distributed.

9.11. (a) In order to make a profit, all insurance companies charge premiums in excess of the product of (a) the probability of the disaster, and (b) the cost of the disaster. Nevertheless, most individuals insure themselves against large losses. Explain this phenomenon.

(b) Some individuals have a predilection for making very risky investments, i.e., gambling. What would be the shape of such an individual's money utility curve?

(c) There are certain individuals who both insure themselves and gamble. What shape would have to be ascribed to such an individual's money utility curve?

9.12. In problem 4.24 we considered the determination of an inventory policy in the face of a deterministic demand. In any realistic situation the demand will be a stochastic variable. Assume, therefore, that the demand values given in problem 4.24 represent the mean values of normally distributed random variables. Assume also that the standard deviation of the monthly demand is 2500 units.

(a) Determine the optimum inventory policy if an additional storage cost of $1.00/unit is incurred whenever the inventory exceeds 50,000. For simplicity consider only the first 6 months of the year. Assume that the expected inventory at the end of June equals the initial inventory.

(b) How would you formulate the above problem if it was required that there be no more than a 5 per cent probability that the 50,000 inventory be exceeded?

9.13. A high-pressure pipeline is being designed to transfer crude oil between two locations. The designer wishes to determine the line size that minimizes the sum of pumping power costs and failure costs.

The pumping power, PP, is given by $PP = (.385 \times 10^{-12} W^3)/(\rho^2 x_1^5)$ where W is the flow in pounds/hour, ρ is the fluid density (50 lb/ft³), and x_1 is the pipe diameter. Although W is nominally fixed at 10.0×10^6 lb/hr, errors in measurement make it a normally distributed random variable with a variance of 1×10^4 lb/hr. Furthermore, fabrication tolerances make the pipe diameter, x_1, a normally distributed random variable, where $\sigma_{(x_1)} = 0.01x_1$. The pumping power correlation itself is not exact and the standard deviation of the correlation is $0.01 \times PP$.

In view of the stochastic nature of the variables, selection of a pump horsepower exactly equal to *PP* has a high probability of resulting in a pump that is actually too small. Selection of a pump size x_2, which cannot meet the requirements, imposes a failure cost of $150,000.

The capital and equivalent pumping costs are given by

$$C = 10{,}000x_1^2 + 50x_2 + 70PP.$$

(a) Derive an appropriate nonlinear programming model, assuming that the oil company's utility is linear with money.
(b) Revise the formulation of (a), assuming that the oil company's utility for money is given by $1 - e^{-(2\times 10^{-7})r}$, where r is the total wealth of the oil company in dollars.
(c) Solve (b) using an appropriate unconstrained minimization technique.

APPENDIX A

Guide To Selection of Optimization Techniques

The plethora of available optimization techniques often makes it difficult to decide which technique is most appropriate for a given problem. Fig. A-1, shown on the following pages, attempts to provide the analyst with some guidance in making this decision.

It should be recognized that this figure represents the experience of the authors and that in some cases there will be disagreement as to the most advantageous routes to follow.

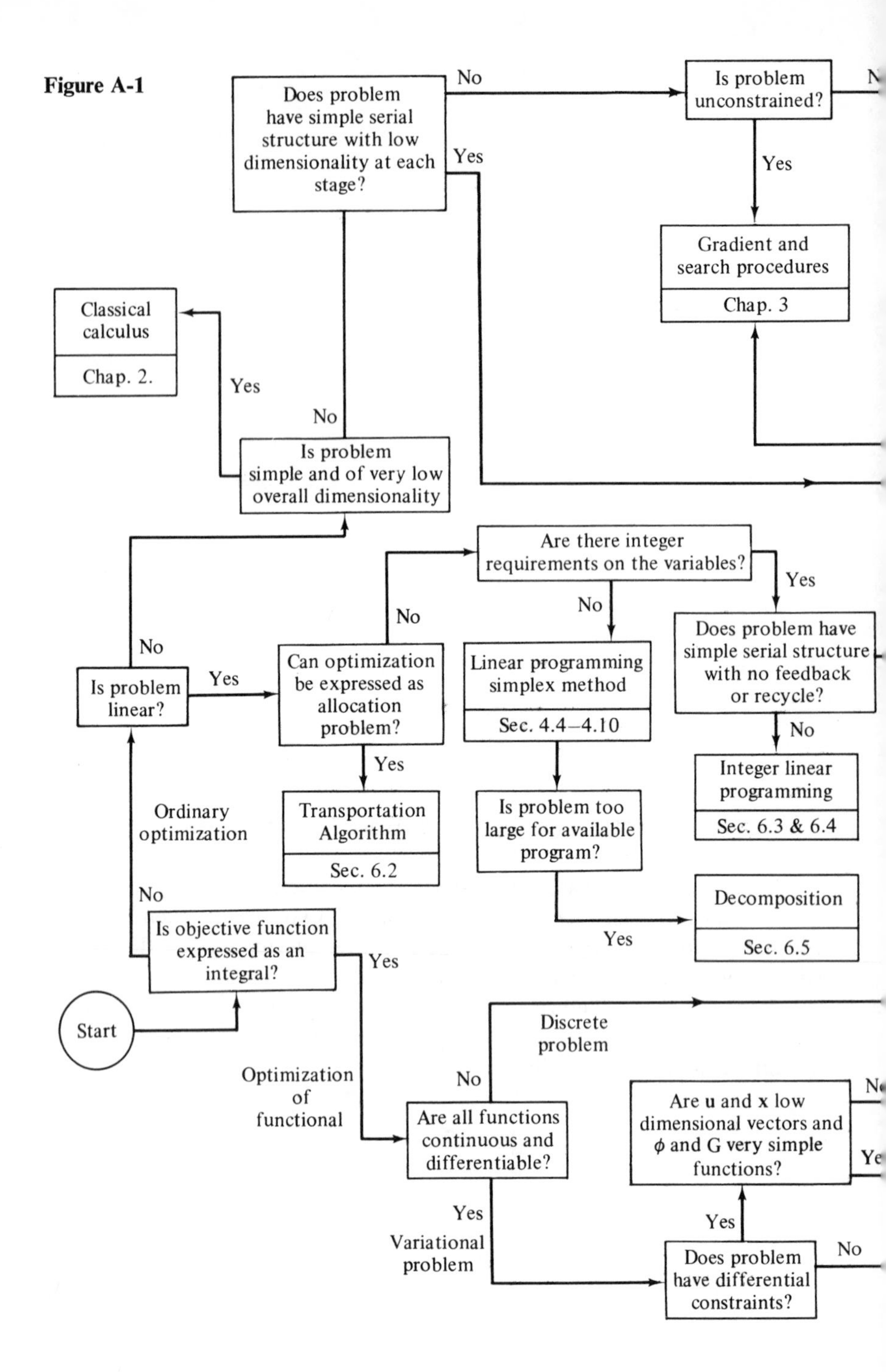
Does problem have simple serial structure with low dimensionality at each stage?
No
Yes
Is problem unconstrained?
Yes
Gradient and search procedures
Chap. 3
Classical calculus
Chap. 2.
Yes
No
Is problem simple and of very low overall dimensionality
Are there integer requirements on the variables?
Yes
No
No
Does problem have simple serial structure with no feedback or recycle?
No
Is problem linear?
Yes
Can optimization be expressed as allocation problem?
Linear programming simplex method
Sec. 4.4–4.10
Yes
Integer linear programming
Sec. 6.3 & 6.4
Transportation Algorithm
Sec. 6.2
Is problem too large for available program?
Ordinary optimization
No
Decomposition
Sec. 6.5
Is objective function expressed as an integral?
Yes
Yes
Start
Discrete problem
Optimization of functional
No
Are all functions continuous and differentiable?
Are u and x low dimensional vectors and φ and G very simple functions?
Yes
Yes
Variational problem
Does problem have differential constraints?
No

Figure A-1

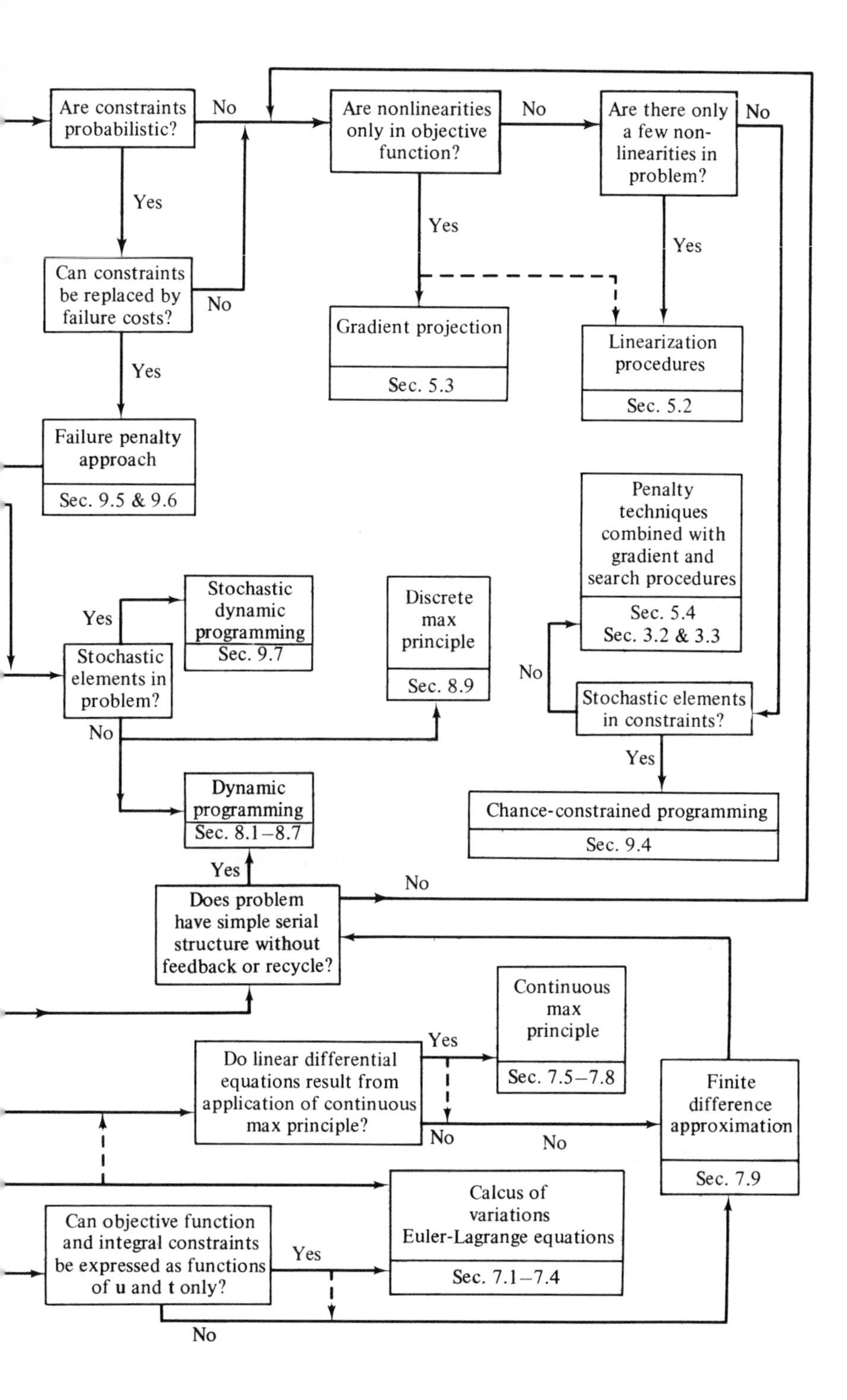

Are constraints probabilistic?
No
Yes
Can constraints be replaced by failure costs?
No
Yes
Failure penalty approach
Sec. 9.5 & 9.6
Are nonlinearities only in objective function?
No
Yes
Gradient projection
Sec. 5.3
Are there only a few non-linearities in problem?
No
Yes
Linearization procedures
Sec. 5.2
Penalty techniques combined with gradient and search procedures
Sec. 5.4
Sec. 3.2 & 3.3
No
Stochastic elements in constraints?
Yes
Chance-constrained programming
Sec. 9.4
Yes
Stochastic dynamic programming
Sec. 9.7
Stochastic elements in problem?
No
Discrete max principle
Sec. 8.9
Dynamic programming
Sec. 8.1–8.7
Yes
Does problem have simple serial structure without feedback or recycle?
No
Continuous max principle
Sec. 7.5–7.8
Yes
Do linear differential equations result from application of continuous max principle?
No
No
Finite difference approximation
Sec. 7.9
Calcus of variations
Euler-Lagrange equations
Sec. 7.1–7.4
Can objective function and integral constraints be expressed as functions of u and t only?
Yes
No

APPENDIX B
ELEMENTS OF MATRIX ALGEBRA

An understanding of linear programming and the nonlinear programming techniques based upon it requires a familiarity with matrix algebra. Matrices and their manipulation form a portion of the subject known as *linear algebra.* We shall not attempt to survey all of this subject area but simply to outline those elements which are essential to optimization theory.

Matrices

A matrix is any rectangular array of elements in which the elements are arranged in rows and columns. The array of numbers.

$$\begin{array}{rrrr} 9 & 1 & 5 & 6 \\ 4 & -3 & 7 & -2 \\ -8 & 5 & 3 & 0 \end{array}$$

is a matrix. This matrix contains three rows and four columns, and therefore we say it is a 3×4 matrix.

More generally, we can represent the elements of a matrix by the symbol a_{ij}, where the first subscript represents the row and the second subscript represents the column in which the element is placed. The entire matrix of all the elements a_{ij} we represent by **A**. Thus the $m \times n$ matrix **A** can be written

$$\mathbf{A} = \begin{bmatrix} a_{11} & a_{12} & a_{13} & a_{14} \cdots a_{1n} \\ a_{21} & a_{22} & a_{23} & a_{24} \cdots a_{2n} \\ \cdots & \cdots & \cdots & \cdots \\ a_{m1} & a_{m2} & a_{m3} & a_{m4} \cdots a_{mn} \end{bmatrix}.$$

It is sometimes desirable to consider only a portion of a matrix. We may *partition* a matrix into smaller segments as shown:

$$\mathbf{B} = \left[\begin{array}{cc|cc} b_{11} & b_{12} & b_{13} & b_{14} \\ b_{21} & b_{22} & b_{23} & b_{24} \\ \hline b_{31} & b_{32} & b_{33} & b_{34} \end{array}\right].$$

Each rectangular portion, separated by the solid lines, is called a *submatrix*. We can rewrite the matrix **B** in terms of only four elements:

$$\mathbf{B} = \begin{bmatrix} \mathbf{B}_{11} & \mathbf{B}_{12} \\ \mathbf{B}_{21} & \mathbf{B}_{22} \end{bmatrix},$$

where each element $\mathbf{B}_{ij}$ is itself a matrix. Thus

$$\mathbf{B}_{11} = \begin{bmatrix} b_{11} & b_{12} \\ b_{21} & b_{22} \end{bmatrix}.$$

We shall be concerned primarily with matrices whose elements are scalar quantities, though the elements may be functions of one or more independent variables.

Vectors

A *vector* is a matrix consisting of a single row or column. Thus the column vector **X** is

$$\mathbf{X} = \begin{bmatrix} x_1 \\ x_2 \\ x_3 \\ \cdot \\ \cdot \\ \cdot \\ x_m \end{bmatrix}$$

and the row vector **W** is

$$\mathbf{W} = [w_1 \quad w_2 \quad w_3 \cdots w_n].$$

In dealing with row or column vectors we obviously need only one subscript to designate a particular element of the vector.

A matrix may be considered to consist of a set of column or row vectors. Thus the matrix **A** may be said to consist of the set of column vectors $\mathbf{P}_1$, $\mathbf{P}_2, \ldots, \mathbf{P}_n$, where

$$\mathbf{P}_j = \begin{bmatrix} a_{1j} \\ a_{2j} \\ a_{3j} \\ \cdot \\ \cdot \\ \cdot \\ a_{mj} \end{bmatrix}.$$

We should not confuse the term vector as used in matrix algebra with the same term used in vector analysis. In vector analysis, a vector represents a quantity having both magnitude and direction. We shall use the term vector to represent any mathematical quantity defined by n parameters. Thus, in matrix algebra, a vector can represent a point in n-dimensional space by having its elements represent the n coordinates of the point with respect to some coordinate axes. The line segment connecting a given point with the origin would have both length and direction, and hence could represent a vector as used in vector analysis. If we were always to designate our line segments as rays from the origin to a given point, then the coordinates of the given point would completely specify the directed line segment. It is only in this limited sense that the matrix-algebra and vector-analysis definitions can be considered to coincide.

Special Matrices

A matrix **A** is said to be square if the number of rows equals the number of columns ($m = n$). A *diagonal matrix* is a square matrix whose off-diagonal elements are zero; i.e.,

$$\mathbf{C} = \begin{bmatrix} c_{11} & 0 & 0 & \cdots & 0 \\ 0 & c_{22} & 0 & \cdots & 0 \\ 0 & 0 & c_{33} & \cdots & 0 \\ \hdashline \\ \hdashline \\ 0 & 0 & 0 & \cdots & c_{mm} \end{bmatrix}.$$

An *identity matrix* or *unit matrix* is a diagonal matrix whose diagonal elements are 1. We represent an identity matrix by the symbol **I**. A 4 × 4 identity matrix is therefore

$$\mathbf{I} = \begin{bmatrix} 1 & 0 & 0 & 0 \\ 0 & 1 & 0 & 0 \\ 0 & 0 & 1 & 0 \\ 0 & 0 & 0 & 1 \end{bmatrix}.$$

The identity matrix has a role in matrix algebra analogous to that played by 1 in elementary algebra.

A *null matrix* is a matrix all of whose elements are zero. We designate it by **0**. A null matrix need not be square. Thus we may have a null vector. For example,

$$\mathbf{0} = \begin{bmatrix} 0 \\ 0 \\ 0 \end{bmatrix}.$$

By the *transpose of a matrix* we mean a new matrix in which the rows and columns of the original matrix have been interchanged. We use the symbol $\mathbf{A}'$ to represent the transpose of **A**. If **A** is as previously defined, then $a'_{ij} = a_{ji}$. If we write the entire matrix, we have

$$\mathbf{A}' = \begin{bmatrix} a_{11} & a_{21} & \cdots & a_{m1} \\ a_{12} & a_{22} & \cdots & a_{m2} \\ a_{13} & a_{23} & \cdots & a_{m3} \\ a_{14} & a_{24} & \cdots & a_{m4} \\ \hdashline \hdashline a_{1n} & a_{2n} & \cdots & a_{mn} \end{bmatrix}.$$

The transpose of an $m \times n$ matrix is therefore an $n \times m$ matrix. Thus the transpose of

$$\mathbf{A} = \begin{bmatrix} 2 & 3 & 4 \\ 1 & 6 & 7 \end{bmatrix}$$

is

$$\mathbf{A}' = \begin{bmatrix} 2 & 1 \\ 3 & 6 \\ 4 & 7 \end{bmatrix}.$$

A *symmetric matrix* is one in which the elements

$$a_{ij} = a_{ji}.$$

A symmetric matrix equals its own transpose ($\mathbf{A} = \mathbf{A}'$). A *skew symmetric matrix* is one in which the elements

$$a_{ij} = -a_{ji}.$$

A skew symmetric matrix equals the negative of its transpose ($\mathbf{A} = -\mathbf{A}'$).

Matrix Addition and Multiplication

There are several mathematical operations that may be used with matrices. We may add two matrices together to form a new matrix. That is,

$$\mathbf{A} + \mathbf{B} = \mathbf{C}.$$

The addition rule which we must follow states that we add corresponding elements of the matrices **A** and **B**. The elements of **C** are thus obtained from

$$a_{ij} + b_{ij} = c_{ij}.$$

It is obvious from this definition that we may only add matrices having the same number of columns and rows. For example,

$$\begin{bmatrix} 1 & 2 \\ 7 & 9 \end{bmatrix} + \begin{bmatrix} 3 & 5 \\ 8 & 4 \end{bmatrix} = \begin{bmatrix} 4 & 7 \\ 15 & 13 \end{bmatrix}.$$

We observe that both the associative and commutative laws hold for matrix addition. That is,

$$(\mathbf{A} + \mathbf{B}) + \mathbf{C} = \mathbf{A} + (\mathbf{B} + \mathbf{C}) \qquad \text{(associative law)},$$

$$\mathbf{A} + \mathbf{B} = \mathbf{B} + \mathbf{A} \qquad \text{(commutative law)}.$$

We may define the operation of scalar multiplication. Here we multiply a scalar, say k, and a matrix; i.e.,

$$k\mathbf{A} = \mathbf{A}k = \mathbf{B}.$$

Each element of **B** is obtained from

$$b_{ij} = ka_{ij}.$$

The distributive law holds for scalar multiplication, and therefore

$$k(\mathbf{A} + \mathbf{B}) = k\mathbf{A} + k\mathbf{B}.$$

We may also multiply two matrices together. A matrix product is represented by

$$\mathbf{AB} = \mathbf{C}.$$

Matrix multiplication is defined only where the number of columns of **A** *equals the number of rows of* **B**. The rule for matrix multiplication requires that we obtain the element c_{ij} in the ith row and jth column of **C** as the sum of the products of the elements of the ith row of **A** multiplied by the corresponding elements in the jth column of **B**. That is,

$$c_{ij} = a_{i1}b_{1j} + a_{i2}b_{2j} + \cdots + a_{im}b_{mj}.$$

For example,

$$\mathbf{DB} = \begin{bmatrix} d_{11} & d_{12} & d_{13} \\ d_{21} & d_{22} & d_{23} \end{bmatrix} \begin{bmatrix} b_{11} & b_{12} \\ b_{21} & b_{22} \\ b_{31} & b_{32} \end{bmatrix} = \begin{bmatrix} c_{11} & c_{12} \\ c_{21} & c_{22} \end{bmatrix},$$

where

$$\begin{aligned} c_{11} &= d_{11}b_{11} + d_{12}b_{21} + d_{13}b_{31}, \\ c_{12} &= d_{11}b_{12} + d_{12}b_{22} + d_{13}b_{32}, \\ c_{21} &= d_{21}b_{11} + d_{22}b_{21} + d_{23}b_{31}, \\ c_{22} &= d_{21}b_{12} + d_{22}b_{22} + d_{23}b_{32}. \end{aligned}$$

As another example, consider

$$[3 \quad 2]\begin{bmatrix} 4 & 6 & 1 \\ -5 & 7 & -2 \end{bmatrix} = [2 \quad 32 \quad -1].$$

We see from the preceding examples that when we multiply an $m \times n$ matrix by an $n \times p$ matrix, our result is an $m \times p$ matrix. Thus if we multiply a $1 \times m$ row vector by an $m \times 1$ column vector, the result is a scalar. The multiplication of a row vector by a column vector is equivalent to taking the dot product of two vectors in vector-analysis notation. Any two vectors **A** and **B** that are mutually perpendicular have a zero dot product. Such vectors are said to be *orthogonal*. Hence whenever we find that

$$\mathbf{A}'\mathbf{B} = \mathbf{B}'\mathbf{A} = 0,$$

we know that **A** and **B** are orthogonal.

It is obvious that if **A** is an $m \times n$ matrix and **B** is an $n \times p$ matrix, we can find the product **AB**, but our multiplication rule prevents us from forming the product **BA**. Thus *matrix multiplication is not commutative.* However, the associative and distributive laws do hold. That is,

$$(\mathbf{AB})\mathbf{C} = \mathbf{A}(\mathbf{BC}) \qquad \text{(associative law)},$$
$$\mathbf{A}(\mathbf{B} + \mathbf{C}) = \mathbf{AB} + \mathbf{AC} \qquad \text{(distributive law)}.$$

If **A** and **B** are both diagonal matrices, then multiplication of these matrices is commutative. That is, we can write

$$\mathbf{AB} = \mathbf{BA} = \mathbf{C}.$$

Furthermore, **C** will also be a diagonal matrix.

It is instructive to multiply **A** by a unit (identity) matrix of appropriate dimensions. For a 3×3 matrix

$$\mathbf{AI} = \begin{bmatrix} a_{11} & a_{12} & a_{13} \\ a_{21} & a_{22} & a_{23} \\ a_{31} & a_{32} & a_{33} \end{bmatrix} \begin{bmatrix} 1 & 0 & 0 \\ 0 & 1 & 0 \\ 0 & 0 & 1 \end{bmatrix} = \begin{bmatrix} a_{11} & a_{12} & a_{13} \\ a_{21} & a_{22} & a_{23} \\ a_{31} & a_{32} & a_{33} \end{bmatrix} = \mathbf{A}.$$

Our matrix multiplication rules are such that multiplication by a unit matrix always leads to a product identical to the original matrix. Furthermore, the same result is obtained whether we premultiply or postmultiply by **I**. We thus may write

$$\mathbf{AI} = \mathbf{IA} = \mathbf{A}.$$

If **A** and **B** are $m \times n$ and $n \times p$ matrices, respectively, then it is easily shown that

$$(\mathbf{AB})' = \mathbf{B}'\mathbf{A}'.$$

In other words, *the transpose of a matrix product equals the product of the transposes.*

Orthogonal Vector Sets

A set of vectors is said to be orthogonal if all the vectors within the set are orthogonal; i.e., if

$$\mathbf{A}_i'\mathbf{A}_j = 0 \qquad \text{for } i \neq j.$$

If the vectors $\mathbf{A}_i$ are also of unit length, the set is said to be *orthonormal.*

In an orthonormal set

$$\mathbf{A}_i'\mathbf{A}_j = \begin{cases} 0 & \text{for } i \neq j, \\ 1 & \text{for } i = j. \end{cases}$$

Determinants

With every square matrix **A** there is associated a scalar quantity, called the *determinant* of **A**, written as $|\mathbf{A}|$ or in the form of an array:

$$|\mathbf{A}| = \begin{vmatrix} a_{11} & a_{12} & \cdots & a_{1m} \\ a_{21} & a_{22} & \cdots & a_{2m} \\ \vdots & & & \\ a_{m1} & a_{m2} & \cdots & a_{mm} \end{vmatrix}.$$

We must not confuse the determinant written in array form with the matrix itself. A matrix can never be represented by a single numerical value. We say that an $m \times m$ matrix has a determinant of *order m*. We define the determinant of **A** as the number obtained from the sum of all possible products in each of which there is exactly one element from each row and column of **A**, each product being assigned a positive or negative sign according to a predetermined rule. In a 2×2 matrix we can obtain the products that are summed to form the determinant by connecting the matrix elements with diagonal line segments. We assign a positive sign to the line segment sloping up and to the left and a negative sign to the other line segment. Thus we have

$$\begin{vmatrix} a_{11} & a_{12} \\ a_{21} & a_{22} \end{vmatrix} = a_{11}a_{22} - a_{12}a_{21}.$$

We may proceed similarly for a 3×3 matrix. We imagine that those line segments which terminate before they include three terms are continued on the other side of the matrix in the next row. We have

$$|\mathbf{A}| = \begin{vmatrix} a_{11} & a_{12} & a_{13} \\ a_{21} & a_{22} & a_{23} \\ a_{31} & a_{32} & a_{33} \end{vmatrix} = a_{11}a_{22}a_{33} + a_{12}a_{23}a_{31} + a_{13}a_{21}a_{32} - a_{13}a_{22}a_{31} - a_{12}a_{21}a_{33} - a_{11}a_{23}a_{32}.$$

Again we have assigned positive signs to the line segments sloping upward and to the left and negative signs to the rest.

We cannot proceed in this manner for larger-size matrices and must make use of a more general procedure. To do so, we shall have to define two new terms. The *minor*, M_{ij}, of element a_{ij} is the determinant obtained from the square matrix **A** by striking out the *i*th row and *j*th column. Thus the minor $|M_{21}|$ of the element a_{21} from the 3 × 3 matrix **A** is

$$|M_{21}| = a_{12}a_{33} - a_{13}a_{32}.$$

We define the *cofactor*, A_{ij}, of element a_{ij} as $(-1)^{(i+j)}|M_{ij}|$. We also call the cofactor A_{ij} the *signed minor* of a_{ij}.

Now that we have defined what we mean by a cofactor, we can make use of the following theorem: *The determinant of a matrix* **A** *is equal to the sum of the products obtained by multiplying the elements of any row (column) by their respective cofactors.* We may therefore choose a column (say column 1) and write for our 3 × 3 matrix **A** that

$$\begin{aligned}|\mathbf{A}| &= a_{11}A_{11} + a_{21}A_{21} + a_{31}A_{31}\\ &= a_{11}\begin{vmatrix} a_{22} & a_{23} \\ a_{32} & a_{33} \end{vmatrix} - a_{21}\begin{vmatrix} a_{12} & a_{13} \\ a_{32} & a_{33} \end{vmatrix} + a_{31}\begin{vmatrix} a_{12} & a_{13} \\ a_{22} & a_{23} \end{vmatrix}\\ &= a_{11}(a_{22}a_{33} - a_{23}a_{32}) - a_{21}(a_{12}a_{33} - a_{13}a_{32}) + a_{31}(a_{12}a_{23} - a_{22}a_{13}).\end{aligned}$$

We see that this result is identical with that previously obtained.

By use of this theorem we are able to decompose a determinant of an $n \times n$ matrix into n sets of determinants of an $(n - 1) \times (n - 1)$ matrix. We may continue this procedure, successively decomposing the determinants that must be evaluated, until we reach a matrix size which is convenient to handle.

When we replace the elements a_{ij} by actual numbers, we may find in some cases that $|\mathbf{A}|$ has a value of zero. A matrix whose determinant is zero is said to be *singular*. We may find that if we obtain the determinant of a portion of a singular matrix that such a determinant is nonzero. The *rank of a matrix* is defined as the order of the largest nonvanishing determinant.

Simple rules can be provided for establishing how the determinant of a matrix is affected by several elementary operations on the matrix:

1. Multiplying a row (or column) of a matrix by a scalar multiplies the determinant by that scalar. Thus if

$$|\mathbf{A}| = \begin{vmatrix} 1 & 2 \\ 3 & 4 \end{vmatrix} = 4 - 6 = -2,$$

   and we multiply the first row by 2, then

$$|\mathbf{B}| = \begin{vmatrix} 2 & 4 \\ 3 & 4 \end{vmatrix} = 8 - 12 = -4.$$

2. Interchanging two adjacent rows (or columns) of a matrix changes the sign of the determinant. If we interchange the rows of $|\mathbf{A}|$, our 2 × 2 matrix

of the previous example, we have

$$|\mathbf{C}| = \begin{vmatrix} 3 & 4 \\ 1 & 2 \end{vmatrix} = 6 - 4 = 2.$$

3. Addition of a multiple of a row (column) to one row (column) does not change the value of the determinant. For example, if we add the first column of **A** to the second column, we have

$$|\mathbf{D}| = \begin{vmatrix} 1 & 3 \\ 3 & 7 \end{vmatrix} = 7 - 9 = -2.$$

4. If one row (column) of a matrix is a multiple of another row (column), the determinant will have a value of zero. For example,

$$|\mathbf{E}| = \begin{vmatrix} 1 & 3 \\ 3 & 9 \end{vmatrix} = 9 - 9 = 0.$$

Inverse Matrices

We have defined matrix multiplication but have not defined any operation similar to division. Although matrix division cannot be defined, we can define an operation analogous to division in some respects. That such an operation would be useful may be seen by considering the matrix equation

$$\mathbf{AB} = \mathbf{I}.$$

If we wished to solve for **B**, we would like to have an operation that would yield a result equivalent to division by **A**. If **A** and **B** are nonsingular square matrices, we may resolve our difficulty through the definition of the *inverse* of matrix **A**.

The inverse, $\mathbf{A}^{-1}$, of a nonsingular square matrix **A** is defined such that

$$\mathbf{AA}^{-1} = \mathbf{A}^{-1}\mathbf{A} = \mathbf{I}.$$

We can now solve our original matrix equation by multiplying both sides by $\mathbf{A}^{-1}$ to obtain

$$\mathbf{A}^{-1}\mathbf{AB} = \mathbf{A}^{-1}\mathbf{I}$$

or

$$\mathbf{B} = \mathbf{A}^{-1}.$$

The inverse of **A** may be evaluated in terms of the cofactors of **A**. We first define the adjoint matrix, **J**, of the $m \times m$ matrix **A** as another $m \times m$ matrix, where the element of the jth row and ith column is the cofactor of the element a_{ij} in the original matrix **A**. That is,

$$\mathbf{J} = \begin{bmatrix} A_{11} & A_{21} & \cdots & A_{m1} \\ A_{12} & A_{22} & \cdots & A_{m2} \\ \hdashline A_{1m} & A_{2m} & \cdots & A_{mm} \end{bmatrix}.$$

If $|\mathbf{A}| \neq 0$, then we may obtain the inverse of **A** from

$$\mathbf{A}^{-1} = \frac{\mathbf{J}}{|\mathbf{A}|}.$$

It is obvious that $\mathbf{A}^{-1}$ does not exist if **A** is singular.

As an illustration, let us obtain the inverse of

$$\mathbf{A} = \begin{bmatrix} 3 & 1 \\ 4 & 2 \end{bmatrix}.$$

We have

$$|\mathbf{A}| = 6 - 4 = 2,$$

$$\mathbf{J} = \begin{bmatrix} 2 & -1 \\ -4 & 3 \end{bmatrix},$$

and

$$\mathbf{A}^{-1} = \begin{bmatrix} 1 & -\frac{1}{2} \\ -2 & \frac{3}{2} \end{bmatrix}.$$

With this result, we can easily verify that $\mathbf{AA}^{-1} = \mathbf{A}^{-1}\mathbf{A} = \mathbf{I}$.

The determination of **J** requires a considerable number of calculations when the matrix is large. We shall see subsequently that we can also obtain the inverse by methods which require less numerical effort.

If we know the inverses of matrices **A** and **B**, we may obtain the inverse of their product, $(\mathbf{AB})^{-1}$, from

$$(\mathbf{AB})^{-1} = \mathbf{B}^{-1}\mathbf{A}^{-1}.$$

Simultaneous Equations: Gauss-Jordan Elimination

Let us consider n simultaneous linear equations in the variables x_1, $x_2, \ldots, x_n$:

$$\begin{aligned} a_{11}x_1 + a_{12}x_2 + \cdots + a_{1n}x_n &= b_1, \\ a_{21}x_1 + a_{22}x_2 + \cdots + a_{2n}x_n &= b_2, \\ \hdashline a_{n1}x_1 + a_{n2}x_2 + \cdots + a_{nn}x_n &= b_n. \end{aligned}$$

We may represent the set of coefficients, a_{ij}, by the matrix **A**, where

$$\mathbf{A} = \begin{bmatrix} a_{11} & a_{12} & \cdots & a_{1n} \\ a_{21} & a_{22} & \cdots & a_{2n} \\ \hdashline a_{n1} & a_{n2} & \cdots & a_{nn} \end{bmatrix}.$$

If we also let

$$\mathbf{B} = \begin{bmatrix} b_1 \\ b_2 \\ \cdot \\ \cdot \\ \cdot \\ b_n \end{bmatrix} \quad \text{and} \quad \mathbf{X} = \begin{bmatrix} x_1 \\ x_2 \\ \cdot \\ \cdot \\ \cdot \\ x_n \end{bmatrix},$$

we see that, by the rules of matrix multiplication, the matrix equation

$$\mathbf{AX} = \mathbf{B}$$

is equivalent to our original set of simultaneous equations.

Since A is a square $n \times n$ matrix, $\mathbf{A}^{-1}$ exists if **A** is nonsingular. If we assume **A** to be nonsingular, we obtain the solution to our set of simultaneous equations by premultiplying both sides of the vector equation by $\mathbf{A}^{-1}$. We find that

$$\mathbf{X} = \mathbf{A}^{-1}\mathbf{B}.$$

All that is needed to obtain the solution is $\mathbf{A}^{-1}$. We have seen that the inverse can be computed from the adjoint matrix **J** and $|\mathbf{A}|$. However, this computation is tedious, and we can achieve the same result more expeditiously by the Gauss–Jordan complete elimination procedure.

Let us write the matrices **A** and **I** and the vector **B** side by side:

$$[\mathbf{A} \mid \mathbf{I} \mid \mathbf{B}].$$

Now let us premultiply each of these matrices by $\mathbf{A}^{-1}$ to obtain

$$[\mathbf{A}^{-1}\mathbf{A} \mid \mathbf{A}^{-1}\mathbf{I} \mid \mathbf{A}^{-1}\mathbf{B}]$$

or

$$[\mathbf{I} \mid \mathbf{A}^{-1} \mid \mathbf{X}].$$

We see that where our **A** matrix originally existed we have an identity matrix, that the inverse of **A** has replaced **I**, and that in place of the original vector **B** we have our desired solution vector **X**.

In the Gauss–Jordan procedure, we perform a set of operations on **A** such that **A** is converted to an identity matrix. Such a set of operations is equivalent to premultiplying **A** by $\mathbf{A}^{-1}$. If we perform the same set of operations on **B**, this is equivalent to premultiplying **B** by $\mathbf{A}^{-1}$, and thus our solution vector **X** is obtained.

Since the elements of the matrices **A** and **B** represent the coefficients of our simultaneous equations, the operations that we may perform on these elements are the same as those we may perform on a set of simultaneous algebraic equations. Thus we may multiply any row by a scalar without altering our equality. Similarly, we may add the elements of one row to the corresponding elements of another row. We therefore perform these operations successively in a manner such that the matrix **A** is reduced to an identity matrix.

The procedure may be most easily understood by considering a simple example. Let us determine the solution and inverse for the following set of equations:

$$\begin{aligned} 2x_1 + 3x_2 + x_3 &= 4, \\ 3x_1 + 5x_2 - 2x_3 &= 10, \\ 2x_1 - 5x_2 + 7x_3 &= 6. \end{aligned}$$

The determinant is not singular, and therefore a solution may be found. We write the coefficients of our equation in matrix form and place them adjacent to an identity matrix. The partitioned matrix, $\mathbf{A}|\mathbf{I}|\mathbf{B}$, obtained is

$$\left[\begin{array}{ccc|ccc|c} 2 & 3 & 1 & 1 & 0 & 0 & 4 \\ 3 & 5 & -2 & 0 & 1 & 0 & 10 \\ 2 & -5 & 7 & 0 & 0 & 1 & 6 \end{array}\right].$$

Since we wish to have a 1 in position a_{11}, we divide our first equation by 2. To obtain zeros in the remaining positions of the first column, we first multiply the revised first line (after the division by 2) by -3 and add it to the second line. We then multiply the revised first line by -2 and add it to the third line. We have

$$\left[\begin{array}{ccc|ccc|c} 1 & \frac{3}{2} & \frac{1}{2} & \frac{1}{2} & 0 & 0 & 2 \\ 0 & \frac{1}{2} & -\frac{7}{2} & -\frac{3}{2} & 1 & 0 & 4 \\ 0 & -8 & 6 & -1 & 0 & 1 & 2 \end{array}\right].$$

We now need to obtain a 1 in position a_{22}. We obtain this by multiplying the second line by 2. The revised second line is then multiplied by 8 and added to the third line to get a zero in position a_{32}. Similarly, we obtain a zero in position a_{12} by multiplying the revised second line by $-\frac{3}{2}$ and adding it to the first line. We then have

$$\left[\begin{array}{ccc|ccc|c} 1 & 0 & 11 & 5 & -3 & 0 & -10 \\ 0 & 1 & -7 & -3 & 2 & 0 & 8 \\ 0 & 0 & -50 & -25 & 16 & 1 & 66 \end{array}\right].$$

All that is left is to revise the third column of the matrix. We obtain a 1 in column 3 by dividing by -50. We then multiply the revised third line by 7 and add it to the second line. Finally, we obtain a zero in position a_{13} by multiplying the revised third line by -11 and adding it to the first line. Our final array is

$$\left[\begin{array}{ccc|ccc|c} 1 & 0 & 0 & -\frac{1}{2} & \frac{13}{15} & \frac{11}{50} & \frac{113}{25} \\ 0 & 1 & 0 & \frac{1}{2} & -\frac{6}{25} & -\frac{7}{50} & -\frac{31}{25} \\ 0 & 0 & 1 & \frac{1}{2} & -\frac{8}{25} & -\frac{1}{50} & -\frac{33}{25} \end{array}\right].$$

Thus we have $x_1 = \frac{113}{25}$, $x_2 = -\frac{31}{25}$, and $x_3 = -\frac{33}{25}$; and the inverse $\mathbf{A}^{-1}$ is given by the middle matrix. The correctness of the result may be verified by substituting the x_i values in the original equation and by carrying out the $\mathbf{AA}^{-1}$ matrix multiplication.

We can see, from the previous example, that the Gauss–Jordan method is a repetitive use of the same set of operations. We can describe our procedure by three simple rules:

1. Normalize the ith row so that $a_{ii} = 1$.
2. Obtain a new jth row through multiplication at the ith row by $-a_{ji}$ and addition of the new row to the jth row. Repeat for each $j \neq i$.
3. Return to step 1 and repeat for each value of i.

These operations cause all elements of the original matrix to vanish except the a_{ii}, which are set equal to 1.

Linear Independence and Bases

Let us look at our original set of simultaneous equations in a slightly different manner and rewrite them as a matrix equation:

$$\begin{bmatrix} 2 & 3 & 1 \\ 3 & 5 & -2 \\ 2 & -5 & 7 \end{bmatrix} \begin{bmatrix} x_1 \\ x_2 \\ x_3 \end{bmatrix} = \begin{bmatrix} 4 \\ 10 \\ 6 \end{bmatrix}.$$

We may represent each column of numbers as a vector. Thus we have

$$\mathbf{P}_1 = \begin{bmatrix} 2 \\ 3 \\ 2 \end{bmatrix}, \quad \mathbf{P}_2 = \begin{bmatrix} 3 \\ 5 \\ -5 \end{bmatrix}, \quad \mathbf{P}_3 = \begin{bmatrix} 1 \\ -2 \\ 7 \end{bmatrix}, \quad \mathbf{P}_0 = \begin{bmatrix} 4 \\ 10 \\ 6 \end{bmatrix},$$

and our matrix equation becomes

$$\mathbf{P}_1 x_1 + \mathbf{P}_2 x_2 + \mathbf{P}_3 x_3 = \mathbf{P}_0.$$

We may thus regard $\mathbf{P}_0$ as a *linear combination* of the vectors $\mathbf{P}_1$, $\mathbf{P}_2$, and $\mathbf{P}_3$. Our solution, x_1, x_2, x_3, is that unique set of numbers which generates this equality. We would not have a solution unless it were possible to express $\mathbf{P}_0$ in terms of the vectors $\mathbf{P}_1$, $\mathbf{P}_2$, and $\mathbf{P}_3$. We can do so because $\mathbf{P}_1$, $\mathbf{P}_2$, and $\mathbf{P}_3$ form a *linearly independent set* in the three-dimensional space of the problem.

A *three-dimensional Euclidean space*, E_3, is defined as the collection of all possible three-dimensional vectors $\mathbf{P}$, where

$$\mathbf{P} = \begin{bmatrix} a_1 \\ a_2 \\ a_3 \end{bmatrix}.$$

In most problems, our space will be greater than three dimensional. We therefore define an *n-dimensional Euclidean space*, E_n, as the collection of all possible n-dimensional vectors $\mathbf{P}$, where

$$\mathbf{P} = \begin{bmatrix} a_1 \\ a_2 \\ \cdot \\ \cdot \\ \cdot \\ a_n \end{bmatrix}.$$

A set of vectors $\mathbf{P}_1, \mathbf{P}_2, \ldots, \mathbf{P}_n$ is said to be *linearly independent* if the expression

$$\alpha_1 \mathbf{P}_1 + \alpha_2 \mathbf{P}_2 + \cdots + \alpha_n \mathbf{P}_n = \mathbf{0}$$

requires that all the scalars α_i be zero. That is,

$$\alpha_1 = \alpha_2 = \alpha_3 \cdots \alpha_n = 0.$$

This means that any one vector $\mathbf{P}_j$ *cannot* be expressed in terms of one or more of the remaining $(n - 1)$ vectors. If any nonzero set of values for the α_i will cause the equality to hold, then the vectors are said to be *linearly dependent*.

If we have a set of n linearly independent vectors in the space E_n, we can express any other vector in E_n as a linear combination of the original n vectors. Because of this property, we call any set of n linearly independent vectors in E_n a *basis* for E_n. These statements imply that any set of $(n + 1)$ vectors in E_n must be linearly dependent. Consider a set of n vectors to which we add an additional vector. From our definition, we know that if the original set was linearly dependent, the augmented set must also be linearly dependent. If the first n vectors are linearly independent, then we can express $\mathbf{P}_{n+1}$ as a linear combination of the first n. That is,

$$\mathbf{P}_{n+1} = \alpha_1\mathbf{P}_1 + \alpha_2\mathbf{P}_2 + \cdots + \alpha_n\mathbf{P}_n, \qquad \text{at least one } \alpha_i \neq 0,$$

or

$$\alpha_1\mathbf{P}_1 + \alpha_2\mathbf{P}_2 + \cdots + \alpha_n\mathbf{P}_n - \mathbf{P}_{n+1} = \mathbf{0}.$$

Hence the set must be linearly dependent, since a set of scalars other than zero causes the linear combination to be zero.

We may illustrate this by a numerical example. Consider the vectors

$$\mathbf{P}_1 = \begin{bmatrix} 1 \\ 2 \end{bmatrix}, \qquad \mathbf{P}_2 = \begin{bmatrix} 5 \\ 7 \end{bmatrix}.$$

We may show that these are linearly independent by writing

$$\alpha_1\mathbf{P}_1 + \alpha_2\mathbf{P}_2 = \mathbf{0}$$

or

$$\alpha_1\begin{bmatrix} 1 \\ 2 \end{bmatrix} + \alpha_2\begin{bmatrix} 5 \\ 7 \end{bmatrix} = \begin{bmatrix} 0 \\ 0 \end{bmatrix},$$

$$\begin{bmatrix} \alpha_1 + 5\alpha_2 \\ 2\alpha_1 + 7\alpha_2 \end{bmatrix} = \begin{bmatrix} 0 \\ 0 \end{bmatrix}.$$

The last vector equation represents two simultaneous algebraic equations:

$$\alpha_1 + 5\alpha_2 = 0,$$
$$2\alpha_1 + 7\alpha_2 = 0.$$

Their solution yields $\alpha_1 = \alpha_2 = 0$.

If we add a third vector,

$$\mathbf{P}_3 = \begin{bmatrix} 0 \\ 4 \end{bmatrix},$$

we can show that the set is no longer linearly independent. We have

$$\alpha_1\mathbf{P}_1 + \alpha_2\mathbf{P}_2 + \alpha_3\mathbf{P}_3 = \alpha_1\begin{bmatrix} 1 \\ 2 \end{bmatrix} + \alpha_2\begin{bmatrix} 5 \\ 7 \end{bmatrix} + \alpha_3\begin{bmatrix} 0 \\ 4 \end{bmatrix} = \begin{bmatrix} 0 \\ 0 \end{bmatrix},$$

$$\begin{bmatrix} \alpha_1 + 5\alpha_2 \\ 2\alpha_1 + 7\alpha_2 + 4\alpha_3 \end{bmatrix} = \begin{bmatrix} 0 \\ 0 \end{bmatrix}.$$

If we again consider the vector equation as two simultaneous equations, we conclude

$$\alpha_1 = -5\alpha_2,$$
$$\alpha_3 = \tfrac{3}{4}\alpha_2.$$

Our equality will then hold with any number arbitrarily assigned to α_1, and therefore the set of vectors is not linearly independent.

If we return to the set of three equations we solved at the beginning of this section, we can call the linearly independent vectors $\mathbf{P}_1$, $\mathbf{P}_2$, and $\mathbf{P}_3$ the *basis* for our solution. If the vectors were not linearly independent, they would not be a basis and we could not express $\mathbf{P}_0$ in terms of them. A solution would then not be possible. Since we can always find a solution to our equation if $\mathbf{A}$ is nonsingular, the nonsingularity of $\mathbf{A}$ must imply the linear independence of the $\mathbf{P}_i$.

Although any set of n linearly independent vectors may serve as a basis for E_n, the unit vectors, $\mathbf{e}_i$, are frequently selected as the basis. These vectors are defined by

$$\mathbf{e}_1 = \begin{bmatrix} 1 \\ 0 \\ 0 \\ 0 \\ \cdot \\ \cdot \\ \cdot \\ 0 \end{bmatrix}, \quad \mathbf{e}_2 = \begin{bmatrix} 0 \\ 1 \\ 0 \\ 0 \\ \cdot \\ \cdot \\ \cdot \\ 0 \end{bmatrix}, \quad \mathbf{e}_3 = \begin{bmatrix} 0 \\ 0 \\ 1 \\ 0 \\ \cdot \\ \cdot \\ \cdot \\ 0 \end{bmatrix}, \ldots, \quad \mathbf{e}_n = \begin{bmatrix} 0 \\ 0 \\ 0 \\ 0 \\ \cdot \\ \cdot \\ \cdot \\ 1 \end{bmatrix}.$$

If this particular basis is selected, the components of the vector $\mathbf{P}_0$ in the expression

$$\mathbf{e}_1x_1 + \mathbf{e}_2x_2 + \cdots + \mathbf{e}_nx_n = \mathbf{P}_0$$

will equal the coefficients $x_1, x_2, \ldots, x_n$ of the basis vectors.

Solution of *m* Linear Equations in *n* Unknowns

We have just considered the solution of a set of n simultaneous equations in n unknowns. However, in linear programming we are concerned with equations that have the form

$$\begin{aligned} a_{11}x_1 + a_{12}x_2 + a_{13}x_3 + \cdots + a_{1n}x_n &= b_1 \\ a_{21}x_1 + a_{22}x_2 + a_{23}x_3 + \cdots + a_{2n}x_n &= b_2 \\ \text{-----------------------}& \\ a_{m1}x_1 + a_{m2}x_2 + a_{m3}x_3 + \cdots + a_{mn}x_n &= b_m, \end{aligned}$$

where m is not necessarily equal to n. In this more general case, the equation set may not have a unique solution. The kinds of solutions we may obtain can be classified by comparing the rank of **A**, denoted $r[\mathbf{A}]$, with the rank of the augmented matrix formed by appending the column vector **B** to the right of matrix **A**. We shall denote the augmented matrix by $[\mathbf{A} \mid \mathbf{B}]$ and its rank by $r[\mathbf{A} \mid \mathbf{B}]$.

We may first observe thar $r[\mathbf{A} \mid \mathbf{B}] \geq r[\mathbf{A}]$, since the cofactors contained in **A** are also contained in $[\mathbf{A} \mid \mathbf{B}]$. When $r[\mathbf{A} \mid \mathbf{B}] > r[\mathbf{A}]$, we find that **B** is linearly independent of the vectors $\mathbf{P}_i$, which compose **A**. We saw in the previous section that if this were so, we could not express **B** in terms of $\mathbf{P}_i$. This is equivalent to stating that there is no solution to our set of equations. We then have an *inconsistent* set of equations.

Another possibility is that $r[\mathbf{A} \mid \mathbf{B}] = r[\mathbf{A}]$. If this is so, and $r[\mathbf{A}] = m = n$, then we have that **A** is a nonsingular $n \times n$ matrix. This is the case considered earlier, and a unique solution is obtained.

Still a third possibility is that $r[\mathbf{A}] = r[\mathbf{A} \mid \mathbf{B}] = l < n$. In this case, there are l independent equations. Therefore, we can solve for any set of l variables in terms of the remaining $(n - l)$ variables. Since any arbitrary values can be assigned to these $(n - l)$ variables, there are an infinite number of possible solutions. It is this situation which is of interest in linear programming.

Although there are an infinite number of solutions, we shall be concerned only with what are called *basic solutions*. To obtain a basic solution, we choose m linearly independent vectors from the $\mathbf{P}_i$ composing **A**, assign these $\mathbf{P}_i$ and corresponding x_i the indices 1 to m, and write our solution as

$$\sum_{i=1}^{m} \mathbf{P}_i x_i = \mathbf{B}.$$

Obviously, only those x_i with indices between 1 and m may have nonzero values. If the number of nonzero x_i in the basis equals m, we have a *non-degenerate* solution. If the number of nonzero x_i is less than m, we have a *degenerate* solution, since this means we express $\mathbf{B}$ in terms of less than m linearly independent vectors. We call the set of m linearly independent vectors that we have chosen the *basis* for our solution, since any other vector can be expressed in terms of these basis vectors. For each set of m linearly independent vectors we choose, we obtain another basic solution. The total possible number of basic solutions is then the number of combinations of n items taken m at a time, or $n!/m![n - m]!$.

Let us consider the simultaneous equations

$$\begin{aligned} 5x_1 + 2x_2 + 3x_3 &= 3, \\ x_2 + 6x_3 &= 4, \end{aligned}$$

and select $\mathbf{P}_1 = \left[\begin{smallmatrix}5\\0\end{smallmatrix}\right]$ and $\mathbf{P}_2 = \left[\begin{smallmatrix}2\\1\end{smallmatrix}\right]$ as the basis for our solution. We may then write

$$\mathbf{P}_1 x_1 + \mathbf{P}_2 x_2 = \begin{bmatrix}5\\0\end{bmatrix} x_1 + \begin{bmatrix}2\\1\end{bmatrix} x_2 = \begin{bmatrix}3\\4\end{bmatrix}.$$

By solving the two simultaneous equations represented by the last vector equation, we obtain ($x_1 = -1, x_2 = 4$) as our solution. Since $\mathbf{P}_3$ does not appear in the basis, $x_3 = 0$.

We may always reduce any given basis to unit vectors by means of the Gauss–Jordan elimination procedure. To reduce the basis ($\mathbf{P}_1, \mathbf{P}_2$) of the previous example to unit vectors $\mathbf{P}_1 = \left[\begin{smallmatrix}1\\0\end{smallmatrix}\right]$, $\mathbf{P}_2 = \left[\begin{smallmatrix}0\\1\end{smallmatrix}\right]$, we write the coefficients of the two simultaneous equations as a partitioned matrix:

$$\left[\begin{array}{ccc|c} 5 & 2 & 3 & 3 \\ 0 & 1 & 6 & 4 \end{array}\right].$$

We now apply the Gauss–Jordan elimination procedure and establish unit vectors in the first two columns. We do so by first dividing the first row by 5. Since we have a zero in the (2, 1) position and a 1 in the (2, 2) position, we need only obtain a zero in the (1, 2) position. This we do by multiplying the second row by $-\frac{2}{5}$ and adding to the first row. We then have

$$\left[\begin{array}{ccc|c} 1 & 0 & -\frac{9}{5} & -1 \\ 0 & 1 & 6 & 4 \end{array}\right].$$

Our basis is now in terms of unit vectors, and the nonzero components of the solution vector, $\mathbf{X} = \begin{bmatrix} -1 \\ 4 \\ 0 \end{bmatrix}$, are in the $\mathbf{P}_0$ column.

Since the vectors $\mathbf{P}_1$, $\mathbf{P}_2$, and $\mathbf{P}_3$ are not linearly independent, we can write $\mathbf{P}_3$ in terms of $\mathbf{P}_1$ and $\mathbf{P}_2$; i.e.,

$$\alpha_1\mathbf{P}_1 + \alpha_2\mathbf{P}_2 = \mathbf{P}_3,$$

where α_1 and α_2 are scalar constants to be determined. These constants are equal to the values in the $\mathbf{P}_3$ column of the final Gauss–Jordan matrix. Hence

$$-\tfrac{9}{5}\mathbf{P}_1 + 6\mathbf{P}_2 = \mathbf{P}_3.$$

We may verify this by substitution of the original values of $\mathbf{P}_1$ and $\mathbf{P}_2$ to obtain

$$-9/5\begin{bmatrix} 5 \\ 0 \end{bmatrix} + 6\begin{bmatrix} 2 \\ 1 \end{bmatrix} = \begin{bmatrix} -9 \\ 0 \end{bmatrix} + \begin{bmatrix} 12 \\ 6 \end{bmatrix} = \begin{bmatrix} 3 \\ 6 \end{bmatrix},$$

which is the original value of the $\mathbf{P}_3$ vector.

Quadratic Forms

A quadratic form in the n variables $x_1, x_2, x_3, \ldots, x_n$ is defined as

$$\mathbf{X}'\mathbf{A}\mathbf{X},$$

where $\mathbf{X}'$ is the row vector $[x_1, x_2, x_3, \ldots, x_n]$ and $\mathbf{A}$ is an $n \times n$ matrix of scalar quantities. The quadratic form will always have a scalar value. Consider the simple case in which

$$\mathbf{X} = \begin{bmatrix} x_1 \\ x_2 \end{bmatrix} \quad \text{and} \quad \mathbf{A} = \begin{bmatrix} 1 & 3 \\ 4 & 5 \end{bmatrix}.$$

Then

$$\begin{aligned} \mathbf{X}'\mathbf{A}\mathbf{X} &= [x_1 \quad x_2]\begin{bmatrix} 1 & 3 \\ 4 & 5 \end{bmatrix}\begin{bmatrix} x_1 \\ x_2 \end{bmatrix} \\ &= \begin{bmatrix} (x_1 + 4x_2)(3x_1 + 5x_2) \end{bmatrix}\begin{bmatrix} x_1 \\ x_2 \end{bmatrix} \\ &= x_1^2 + 7x_1x_2 + 5x_2^2. \end{aligned}$$

A quadratic form is classified by considering whether it may be positive or negative. A quadratic form is said to be *positive definite* if $\mathbf{X}'\mathbf{A}\mathbf{X} > 0$ for every $\mathbf{X}$ except $\mathbf{X} < \mathbf{0}$. It is *positive semidefinite* if $\mathbf{X}'\mathbf{A}\mathbf{X} \geq 0$ for every $\mathbf{X}$, and there exist $\mathbf{X} \neq \mathbf{0}$ for which $\mathbf{X}'\mathbf{A}\mathbf{X} = 0$. We say that a quadratic form is *negative definite* if $-\mathbf{X}'\mathbf{A}\mathbf{X}$ is positive definite. Similarly, we say it is *negative semidefinite* if $-\mathbf{X}'\mathbf{A}\mathbf{X}$ is positive semidefinite. A quadratic form is called *indefinite* if $\mathbf{X}'\mathbf{A}\mathbf{X}$ may take on both positive and negative values. In the example shown above, we can choose values of x_1 and x_2 that will lead to positive or negative values for $(x_1^2 + 7x_1x_2 + 5x_2^2)$, therefore, $\mathbf{X}'\mathbf{A}\mathbf{X}$ is indefinite.

Quadratic forms are particularly useful for the determination of the convexity or concavity of differentiable functions. To determine whether a differentiable function $\phi(y_1, y_2, \ldots, y_n)$ is convex, we make use of the matrix whose elements are the second partial derivatives

$$\frac{\partial^2 \phi}{\partial y_i \, \partial y_j}.$$

We call such a matrix a *Hessian matrix*. For a function of three variables, the Hessian matrix $\mathbf{H}$ would be

$$\mathbf{H} = \begin{bmatrix} \dfrac{\partial^2 \phi}{\partial y_1^2} & \dfrac{\partial^2 \phi}{\partial y_1 \, \partial y_2} & \dfrac{\partial^2 \phi}{\partial y_1 \, \partial y_3} \\[2ex] \dfrac{\partial^2 \phi}{\partial y_2 \, \partial y_1} & \dfrac{\partial^2 \phi}{\partial y_2^2} & \dfrac{\partial^2 \phi}{\partial y_2 \, \partial y_3} \\[2ex] \dfrac{\partial^2 \phi}{\partial y_3 \, \partial y_1} & \dfrac{\partial^2 \phi}{\partial y_3 \, \partial y_2} & \dfrac{\partial^2 \phi}{\partial y_3^2} \end{bmatrix}.$$

We would then have

$$\mathbf{X}'\mathbf{H}\mathbf{X} = \frac{\partial^2 \phi}{\partial y_1^2} x_1^2 + 2\frac{\partial^2 \phi}{\partial y_1 \, \partial y_2} x_1 x_2 + 2\frac{\partial^2 \phi}{\partial y_1 \, \partial y_3} x_1 x_3 + \frac{\partial^2 \phi}{\partial y_2^2} x_2^2 + 2\frac{\partial^2 \phi}{\partial y_2 \, \partial y_3} x_2 x_3 + \frac{\partial^2 \phi}{\partial y_3^2} x_3^2.$$

Note that since $(\partial^2\phi/\partial y_i \, \partial y_j) = (\partial^2\phi/\partial y_j \, \partial y_i)$, we have only six terms in our result rather than nine.

A differentiable function $\phi(y_1, y_2, \ldots, y_n)$ is *convex* over the set E_n if its Hessian matrix is positive semidefinite for all values of $\mathbf{Y}$ in E_n. It is *strictly convex* over E_n if its Hessian matrix is positive definite for all values of $\mathbf{Y}$. It is *concave* over the set E_n if its Hessian matrix is negative semidefinite, and *strictly concave* if its Hessian matrix is negative definite over all values of $\mathbf{Y}$ in E_n.

Consider the function $\phi = y_1^2 + y_2^2$ over the set of all real numbers. We would then have that

$$\mathbf{X'HX} = 2x_1^2 + 0[x_1x_2] + 2x_2^2$$

for all values of **X**, and therefore **H** is positive definite. We conclude that the function considered is strictly convex.

If our function had been linear (e.g., $\phi = a_1y_1 + a_2y_2 + a_3y_3 + a_4y_4$), we would have found $\mathbf{X'HX} = 0$ for all values of **X**. Our quadratic form is then both positive and negative semidefinite. A linear function thus satisfies the conditions of both convexity and concavity.

PROBLEMS

A.1. Perform the following matrix operations:

(a) $\begin{bmatrix} -1 & 3 & 5 \\ 2 & 1 & 7 \\ 3 & 4 & -3 \end{bmatrix} + \begin{bmatrix} 2 & 8 & 4 \\ 1 & -3 & 6 \\ -4 & 5 & 7 \end{bmatrix}$

(b) $2\mathbf{A} - 3\mathbf{B} + \mathbf{C}$ where $\mathbf{A} = \begin{bmatrix} -1 & 2 & 3 \\ 5 & 6 & 7 \end{bmatrix}$, $\mathbf{B} = \begin{bmatrix} 2 & 9 & 4 \\ 8 & -5 & 3 \end{bmatrix}$,

$\mathbf{C} = \begin{bmatrix} 0 & 2 & 1 \\ 3 & -4 & 1 \end{bmatrix}$

(c) $[4 \quad -2 \quad 7] \cdot \begin{bmatrix} 8 & 3 & 9 \\ -2 & 4 & 5 \\ -1 & 5 & 6 \end{bmatrix}$

(d) $\begin{bmatrix} 8 & 2 \\ 9 & 6 \end{bmatrix} \cdot \begin{bmatrix} 7 & 4 \\ 1 & 3 \end{bmatrix}$

(e) $\begin{bmatrix} 8 & 2 \\ 9 & 6 \end{bmatrix} \cdot \begin{bmatrix} 7 & 4 \\ 1 & 3 \end{bmatrix}'$ (f) $\begin{bmatrix} 7 & 4 \\ 1 & 3 \end{bmatrix} \cdot \begin{bmatrix} 8 & 2 \\ 9 & 6 \end{bmatrix}$

(g) $\mathbf{AB'} + \mathbf{BA'}$, where **A** and **B** are as designated in part (b).

A.2. Compute the value of the largest nonvanishing determinant of each of the following matrices:

(a) $\begin{bmatrix} 3 & 1 & 2 \\ 6 & 5 & 9 \\ 8 & 4 & 1 \end{bmatrix}$

(b) $\begin{bmatrix} 5 & 6 & 1 & 5 \\ 7 & 8 & -2 & 1 \\ -3 & 8 & 4 & 3 \\ 2 & 9 & 7 & 4 \end{bmatrix}$

(c) $\begin{bmatrix} 1 & 0 & 2 & 4 & -5 \\ 2 & 1 & 4 & 0 & 1 \\ 4 & -3 & 8 & 6 & 9 \\ 0 & 2 & 0 & 5 & 1 \\ 3 & 8 & 6 & 2 & -2 \end{bmatrix}$

(d) What is the rank of the matrices of parts (a), (b), and (c)?

A.3. Verify that pre- or postmultiplication of a multi-row matrix by a vector always yields a vector.

A.4. Determine the inverses of the matrices of problem A.2 by
(a) Use of the adjoint matrix.
(b) Use of the Gauss–Jordan elimination procedure.

A.5. Solve the following sets of simultaneous equations using the Gauss–Jordan elimination procedure. Check your results by substitution into the original equations.

(a)
$$\begin{aligned} x_1 + 2x_2 + 5x_3 &= 9 \\ 2x_1 + 3x_2 - x_3 &= 7 \\ 4x_1 - 2x_2 + 3x_3 &= 6 \end{aligned}$$

(b)
$$\begin{aligned} 5x_1 + 3x_2 + 7x_3 - x_4 &= 1 \\ 3x_1 + x_2 + 5x_3 - 7x_4 &= 0 \\ 2x_1 - 2x_2 + 4x_3 + 6x_4 &= 5 \\ x_1 + x_2 + 3x_3 + 2x_4 &= 4 \end{aligned}$$

A.6. The well-known *Cramer's rule* states that "A system of linear equations, $\mathbf{AX} = \mathbf{B}$. whose determinant $|\mathbf{A}|$ is nonzero, has the solution

$$x_1 = \frac{d_1}{|\mathbf{A}|}, x_2 = \frac{d_2}{|\mathbf{A}|}, \ldots, x_n = \frac{d_n}{|\mathbf{A}|},$$

where d_i is the determinant obtained from the matrix $\mathbf{A}$ by replacing the ith column of $\mathbf{A}$ with the column vector $\mathbf{B}$." Verify this rule by showing that it is correct for any 3×3 matrix having a nonzero determinant.

A.7. Solve the following simultaneous equations using the unit vectors

$$\mathbf{P}_1 = \begin{bmatrix} 1 \\ 0 \\ 0 \end{bmatrix}, \quad \mathbf{P}_2 = \begin{bmatrix} 0 \\ 1 \\ 0 \end{bmatrix}, \quad \mathbf{P}_3 = \begin{bmatrix} 0 \\ 0 \\ 1 \end{bmatrix}$$

as a basis for the solution:

$$\begin{aligned} x_1 + 3x_2 + 4x_3 + 2x_4 &= 2 \\ 5x_1 + 2x_2 + 3x_3 &= 3 \\ 6x_1 + 3x_2 - x_3 - x_4 &= 1 \end{aligned}$$

A.8. Compute the quadratic forms of the following matrices:

(a) $\begin{bmatrix} 4 & 5 \\ -1 & 3 \end{bmatrix}$ (b) $\begin{bmatrix} 1 & 2 & 3 \\ 4 & 5 & 6 \\ 7 & 8 & 9 \end{bmatrix}$ (c) $\begin{bmatrix} -2 & 3 & 5 & 4 \\ 7 & 8 & -7 & 1 \\ 1 & 0 & 4 & 3 \\ 5 & 2 & 6 & -2 \end{bmatrix}$

A.9. Determine the convexity or concavity of the following functions by examining their quadratic forms:

(a) $3x_1^2 - 2x_1 + 5x_1^3$
(b) $-3x_1^2 - 2x_1 + 5x_1^3$
(c) $x_1x_2x_3 + x_1^2$
(d) $-4x_1^3 + 3x_2 - e^{-x_2}$
(e) $\log(x_1 + x_2 + x_3)$

A.10. Show that the quadratic form of the Hessian matrix can be written as

$$\sum_{i=1}^{n} \sum_{j=1}^{n} \left(\frac{\partial^2 \phi}{\partial y_i \, \partial y_j}\right) x_i x_j.$$

A.11. If **A** is a skew symmetric matrix, show that $a_{ii} = 0$ for all i.

A.12. A triangular matrix is defined as a matrix wherein the nonzero elements form a triangular array. Thus a 4×4 square triangular matrix is symbolized by

$$\mathbf{B} = \begin{bmatrix} b_{11} & b_{12} & b_{13} & b_{14} \\ 0 & b_{22} & b_{23} & b_{24} \\ 0 & 0 & b_{33} & b_{34} \\ 0 & 0 & 0 & b_{44} \end{bmatrix}.$$

(a) Show that the determinant of a square triangular matrix is given by the product of the diagonal elements. (That is, $|\mathbf{B}| = b_{11}b_{22}b_{33}b_{44}$.)
(b) Indicate how (a) can lead to an efficient algorithm for computation of determinants.

INDEX

C

D

H

I

N

O

P

Q

R

S

T

U

V

W

Z